ALLZEIT WACH
1842

E. Gwinner (Ed.)

Bird Migration

Physiology and Ecophysiology

With 123 Figures and 20 Tables

Springer-Verlag
Berlin Heidelberg New York
London Paris Tokyo
Hong Kong Barcelona

Professor Dr. EBERHARD GWINNER
Max-Planck-Institut für Verhaltensphysiologie
Vogelwarte
D-8138 Andechs, FRG

ISBN-13: 978-3-642-74544-7 e-ISBN-13: 978-3-642-74542-3
DOI: 10.1007/978-3-642-74542-3

Library of Congress-Cataloging-in-Publication Data. Bird migration : physiology and ecophysiology / E. Gwinner, editor. p. cm. ISBN 0-387-50855-4 (U.S. : alk. paper) 1. Birds–Migration. 2. Birds–Physiology. 3. Birds–Ecology. I. Gwinner, Eberhard, 1938– . QL698.9.B57 1990 598.252'5–dc20 90-34465 CIP

Softcover reprint of the hardcover 1st edition 1990

Typesetting: International Typesetters Inc., Makati, Philippines
2131/3145(3011)-543210 – Printed on acid-free paper

The articles of this volume are dedicated
to the memory of DONALD S. FARNER (1915–1988)
in appreciation of his numerous contributions
to bird migration research

Contents

Contributors

Alerstam, T., Department of Ecology, University of Lund, Ecology Building, S-22362 Lund, Sweden

Bairlein, F., Department of Zoology, Physiological Ecology Section, University of Köln, Weyertal 119, D-5000 Köln 41, FRG

Berthold, P., Max-Planck-Institut für Verhaltensphysiologie, Vogelwarte, D-7760 Radolfzell-Möggingen, FRG

Biebach, H., Max-Planck-Institut für Verhaltensphysiologie, Vogelwarte, D-8138 Andechs, FRG

Bruderer, B., Schweizerische Vogelwarte, CH-6204 Sempach, Switzerland

Butler, P. J., School of Biological Sciences, University of Birmingham, P. O. Box 363, Birmingham B15 2TT, UK

Davidson, N. C., Nature Conservancy Council, Northminster House, Peterborough PE1 1UA, UK

Dolnik, V. R., Zoological Institute of the USSR, Academy of Sciences, Leningrad 199034, USSR

Drent, R., University of Groningen, Zoological Laboratory, P.O.Box 14, NL-9750 AA Haren (Gr.), The Netherlands

Evans, P. R., Department of Biological Sciences, University of Durham, South Road, Durham DH1 3LE, UK

Gwinner, E., Max-Planck-Institut für Verhaltensphysiologie, Vogelwarte, D-8138 Andechs, FRG

Herter, D. R., LGL Alaska Research Associates, Inc., 505 W. Northern Lights Blvd. Anchorage, Alaska 99503, USA

Jehl, J. R., Jr., Sea World Research Institute, Hubbs Marine Research Center, 1700 South Shores Road, San Diego, California 92109, USA

Jenni, L., Schweizerische Vogelwarte, CH-6204 Sempach, Switzerland

Johnson, S. R., LGL Ltd., Environmental Research Associates, 9768 Second Street, Sidney, B. C. V8L 3Y8, Canada

Ketterson, E. D., Department of Biology, Jordan Hall, Indiana University, Bloomington, Indiana 47405, USA

Leisler, B., Max-Planck-Institut für Verhaltensphysiologie, Vogelwarte, D-7760 Radolfzell-Möggingen, FRG

Lindström, Å., Department of Ecology, University of Lund, Ecology Building, S-22362-Lund, Sweden

Martin, G.R., University of Birmingham, School of Biological Sciences and School of Continuing Studies, P.O. Box 363, Birmingham B15 2TT, UK

Mattocks, P.W., Jr., 915 E Third Ave., Ellensburg, Washington 98926, USA

Nachtigall, W., Fachrichtung Zoologie, Universität des Saarlandes, D-6600 Saarbrücken, FRG

Nolan, V., Jr., Department of Biology, Jordan Hall, Indiana University, Bloomington, Indiana 47405, USA

O'Connor, R.J., Department of Wildlife, University of Maine, 240 Nutting Hall, Orono, Maine 04469, USA

Pearson, D.J., Browses, Sibton, Saxmundham, Suffolk, UK

Piersma, T., Netherlands Institute of Sea Research, P.O. Box 59, NL-1790 AB Den Burg, Texel, The Netherlands

Ramenofsky, M., Department of Zoology, NJ-15, University of Washington, Seattle, Washington 98195, USA

Rayner, J.M.V., Department of Zoology, University of Bristol, Woodland Road, Bristol BS8 1UG, UK

Richardson, W.J., LGL Ltd., Environmental Research Associates, P.O. Box 280, King City, Ontario L0G 1K0, Canada

Schwabl, H., Max-Planck-Institut für Verhaltensphysiologie, Vogelwarte, D-8138 Andechs, FRG

Silverin, B., Department of Zoology, University of Gothenburg, Box 25059, S-40031 Gothenburg, Sweden

Terrill, S., Siena College, Division of Science, Route 9, Loudonville, N.Y. 12211, USA

Walsberg, G.E., Department of Zoology, Arizona State University, Tempe, Arizona 85287-1501, USA

Williams, J.M., Swarthmore College, Department of Biology, Swarthmore, Pennsylvania 19081, USA

Williams, T.C., Swarthmore College, Department of Biology, Swarthmore, Pennsylvania 19081, USA

Wingfield, J.C., Department of Zoology, NJ-15, University of Washington, Seattle, Washington 98195, USA

Woakes, A.J., School of Biological Sciences, University of Birmingham, P.O. Box 363, Birmingham B15 2TT, UK

Introduction

E. GWINNER[1]

The phenomenon of bird migration with its large scale dimensions has attracted the attention of naturalists for centuries. Worldwide billions of birds leave their breeding grounds every autumn to migrate to areas with seasonally more favorable conditions. Many of these migrants travel only over a few hundred kilometers but others cover distances equivalent to the circumference of the earth. Among these long-distance migrants are several billion birds that invade Africa every autumn from their West and Central Palaearctic breeding areas. In the Americas and in Asia the scope of bird migration is of a similar magnitude.

Just as impressive as the numbers of birds are their achievements. They have to cope with the enormous energetic costs of long-distance flying, particularly while crossing oceans and deserts that do not allow replenishment of depleted fat reserves. They have to appropriately time the onset and end of migrations, both on a daily and annual basis. And finally, they have to orient their migratory movements in space to reach their species- or population-specific wintering and breeding grounds, irrespective of the variable climatic conditions along their migratory routes.

Since Gustav Kramer's pioneering discovery in 1950 of sun compass orientation in migratory birds, a great deal of research has been devoted to this latter set of problems of spatial orientation. Progress has been fast and exciting in this area and several symposia (e.g. in Wallops Island 1981, in Tübingen 1977, in Tirrenia 1981) have been predominantly devoted to problems of orientation in space. The proceedings of these symposia (Galler et al. 1972; Schmidt-Koenig and Keeton 1978; Papi and Wallraff 1982) give good overviews of the states of avian orientation research at the respective times and can be consulted by both students and senior scientists as a quick and competent source of information. In contrast, other areas of bird migration research have developed only slowly and correspondingly have been treated much less intensively or not at all in symposia, a notable exception being the Falsterbo meeting in 1977 (Alerstam et al. 1978; see also the review articles in Rankin 1985). In recent years, however, the situation has begun to change and some substantial progress has been made in the understanding of several physiological and ecophysiological aspects of bird migration. It seemed timely, therefore, to bring researchers working in these areas together at a meeting on these themes. The symposium was organized by asking experts to

[1]Max-Planck-Institut für Verhaltensphysiologie, Vogelwarte, D-8138 Andechs, FRG

E. Gwinner (Ed.) Bird Migration

review their fields and assess our present understanding and the most promising directions future research should take. Following a planning phase of about 2 years, the meeting, attended by about 90 participants, took place on October 18-20, 1988 in the Evangelische Akademie in Tutzing. While, due to space limitations, the contents of the 20 poster presentations could unfortunately not be considered for publication, the 26 review papers that represented the core of the symposium are presented in this book.

Chapter 1 contains articles on the *Patterns of Migration* across certain particularly inhospitable areas of the world like oceans (Williams and Williams), the Arctic (Johnson and Herter), the Alps (Bruderer and Jenni) and East Africa (Pearson). It also reviews the effects of weather conditions on migratory timing (Richardson) and the phenomenon of molt migration (Jehl). The pictures that have emerged here are partly based on intensive observations and trappings of birds on migration, and partly on radar studies which continue to reveal valuable information about the birds' migratory performance. Studies of this kind set the stage for the physiological and ecophysiological analyses dealt with in subsequent chapters. Due to spatial limitations the treatment of migratory patterns had unfortunately to be selective rather than comprehensive so the reader must be referred to some other books and review articles for further information (e.g. Africa: Curry-Lindahl 1981; Asia: McClure 1974; America: Keast and Murton 1980; Rappole et al. 1983).

Chapter 2 covers *Ecological and Behavioral Aspects of Migration* and particularly concentrates on the interactions between birds and their environment in the wintering and breeding area. Depending on species and circumstances, migratory birds may either show a high degree of fidelity to specific wintering areas or move around in their general wintering range. The environmental factors as well as the internal mechanisms that play a role in this phenomenon are discussed in the papers by Ketterson and Nolan and by Terrill. In populations of "partial migrants" some individuals migrate while others spend the winter in their breeding area and it has long been debated whether environmental factors or differences in the genetic constitution are responsible for these behavioral differences. The paper by Schwabl and Silverin critically evaluates the evidence supporting either of these hypotheses and emphasizes the possible role of behavioral interactions in partial migration. The fascinating problem of how migratory birds select habitats in the wintering area and how they interfere with resident birds has received little attention until now, so the summary account by Leisler should be helpful in guiding future research. In a final presentation in this chapter O'Connor discusses evidence suggesting that the migratory status of a species also has effects on its ecology and eventually its reproductive strategies during the breeding season. The information accumulated in this chapter is not only of theoretical interest, but is also relevant in the context of conservation efforts, because effective protection of migratory birds depends on the understanding of their ecological requirements, not only at home but also along their migratory routes and in the wintering area.

Chapter 3 deals with *Physiological Adaptations to Migration*. Since fat serves as the major fuel for migratory birds, the understanding of how birds store and metabolize fat is of fundamental significance. Ramenofsky reviews the current information on avian lipid biochemistry, particularly during premigratory fat deposition and during migratory fat utilization. This paper is complemented by Bairlein's treatment of nutritional adaptations and food selection by migrants which correspond with the birds' migratory requirements. The endocrine mechanisms controlling migration are still not well understood, but with the advent of new techniques, particularly radioimmunoassays, rapid development can be expected in the near future. The review by Wingfield and Schwabl provides a good starting point for future research. The conceptual framework developed during the last 3 decades in biological rhythms research, particularly circadian rhythms, has prompted intensive work on circannual rhythms and revealed the significance of endogenous programs for seasonal timing processes. The merits as well as the possible limitations of this concept are discussed by Gwinner, while Berthold reviews the evidence for genetic factors as a basis for circannual programs and other seasonal timing mechanisms. Many normally diurnal birds migrate at night, so the problem emerges of how such birds cope with the visual problems of nocturnal migrations. Martin illustrates the ubiquity of nocturnal migration and defines the unanswered questions posed by this phenomenon.

Flight is the major means of locomotion during migration, so an analysis of its mechanics and physiology is of the utmost significance. Calculations based on physiological aerodynamic models and sophisticated physiological measurements of birds flying free or in wind tunnels have contributed substantially to our understanding of many phenomena. In Chapter 4 on *Avian Flight* the papers by Rayner, Butler and Woakes, and Nachtigall give impressive accounts of these achievements. A critical evaluation of some predictions based on aerodynamic theory can be found in the paper by Walsberg in Chapter 5.

The papers in the final Chapter 5 on *Strategies and Tactics of Migration* use ecophysiological approaches to understand how migratory birds, with their particular physiological makeup, cope with the particular environmental conditions to which they are exposed. Alerstam and Lindström show that optimization models can be applied to evaluate migratory patterns, while Biebach interprets the unusual migration patterns of birds flying across the inhospitable habitats of the Sahara desert in light of the constraints imposed on these birds by their physiology. A similar approach is used by Dolnik in his treatment of bird migration across the deserts and mountains of middle Asia and Kasakhstan – the first English summary of an impressive large-scale study in Russia. The final papers by Evans and Davidson and by Drent and Piersma deal with well investigated cases of shore-bird migration relating migratory performance to energy and time budgets, while Walsberg emphasizes the need for more studies on en-route foraging as a basis for understanding migratory energetics.

I am grateful to the Max-Planck-Gesellschaft and the Deutsche Forschungsgemeinschaft for sponsoring the symposium and to the Evangelische Akademie Tutzing, whose competent management in a beautiful setting provided the relaxed atmosphere that renders fruitful scientific discussions possible. The Vogelpark Walsrode and the Keller-Verlag, Starnberg, supplied additional funds that enabled us to organize the meeting in a generous fashion. Their help is gratefully acknowledged.

I also thank K. Able, T. Alerstam, P. Berthold, H. Biebach, D.M. Bryant, R. Drent, D.S. Farner, J. King, B. Leisler, F. Moore, H. Schwabl and J.C. Wingfield for valuable comments and suggestions on the scientific program during the planning stage of the meeting and J. Aschoff, A. Chandola-Saklani, F.W. Merkel, V. Nolan, K. Schmidt-Koenig and H.G. Wallraff for their efficient performance as chairpersons during the symposium. H. Biebach, J. Dittami, D. Janik and J. King helped editing the manuscripts. Their contributions as well as those of several anonymous reviewers are greatly appreciated. Last but not least I wish to thank the staff of the Vogelwarte at the Max-Planck-Institut für Verhaltensphysiologie, particularly Ch. v. Goetz, I. Schwabl and I. Tarabichi who gave their best in preparing the meeting, as well as my wife Helga for her support and advice throughout.

References

Alerstam TH, Enkell PH, Ulfstrand S (eds) (1978) Current bird migration research. Oikos 30:165–176

Curry-Lindahl K (1981) Bird migration in Africa. Academic Press, London

Galler S, Schmidt-Koenig K, Jakobs GJ, Belleville RE (eds) (1972) Animal orientation and navigation. NASA, Washington, DC

Keast A, Morton ES (eds) (1980) Migrant birds in the Neotropics: ecology, behavior, distribution, and conservation. Smithsonian Institution Press, Washington, DC

McClure HE (1974) Migration and survival of the birds of Asia. US Army Medical Component, SEATO, Med Projects, Bangkok, Thailand

Papi F, Wallraff HG (eds) (1982) Avian navigation. Springer, Berlin Heidelberg New York

Rankin MA (ed) (1985) Migration: mechanisms and adaptive significance. Contributions in Marine Science, Suppl 27, University of Texas Marine Science Institute, Port Aransas, Texas

Rappole JH, Morton ES, Lovejoy TE, Ruos JL (1983) Nearctic avian migrants in the Neotropics. Publ US Dept Interior, Fish and Wildlife Service

Schmidt-Koenig K, Keeton WT (eds) (1978) Animal migration, navigation, and homing. Springer, Berlin Heidelberg New York

Chapter 1 Patterns of Migration

The Orientation of Transoceanic Migrants

T.C. Williams and J.M. Williams[1]

1 Introduction

Land birds which migrate over oceans offer one of the rare opportunities in nature of testing the limits of avian orientation. Land birds, with a few exceptions, are unable to replenish their energy stores at sea and, thus, must continue their flights over the water under all conditions. They cannot stop, as birds often do over land, when conditions are unfavorable.

This review will concentrate on the movements of land birds making over-water flights of 1000 km or more. Sea birds also make long overwater flights and return accurately to remote islands such as Ascension in the Atlantic (Ashmole 1963) or Midway in the Pacific (Kenyon and Rice 1958; Fisher 1971), but the accuracy of orientation and the mechanisms by which long-distance homing flights are accomplished have not, to our knowledge, been elucidated. When migratory routes of sea birds are established by seasonal observation of species distributions or radar, they often reflect the occurrence of good feeding conditions for the birds at sea rather than an efficient route between two points (Grimes 1977; Harrison 1983).

2 Methods

Methods used to establish the movements of land birds over open oceans include seasonal distribution of species, banding, radar, and nocturnal visual techniques such as watching birds fly across the face of a full moon or through a light beam. Seasonal concentrations of migrants at coastal areas and their subsequent appearance on a distant shore support long overwater flights by land birds, such as the blackpoll warbler (*Dendroica striata*) (Nisbet 1970), although the data can be interpreted in other ways (Murray 1965). Observations of the seasonal occurrence of large numbers of land bird migrants on islands far from continental coasts or on ships at sea are strongly suggestive of overwater migration but must be interpreted with caution as they may represent not true migrants but rather vagrants or birds unable to continue their migration (Wingate 1973; Pratt et al. 1987). Birds which land on ships are often immatures, have reduced energy

[1]Swarthmore College, Department of Biology, Swarthmore, PA 19081, USA

E. Gwinner (Ed.) Bird Migration

reserves, or are species not known to be transoceanic migrants (Scholander 1955; McClintock et al. 1978; Williams and Williams 1978). Gauthreaux (1971) reports that all birds collected from oil drilling platforms in the Gulf of Mexico were severely emaciated. Radar observations at continental sites, on islands, and from ships at sea all indicate that some bird migrations pass overhead without being detected by observers (Drury and Keith 1962; Casement 1966; Eastwood 1967; McClintock et al. 1978; Larkin et al. 1979; Williams et al. 1981; Williams and Williams 1988).

Banding data unequivocally establishes two points on a migratory flight but the route between these two points is unknown. Even massive banding efforts over long periods of time for a highly visible species such as white stork (*Ciconia ciconia*) often do not clearly distinguish between an overwater migration as opposed to movement via continental coastlines (Kania 1985).

Radar provides more direct evidence of transoceanic migration, but any one radar can only follow birds over a small portion of a long overwater flight. Radars can reveal the relative density of migration, the altitude (with some radars), the track direction, and the ground speed of birds. If wind velocity can be determined at the altitude of birds, airspeed and heading of migrants can be calculated by addition of the ground speed and track vector to the negative of the wind vector. Little taxonomic information on migrant birds is available from the radar echo. Size and persistence of the echo when coupled with calculated airspeeds can be used to separate two classes of bird echoes, "passerine" and "shorebird or waterfowl". Further taxonomic distinctions on radar characteristics alone are unreliable due to extensive overlap in airspeed, average echoing area, and radar signature of birds (Eastwood 1967; Bruderer 1977; Larkin 1980; Williams and Williams 1980; Vaughn 1985). Temporal association between observed radar activity and the change in numbers of migrants seen on the ground can assist in identification of large taxonomic groups with major differences in time of migration, such as shorebirds (Charadriiformes) and waterfowl (Anseriformes). The absence of certain species from ground counts, however, cannot be taken to imply their absence from migrations seen passing overhead or radar (Williams et al. 1981; Williams and Williams 1988).

Watching birds cross the face of a full moon or flying through a light beam has been used to study birds departing the North American coast (Lowery and Newman 1966) and crossing the Gulf of Mexico (Gauthreaux 1971). These techniques are limited to altitudes of less than 1 km (Gauthreaux 1970) and are not suitable for shipboard use.

3 Geography of Transoceanic Migrations

Overwater flight per se appears to pose no barrier to land birds. Visual observations and radar confirm that large numbers of birds regularly cross several hundred km of open water such as the North Sea (Eastwood 1967), the Mediterranean (Casement 1966; Eastwood 1967; Thake 1982; Wood 1982), and the Great Lakes (Giltz and Burtt 1974). In general, overwater flight appears to occur wherever such flight represents a saving of energy and is within the flight capability of the birds. In most cases this occurs where winds are favorable and distances are shorter than a coastal migration (Williams et al. 1977a; Richardson 1979; Alerstam 1981; Alerstam et al. 1986; Williams and Williams 1988). A review of the current literature suggests that transoceanic migration of land birds is both more common and more extensive than previously thought. As shown diagrammatically in Fig. 1, land birds fly over almost all the major oceans of the world and some of their overwater flights may exceed 7000 km. In the following summary of the primary long-distance transoceanic migrations of land birds, all data are from species distribution and banding studies except where indicated.

3.1 North Atlantic

Several kinds of birds including loons (Gaviiformes), geese and ducks (Anseriformes), shorebirds and terns (Charadriiformes), and passerines (Passeriformes) migrate across the northern Atlantic Ocean, from the high Canadian arctic, over the Greenland icecap, across Iceland which is used heavily as a staging area during spring migration, and on to the Soviet Union, Scandinavia, western Europe or western and southern Africa. It is probable that these routes are used during both spring and fall migrations and involve long, nonstop

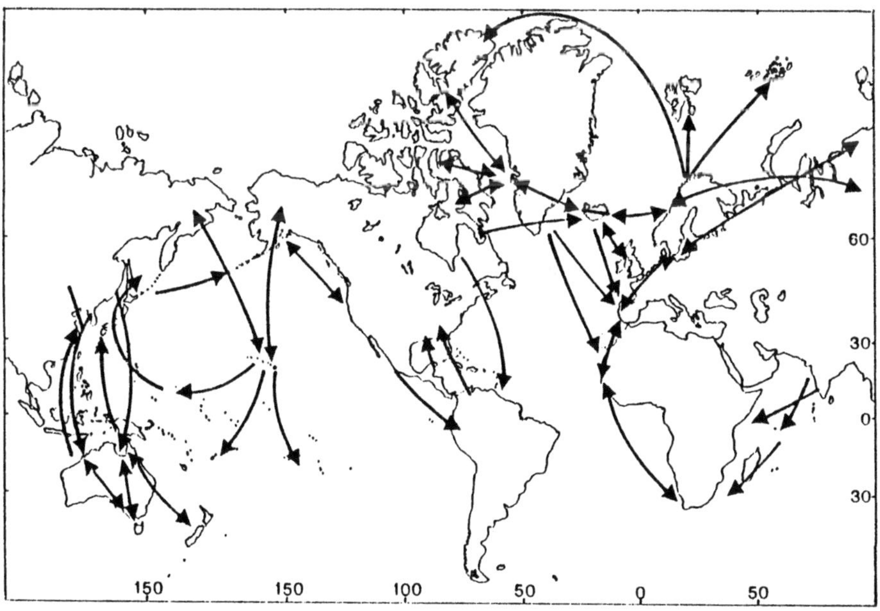

Fig. 1. Principal transoceanic migration routes. *Northward pointing arrows* schematically indicate spring migrations, *southward pointing arrows* indicate fall migrations. Actual migrations are usually broad front, hundreds or thousands of kilometers wide

flights of 3000 to 4000 km, much of which are over water. These migrations include white-fronted (*Anser albifrons*) and brent (*Branta bernicla*) geese, several species of high arctic shorebirds including ringed plover (*Charadrius hiaticula*), ruddy turnstone (*Arenaria interpres*), red knot (*Calidris canutus*), sanderling (*C. alba*), purple sandpiper (*C. maritima*) and dunlin (*C. alpina*), the arctic tern (*Sterna paradisaea*), and some passerines such as common wheatear (*oenanthe oenanthe*), lapland bunting (*Calcarius lapponicus*), and snow bunting (*Plectrophenax nivalis*) (Moreau 1972; Alerstam 1981; Morrison 1984; Alerstam et al. 1986; Dick et al. 1987; Piersma et al. 1987). In addition, spring migrations of red knot (*Canutus canutus islandica*) may also include flights from staging areas in northern Norway to Spitsbergen and Franz Josef Land (Davidson et al. 1986). Alternatively, these migrants may go directly from northern Norway to Ellesmere Island, flying to the north of Greenland, a distance of about 2000 km (Davidson et al. 1986; Uttley et al. 1987).

The average direction of spring migration observed by radar at Iceland (300°, GA Gudmundsson pers. comm.) was similar to the average direction for transglacial migrants in southern Greenland (301°, Alerstam et al. 1986) suggesting broad front, probably nonstop, migration from Iceland over the Greenland icecap. Radar observations of fall flights also indicate direct nonstop flights over Greenland ice with some migrants continuing their flights over the Denmark Strait (Alerstam et al. 1986). Radar studies in northeastern England (Evans 1966) and northeast Scotland (Bourne and Lucas 1981) revealed massive southward broad front migrations to the Brittish Isles.

3.2 Eastern Atlantic

Many of the birds involved in the migrations across the northern Atlantic fly south to parts of Europe where they winter or stop in staging areas before they continue to western and southern Africa (Engelmoer et al. 1984; Piersma 1987; Piersma et al. 1987). The route between western and southern Africa is not well documented (Engelmoer et al. 1984), although species such as bar-tailed godwit (*Limosa lapponica*) and red knot may fly nonstop over the water between the westernmost part of Africa and southern Africa, a distance of about 6000 km (Piersma et al. 1987).

Hilgerloh (1988) concludes from both radar and moon-watching data that in autumn passerines migrate in a broad front southwest down the Iberian peninsula and continue their flight 1000 km or more over the eastern Atlantic, landing on the Canary Islands or the adjacent west African coast.

3.3 Western Atlantic

During fall migration in the western Atlantic several species of birds take an overwater route from the northeast coast of North America south to the Caribbean and South America, a distance of about 3000 to 4000 km (Drury and Keith

1962; Williams and Williams 1978). Both shorebirds and passerines follow this route; some of the more common migrants are American golden-plover (*Pluvialis dominica*), ruddy turnstone, Hudsonian godwit (*L. haemastica*), red knot (*C. canutus rufa*), semipalmated sandpiper (*C. pusilla*), and blackpoll warbler (Hager 1966; Nisbet 1970; Richardson 1979; Morrison 1984; Williams 1984). During spring migration most of the migrants follow a land-based route north, rather than returning over water (Eastwood 1967; Nisbet and Drury 1967; Harrington and Morrison 1979).

Southward departures from the North American coast have been monitored by radar at several sites (Drury and Keith 1962; Drury and Nisbet 1964; Williams et al. 1977a, 1977b; Richardson 1979, 1980). Radar observations of en route migration were made as birds passed over Bermuda (Ireland and Williams 1974), and from ships at sea (McClintock et al. 1978; Williams and Williams 1978; Larkin et al. 1979). Radars at Antigua (Hilditch et al. 1973; Williams 1985), Barbados (Williams et al. 1977a), and Puerto Rico (Richardson 1976) revealed that the great majority of Atlantic migrants continue their nonstop flights over the Caribbean at altitudes of up to 6 km. Radar on the island of Tobago, near South America, indicated birds descending for landing (Williams et al. 1977a). Many of these North Atlantic observations were made simultaneously and thus waves of migrants could be followed as they crossed the Atlantic (Williams et al. 1977a). Flight times for early October migrants (presumably passerine) were 18 h to Bermuda, 64 to 70 h to the Caribbean and 80 to 90 h to South America (Williams et al. 1978). Shorebird echoes were detected in August and September and the low airspeeds of most migrants in October were attributed to passerines (Richardson 1976; Williams 1985).

Radar showed a low density return flight of spring migrants moving NW over Puerto Rico in late spring (Richardson 1974), and small numbers of shorebird echoes approaching the North American coast from the open Atlantic (Nisbet and Drury 1967; Richardson 1971). Our radar observations at Antigua and Bermuda in early spring, and spring radar observations at Charleston, SC failed to reveal northward movements over the Atlantic (pers. comm. JM and TC Williams and SA Gauthreaux Jr).

3.4 Gulf of Mexico

During the spring, migrants regularly cross the Gulf of Mexico from the Yucatan Peninsula to southern Louisiana, a distance of about 1200 km over water. Radar studies show that many of these migrants fly an additional 40 to 120 km north of the Louisiana coast before landing (Gauthreaux 1971). Most of the migrants are passerines and some shorebirds (Gauthreaux 1971), but also include the ruby-throated hummingbird (*Archilochus colubris*) (Alerstam 1981). Few birds cross the Gulf during fall migration apparently due to the prevailing southerly winds (Able 1972).

3.5 Arctic and Northern Pacific

Radar has shown both the Bering and Beaufort Seas to be sites of active bird movements during much of the spring and summer (Flock 1973; Richardson and Johnson 1981; Johnson and Richardson 1982). Ground counts indicate that the migrants include a wide spectrum of water birds and some shorebirds (Richardson and Johnson 1981; Johnson and Richardson 1982) as well as some passerines (Tove 1988). The inability of radar to distinguish shorebirds from waterfowl makes it difficult to be certain of land bird migration over water (Richardson and Johnson 1981), although Tove (1988), on the basis of timing of arrivals at the Aleutian Islands, concludes that shorebirds and some passerines may have crossed 1600 to 2000 km of open water from the central Kurile Islands.

3.6 Eastern Pacific

Several species of birds cross the Gulf of Alaska between coastal Alaska and California, or points further south, during fall migration and probably also spring migration. Brent geese cross about 4000 km of water from the Alaska Peninsula to southern California (Alerstam 1981), and various species of shorebirds, including western sandpiper (*C. mauri*) and dunlin also fly across the Gulf (Morrison 1984). Visual observations from ships in the Gulf of Alaska revealed some passerines, mostly warblers (Parulidae) and sparrows (Fringillidae), heading SE from Alaska toward Oregon and California, and numerous shorebirds, primarily plovers (Charadriidae), turnstones, dowitchers and sandpipers (all Scolopacidae) heading in a more southerly direction (115°–220°) toward the Pacific islands (Kloeckner et al. 1982). Radar observations from these ships were ineffective due to high seas in the Gulf.

The southward migration of Wilson phalarope (*Phalaropus tricolor*) between North and South America appears to inlude an overwater flight of about 4000 km from the southwestern United States or Mexico to Ecuador or Peru (Jehl 1988). It is estimated that these phalarope are capable of making nonstop flights of 7000 km based on premigratory weight gain (Jehl 1988).

3.7 Central Pacific

Kloeckner et al. (1982) and Williams et al. (1986) concluded on the basis of both radar and visual observations that the Hawaiian Islands represent a terminus of migration in the mid-Pacific where several species of birds which have flown from the Arctic stop to feed before continuing on to more distant islands in the southern Pacific. Shorebirds, which include Pacific golden-plover (*P. fulva*), wandering tattler (*Heteroscelus incanus*), ruddy turnstone, and sanderling are the most common migrants on the main Hawaiian Islands (Kloeckner et al. 1982; Pratt et al. 1987). Many Pacific golden-plover overwinter in Hawaii, establishing winter territories which they defend year after year (Johnson et al., 1989). Some of these plover are currently in their tenth winter on the same site on Oahu (OW Johnson pers. comm.). Ducks, primarily Northern pintail (*Anas acuta*) and northern shoveler (*A. clypeata*), also migrate each year to the Hawaiian Islands where they overwinter (Kloeckner et al. 1982).

3.8 Western Pacific

Several species of shorebirds which breed in northeastern Asia, Siberia, and western Alaska overwinter in southeast Asia, on islands in the western and southern Pacific, and in Australia and New Zealand (Portenko 1981; Pyle and Engbring 1985; Lane 1987a; Parish et al. 1987). Although many migrants move through Korea and eastern China (McClure 1974; Piersma 1985; Cheng 1987), several species including Pacific golden-plover, black-tailed godwit (*L. limosa*), bar-tailed godwit, ruddy turnstone, great knot (*C. tenuirostris*), and red knot (*C. canutus rogersi*) may migrate directly over the western Pacific (Xu and Purchase 1983; Zhang 1985, unpublished work; Lane 1987a; Parish et al. 1987). There is no evidence that passerines follow this overwater route, but barn swallows (*Hirundo rustica*) fly over the Pacific as far as the Philippines (McClure 1974).

Recent observations in eastern China at Chenlushan Island in the Yellow Sea (Zhou 1985) and at Beidaihe, Hebei Province (Williams et al. 1986; Bakewell 1987) have found that the majority of migrants are passerines and shorebirds and that spring migration along the coast is heavier than that in the fall. Radar observations at Haizhou Bay revealed relatively small numbers of birds, probably shorebirds, departing the Asian coast to the SE and SSE in autumn. The density of these migrants was on the order of 100 times less than the density of migrants moving SW parallel to the general trend of the coast. Spring migrations were denser than those in the fall at Haizhou Bay; no significant numbers of birds were detected arriving from the Pacific (from the E or SE) in spring (Mao 1984, 1985; Mao and Zhou 1987). Radar observations during spring migration at Hong Kong showed both passerine and shorebird echoes arriving from the SW to SSE which suggested flights of 1300 to 2200 km over the South China Sea. Some spring migrants were detected at up to 3.6 km altitude above Hong Kong suggesting that, like spring migrants in Louisiana, they continued their flight inland before landing. During fall migration, movements to the SW were recorded in September and October (Myres and Apps 1973; Melville 1980).

Radar studies from August to November on the island of Guam in the western Pacific showed migrants passing over the island to the SSE. Small numbers of several species of shorebirds, mostly Pacific golden-plover and ruddy turnstone, regularly stop on the island (Williams and Grout 1985). It is likely that some of these birds stay on the island for the winter, as in Hawaii (Johnson et al., 1989), while others continue on to more distant islands (Williams and Williams 1988).

Several migration studies of shorebirds have been conducted in the Port Hedland-Broome region of northwestern Australia, some of which included the use of radar. The coastal area in northwest Australia is used by over a half million shorebirds as a staging area during the southward migration and by about half that number during the northward migration (Lane and Jessop 1985a; Lane 1987a). Radar studies in March and April revealed most migrants departed the coast in a northwesterly direction (310°–330°) toward Borneo about 2000 km away or the south coast of China nearly 5000 km away (Lane and Jessop 1985b; Lane 1987b; Minton 1987). Many migrants also pass through the Carpentaria area in northeastern Australia. This may be a staging area for birds which go on to New Zealand, as well as for those continuing to southern Australia (Parish et al.1987; Barter et al. 1988).

Shorebirds in the Pacific may be making longer overwater flights than once thought. Larger birds such as Pacific golden-plover, black-tailed and bar-tailed godwit, and great knot are capable of flights from 6000 to 7500 km, depending on body weight (Barter 1986; Lane 1987a). Even red knot weighing 200 g or more may be capable of making nonstop flights between New Zealand or southern Australia and the south coast of China, a distance of 7500 km (Barter et al. 1988).

3.9 Indian Ocean

Some shorebirds, which breed in the eastern Palaearctic and winter in southern Africa, fly in a southeasterly direction over the Indian Ocean crossing the Iran-Pakistan coast. Some of these migrants, particularly large numbers of curlew sandpiper (*C. ferruginea*), stop in the Seychelles (Feare and High 1977). Other migrants, including passerines, bee-eaters (Meropidae), and falcons (Falconidae) fly a distance of about 3000 km from India to East Africa (Alerstam 1981). We know of no radar studies in this area.

4 Orientation

One might expect that migrants would show a change in orientation as they crossed a landmark as significant as the coastline of a continent, and diurnal, visual observations of coastal migrants, especially waterfowl, often do show flight with reference to a shoreline (Alerstam et al. 1974; Able 1980). Visual observations, however, can detect only a small percentage of all migrants, especially at continental sites where most migration occurs at night (Eastwood 1967). In contrast to the relatively small numbers of birds which can be observed visually, radar has revealed that the great majority of birds move in a broad-front migration and pass over continental coastlines and oceanic islands without deviation (Drury and Keith 1962; Eastwood 1967; Williams et al. 1977a; Richardson 1979; Alerstam 1981; Williams and Williams 1988). With the exception of raptors using updrafts along a coast (Richardson 1975; Kerlinger and Gauthreaux 1984), only a few radar studies have observed the use of a shoreline for orientation (Eastwood 1967; Alerstam et al. 1974, 1986; Able 1980).

The most extensive radar studies of transoceanic migration have been made for autumnal (southward) movements over the western North Atlantic (Williams et al. 1977a; Richardson 1980; Williams 1985). Weather conditions stimulating departure from North America are critical to a successful transatlantic flight. The passage of a cold front over the coast triggers departure with favorable winds and is associated with predictable weather conditions for the duration of the flight, in particular the absence of tropical storms (Williams et al. 1977a). Once over the sea the birds soon reach the frontal system which triggered their migration. Transoceanic migrants penetrate the front and continue moving SSE. When these birds encounter the NE trade winds near 25° N, their tracks are drifted to the S and SSW as they pass over the Caribbean. Groups of migrants which fail to penetrate the frontal system show near random distributions of headings on radar and are often attracted to ships (McClintock et al. 1978; Williams and Williams 1978; Larkin et al. 1979). Williams and Williams (1978) concluded that such behavior over the western North Atlantic was associated with large numbers of coastal migrants, primarily fringillids, not adapted to transoceanic conditions, which were blown out to sea by strong northwest winds and stranded.

We have suggested that birds moving from North to South America over the Atlantic maintain a simple compass heading for the duration of their flight (Williams and Williams 1978). Average headings computed from radar tracks and winds measured at the altitude of migrants were clustered tightly around 155° at Bermuda, Miami, Antigua, Barbados, and Tobago (Williams 1985). As shown in Fig. 2, the shift in track from SSE at the North American coast to S or SSW over Caribbean islands is explained as drift by winds from a constant heading (Williams 1985). Richardson (1979, 1980) observed large numbers of shorebirds departing Nova Scotia on headings too far east to reach South America unless they redirected their flight en route. Stoddard et al. (1983) showed that constant heading was a tenable strategy for birds departing the coast at least as far north

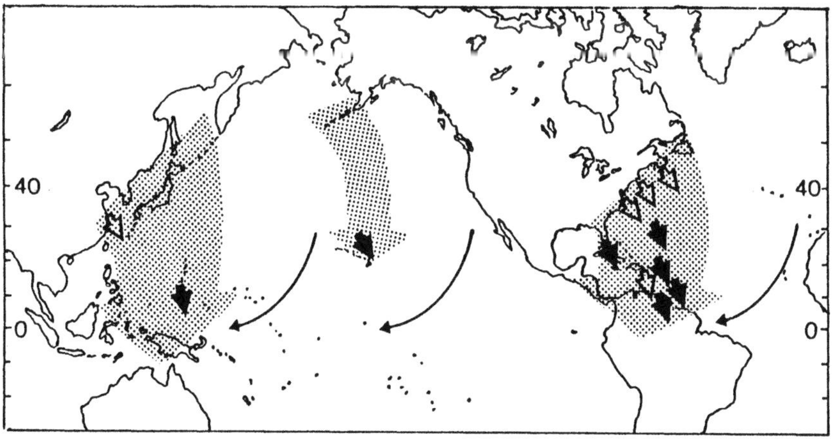

Fig. 2. Track, heading, and trade winds for three southward transoceanic bird migrations. Estimated extent of the heaviest broad-front migrations (track) is indicated by *gray arrows*. *Fine black arrows* indicate NE trade winds. Headings calculated from radar tracks and winds aloft at altitude of migrants are indicated by *solid arrows*. Estimated headings are indicated by *open arrows*. Radar observations from ships at sea are not included

as Nova Scotia. The range of successful headings, however, became increasingly restricted at northern latitudes. Acceptance of the constant compass heading hypothesis thus requires either significant mortality for birds observed departing Nova Scotia on headings to the E and SE or a Palaearctic-African destination for these migrants as demonstrated for more northern departures (Alerstam et al. 1986). Richardson (1976) estimated headings of migrants at Puerto Rico to be about 10° south of those observed on the more eastern islands. Stoddard et al. (1983) demonstrated that as a group of birds with normally distributed headings moved south to the Caribbean, observed headings should shift clockwise for more westward observation sites.

Radar observations on Guam, Mariana Islands, and at Haizhou Bay, China, showed many similarities with observations from the western North Atlantic. In both cases a large number of birds moved SW along the general continental coastline while a smaller number moved S to SE over the sea flying thousands of kilometers from land. Coastal migrants were strongly affected by local conditions; the passage of a cold front was the most potent stimulus for autumnal migration (Mao 1985; Mao and Zhou 1987). Transoceanic migrants appeared to maintain constant compass headings without regard to local conditions; they passed over islands en route without showing compensation for drift by winds (Williams and Williams 1988). Both timing of migration and airspeeds indicated large numbers of passerines making flights over the Atlantic while only shorebirds appeared to use the route over the Pacific (Williams and Williams 1988; Williams and Mao in press).

Observations at Hawaii (Kloeckner et al. 1982; Williams et al. 1986) revealed no birds passing high over the islands as in the Atlantic; most birds detected with radar were landing and the remainder had recently taken off. Birds approaching Hawaii showed compensation for wind drift as they prepared to land. A simulation of flight from the Aleutian Islands to Hawaii under the observed range of wind conditions en route showed that constant compass orientation was sufficient to bring migrants to the chain of Hawaiian Islands even under adverse weather conditions in the Gulf of Alaska (Kloeckner et al. 1982).

In summary, transoceanic migrants show three different phases of flight behavior, each characterized by different orientation techniques. At takeoff transoceanic migrants, like continental migrants, are sensitive to local wind and weather conditions (Richardson 1978). Birds north of the tropics respond to synoptic weather patterns. Birds departing from tropical areas avoid unfavorable wind conditions and may respond to other weather factors (Lowery 1951; Gauthreaux 1971; Richardson 1974; Lane and Jessop 1985b).

Once transoceanic flight is initiated, migrants appear to enter a programmed phase of flight in which they are insensitive to local wind and weather conditions or to landmarks. While continental radar studies indicate at least partial compensation for wind drift by migrant birds, migrants over the water show little or no compensation for wind drift (Able 1980; Alerstam 1981). Raptors over the sea may be an exception as Kerlinger (1984) finds compensation for drift at wind speeds less than 5 m/s; at higher wind speeds the birds avoid overwater flight. Unlike continental migrants which land after 5 to 10 h of flight and reassess weather conditions and their orientation on at least a daily basis (Eastwood 1967; Able 1980), transoceanic migrants may fly for 100 h without any apparent reference to local conditions or topography. As shown in Fig. 2, the orientation of transoceanic migrants may be largely explained by maintenance of a simple compass heading. Radar observations of migrants far from continents (where all birds will presumably be in the programmed phase of flight) do not show significant differences in the orientation of adult and juvenile autumnal shorebirds or between shorebirds and passerines (Richardson 1976; Williams 1985; Williams and Williams 1988). This uniformity of behavior for transoceanic migrants contrasts with continental radar studies which typically show a great diversity of migratory directions. Transoceanic migration thus appears to be strongly selected by present wind patterns and geography to produce a single, reliable solution based on genetically programmed behavior.

During the final phase of transoceanic flight, migrants again become sensitive to local wind and weather conditions and to landmarks as they descend for landfall. This was clearly seen at Hawaii (Kloeckner et al. 1982; Williams et al. 1986) and is consistent with the observations of Gauthreaux (1971) on the northern Gulf of Mexico. Thus the programmed behavior of transoceanic migrants appears to have both a direction and a distance or time factor.

5 Relationship to Experimental Work

Abundant evidence from experimental studies now supports the existence of vector orientation, a genetically encoded set of instructions for direction and duration of at least the first year migration of several species (Gwinner and Wiltschko 1978; Wiltschko and Wiltschko 1988; Berthold 1988; Gwinner, this Vol.). Studies of transoceanic migration not only offer field observations of vector orientation in action, but also suggest limiting conditions and additional behaviors needed to make the system fully operational.

Weather conditions during the flight play an important role in the overall success of vector orientation as a strategy for migrants. Predictability of weather is important both in terms of the accuracy of orientation and in mortality of migrants. Radar studies in the western North Atlantic have shown that selection of proper cues for initiation of transoceanic flights can reliably predict weather conditions over a total flight distance of 3500 km (Williams et al. 1977a). Not only do transatlantic migrants avoid tropical storms in this way but they also fly with favorable winds for much of their journey (Stoddard et al. 1983). Simulations of flight using vector orientation are much more informative if they use en route wind conditions known to have prevailed during observed migrations. In our experience, use of average winds tends to increase the variance of simulated tracks, to increase simulated flight time, and to shift simulated destination relative to the same simulation using observed winds.

Studies of transoceanic migrants show that taxonomic groups of migrants differ with respect to their use of vector orientation. Transoceanic migrants and continental migrants in North America differ with respect to environmental cues for the initiation of migration and in their response to islands, ships at sea, or local weather conditions.

Vector orientation and selection of proper weather cues for the initiation of the flight appear sufficient to allow the majority of migrants to make landfall reliably on distant coasts or island chains. Migrants must then use additional orientation techniques such as systematic search along coastlines, long-range visual orientation on cloud formations over islands, or perhaps bicoordinate navigation to locate traditional en route staging areas and breeding or wintering areas.

Acknowledgements. The preparation of this review was supported by National Science Foundation grant ATM-87-13981 (ROA) to Brown University, and by Swarthmore College through its Flack award and Faculty Research and Travel Funds.

References

Able KP (1972) Fall migration in coastal Louisiana and the evolution of migration patterns in the Gulf region. Wilson Bull 84:231–242

Able KP (1980) Mechanisms of orientation, navigation, and homing. In: Gauthreaux SA Jr (ed) Animal migration, orientation, and navigation. Academic Press, London, pp 283–373

Alerstam T (1981) The course and timing of bird migration. In: Aidley DJ (ed) Animal migration. Cambridge University Press, Cambridge, pp 9–54

Alerstam T, Bauer C, Roos G (1974) Fält-och radarstudier av östersjöejdrarnas *Somateria mollissima* vårsträck. Var Fagelvarld 33:15–27

Alerstam T, Hjort C, Högstedt G, Jönsson PE, Karlsson J, Larsson B (1986) Spring migration of birds across the Greenland inlandice. Medd Grønl Biosci 21:3–38

Ashmole NP (1963) The biology of the wideawake or sooty tern on Ascension Island. Ibis 103b:297–364

Bakewell D (1987) Autumn wader migration at Beidaihe, Hebei Province, China. Interwader Newsl 8:16–17

Barter M (1986) Great knots partly undone. Stilt 9:5–20

Barter M, Jessop A, Minton C (1988) Red knot (*Calidris canutus rogersi*) in Australia. Part I. Sub-species confirmation, distribution and migration. Stilt 12:29–32

Berthold P (1988) The control of migration in European warblers. Proc 19 Intern Ornithol Congr, pp 215–249

Bourne WRP, Lucas A (1981) A radar film of migration in northeast Scotland. Ibis 123:581

Bruderer B (1977) Beitrag der Radar-Ornithologie zu Fragen der Orientierung, der Zugphysiologie und der Umweltabhängigkeit des Vogelzuges. Vogelwarte 29:83–91

Casement MB (1966) Migration across the Mediterranean observed by radar. Ibis 108:461–491

Cheng T (1987) A synopsis of the avifauna of China. Science, Beijing Parey Berlin

Davidson NC, Strann KB, Crockford NJ et al. (1986) The origins of knots *Calidris canutus* in arctic Norway in spring. Ornis Scand 17:175–179

Dick WJA, Piersma T, Prokosch P (1987) Spring migration of the Siberian knots *Calidris canutus canutus*: results of a co-operative Wader Study Group project. Ornis Scand 18:5–16

Drury WH, Keith JA (1962) Radar studies of songbird migration in coastal New England. Ibis 104:449–489

Drury WH, Nisbet ICT (1964) Radar studies of orientation of songbird migrants in southeastern New England. Bird Banding 35:69–119

Eastwood E (1967) Radar ornithology. Methuen, London

Engelmoer M, Piersma T, Altenburg W, Mes R (1984) The Banc d'Arguin (Mauritania). In: Evans PR,

Goss-Custard JD, Hale WG (eds) Coastal waders and wildfowl in winter. Cambridge University Press, Cambridge, pp 293–310
Evans PR (1966) Migration and orientation of passerine night migrants in northeast England. J Zool 150:319–369
Feare CJ, High J (1977) Migrant shorebirds in the Seychelles. Ibis 119:323–338
Fisher HI (1971) Experiments on homing in Laysan albatrosses *Diomedea immutabilis*. Condor 73:389–400
Flock WL (1973) Radar observations of bird movements along the arctic coast of Alaska. Wilson Bull 85:259–275
Gauthreaux SA Jr (1970) Weather radar quantification of bird migration. BioScience 20:17–20
Gauthreaux SA Jr (1971) A radar and direct visual study of passerine spring migrants in southern Louisiana. Auk 88:343–365
Giltz M, Burtt HE (1974) Observations of the behavior, flight patterns, and movements of starlings and blackbirds which may endanger aircraft. In: Gauthreaux SA Jr (ed) Proc Conf Biol Aspects Bird/Aircraft Collision Problem. Dept Zoology Clemson University, Clemson, SC
Grimes LG (1977) A radar study of tern movements along the coast of Ghana. Ibis 119:28–36
Gwinner E, Wiltschko W (1978) Endogenously controlled changes in migratory direction of the garden warbler, *Sylvia borin*. J Comp Physiol 125:267–273
Hager JA (1966) Nesting of the Hudsonian godwit at Churchill, Manitoba. Living Bird 5:1–43
Harrington BA, Morrison RIG (1979) Semipalmated sandpiper migration in North America. Stud Avian Biol 2:83–100
Harrison P (1983) Seabirds, an identification guide. Houghton Mifflin, Boston
Hilditch CDM, Williams TC, Nisbet ICT (1973) Autumnal bird migration over Antigua, WI. Bird-Banding 44:171–179
Hilgerloh G (1988) Radar observations of passerine transsaharan migrants in southern Portugal. Ardeola 35:221–232
Ireland LC, Williams TC (1974) Radar observations of bird migration over Bermuda. In: Gauthreaux SA Jr (ed) Proc Conf Biol Aspects Bird/Aircraft Collision Problem. Dept Zool Clemson University, Clemson, SC
Jehl JR Jr (1988) Biology of the eared grebe and Wilson's phalarope in the nonbreeding season: a study of adaptations to saline lakes. Stud Avian Biol 12:1–74
Johnson OW, Morton ML, Bruner PL, Johnson PM (1989) Fat cyclicity, predicted migratory flight ranges, and features of wintering behavior in Pacific golden-plovers 91:156–177 Condor
Johnson SR, Richardson WJ (1982) Waterbird migration near the Yukon and Alaskan coast of the Beaufort Sea. II. Moult migration of seaducks in summer. Arctic 35:291–301
Kania W (1985) Results of bird ringing in Poland. Migrations of white stork *Ciconia ciconia*. Acta Ornithol 21:1–14
Kenyon KW, Rice DW (1958) Homing of the Laysan albatrosses. Condor 60:3–6
Kerlinger P (1984) Flight behavior of sharp-shinned hawks during migration. II: over water. Anim Behav 32:1029–1034
Kerlinger P, Gauthreaux SA Jr (1984) Flight behavior of sharp-shinned hawks during migration. I: over land. Anim Behav 32:1021–1028
Kloeckner PD, Williams JM, Williams TC (1982) Radar and visual observations of transpacific migrants. Elepaio 42:77–80
Lane B (1987a) Shorebirds in Australia. Nelsons, Melbourne
Lane B (1987b) Northern Australia expeditions, 1986. Stilt 11:3–5
Lane B, Jessop A (1985a) Report on the 1985 North-west Australia Wader Studies Expedition. Stilt 6:2–16
Lane B, Jessop A (1985b) Tracking of migrating waders in north-western Australia using meteorological radar. Stilt 6:17–28
Larkin RP (1980) Transoceanic bird migration: evidence for detection of wind direction. Behav Ecol Sociobiol 6:229–232
Larkin RP, Griffin DR, Torre-Bueno JR, Teal J (1979) Radar observations of bird migration over the western North Atlantic Ocean. Behav Ecol Sociobiol 4:225–264
Lowery GH Jr (1951) A quantitative study of the nocturnal migration of birds. Univ Kans Publ Mus Nat Hist 3:361–472

Lowery GH Jr, Newman RJ (1966) A continentwide view of bird migration on four nights in October. Auk 83:547–586
Mao Y (1984) Observation of bird migration by radar in Haizhou Bay, China. MS Thesis, East China Normal Univ, Shanghai
Mao Y (1985) Wader migration in Haizhou Bay, eastern China. Interwader Newsl 6:10–11
Mao Y, Zhou B (1987) Observations on bird migration by radar in Haizhou Bay. Acta Zool Sin 33:277–284
McClintock CP, Williams TC, Teal JM (1978) Autumnal bird migration observed from ships in the western North Atlantic Ocean. Bird-Banding 49:262–275
McClure HE (1974) Migration and survival of the birds of Asia. US Army Med Comp, Seato Med Proj, Bangkok
Melville D (1980) Bird migration through Hong Kong observed by radar and its importance for birdstrike control. Report to Birdstrike Research Unit, Agriculture and Fisheries Dept, Hong Kong
Minton C (1987) Report on visit to NW Australia 21 March to 5 April 1987. Stilt 11:6–12
Moreau RE (1972) The Palaearctic-African bird migration systems. Academic Press, London
Morrison RIG (1984) Migration systems of some new world shorebirds. In: Burger J, Olla BL (eds) Shorebirds, migration and foraging behavior. Plenum, New York (Winn HE (ed) Behavior of marine animals, vol 6) pp 125–202
Murray B (1965) On the autumn migration of the blackpoll warbler. Wilson Bull 77:122–133
Myres MT, Apps RF (1973) Migration of birds over the south coast of China recorded by radar. Nature 241:552
Nisbet ICT (1970) Autumn migration of the blackpoll warbler. Evidence for long flight provided by regional survey. Bird-Banding 41:207–240
Nisbet ICT, Drury WH (1967) Orientation of spring migrants studied by radar. Bird-Banding 38:173–186
Parish D, Lane B, Sagar P, Tomkovitch P (1987) Wader migration systems in east Asia and Australia. Wader Study Group Bull 49 Suppl, IWRB Special Publ 7:4–14
Piersma T (1985) Abundance of waders in the Nakdong Estuary, South Korea, in September 1984. Wader Study Group Bull 44:21–26
Piersma T (1987) Hink, stap of sprong? De reisbeperkingen van arctische steltlopers door voedselzoeken, vetopbouw en vliegsnelheid. Limosa 60:185–194
Piersma T, Beintema AJ, Davidson NC, Munster OAG, Pienkowski MW (1987) Wader migration systems in the east Atlantic. Wader Study Group Bull 49 Suppl, IWRB Special Publ 7:35–56
Portenko LA (1981) Birds of the Chukchi Peninsula and Wrangel Island. Amerind, New Delhi
Pratt HD, Bruner PL, Berrett DG (1987) The birds of Hawaii and the tropical Pacific. Princeton University Press, Princeton
Pyle P, Engbring J (1985) Checklist of the birds of Micronesia. Elepaio 46:57–68
Richardson WJ (1971) Spring migration and weather in eastern Canada: a radar study. Am Birds 25:684–690
Richardson WJ (1974) Spring migration over Puerto Rico and the western Atlantic, a radar study. Ibis 116:172–193
Richardson WJ (1975) Autumn hawk migration studied with radar. In: Harwood H (ed) Proc N Am Hawk Migration Conf. Syracuse 1974, pp 47–58
Richardson WJ (1976) Autumn migration over Puerto Rico and the western Atlantic: a radar study. Ibis 118:309–331
Richardson WJ (1978) Timing and amount of bird migration in relation to weather: a review. Oikos 30:224–272
Richardson WJ (1979) Southeastward shorebird migration over Nova Scotia and New Brunswick in autumn: a radar study. Can J Zool 57:107–124
Richardson WJ (1980) Autumn landbird migration over the western Atlantic Ocean as evident from radar. Proc 17 Intern Ornithol Congr Berlin 1978
Richardson WJ, Johnson SR (1981) Waterbird migration near the Yukon and Alaska coast of the Beaufort Sea. I. Timing, routes and numbers in spring. Arctic 34:108–121
Scholander SI (1955) Land birds over the western North Atlantic. Auk 72:225–239

Stoddard PK, Marsden JE, Williams TC (1983) Computer simulation of autumnal bird migration over the western North Atlantic. Anim Behav 31:173–180
Thake MA (1982) Kestrel migration through Malta in autumn in relation to weather. Riv Ital Ornithol 52:178–186
Tove MH (1988) Observations of Asiatic migrants in the western Aleutians. J Field Ornithol 59:101–109
Uttley JD, Thomas CJ, Davidson NC, Strann KB, Evans PR (1987) The spring migration system of Nearctic knots *Calidris calidris islandica*: a re-appraisal. Wader Study Group Bull 49 Suppl, IWRB Special Publ 7:80–84
Vaughn CR (1985) Birds and insects as radar targets: a review. Proc IEEE 73:205–227
Williams JM (1984) That divine impulse. Living Bird Quart 3:8–12
Williams JM, Grout PC (1985) Migrants on Guam, Fall 1983. Elepaio 46:41–44
Williams MD, Bakewell DN, Carey GJ, Holloway SJ (1986) On the bird migration at Beidaihe, Hebei province, China, during spring 1985. Forktail 2:3–20
Williams TC (1985) Autumnal bird migration over the windward Caribbean Islands. Auk 102:163–167
Williams TC, Mao Y (in press) A comparison of radar observations of bird migration at Haizhou Bay, China and Guam, Marianas. Auk
Williams TC, Williams JM (1978) Orientation of transatlantic migrants. In: Schmidt-Koenig K, Keeton WT (eds) Proc Symp Animal migration, navigation, and homing, Univ of Tübingen, Aug 17–20, 1977. Springer, Berlin Heidelberg New York, 462 pp (Proceedings in the Life Sciences) pp 239–251
Williams TC, Williams JM (1980) A Peterson's guide to radar ornithology? Am Birds 34:738–741
Williams TC, Williams JM (1988) Radar and visual observations of autumnal (southward) shorebird migration on Guam. Auk 105:460–466
Williams TC, Williams JM, Ireland LC, Teal JM (1977a) Bird migration over the western North Atlantic Ocean. Am Birds 31:251–267
Williams TC, Berkeley P, Harris V (1977b) Autumnal bird migration over Miami studied by radar: a possible test of the wind drift hypothesis. Bird-Banding 48:1–10
Williams TC, Williams JM, Ireland LC, Teal JM (1978) Estimated flight time for transatlantic autumnal migrants. Am Birds 32:275–280
Williams TC, Marsden JE, Lloyd-Evans TL, Krauthamer V, Krauthamer H (1981) Spring migration studied by mistnetting, ceilometer, and radar. J Field Ornithol 52:177–190
Williams TC, Williams JM, Kloeckner PD (1986) Airspeed and heading of autumnal migrants over Hawaii. Auk 103:634–635
Wiltschko W, Wiltschko R (1988) Magnetic orientation in birds. Curr Ornithol 5:67–121
Wingate DB (1973) A checklist and guide to the birds of Bermuda. Island, Bermuda
Wood B (1982) Weights and migration strategy of blackcaps *Sylvia atricapilla* wintering in Tunisia. Ibis 124:66–72
Xu W, Purchase D (1983) On the migration of banded waders between China and Australia. Mem Beijing Nat Hist Mus 23:1–12
Zhou X (1985) Observation of birds migration at the Chenlushan Island in Yellow Sea. MS Thesis, East China Normal University, Shanghai

Bird Migration in the Arctic: A Review

S.R. JOHNSON and D.R. HERTER[1]

1 Introduction

The migration of birds to and from the Arctic is one of the most remarkable events in the natural world. Though the Arctic is a hostile environment from October through May, 24 h of daylight and a superabundance of food during the brief summer provides favorable breeding conditions for millions of birds. About 135 species of birds migrate there to nest. Some make the longest migrations known to science. The Arctic tern (*Sterna paradisaea*) is the classic example; it makes a round-trip migration of over 30 000 km from the Antarctic to the Arctic (Salomonsen 1967a).

Arctic migrants comprise about 95% of the total number of bird species found in the Arctic. However, these migrants are only a fraction (< 10%) of the total number that regularly occur in temperate North America and Eurasia (Moreau 1972; Cramp and Simmons 1977, 1980, 1983, 1985; AOU 1983).

Very large single-species concentrations of birds occur in the Arctic in summer. The eastbound spring flights of eider ducks (*Somateria* spp.) and oldsquaws (*Clangula hyemalis*) past Point Barrow, Alaska involve millions of birds (Woodby and Divoky 1982; Johnson 1985). The vast flocks of shearwaters (*Puffinus* spp.) that migrate to rich feeding grounds in Arctic seas number in the millions (Salomonsen 1967b; Brown et al. 1975; Guzman and Myres 1982; Troy and Johnson 1987). The immense concentrations of dovekies (*Alle alle*) that swarm over spring staging areas in northern Baffin Bay (Renaud et al. 1982) and the approximately 20 million Arctic migrant shorebirds and waterfowl that stage in the Copper-Bering River delta system in Alaska (Islieb 1979; Senner 1979; Senner et al. 1981) and the Wadden Sea (Pienkowski and Evans 1984; Piersma et al. 1987) represent some of the world's most impressive bird concentrations.

Migrations in the Arctic are often characterized by a sudden arrival of nesting birds in spring, and an equally sudden departure in the fall. A majority of birds breeding in the Arctic are marine and coastal species that move inland to nest on the open Arctic tundra. Because circumpolar tundra habitats generally lie to the south of east-west oriented coastlines, migrations of waterbirds at high latitudes

[1]LGL Limited, Environmental Research Associates, 9768 Second Street, Sidney, British Columbia V8L 3Y8, Canada and LGL Alaska Research Associates, Inc., 505 W. Northern Lights Blvd., Anchorage, Alaska 99503, USA

E. Gwinner (Ed.) Bird Migration

(> 70° N) generally are along an east-west axis. The origins of these migration routes are no doubt highly traditional and probably have evolved in response to climatic changes 2–3 million years ago in the Early Pleistocene.

2 Evolution of Arctic Bird Migration

Bird migration may have originated as early as the Eocene, about 50 million years ago, shortly after modern forms of birds first appeared (Feduccia 1980). Many species alive 2 or 3 million years ago would probably be easily identifiable today, and some of these species no doubt migrated from one region to another (Gauthreaux 1982). Many of the species now present in the Arctic appear to have originated in the Old World. Studies by Mayr and Amadon (1951) and Mayr (1960) showed that large proportions of passerine species present in North America are of Palearctic origin, especially at far northern latitudes where well over half originated in the Old World.

The Pleistocene glaciations, which lasted until only 8–12 000 years ago, no doubt had a profound effect on migrations to the Arctic. In the Nearctic, ice sheets separated refugia at high latitudes from ice-free areas to the south (Irving 1972; Hopkins et al. 1982). But the Beringian refugium in Alaska was never separated from the Old World. The unglaciated land bridge connecting present-day Alaska to East Siberia was contiguous with ice-free portions of Asia and regions farther south (Moreau 1972; Denton and Hughes 1981; Hamilton and Thorson 1986; Fig. 1).

Beringia was a vast area of steppe-tundra habitat occupied by large and small mammals (Guthrie 1982) and perhaps as many as 15 to 17 non-migratory species of birds (Irving 1972). At its southern extension in present-day Alaska-Yukon, Beringia was probably separated from ice-free tundra and boreal habitats to the south by about 1500 km of ice, though some workers argue that at times an ice-free corridor may have existed between these two regions (Reeves 1973). Ice-free areas along the Pacific coast and on the northern tips of Ellesmere and adjacent islands, northern Greenland and northern Norway also may have provided habitat for birds (Macpherson 1965; Leech 1966; Brassard 1971; Fladmark 1978; Warner et al. 1982; Hjørt and Björk 1984; England 1987; Vorren et al. 1988).

Nevertheless, there has been a general reluctance by ornithologists to accept the possibility of migration from New World wintering areas to the Arctic during the Pleistocene. According to Irving (1960, 1972), Nearctic birds could not have migrated across the vast expanses of ice to reach refugia at high latitudes in Alaska and Canada. Similarly, Mead (1983) stated that there is today "... a very obvious route for shore birds from Arctic Canada and Greenland to the European and North African coasts, but recent glaciations will certainly have extinguished any ancestral bird populations that could have been using this route, especially over the time-scale that the continents have been moving in."

It has been more generally accepted that most birds gradually colonized newly exposed habitats, and expanded their breeding ranges in the Arctic during the slow process of climatic amelioration (Williamson 1975) and diminution of

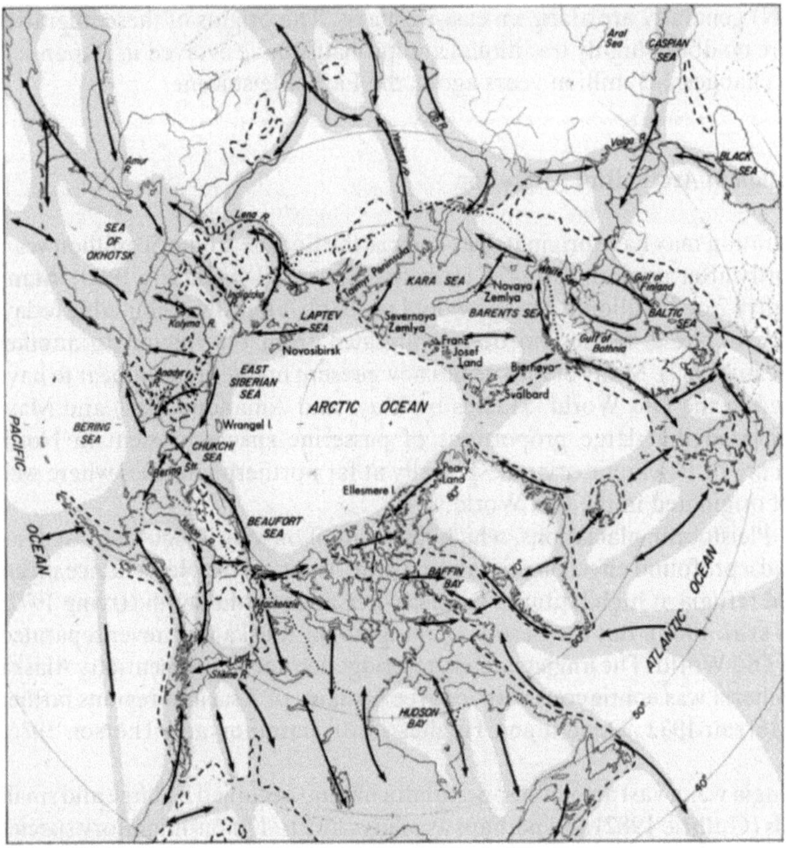

Fig. 1. Map showing major bird migration corridors to and from the Arctic. We have defined the Arctic as northern high latitude tundra and marine habitats that are snow and ice covered for most of the year (Remmert 1980), generally north of the tree line (*dotted line*). The *dashed line* indicates the maximum extent of ice about 20 000 years ago (Macpherson 1965; Denton and Hughes 1981; England 1987)

the ice sheets, which began about 15–18 000 years ago (Denton and Hughes 1981). Nørrevang (1963) and Hale (1984) discussed the distribution of the ice sheets and refugia in northern Europe and the postglacial climatic development that probably led to "recent" colonizations and migrations of birds to and from Arctic regions in the western Palearctic.

Based on present-day examples, however, it seems reasonable that many New World species may have traveled great distances over water or ice to reach Nearctic refugia (see Prater 1981). Similar or identical long-distance non-stop migrations, usually with the help of tailwinds (Richardson 1978; Reiff et al. 1986), are not uncommon today, particularly among waterfowl, shorebirds, cranes, and a few passerines (Snow 1953; Salomonsen 1967b; Thompson 1973; Wilson 1981; Kessel 1984; Owen and Gullestad 1984; Pienkowski and Evans 1984; Meltofte 1985; Mercier 1985; Davidson et al. 1986; Piersma 1987). Even today several species of birds nest only in what was formerly Beringia, and migrate solely to the New World (e.g., *Limnodromus scolopaceus, Calidris mauri, Aphriza virgata, Heteroscelus incanus, Arenaria melanocephala*).

Tailwinds probably were important in assisting migrations across ice sheets. Alerstam et al. (1986) discussed the importance of tailwinds in the long-distance migrations of birds over the Greenland inland ice, and Owen and Gullestad (1984) noted that tailwinds assist Svalbard barnacle geese in their 2400 km migration from Bjørnøya to Britain; other examples of long-distance migration assisted by tailwinds are discussed below. Various models of paleoclimatic conditions in Beringia (Barry 1982; Lamb and Woodroffe 1970; Gal-Chen 1982) during the glacial maximum about 18 000 years ago indicate that winds were favorable for seasonal long-distance migrations. Although there are no reconstructions for "spring" or "autumn", the models indicate southerly flows entering the Arctic from the Pacific across Alaska during July, and north and northeasterly airflows during January (Williams et al. 1974 in Barry 1982). Thus, given our current understanding of long-distance bird migrations, the distribution and number of Pleistocene refugia, the possible existence of an ice-free corridor to Beringia, and climate and weather patterns in the Nearctic (especially wind patterns in the late Pleistocene), it is probable that many birds migrated to the Arctic throughout the Pleistocene.

3 A Review of Some Arctic Migration Studies

3.1 Palearctic

The first detailed studies of bird migration were conducted in Europe. The Danish ornithologist H.C. Mortensen first attached tarsus bands to birds in 1890 and subsequently determined the migration routes of northern pintails (*Anus acuta*) and several other species that nest in the Old World Arctic (Jespersen and Tåning 1950). The banding studies carried out from 1901 to 1944 at Rossitten on the Baltic coast (Thienemann 1927) were the first long-term investigations of migration using banding techniques (Dol'nik 1969). These and later large-scale banding studies at Rybachi by the Zoological Institute of the USSR Academy of Sciences (Lusis et al. 1961; Paevskii 1973; Kumari 1976) have determined the migration routes of many species, mostly passerines, to and from the Old World Arctic.

Løvenskiold (1964) reviewed the ornithology of the Svalbard Archipelago and first identified Bjørnøya (Bear Island) as an important resting area for Svalbard-bound migrants enroute from northern Norway. He also reviewed new records of birds migrating to Svalbard in relation to their status on Bjørnøya and in northern Norway. His extensive annotated bibliography (655 citations covering the years 1598 to 1958) included virtually all early records of bird migration near Greenland, Iceland, Svalbard, Bjørnøya, and Novaya Zemlya. Later, Haftorne (1971) reviewed the distribution, abundance, and migrations of birds in Arctic Norway. Morrison (1977) summarized current research on Arctic nesting shorebirds that migrate through Britain and western Europe, with special emphasis on the red knot (*Calidris canutus*) which nests in High Arctic Greenland and Canada. More recently, Pienkowski and Evans (1984) reviewed the migration of shorebirds in the western Palaearctic. Leslie and Lessels (1978) described the age-segregated autumn migrations of dunlin (*Calidris alpina*) through Arctic Norway, and Greenwood (1984) summarized the migrations of all subspecies of dunlin and provided an overview of migration patterns in the Holarctic. Davidson et al. (1986) showed the importance of northern Norway as a staging area for *C. canutus* that migrate to northern Greenland, Ellesmere Island, and Svalbard. Holgersen (1960) reported on the migration routes and staging on Bjørnøya of Svalbard pink-footed geese (*Anser brachyrhynchus*), and recently Gullestad et al. (1984) and Owen and Gullestad (1984) discussed the importance of Helgeland and Bjørnøya as critical spring and fall staging areas, respectively, for migrant Svalbard barnacle geese. Ebbinge et al. (1982) reviewed the importance of spring staging areas for all Arctic nesting geese wintering in western Europe, and very recently Piersma et al. (1987) resynthesized all information on shorebird migration systems (routing, timing, volume, staging areas) in the East Atlantic flyway.

Hildén (1971), Bergman (1974, 1976, 1977), and Bergman and Donner (1964) used radar and visual techniques to study spring and fall migration, and the various factors which influence the direction, speed, altitude, and timing of movements of Arctic nesting waterfowl and gulls between the Baltic (Gulf of Bothnia and Gulf of Finland) and the White Sea. Bianchi (1983) described the times and routes of the large passages of waterbirds that migrate in spring along the east side of the White Sea, and Kokhanov (1983) investigated apparent weather-related variations in the summer (molt) and autumn migrations of sea ducks through the White Sea.

Belopol'skii (1957) and his colleagues at the Seven Islands Bird Sanctuary in the Barents Sea conducted a number of important studies concerning the migrations of colonial gulls, terns, alcids, and seaducks from the 1930s through 1950s. Their investigations of the migration routes, physiological conditions of migrants, and the timing of arrival at colonies were the first of their kind in the Soviet Arctic. McClure (1974), Cramp and Simmons (1977, 1983), Owen (1980), and Gibson (1981) described the spring migration of birds bound for Arctic Alaska and eastern Siberia along the East Asia flyway. Vronsky (1985) studied the spring migration of birds in the lower Yenisey River valley and found that this area and the Yenisey Bay coast are the "... main path of migration of some avian species to the north on the Taimyr." Summers et al. (1987) reviewed existing information on shorebird migration systems in western Asia and the USSR, with emphasis on the Taimyr region. Kistchinski (1979) reviewed the migrations of birds of northern Asia, including a detailed treatment of the migration of Wrangel Island snow geese (*Chen caerulescens*). Uspenskii (1969) discussed the possibility of polar migrations, primarily by geese, shorebirds, larids, and passerines, between Novosibirsk and Severnaya Zemlya-Franz Josef Land to the northwest, and the remote possibility of a transcontinental migration to the Soviet Central Arctic from Alaska and Greenland, although he found little evidence of the latter.

3.2 Nearctic

In Alaska, Murdoch (1885) and Palmer (1889) conducted early studies of bird migration in the Bering Strait and the eastern Chukchi Sea, an area of convergence for both New and Old World migrants. Later, Swarth (1934) studied bird migrations in this same area and discussed the timing and volume of migration past the Yukon-Kuskokwim Delta (Nunivak Island) and through the northern Bering Sea. Bailey (1948) discussed existing knowledge of bird migrations in Arctic Alaska and far east Siberia, with special references to the timing and volume of passage through the Bering Strait region, past Point Barrow and along the Beaufort Sea coast to the Canada border. Most of his work was conducted in the vicinity of Bering Strait and Point Barrow in the 1920s. Anderson (1913, 1915, 1937) reported on bird observations (including the timing and directions of migrations) and bird collections made during explorations in the Chukchi and Beaufort seas by Stefansson (1908–1912) and during the Canadian Arctic Expedition to the Beaufort Sea (1914–1916), and Brooks (1915) and Dixon (1943) recorded the directions, timing, and volume of spring and summer migrations of waterbirds along the Beaufort Sea coast of Alaska.

Studies began later in northwestern Canada. Cooke (1915) first described the importance of the Mackenzie River valley in the Northwest Territories as a major migration corridor to the Arctic for migrants crossing the continent from the Gulf of Mexico and the east coast of the United States, and Sladen (1973) described the spring migration of tundra swans from Chesapeake Bay, through the Great Lakes, across the Canadian prairie provinces and down the Mackenzie River valley to the Arctic coasts of Alaska and western Canada. He pioneered the extensive use of colored neck-collars on geese and swans to more easily determine their migrations and flyway affiliations (Sladen and Kistchinskii 1977).

General reviews by Gabrielson and Lincoln (1959), Irving (1960, 1972), Johnson et al. (1975), and Kessel and Gibson (1980) included information on the status, timing, routing, and rates of passage of Alaskan Arctic migrants, and Richardson and Johnson (1981), Johnson and Richardson (1982), and Woodby and Divoky (1982) summarized existing information on the timing, routing, and volume of waterbird migration along the Alaskan and Canadian Beaufort Sea coasts.

Arctic whaling and then the search for Sir John Franklin, lost in 1845 while searching for the Northwest Passage, stimulated a pulse of exploration in the Canadian Arctic Archipelago. Until this busy period of Arctic exploration the reports by Sir John Richardson (1836, 1852) were the earliest accounts by a naturalist of bird migration in the Canadian Eastern Arctic. Ellis and Evans (1960) described the timing and volume of bird migration in the Foxe Basin area southwest of Baffin Island. Kumlien (1879) noted both Palearctic and Nearctic migrant plovers, *Charadrius hiaticula* and *C. semipalmatus*, respectively, nesting in this region. Almost 80 years later during the Baird expedition, Wynne-Edwards (1952) studied birds and bird movements along the same stretch of coast on southeast Baffin Island. Salomonsen (1950) recognized the potential importance of Iceland as a critical staging area for migrants moving to and from High Arctic

Greenland and Canada, and this has been confirmed by several researchers in recent years (Morrison 1975, 1977, 1984; Maltby-Prevett et al. 1975; Wilson 1981; Meltofte 1985; Alerstam et al. 1986; Morrison and Myers 1987; Uttley et al. 1987; Whitfield and Magnusson 1987). Salomonsen (1950, 1967b) and Alerstam et al. (1986) also confirmed earlier observations by Helms (1926) of a direct non-stop migration by several species over the Greenland ice cap. Meltofte (1976) showed that birds migrating to Peary Land, North Greenland, had two separate migration routes to the area, one along the west coast of Greenland and another along the east coast. Schanning (1933); Degerbøl and Møhl-Hansen (1935), and Hørring (1939) conducted early studies of birds and their movements in East Greenland.

4 Patterns of Migration to the Arctic

4.1 Long-Distance Migrants

Many Arctic species make long-distance migrations to and from wintering grounds in the southern hemisphere (Cramp and Simmons 1977, 1983; AOU 1983; Morrison and Myers 1987; Piersma et al. 1987). Like the Arctic tern, the peregrine falcon (*Falco peregrinus*), many shorebirds (*Numenius phaeopus, N. borealis, Limosa lapponica, Charadrius hiaticula, C. semipalmatus,* and *C. mongolus, Pluvialis dominica, P. squatarola, Arenaria interpres, Calidris* spp., *Tringites subruficollis, Philomachus pugnax, Aphriza virgata, Heteroscelus incanus, Phalaropus fulicaria, P. lobatus*), the three jaegers (*Stercorarius* spp.), and Sabine's gull (*Xema sabini*) also winter well into the southern hemisphere and migrate over at least 80° of latitude to nest in the Arctic (Cramp and Simmons 1977, 1983; AOU 1983; Fig. 2).

Nilsson (1858) first proposed that wintering areas of allopatric populations are located in the same latitudinal sequence as their nesting areas; movements between these areas follow a pattern called "chain migration" (Lundberg and Alerstam 1986). Palmén (1874) described "leap-frog" migration, which Mead (1983) characterized as "... the northernmost breeding populations of a species actually wintering further south than any others." Salomonsen (1955) examined various examples of chain and leap-frog migrations in the Arctic and discussed their evolutionary significance. Lundberg and Alerstam (1986) discussed the selective forces that could promote the evolution of these types of migrations and Alerstam and Högstedt (1980, 1985) and Pienkowski et al. (1985) further examined the mechanism, its significance, and possible alternatives to the leap-frog hypothesis (see also Drent and Piersma, this Vol.).

Migration routes of long-distance migrants are similar in both the Nearctic and Palearctic. Shorebirds in general stage at widely separated coastal sites on both shores of the North Atlantic and North Pacific. A few species on each continent are seen more commonly in the interior [e.g., Baird's sandpiper (*Calidris bairdii*), buff-breasted sandpiper, Eurasian dotterel (*Charadrius morinellus*)] (Morrison and Myers 1987; Cramp and Simmons 1983), although many

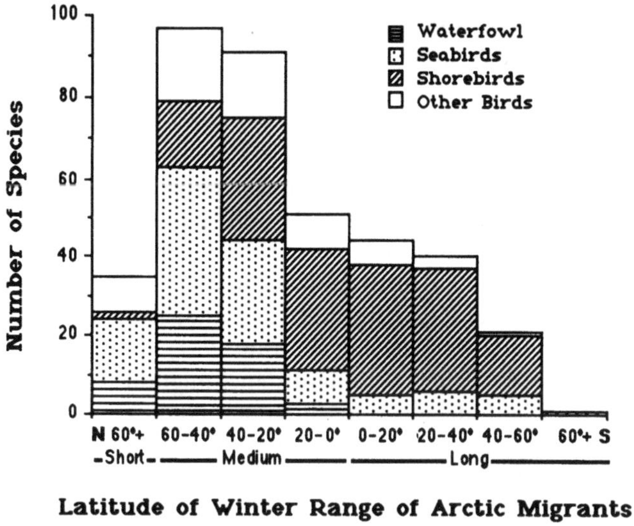

Fig. 2. Latitudes from which Arctic nesting birds (short-, medium-, and long-distance migrants) initiate spring migration. A few species may winter over a broad range of latitudes and therefore may be included in several categories. (Based on information from Cramp and Simmons 1977, 1983; Cramp 1985; AOU 1983; Harrison 1985; Hayman et al. 1986)

coastal-staging species also undertake long non-stop flights over the interior and over oceans. Once in the Arctic, spring migrants disperse to breeding areas and few large concentrations are seen. Arctic nesting larids, jaegers, and red phalaropes generally follow trans-oceanic routes at southern latitudes, but at higher latitudes visible migrations of these species are often noted along coastlines (Johnson 1975; Johnson and Herter 1989; Richardson and Johnson 1981; Orr et al. 1982; Haney and Stone 1988). In general, however, trans-oceanic migrants are inconspicuous until they arrive near tundra breeding grounds. The final leg of spring migration for Sabine's gulls, long-tailed and parasitic jaegers, and several other species may be at high altitudes over northern interior regions, thence along river valleys and other physiographic features to tundra nesting grounds (Dean et al. 1976; Richardson and Johnson 1981; AOU 1983).

Other Arctic shorebirds, e.g., *Calidris ruficollis, C. ferruginea, C. acuminata,* and *Limicola facinellus, Calidris tenuirostris, C. temminckii, C. subminuta, Philomachus pugnax* and *Heteroscelus brevipes*, are generally restricted to Old World arctic migration systems, except in the zone of convergence near Bering Strait where variable numbers of Old and New World species regularly cross to adjacent continents to nest (AOU 1983; Cramp and Simmons 1977, 1983; Pienkowski and Evans 1984; Hayman et al. 1986). In the North Atlantic Basin a similar zone of convergence for Old and New World migrants includes Iceland, Greenland, and Arctic Canada (Owen 1980; AOU 1983; Morrison 1984). A few species of passerines also undergo migrations to Arctic regions from the southern hemisphere. Northern wheatears (*Oenanthe oenanthe*), yellow and white wagtails (*Motacilla flava* and *M. alba*), and pechora and red-throated pipits (*Anthus gustavi, A. cervinus*) migrate between southern Africa and Australasia to nest on Arctic tundra in the Palearctic and Nearctic, distances of over 10 000 km (AOU 1983).

Some seabirds, such as the shearwaters (*Puffinis tenuirostris, P. griseus*, and *P. gravis*) nest in the southern hemisphere during the Austral summer and migrate over 10 000 km to Arctic seas during the Austral winter (Cramp and Simmons 1977; Salomonsen 1979a,b; Guzman and Myres 1982; AOU 1983). These migrations take them over unproductive tropical seas where they apparently do not feed (Guzman and Myres 1982), so these birds must accumulate substantial fat reserves in arctic and subarctic seas for their long return migration to the southern hemisphere (Guzman and Myres 1982; Troy and Johnson 1987).

It is notable that many of the exceptionally long-distance migrants, i.e., birds that migrate over 120–140° of latitude, are waterbirds that take trans-oceanic routes or routes along coasts or through inland waterway systems. Some of the more terrestrial waterbirds, such as shorebirds, are very dependent on specific stopover or staging areas where they refuel for subsequent stages in their migration. Work by Alerstam et al. (1986) has shown how the timing and routes of migration and gaps in the circumpolar breeding distributions of some Arctic birds may have evolved in relation to the locations of major staging areas along different flyway systems. Such constraints on migration could also facilitate racial segregation in some Arctic migrants. In fact, Irving (1960, 1972) had earlier speculated that the discrete nature of the routes of some migrations to the Arctic, such as along the Mackenzie River or the Yukon River drainages (Irving 1961; West and Peyton 1980), may serve as partial isolating mechanisms that may help maintain distinct breeding populations.

4.2 Medium-Distance Migrants

This group includes many birds that migrate across 40–60° of latitude (3500–6500 km) to nest in the Arctic (Fig. 2). Most species are waterbirds. Included are most loons (*Gavia* spp.), arctic plovers, curlews, godwits, and the *Calidris* sandpipers, several gulls, most arctic waterfowl (except seaducks), some arctic passerines such as *Anthus spinoletta, Eremophila alpestris, Calcarius lapponicus, Plectrophenax nivalis*, several raptors (*Buteo lagopus, Circus cyaneus, Asio flammeus*), and sandhill crane (*Grus canadensis*). Like the long-distance migrants, many of the waterbirds in this group make non-stop migrations and therefore depend on staging areas along migration routes where food is abundant and fat can be accumulated for subsequent stages of migration and for nesting.

Traditional migration pathways for most species, especially the waterfowl, are well known (West et al. 1968; Bellrose 1980; Cramp and Simmons 1977; Owen 1980; AOU 1983) because they occur within the heavily populated temperate zone, they tend to be less oceanic than seabirds and shorebirds, and because spring and fall migrations of most waterfowl involve large flocks that are visible during daylight hours (Palmer 1976; Owen 1980). As with the long-distance migrants, stopover or staging areas for medium-distance migrants appear to be highly traditional and physiologically important (Senner et al. 1981; USFWS 1981; Ebbinge et al. 1982; Pienkowski and Evans 1984; Piersma et al. 1987). Staging areas for northbound migrants tend to be in temperate latitudes close to the wintering grounds, while those used by southbound migrants tend to be in arctic or boreal zones close to the breeding grounds. This allows for fat storage prior to the long non-stop legs of spring and fall migration (Drent et al. 1978; Owen 1980; Ebbinge et al. 1982).

Many seabirds that nest and overwinter at high latitudes, such as many gulls, northern gannet (*Morus bassanus*), and several large alcids (*Uria aalge, Alca torda, Fratercula arctica*, and *F. cirrhata*) also migrate to temperate regions during winter (Harrison 1985). Most of the visible migration of these species is coastal or nearshore and generally does not involve concentrations of diurnal migrants. In contrast, migrant shorebirds are visible in spectacular concentrations at staging sites along both coasts of the North Atlantic and North Pacific; *Calidris alpina* and *C. mauri* number in the millions at some spring staging sites (Senner et al. 1981; Butler et al. 1987; Morrison and Myers 1987; Piersma et al. 1987). Some dunlins, especially Alaskan and Siberian races, molt near their Arctic breeding grounds (Holmes 1966; Cramp and Simmons 1983); consequently, noticeable concentrations of this species also occur along Arctic coastlines prior to fall migration.

4.3 Short-Distance Migrants

Birds in this group normally migrate across 10–20° of latitude (about 1000–2000 km) and are often forced from very high latitudes by ice or snow cover and the unavailability of food. Several migrate to areas above 60° N latitude (Fig. 2). They are primarily marine waterfowl (*Chen canagica, Somateria* spp., *Clangula hyemalis*), northern fulmar (*Fulmarus glacialis*), most alcids (*Alca torda, Uria* spp., *Aethia* spp., *Alle alle, Fratercula* spp.), several high Arctic gulls (*Larus hyperboreus, L. glaucoides, Rhodostethia rosea, Pagophila eburnea*), gyrfalcon (*Falco rusticola*), and snowy owl (*Nyctea scandiaca*). Other species, such as black guillemot (*Cepphus grylle*), ptarmigan (*Lagopus* spp.), common raven (*Corvus corax*), and hoary redpoll (*Carduelis hornemanni*) either migrate very short distances (less than 10° latitude) or not at all (Cramp and Simmons 1977; AOU 1983). These birds are well-adapted for obtaining food and surviving in cold northern seas (seaducks, *Fulmaris glacialis*, alcids, gulls) or in tundra habitats (*Falco, Lagopus, Nyctea, Corvus, Carduelis*) (Gessaman 1972; West 1972a,b; Johnson and West 1975; Platt 1976). Some of the resident terrestrial species make short-distance migrations to boreal regions south of the Arctic tundra. These migrations may be irruptive, i.e., may be related to lemming population cycles or seed production cycles (Uspenskii 1969; Irving 1972). Population cycles in ptarmigan can result in migrations of tens of thousands of birds from Arctic tundra to forest habitats farther south (Salomonsen 1950; Irving et al. 1967).

Only a few species of birds remain in the Arctic throughout the winter. Ptarmigan can survive if sufficient willow stems remain exposed above the snow (Irving et al. 1967), and some gyrfalcons remain if ptarmigan or other prey are available (Platt 1976; Palmer 1988). Snowy owls may remain in High Arctic areas such as northern Greenland and Novaya Zemlya if lemming and ptarmigan populations are sufficient to sustain them (Uspenskii 1969; Cramp 1985). Ravens also overwinter in the Arctic where they scavenge from humans and mammalian predators (Uspenskii 1969). Small granivorous passerines such as *Carduelis hornemanni* and *Plectrophenax hyperboreus* may overwinter in sheltered tundra areas where seeds are abundant (Irving 1960, 1972; West 1972b).

In some circumstances waterfowl and gulls may remain at high latitudes to feed in patches of open water long after most other fall migrants have left (Uspenskii 1969). These delayed migrants are forced out of the Arctic as the last ice-free areas freeze in early winter, thus they sometimes must cross over habitats that have been ice- or snow-covered for months. Haila (1980) discussed the ecological significance of delayed migration in arctic waterfowl and suggested that there is an adaptive advantage in birds trying to remain at high latitudes near nesting areas (Haila et al. 1986).

Many seaducks, alcids, and kittiwakes and other gulls overwinter at or south of the ice front in the Bering Sea (Irving et al. 1970), and at least 1–2 million migrate north through the Chukchi, Beaufort, and East Siberian seas in April and May as soon as lead systems develop in the Arctic pack ice. This migration is on a broad front mostly offshore, but there is also a strong coastal component (Thompson and Person 1963; Johnson 1971; Portenko 1972; Flock 1973; Searing et al. 1975; Richardson and Johnson 1981; Woodby and Divoky 1982). A very similar situation exists in the North Atlantic where seabirds and waterfowl, overwintering in polynyas (Brown and Nettleship 1981) and along the ice front off Greenland, Newfoundland, Jan Mayen, and Norway, make relatively short-distance migrations to and from nesting grounds at higher latitudes (Tuck 1971; Brown et al. 1975; Salomonsen 1967b; Hjørt 1976; McLaren 1982).

Most short-distance seabird migrations are relatively unremarkable, but the migration of *Alle alle* to spring staging areas in northwest Baffin Bay involves millions of birds (Renaud et al. 1982). A very large proportion of the world population of *Rodostethia rosea* migrates to fall staging locations near Point Barrow, Alaska. Several species form large wintering concentrations in polynyas and lead systems in the pack ice (Irving et al. 1970; Brown and Nettleship 1981), but movements to and from these areas usually involve gradual shifts as the ice front advances and retreats (Brown et al. 1975); one exception is the relatively rapid and concentrated southward migration of ivory gulls along the east coast of Greenland in late fall (Hjørt 1976). Juvenile thick-billed murres (*Uria lomvia*) are flightless when they leave their nesting colonies, some at 74° N. They nevertheless migrate to overwintering areas south of the ice front by swimming (Salomonsen 1967b, 1979a), sometimes faster than 40 km/day (Gaston 1982).

Short-distance overland movements by Arctic birds between the Baltic, White, and Barents Seas have been well studied (Hildén 1971; Yakobi and Yigi 1972; Bergman 1974, 1976, 1977; Kumari 1980). Similar migrations in Siberia (Dement'ev and Gladkov 1951, 1952; Uspenskii 1969; Portenko 1972; AOU 1983) and Alaska (Cade 1955; Irving 1960; Johnson et al. 1975; Richardson and Johnson 1981) remain poorly understood. Overland migrants often follow rivers, such as the Amur, Yenisey, Lena, Indigirka, Kolyma, and Anadyr in Siberia, and the Kobuk, Noatak, and Colville in Alaska (AOU 1983; Dement'ev and Gladkov 1951, 1952; Cramp and Simmons 1983) (Fig. 1).

4.4 Molt Migration

Molt migrations of post-breeding or non-breeding waterfowl from Arctic or sub-Arctic nesting areas to specific molting areas were reviewed by Salomonsen (1968). Some notable examples of molt migrations in the Nearctic are the southwestward molt migration of eiders past Point Barrow, Alaska (Thompson and Person 1963; Johnson 1971; Timson 1976), and northward migrations of geese to specific arctic molting resorts (King 1970; Derksen et al. 1979; Abraham 1980; Davis et al. 1985). Salomonsen (1968) and Owen (1980) described similar movements by geese in the Palaearctic (also see Jehl, this Vol.). Molt migrations, especially by Arctic and sub-Arctic nesting geese, are relatively short and are frequently via the most direct route, often due north (Salomonsen 1968; Owen 1980). Notable exceptions are the 1500 km eastward molt migrations of eiders from the Canadian eastern Arctic to west Greenland (Salomonsen 1968, 1979b), the eastward molt migration of seaducks in East Siberia (Portenko 1972), the westward molt migrations of seaducks in the Beaufort Sea (Johnson and Richardson 1982) and White Sea (Salomonsen 1968; Kokhanov 1983), and the southwestward molt migration of scoters to the Baltic (Zhalakyavichus 1982).

The post-breeding molt by Arctic waterfowl is an energy-demanding process and the evolution of a special molt migration no doubt has significant advantages. Feather replacement requires a high protein diet, and many geese are able to replace feathers in food-rich molting areas without losing much weight (Newton 1968; Ankney 1979). Thus, Owen and Ogilvie (1979) proposed the "food quality" hypothesis to explain molt migration in the Arctic. Because the early growth of grasses and sedges is more digestible and higher in protein than later growth, and because the "spring flush" is later at higher latitudes, by migrating north, post-breeding geese would have access to highly digestible plants that are rich in protein. Salomonsen (1968) and others present alternative hypotheses to explain the molt migration of other waterbirds, but the food quality hypothesis best explains Arctic molt migrations of geese (see also Jehl, in this Vol.).

5 Limitations and Adaptations

5.1 Weather and Food

There is a large body of information (see Richardson 1978 and Richardson, this Vol.) showing how dates of arrival of many sub-Arctic and some Arctic birds are correlated with the earliness or lateness of the season. However, Meltofte (1985) and Elander and Blomqvist (1986) suggest that spring migration and arrival schedules of High Arctic migrants have evolved to more closely coincide with the average date when conditions are favorable on the nesting grounds, rather than to annual differences in temperatures along spring migration routes. In support of this idea, Davidson et al. (1986) found that the migrations of Siberian knots through northern Norway were remarkably similar in years when weather was very different, and Fox et al. (1986) found that over several years male Lapland longspurs arrived at nesting sites in West Greenland at about the same time each year regardless of weather conditions. Uspenskii (1969) found that many Arctic species arrived at nesting grounds in the Soviet Arctic with remarkable regularity from one year to the next. Such rigid schedules on the final few legs of spring migration are particularly critical for High Arctic species because of the short nesting season (45–65 days for many species) and the need to begin nesting as early as possible after arrival.

Throughout the Arctic virtually all breeding adults have arrived by the middle of June. Individuals that arrive on the breeding grounds late, or with marginal energy reserves, or that find nesting habitat covered with snow, may fail in their nesting attempt (Newton 1977; Green et al. 1977; Boyd 1987; Meltofte 1985), or may conduct a reverse migration out of the area (Meltofte 1979; deKorte et al. 1981). The jaegers, especially *Stercorarius pomarinus*, conduct such a reverse migration out of the Arctic during years when their food supply, mostly lemmings, is depressed (Parmelee et al. 1967; Maher 1974; Richardson et al. 1975; Richardson and Johnson 1981).

The migration routes and locations of staging areas for Arctic birds no doubt are traditional and have evolved over long periods, primarily in relation to food availability and climatic fluctuations. However, migration routes and wintering areas of some species, e.g. Palaearctic tundra swans and some Nearctic brant, have changed rather markedly in recent decades in relation to weather and climate (Palmer 1976; Martin 1983; Erskine 1988), and it is argued that such flexibility provides an adaptive advantage to survive changing climatic conditions and reduced food supplies (Martin 1983). Similarly, during years of drought when nesting habitat is unavailable at mid continent locations, large numbers of some prairie species (e.g., *Anas acuta*) extend their spring migration to "prairie" habitats in the Arctic (Hansen and McKnight 1964; Henny 1973; Derksen and Eldridge 1980) and may successfully nest there.

5.2 Time and Energy

The short Arctic summer coupled with the energy required for birds to migrate to the Arctic are severe constraints on reproduction. Individuals that migrate on schedule and arrive early and in good physical condition have the best chance of producing young (Ryder 1970; Newton 1977; Owen 1980; Alerstam and Högstedt 1985; Pienkowski et al. 1985; Davidson and Pienkowski 1987). Consequently, the number, locations, and condition of productive spring staging areas where Arctic migrants build up fat reserves and rest may significantly influence their reproductive success when they reach breeding grounds in the Arctic (Wypkema and Ankney 1979; Senner et al. 1981; Ebbinge et al. 1982;

Pienkowski and Evans 1984, 1985; Alerstam and Högstedt 1985; Alerstam et al. 1986; Piersma et al. 1987). Summers and Underhill (1987) have also shown, however, that in some parts of the Arctic the reproductive success of migrant geese and shorebirds is less related to fat reserves and body condition on arrival, than it is to microtine abundance and prey switching by foxes and other predators.

Piersma (1987) and Dick et al. (1987) describe how shorebirds that use different "hop, skip, or jump" strategies to migrate over long distances (e.g., West Africa to the Wadden Sea) may have different food and energy requirements. Short hops are apparently energetically more efficient than covering the same distance in one long flight because of the added costs of carrying extra fuel (fat) in the long "jumps". Nevertheless, many Arctic migrants make the very long flights, probably because of limited availability of high-quality staging areas along the flyways, and because tailwinds at high altitudes can substantially reduce energy expenditure (Hubbard and Flock 1974; Pennycuick 1978; Dick et al. 1987).

Arctic birds show several physiological adaptations for migration. In particular, higher than predicted basal metabolic rates in Arctic gulls (Ellis 1984), shorebirds (Kersten and Piersma 1987), and geese (Drent et al. 1978) facilitate the accelerated rates of feeding necessary for premigratory fattening and the subsequent long-distance flights. Shorebirds may increase their body mass during staging stops by 1–2 and possibly up to 5 g fat/day, depending on the season and the quality of the feeding habitat. Theoretical rates of spring fattening in Siberian knots (up to 5 g/day, Wadden Sea, Piersma 1987) are especially remarkable considering that on average a 1-g increase in body weight requires at least a 2 g increase in food intake (Kersten and Piersma 1987). Average rates of fall fattening in migrant red-necked phalaropes are about 1 g/day (Bay of Fundy, Mercier 1985), which is even more remarkable considering the small size of this bird. In many cases the total body mass may increase by up to 40–50% at some staging areas (Thompson 1973; Summers and Waltner 1978; Mercier 1985; Piersma et al. 1987). Such high proportions of body fat are needed to fuel the birds' very long-distance migrations. Alerstam (1985) calculated that Arctic terns staging off Norway and Britain would require a fuel ratio of 30–40% body weight before conducting a direct flight of 3000–5000 km to their next staging area off West Africa. Ebbinge et al. (1982) found that brant increased body mass by 10–20 g/day during May at their spring staging areas in the Wadden Sea; this represented a 15–20% increase in mass during the month prior to their 4000 km migration to the Taimyr Peninsula, USSR. Mercier (1985) calculated that Arctic red-necked phalaropes, semipalmated sandpipers, and other fall migrant shorebirds staging in the Bay of Fundy would need to carry up to 40% fat to make the flight to their presumed next major staging area in Central or South America.

Theoretically, if spring migration was delayed too long, or if birds were unable to store enough fat during spring, they would arrive late or in poor condition on the nesting grounds and thus might not reproduce (Ryder 1970; Newton 1977; Owen 1980). Non-breeding by Arctic birds is common (Bird and Bird 1940; Barry 1962; Uspenskii 1969; Owen 1980; Meltofte 1981), and has been most often described for species such as Arctic geese, shorebirds, jaegers, gulls, and terns that make substantial migrations to the High Arctic.

Thus, food availability, foraging activity, staging time, and fattening rates by Arctic migrants are all interrelated in the adaptation for migrating great distances to the Arctic. Slow rates of fattening that delay migration at early staging sites affect fattening rates and schedules at later sites, i.e., the "domino effect" of Piersma (1987), and migrants that arrive late or in poor condition may not reproduce at all. The behavioral and physiological adaptations shown by Arctic migrants, especially the long-distance migrants, are remarkable and no doubt have evolved in relation to many environmental factors over the past 100 000 years, most notably climatic conditions and food availability. Whether they can also adapt to man-induced environmental changes (Ebbinge et al. 1982; Davidson and Pienkowski 1987) may be their greatest challenge.

References

Abraham KF (1980) Moult migration of lesser snow geese. Wildfowl 34:89–93

Alerstam T (1985) Strategies of migratory flight, illustrated by arctic and common terns, *Sterna paradisaea* and *Sterna hirundo*. In: Rankin MA (ed) Migration: mechanisms and adaptive significance. Contrib Mar Sci (Suppl) 27:580–603

Alerstam T, Högstedt G (1980) Spring predictability and leap-frog migration. Ornis Scand 11:196–200

Alerstam T, Högstedt G (1985) Leap-frog arguments: reply to Pienkowski, Evans and Townshend. Ornis Scand 16:71–74

Alertstam T, Hjørt C, Högstedt G, Jönsson PE, Karlsson J, Larsson B (1986) Spring migration of birds across the Greenland inland ice. Med Om Grønl Biosci 21:1–38

AOU (American Ornithologists Union) (1983) Checklist of North American birds, 6th Ed. Allen, Lawrence

Anderson RM (1913) Report on the natural history collections of the expedition (1908–1912). In: Stefansson V (ed) My life with the eskimo. MacMillan, New York

Anderson RM (1915) Canadian Arctic expedition 1913, 1914. Summ Rep Geol Surv, Dep Mines 1914. Sess Pap 26:163–166

Anderson RM (1937) Mammals and birds of the western Arctic district, NWT, Canada. Canada's Western Northland, Parks and Forest Branch, Kings Printer, Ottawa

Ankney CD (1979) Does the wing molt cause nutritional stress in lesser snow geese? Auk 96:68–72

Bailey AM (1948) Birds of Arctic Alaska. Colo Mus Nat Hist Pop Ser 8, 317 pp

Barry RG (1982) Approaches to reconstructing the climate of the steppe-tundra biome. In: Hopkins DM, Matthews JV Jr, Schweger CE, Young SB (eds) Paleoecology of Beringia. Academic Press, London

Barry TW (1962) Effect of late seasons on Atlantic brant reproduction. J Wildl Manage 25:19–25

Bellrose FC (1980) Duck, geese and swans of North America. Stackpole Books, Harrisburg

Belopol'skii LO (1957) Ecology of sea colony birds of the Barents Sea. Israel Prog Sci Transl. 1961, Jerusalem

Bergman G (1974) The spring migration of the long-tailed duck and the common scoter in western Finland. Ornis Fenn 51:129–145

Bergman G (1976) Vogelzugstudien und Migrationsprobleme in Finnland. Ornis Fenn 53:107–114

Bergman G (1977) Finnish radar investigations on migration of waterfowl between the Baltic and the White Sea. 24 Rassegna Internazionale Elettronica Nucleare ed Aerospaziale, Rome

Bergman G, Donner KO (1964) An analysis of the spring migration of the common scoter and the long-tailed duck in southern Finland. Acta Zool Fenn 105:1–59

Bianchi V (1983) On the passage of waterfowl across the White Sea. Commun Baltic Comm Study Bird Migr 16:3–13

Bird CG, Bird EG (1940) Some remarks on non-breeding in the Arctic, especially in northeast Greenland. Ibis 14:671–678

Boyd H (1987) Do June temperatures affect the breeding success of dark-bellied brent geese *Branta b. bernicla?* Bird Study 34:155–159

Brassard GR (1971) The mosses of northern Ellesmere Island, Arctic Canada. I. Ecology and phytogeography, with an analysis for the Queen Elizabeth Islands. Bryologist 74:233–281

Brooks WS (1915) Notes on birds from east Siberia and arctic Alaska. Bull Mus Comp Zool 59:361–413

Brown RGB, Nettleship DN (1981) The biological significance of polynyas to arctic colonial seabirds. In: Stirling I, Cleator H (eds) Polynyas in the Canadian Arctic. Can Wildl Serv Occas Pap 45:59–66

Brown RGB, Nettleship DN, Germain P, Tull CE, Davis T (1975) Atlas of eastern Canadian seabirds. Can Wildl Serv, Ottawa, 220 pp

Butler RW, Kaiser GW, Smith GEJ (1987) Migration chronology, length of stay, sex ratio, and weight of western sandpipers, (*Calidris mauri*) on the south coast of British Columbia. J Field Ornithol 58:103–111

Cade TJ (1955) Records of the black brant in the Yukon Basin and the question of a spring migration route. J Wildl Manage 19:321–324

Cooke WW (1915) Bird migration in the Mackenzie Valley. Ibis 32:442–459

Cramp S (1985) Handbook of the birds of Europe and Middle East and North Africa, Vol 4. Oxford University Press, Oxford

Cramp S, Simmons KEL (1977) Handbook of the birds of Europe and Middle East and North Africa, Vol 1. Oxford University Press, Oxford

Cramp S, Simmons KEL (1980) Handbook of the birds of Europe and Middle East and North Africa, Vol 2. Oxford University Press, Oxford

Cramp S, Simmons KE (1983) Handbook of the birds of Europe and Middle East and North Africa, Vol 3. Oxford University Press, Oxford

Cramp S, Simmons KEL (1985) Handbook of the birds of Europe and Middle East and North Africa, Vol 4. Oxford University Press, Oxford

Davidson NC, Pienkowski MW (1987) The conservation of international flyway populations of waders. Wader Study Group Bull 49 Suppl/IWRB Spec Publ 7:1–151

Davidson NV, Strann KB, Crockford NJ et al. (1986) The origins of knots *Calidris canutus* in arctic Norway in spring. Ornis Scand 17:175–179

Davis RA, Jones RN, MacInnes CD, Pakulak AJ (1985) Molt migration of large Canada geese on the west coast of Hudson Bay. Wilson Bull 97:296–305

Dean FC, Valkenburg P, Magoun AJ (1976) Inland migration of jaegers in northeastern Alaska. Condor 78:271–273

Degerbøl M, Møhl-Hansen U (1935) The Scoresby Sound committee's 2nd East Greenland expedition in 1932 to King Christian IX's Land. Birds. Med Om Grønl Biosci 104:1–30

Dement'ev GP, Gladkov NA (1951) Birds of the Soviet Union, Vol 3. Israel Prog Sci Trans, 1969, Jerusalem

Dement'ev GP, Gladkov NA (1952) Birds of the Soviet Union, Vol 4. Israel Prog Sci Trans, 1967, Jerusalem

Denton GH, Hughes TJ (eds) (1981) The last great ice sheets. Wiley, New York

Derksen DV, Eldridge WD (1980) Drought-displacement of pintails to the arctic coastal plain, Alaska. J Wildl Manage 44:224–229

Derksen DV, Weller MW, Eldridge WS (1979) Distributional ecology of geese molting near Teshekpuk Lake, National Petroleum Reserve-Alaska. In: Jarvis RL, Bartonek JC (eds) Management and biology of Pacific flyway geese. OSU Bookstores, Corvallis, pp 189–207

Dick WJA, Piersma T, Prokosch P (1987) Spring migration of the Siberian knots *Calidris canutus canutus*: results of a co-operative Wader Study Group project. Ornis Scand 18:5–16

Dixon JS (1943) Birds observed between Point Barrow and Herschel Island on the arctic coast of Alaska. Condor 45:49–57

Dol'nik VR (1969) Mysterious migration of birds. Transl by US Dep Comm, Washington DC

Drent RH, Ebbinge B, Weijand B (1978) Balancing the energy budgets of arctic-breeding geese throughout the annual cycle: a progress report. Verh Ornithol Ges Bayern 23:239–263

Ebbinge B, St. Joseph A, Prokosch P, Spaans B (1982) The importance of spring staging areas for arctic-breeding geese wintering in western Europe. Aquila 89:249–258

Elander M, Blomqvist S (1986) The avifauna of central northeast Greenland, 73° 15′ N – 74° 05′ N, based on a visit to Myggbukta, May – July 1979. Med Om Grønl Biosci 19:1–44

Ellis HI (1984) Energetics of free ranging seabirds. In: Whittow GC, Rahn H (eds) Seabird energetics. Plenum, New York, pp 203–234

Ellis DV, Evans J (1960) Comments on the distribution and migration of birds in Foxe Basin, Northwest Territories Can Field-Nat 74:59–70

England J (1987) Glaciation and evolution of the Canadian high arctic landscape. Geology (Boulder) 15:419–424

Erskine AJ (1988) The changing patterns of brant migration in eastern North America. J Field Ornithol 59:110–119

Feduccia A (1980) The age of birds. Harvard, Cambridge

Fladmark KR (1978) The feasibility of the northwest coast as a migration route for early man. In: Bryan AL (ed) Early man in America from a circum-Pacific perspective. Occas Pap 1, Dep Anthro, Univ. Alberta, Edmonton, pp 119–128

Flock WL (1973) Radar observations of bird movements along the arctic coast of Alaska. Wilson Bull 85:259–275

Fox AD, Francis IS, Madsen J, Stroud JM (1986) The breeding biology of the Lapland bunting *Calcarius lapponicus* in West Greenland during two contrasting years. Ibis 129:541–552

Gabrielson IN, Lincoln FC (1959) The birds of Alaska. Stackpole, Harrisburg

Gal-Chen T (1982) Approaches to mathematical modeling of the steppe-tundra paleoclimate. In: Hopkins DM, Matthews JV Jr, Schweger CE, Young SB (eds) Paleoecology of Beringia. Academic Press, London

Gaston AJ (1982) Migration of juvenile thick-billed murres through Hudson Strait in 1980. Can Field-Nat 96:30–34

Gauthreaux SJ (1982) The ecology and evolution of avian migration systems. In: Farner DS, King JR, Parkes KC (eds) Avian biology, Vol. 6. Academic Press, London, pp 93–168

Gessaman JA (1972) Bioenergetics of the snowy owl (*Nyctea scandiaca*). Arctic Alpine Res 4:223–238

Gibson DD (1981) Migrant birds at Shemya Island, Aleutian Islands, Alaska. Condor 83:65–77

Green GH, Greenwood JJD, LLoyd CS (1977) The influence of snow conditions on the date of breeding of wading birds in north-east Greenland. J Zool (Lond) 183:311–328

Greenwood JG (1984) Migration of dunlin *Calidris alpina*: a worldwide overview. Ringing & Migr 5:35–39

Gullestad N, Owen M, Nugent MJ (1984) Numbers and distribution of barnacle geese *Branta leucopsis* on Norwegian staging islands and the importance of the staging area to the Svalbard population. Nor Polarinst Skr 181:57–65

Guthrie RD (1982) Mammals of the Mammoth Steppe as paleoenvironmental indicators. In: Hopkins DM, Matthews JV Jr, Schweger CE, Young SB (eds) Paleoecology of Beringia. Academic Press, London

Guzman J, Myres MT (1982) Ecology and behavior of southern hemisphere shearwaters (genus *Puffinus*), when over the outer continental shelf of the Gulf of Alaska and Bering Sea during the northern summer (1975-76). In: Outer Cont Shelf Envir Assess Prog Fin Rep Princ Invest 54, pp 571–682

Haftorne S (1971) Norges fugler. Universitetsforlaget, Trondheim

Haila Y (1980) Delayed autumn migration of waterfowl in the northern Baltic. Ornis Fenn Suppl 3:41–43

Haila Y, Tiainen J, Vepsäläinen K (1986) Delayed autumn migration as an adaptive strategy of birds in northern Europe: evidence from Finland. Ornis Fenn 63:1–9

Hale GW (1984) The changing face of European winter areas. In: Goss-Custard JD, Hale WG (eds) Coastal waders and wildfowl in winter. Cambridge University Press, Cambridge

Hamilton TD, Thorson RM (1986) Glaciation in Alaska – the geologic record. Alaska Geol Soc, Anchorage

Haney JC, Stone AE (1988) Littoral foraging by red phalaropes during spring in the northern Bering Sea. Condor 90:723–726
Hansen HA, McKnight DE (1964) Emigration of drought-displaced ducks to the Arctic. Trans N Am Wildl Nat Resour Conf 29:119–127
Harrison P (1985) Seabirds, an identification guide. Reed, Auckland
Hayman P, Marchant J, Prater T (1986) Shorebirds, an identification guide to the waders of the world. Houghton Mifflin, Boston
Helms O (1926) The birds of Angmagssalik. Med Om Grønl Biosci 58:205–275
Henny CJ (1973) Drought displaced movement of North American pintails into Siberia. J Wildl Manage 37:23–29
Hildén O (1971) Occurrence, migration and colour phases of the Arctic skua (*Stercorarius parasiticus*) in Finland. Ann Zool Fenn 8:223–230
Hjørt C (1976) An observation of ivory gull *Pagophila eburnea* migration along the east Greenland current. Dan Ornithol Foren Tidsskr 70:72–73
Hjørt C, Björk S (1984) A re-evaluated glacial chronology for northern East Greenland. Geol Foren Stockn Forh 105:235–243
Holgersen H (1960) Wanderungen und Winterquartiere der Spitzbergen Kurzschnabelgänse. Proc Int Ornithol Congr 12:310–316
Holmes RT (1966) Molt cycle of the red-backed sandpiper (*Calidris alpina*) in western North America. Auk 83:517–533
Hopkins DM, Matthews JV Jr, Schweger CE, Young SB (1982) Paleoecology of Beringia. Academic Press, London
Hørring R (1939) 6 og 7 Thule expedition til Sydøstrgrønland 1931–1933, birds. Med Om Grønl Biosci 108:1–44
Hubbard JD, Flock WB (1974) Radar observation of migratory waterfowl at Cold Bay, Alaska. AFWL-DE-TN-74-008. Rep for US Air Force Weapons Lab, Kirkland AFB, NM
Irving L (1960) Birds of Anaktuvuk Pass, Kobuk, and Old Crow. US Nat Mus Bull 217, Washington DC
Irving L (1961) The migration of Lapland longspurs to Alaska. Auk 78:327–342
Irving L (1972) Arctic life of birds and mammals, including man. Zoophysiology and Ecology 2. Springer, Berlin Heidelberg New York, 192 pp
Irving L, West GC, Peyton LJ, Paneak S (1967) Migration of willow ptarmigan in Arctic Alaska. Arctic 20:77–85
Irving L, McRoy CP, Burns JJ (1970) Birds observed during a cruise in the ice-covered Bering Sea in March 1968. Condor 72:110–112
Islieb ME (1979) Migratory shorebird populations on the Copper River delta and eastern Prince William Sound, Alaska. Stud Avian Biol 2:125–129
Jespersen P, Tåning ÅV (1950) Studies in bird migration, the collected papers of H. Chr. C. Mortensen. Dan Ornithol Foren, Copenhagen
Johnson LL (1971) The migration, harvest, and importance of waterfowl at Barrow, Alaska. M.S. Thesis University Alaska, Fairbanks USA
Johnson SR (1975) Spring movements and abundance of birds at Northwest Cape, St. Lawrence Island, Bering Sea, Alaska. Syesis 9:31–44
Johnson SR (1985) Adaptations of the long-tailed duck (*Clangula hyemalis* L.) during the period of molt in Arctic Alaska. Proc Int Ornithol Congr 18:530–540
Johnson SR, Richardson WJ (1982) Waterbird migration near the Yukon and Alaskan coast of the Beaufort Sea. II. Moult migration of seaducks in summer. Arctic 35:292–301
Johnson SR, West GC (1975) Growth and development of heat regulation in nestlings, and metabolism of adult common and thick-billed murres. Ornis Scand 6:109–115
Johnson SR, Herter DR (1989) The birds of the Beaufort Sea. BP, Anchorage USA
Kersten M, Piersma T (1987) High levels of energy expenditure in shorebirds; metabolic adaptations to an energetically expensive way of life. Ardea 75:175–187
Kessel B (1984) Migration of sandhill cranes, *Grus canadensis*, in east-central Alaska, with routes through Alaska and western Canada. Can Field-Nat 98:279–292
Kessel B, Gibson DD (1980) Status and distribution of Alaska birds. Stud Avian Biol 1, COS, Los Angeles

King JG (1970) The swans and geese of Alaska's Arctic slope. Wildfowl 21:11–17
Kistchinski, AA (1979) Migrations of birds of eastern Europe and northern Asia. Nauka, Moscow
Kokhanov V (1983) Peculiarities of the summer migration of the common scoter in different regions of the White Sea. Commun Baltic Comm Study Bird Mig 16:14–23
Korte J de, Bosman CAW, Meltofte H (1981) Observations on waders (Charadriidae) at Scoresby Sund, East Greenland. Med Om Grønl Biosci 7:1–21
Kumari E (1976) On the study of bird migration in the Soviet Union. Ornis Fenn 53:119–121
Kumari E (1980) Characteristics of seaduck movements in the Baltic. Ornis Fenn Suppl 3:39–41
Kumlien L (1879) Contribution to the natural history of Arctic America, made in connection with the Howgate Polar Expedition, 1877–1878, Gov Print Off, Washington DC
Lamb HH, Woodroffe A (1970) Atmospheric circulation during the last ice age. Q Res 1:29–58
Leech RE (1966) The spiders (Araneida) of Hazen Camp 81° 49′N, 71° 18′W. Quaest Entomol 2:153–212
Leslie R, Lessels CM (1978) The migration of dunlin *Calidris alpina* through northern Scandinavia. Ornis Scand 9:84–86
Løvenskiold H (1964) Avifauna Svalbardensis. Nor Polarinst Skr 129, 460 pp
Lundberg S, Alerstam T (1986) Bird migration patterns: conditions for stable geographical population segregation. J Theor Biol 123:403–414
Lusis YY, Spurin ZD, Taurins EJ, Vilka EK (1961) Birds of the Baltic region: ecology and migrations. Israel Prog Sci Transl, 1968, Jerusalem
Macpherson AH (1965) The origin of diversity in mammals of the Canadian arctic tundra. Syst Zool 14:153–173
Maher W (1974) Ecology of pomarine, parasitic and long-tailed jaegers in northern Alaska. Pac Coast Avifauna 37, 148 pp
Maltby-Prevett LS, Boyd H, Heyland JD (1975) Observations in Iceland and north-west Europe of brant from the Queen Elizabeth Islands, NWT, Canada. Bird-Banding 46:155–170
Martin J-L (1983) Adaptive significance of changes in migratory routes and wintering grounds: two examples. Ornis Fenn Suppl 3:54–55
Mayr E (1960) The emergence of evolutionary novelties. In: Tax S (ed) The evolution of life. University Chicago Press, Chicago, pp 349–380
Mayr E, Amadon D (1951) A classification of recent birds. Am Mus Novit 14:1–42
McClure HE (1974) Migration and survival of the birds of Asia. Appl Sci Res Corp Thail, Bangkok
McLaren PL (1982) Spring migration and habitat use by seabirds in eastern Lancaster Sound and western Baffin Bay. Arctic 35:88–111
Mead C (1983) Bird migration. Facts on File Pub, New York
Meltofte H (1976) Ornithological observations in southern Peary Land, North Greenland, 1973. Med Om Grønl Biosci 205:1–57
Meltofte H (1979) The population of waders *Charadriidae* at Danmarks Havn, northeast Greenland, 1975. Dan Ornithol Foren Tidsskr 73:69–94
Meltofte H (1981) Ornithological observations in northeast Greenland between 74° 30′ and 76° 00′ N. lat., 1976. Med Om Grønl Biosci 3:1–53
Meltofte H (1985) Populations and breeding schedules of waders, Charadrii, in high arctic Greenland. Med Om Grønl Biosci 16:1–43
Mercier FM (1985) Fat reserves and migration of red-necked phalaropes (*Phalaropus lobatus*) in the Quoddy region, New Brunswick, Canada. Can J Zool 63:2810–2816
Moreau RE (1972) The Palaearctic-African bird migration systems. Academic Press, London
Morrison RIG (1975) Migration and morphometrics of European knot and turnstone in Ellesmere Island, NWT, Canada. Bird-Banding 46:290–301
Morrison RIG (1977) Migration of Arctic waders wintering in Europe. Polar Rec 18:475–486
Morrison RIG (1984) Migration systems of some New World shorebirds. In: Burger J, Olla BL (eds) Shorebirds, migration and foraging behavior, Behav Mar Anim, vol 6. Plenum, New York, pp 125–202
Morrison RIG, Myers JP (1987) Wader migration systems in the New World. Wader Study Group Bull 49 Suppl/IWRB Spec Publ 7:57–69
Murdoch J (1885) Birds. In: Report of the International Polar Expedition to Point Barrow, Alaska. US Gov Print Office, Washington, pp 104–128

Newton I (1968) The temperatures, weights and body composition of molting bulfinches. Condor 70:323-332
Newton I (1977) Timing and success of breeding in tundra-nesting geese. In: Stonehouse B, Perrins CM (eds) Evolutionary ecology. University Park Press, London
Nilsson S (1858) Skandinavisk fauna. Foglarna. Gleerups, Lund
Nørrevang A (1963) Considerations of avifaunal connections across the North Atlantic. Dan Ornithol Foren Tidsskr 57:99-109
Orr CD, Ward MP, Williams NA, Brown RGB (1982) Migration patterns of red and northern phalaropes in southwest Davis Strait and in the northern Labrador Sea. Wilson Bull 94:303-312
Owen M (1980) Wild geese of the world. Batsford, London
Owen M, Gullestad N (1984) Migration routes of Svalbard barnacle geese *Branta leucopsis* with a preliminary report on the importance of the Bjørnøya staging area. Nor Poarinst Skr 181:67-77
Owen M, Ogilvie MA (1979) Wing molt and weights of barnacle geese in Spitsbergen. Condor 81:42-52
Paevskii VA (1973) Atlas of bird migrations according to banding data at the Courland Spit. In: Bykhovskii BE (ed) Birds migrations, ecological and physiological factors. Wiley, New York
Palmén JA (1874) Om foglarnes flyttningsvägar. Frenckell, Helsinki
Palmer RS (1976) Handbook of North American birds, vols 2-3, Waterfowl. Yale University Press, New Haven
Palmer RS (1988) Handbook of North American birds, vol 5, Raptors. Yale University Press, New Haven
Palmer W (1889) The avifauna of the Pribilof Islands. In: The fur seals and fur seal islands of the North Pacific Ocean. Gov Print Office, Washington DC, pp 355-431
Parmelee DF, Stephens HA, Schmidt RH (1967) The birds of southeastern Victoria Island and adjacent small islands. Natl Mus Can Bull 222:1-173
Pennycuick CJ (1978) Fifteen testable predictions about bird flight. Oikos 30:165-176
Pienkowski MW, Evans PR (1984) Migratory behavior of shorebirds in the western Palaearctic. In: Burger J, Olla BL (eds) Shorebirds, migration and foraging behavior. Behav Mar Anim, vol 6. Plenum, New York, pp 73-124
Pienkowski MW, Evans PR (1985) The role of migration in the population dynamics of birds. In: Sibly RM, Smith RH (eds) Behavioural ecology, the ecological consequences of adaptive behaviour. Blackwell, Oxford, pp 331-352
Pienkowski MW, Evans PR, Townshend DJ (1985) Leap-frog and other migration patterns of waders: a critique of the Alerstam and Högstedt hypothesis, and some alternatives. Ornis Scand 16:61-70
Piersma T (1987) Hink, stap of sprong? Reisbeperkingen van arctische steltlopers door voedselzoeken, vetopbouw en vliegsnelheid. Limosa 60:185-194
Piersma T, Bientema AJ, Davidson NC, Munster OAG, Pienkowski MW (1987) Wader migration systems in the East Atlantic. Wader Study Group Bull 49 Suppl/IWRB Spec Publ 7:35-56
Platt JB (1976) Gyrfalcon nest site selection and winter activity in the western Canadian Arctic. Can Field-Nat 90:338-345
Portenko LA (1972) Birds of the Chukchi Peninsula and Wrangel Island. Amerind, New Delhi, 1981 (in English)
Prater AJ (1981) Estuary birds of Britain and Ireland. Poyser, Calton
Reeves BOK (1973) The nature and age of the contact between the Laurentide and Cordilleran ice sheets in the western interior of North America. Arctic & Alpine Res 4:41-61
Reiff J, Forbes GS, Spieksma FTM, Reynders JJ (1986) African dust reaching northwestern Europe: a case study to verify trajectory calculations. J Climat Appl Meteorol 25:1543-1567
Remmert H (1980) Arctic animal ecology. Springer, Berlin Heidelberg New York
Renaud WE, McLaren PL, Johnson SR (1982) The Dovekie, *Alle alle*, as a spring migrant in eastern Lancaster Sound and western Baffin Bay. Arctic 35:118-125
Richardson J (1836) Zoological remarks, App 1. In: Back G (ed) Narrative of the Arctic Land Expedition. Murray, London, pp 475-522
Richardson J (1852) Arctic expedition in search of Sir John Franklin. Harpers, New York
Richardson WJ (1978) Timing and amount of bird migration in relation to weather: a review: Oikos 30:224-272

Richardson WJ, Johnson SR (1981) Waterbird migration near the Yukon and Alaskan coast of the Beaufort Sea. I. Timing, routes and numbers in spring. Arctic 34:108–121

Richardson WJ, Morrell MR, Johnson SR (1975) Bird migration along the Beaufort Sea coast: radar and visual observations in 1975. Beauf Sea Tech Rep 3c. Dep Envir, Victoria, Canada

Ryder JP (1970) A possible factor in the evolution of clutch size in Ross' goose. Wilson Bull 82:5–13

Salomonsen F (1950) Grønlands fugle, the birds of Greenland. Munksgaard, Copenhagen

Salomonsen F (1955) The evolutionary significance of bird migration. Biol Medd K Dan Vidensk Selsk 22:1–62

Salomonsen F (1967a) Migratory movements of the Arctic tern (*Sterna paradisaea*) in the Southern Ocean. Biol Medd Dan Vidensk Selsk 24

Salomonsen F (1967b) Fuglene på Grønland. Rhodos, Copenhagen, p 341

Salomonsen F (1968) The moult migration. Wildfowl 19:5–24

Salomonsen F (1979a) Ornithological and ecological studies in S.W. Greenland (59°46′–62°27′ N. Lat.). Med Om Grønl Biosci 204

Salomonsen F (1979b) Trettende foreløbige liste over genfundne grønlandske ringfugle. Dan Ornithol Foren Tidsskr 73:191–206

Schanning HTL (1933) Zoological results of the Norwegian scientific expeditions to East Greenland. I. A contribution to the bird fauna of East Greenland. Skr Svalbard Ishavet 49:2–24

Searing GF, Kuyt E, Richardson WJ, Barry TW (1975) Seabirds of the southeastern Beaufort Sea: aircraft and ground observations in 1972 and 1974. Beaufort Sea Tech Rep 3b. Dep Environ, Victoria

Senner SE (1979) An evaluation of the Copper River delta as critical habitat for migrating shorebirds. Stud Avian Biol 2:131–145

Senner SE, West GC, Norton DW (1981) The spring migration of western sandpipers and dunlins in southcentral Alaska: numbers, timing and sex ratios. J Field Ornithol 52:271–284

Sladen WJL (1973) A continental study of whistling swans using neck collars. Wildfowl 24:8–14

Sladen WJL, Kistchinskii AA (1977) Some results from circumpolar marking programs on northern swans and snow geese. Proc 13th Congr Game Biol, Atlanta, pp 498–507

Snow DW (1953) Migration of the Greenland Wheatear. Ibis 95:376–378

Summers RW, Underhill LG (1987) Factors related to breeding production of brent geese *Branta b. bernicla* and waders (Charadrii) on the Taimyr Peninsula. Bird Study 34:161–171

Summers RW, Waltner M (1978) Seasonal variations in the mass of waders in southern Africa with special reference to migration. Ostrich 50:21–39

Summers RW, Underhill LG, Pearson DJ, Scott DA (1987) Wader migration systems in southern and eastern Africa and western Asia. Wader Study Group Bull 49 Suppl/IWRB Spec Publ 7:15–34

Swarth HS (1934) Birds of Nunivak Island, Alaska. Pac Coast Avifauna 22:1–64

Thienemann J (1927) Rossitten. J Neumann, Neudamm

Thompson MC (1973) Migratory patterns of ruddy turnstones in the Central Pacific region. Living Bird 12:5–23

Thompson DQ, Person RA (1963) The eider pass at Point Barrow, Alaska. J Wildl Manage 27:348–356

Timson RS (1976) Late summer migration at Barrow, Alaska. Environ Assess Alaskan Cont Shelf, Final Rep Princ Invest I:354–400

Troy DM, Johnson SR (1987) Marine birds. In: Truett JC (ed) Environmental characterization and biological utilization of the North Aleutian Shelf nearshore zone. LGL Ecol Res Assoc, Inc rep for US Dep Comm, Nat Ocean Atmosph Admin, Anchorage, pp 322–422

Tuck LM (1971) The occurrence of Greenland and European birds in Newfoundland. Bird-Banding 42:184–209

US Fish and Wildlife Service (1981) The Platte River ecology study. Spec Res Rep, North Prairie Wildl Res Center, Jamestown

Uspenskii SM (1969) Life in high latitudes, a study of bird life. Amerind, 1984, Delhi

Uttley JD, Thomas CJ, Davidson NC, Strann K-B, Evans PR (1987) The spring migration system of nearctic knots *Calidris canutus islandica*: a reappraisal. Wader Study Group Bull 49 Suppl/IWRB Spec Publ 7:80–84

Vorren TO, Vorren D-D, Alm T, Gulliksen S, Løvlie R (1988) The last deglaciation (20,000 to 11,000 B.P.) on Andøya, northern Norway. Boreas 17:41–77

Vronsky NV (1985) Flights and migrations of birds on West Taimyr. Zool Zh 64:78–86

Warner BG, Mathewes R, Clague JJ (1982) Ice-free conditions on the Queen Charlotte Islands, British Columbia, at the height of Late Wisconsin glaciation. Science 218:675–677
West GC (1972a) Seasonal differences in resting metabolic rate of Alaskan ptarmigan. Comp Biochem Physiol A Comp Physiol 42:867–876
West GC (1972b) Effect of acclimation and acclimatization on the resting metabolic rate of the common redpoll. Comp Biochem Physiol A Comp Physiol 43:253–310
West GC, Peyton LJ (1980) Fatty acids of depot lipids in migrating Lapland longspurs. J Field Ornithol 51:138–143
West GC, Peyton LJ, Irving L (1968) Analysis of spring migration of Lapland longspurs to Alaska. Auk 85:639–653
Whitfield JP, Magnusson J (1987) The migration of high arctic waders through Melrakkasletta, north-east Iceland. Wader Study Group Bull 49 Suppl/IWRB Spec Publ 7:85–89
Williams J, Barry RG, Washington WM (1974) Simulation of the atmospheric circulation using the NCAR global circulation model with Ice Age boundary conditions. J Appl Meteorol 13:305–317
Williamson K (1975) Birds and climatic change. Bird Study 22:143–164
Wilson JR (1981) The migration of High Arctic shorebirds through Iceland. Bird Study 28:21–32
Woodby DA, Divoky GJ (1982) Spring migration of eiders and other waterbirds at Point Barrow, Alaska. Arctic 35:403–410
Wynne-Edwards VC (1952) Zoology of the Baird Expedition (1950), I. The birds observed in central and southeast Baffin Island. Auk 69:353–391
Wypkema RCP, Ankney CD (1979) Nutrient reserve dynamics of lesser snow geese staging at James Bay, Ontario. Can J Zool 57:213–219
Yakobi VE, Yigi AI (1972) Radar and visual observations of the common scoter moult migration. Inform Baltic Comm Stud Bird Mig 7:118–137
Zhalakyavichus M (1982) Radar observations of migration for moulting by common scoter in the Lithuanian SSR. Ornith Studies USSR, Vol 2. Zool Inst, USSR Acad Sci, Moscow, pp 274–284

Palaearctic Passerine Migrants in Kenya and Uganda: Temporal and Spatial Patterns of Their Movements

D.J. Pearson[1]

1 Introduction

1.1 Palaearctic Migration to and from Southern Africa

Passerines reaching tropical Africa from the West and mid-Palaearctic initially enter marginal savanna habitats. These are relatively rich in resources during the June-September rainy season, but then become progressively desiccated (Moreau 1972). This contrasts with the situation in Southeast Asia and Central America, where migrants from the East Palaearctic and Nearctic respectively find more humid environments. In Africa, therefore, migrants tend to penetrate further south than in Asia/Australasia or tropical America. Forty-four passerine species commonly reach the equator, and 21 of these extend their migrations further to winter mainly or in part at 5°–25° S (Pearson et al. 1988), thus effectively following the intertropical rain belt and benefitting from humid conditions during most of their stay on the continent. These migrations between the Palaearctic and southern Africa, typically some 6000–10 000 km, include some of the longest passerine journeys known. A few species clearly cross the Congo basin (Chapin 1953; Moreau 1972), but the majority appear to migrate through the savanna dominated, 1400-km-wide corridor between easternmost Zaire and Kenya.

1.2 Origin of the Birds Involved

Representatives of a few species from west, central and north Europe cross forested West Africa to reach wintering areas from south Zaire and Zambia to Botswana and South Africa. Examples include the swallow, *Hirundo rustica*, the spotted flycatcher, *Muscicapa striata*, the collared flycatcher, *Ficedula albicollis*, the garden warbler, *Sylvia borin*, the icterine warbler, *Hippolais icterina*, and the willow warbler, *Phylloscopus trochilus* (Zink 1971; Dowsett et al. 1988). However, the majority of passerines wintering in southern Africa arrive and depart through east and northeast Africa (Pearson et al. 1988). Most of these are from USSR breeding areas at 30°–80° E, but in a few species European birds move southeast

[1]Browses, Sibton, Saxmundham, Suffolk, UK
[2]Seasonal terms are used here in the sense of a northern hemisphere.

E. Gwinner (Ed.) Bird Migration

to join representatives from Asia in the Middle East, so that all birds migrate together through eastern Africa, returning via a similar route in spring.[2] This pattern is seen in the sprosser, *Luscinia luscinia*, the marsh warbler, *Acrocephalus palustris*, the river warbler, *Locustella fluviatilis*, the olive-tree warbler, *Hippolais olivetorum*, the barred warbler, *Sylvia nisoria*, the red-backed shrike, *Lanius collurio*, and the lesser grey shrike, *L. minor* (Moreau 1972; Zink 1973; Dowsett et al. 1988).

1.3 Recent Studies in Kenya and Uganda

Systematic observations on the seasonality and migrations of Palaearctic passerines have been made since 1964 at a number of Kenya and Uganda sites (e.g. Pearson 1972; Pearson and Backhurst 1976a, b; Pearson et al. 1979; Backhurst and Pearson 1984; Boothroyd 1987; Pearson et al. 1988; see also Acknowledgements for observers who have supplied unpublished records). In some cases these have been supplemented by ringing and associated weight and moult studies. Complementary data from Rwanda have been given by Vande weghe (1979; *in litt.*). Migration is best known in Uganda at Kampala (Pearson 1972) and at Lake Edward in the southwest (M.P.L. Fogden *in litt.*; J.M. Lock *in litt.*), and in Kenya in central Nyanza (P.L. Britton *in litt.*), around Nairobi (e.g. Pearson and Backhurst 1976b) and in Tsavo (for localities see Fig. 1). In the Ngulia Hills in Tsavo National Park (West), southward-bound night migrants are attracted and grounded in mist, often in thousands, at the floodlights of a game viewing lodge during moonless periods in November and December. Over 80 000 Palaearctic passerines have been caught and ringed here since 1972 (Pearson and Backhurst 1976a; Backhurst and Pearson 1977, 1984, 1988).

The timing and composition of the main passage movements through the two countries have been well established. Twenty-eight species feature commonly (see Table 1). Major differences exist from one locality to another, emphasizing a marked channelling associated with major topographical features. Pronounced species differences between the two passage seasons indicate loop migrations and/or overflying.

2 Southward Passage

This movement is protracted. A few species reach the East African equator by September, but numbers peak in November. Some birds arrive as late as January.

2.1 Species Distributions and Routes

Migration appears to be concentrated along two major corridors, one through Uganda and the Lake Victoria basin and the other through the plateau country east of the Kenya highlands. Little passage is evident within 50 km of the coast, and migration through the Kenya highlands and rift valley appears limited to a

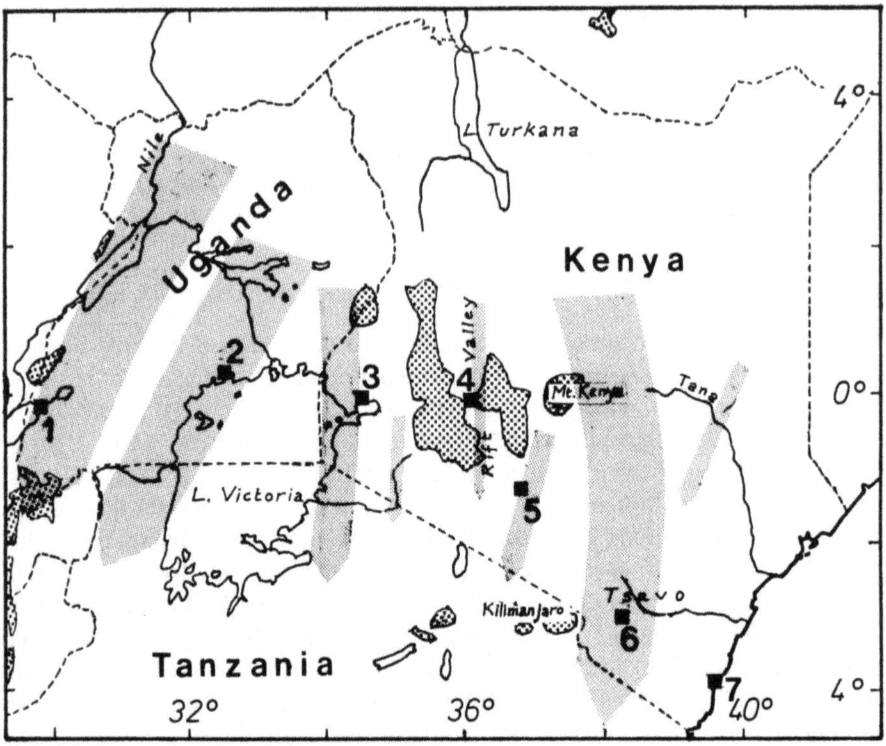

Fig. 1. Map of Kenya and Uganda, showing geographical features mentioned in the text. Observation and ringing localities are as follows: *1* Lake Edward; *2* Kampala; *3* central Nyanza; *4* Nakuru (rift valley); *5* Nairobi; *6* Ngulia (Tsavo); *7* Mombasa. *Arrows* indicate concentrations of southward passage movement. ▩ Land above 2700 m

few species (Fig. 1). The main eastern movement passes immediately east of Mt. Kenya and then east of Kilimanjaro on a front no more than about 150 km wide.

The abundances of species at various sites (Table 1) emphasize that the composition of the two main arms of migration is quite different. Uganda passage is dominated by willow warblers, garden warblers and red-backed shrikes, and includes many yellow wagtails (*Motacilla flava*) and sand martins (*Riparia riparia*) as well as reed warblers (*Acrocephalus scirpaceus*). The eastern Kenya movement by contrast is dominated by marsh warblers, whitethroats (*Sylvia communis*) and sprossers, and also includes iranias (*Irania gutturalis*), river warblers, Basra reed warblers (*Acrocaphalus griseldis*) and rufous bush chats (*Cercotrichas galactotes*); none of these occur commonly in Uganda and some are unrecorded there. Garden warblers, willow warblers and red-backed shrikes contribute relatively small numbers to the eastern passage, and reed warblers and sedge warblers (*Acrocephalus schoenobaenus*) are practically absent.

Most southward migrants occur at altitudes between 500 and 1600 m, in scrub, thicket and small trees, in bushed and wooded savanna and areas of marginal cultivation. The few species recorded on passage through the Kenya

Table 1. Palaearctic passerine passage migrants in Kenya and Uganda, showing their relative abundance at different sites[a]

		*Southward passage**							*Northward passage*						
		A	B	C	D	E	F	G	A	B	C	D	E	F	G
I	Southward passage mainly through Uganda (some species more eastern in spring)*														
	Riparia riparia	3	3	3	3	1	1	1	3	3	3	2			
	Saxicola rubetra (T)	2	2	1					2	1	1				
	Acrocephalus scirpaceus	2	2	2		1			1	1	1	1	2	1	
	A. schoenobaenus	2	1	2	2	2			2	2	3	3	3	3	2
	Phylloscopus trochilus	3	3	3	3	2	2	1	3	3	3	3	3	3	2
	Sylvia borin	2	3	3	1	2	1		2	3	2	1	2	1	
	Anthus trivialis	2	2	2	1	1	1		2	1	2	1	1		
	Motacilla flava	3	3	3	3	2	1	1	3	3	3	3	2		
	*Lanius collurio**	3	3	2	1	1	2		1	1	1	2	2	3	3
	*Lanius minor**	1	1						2	2	2	2	2	2	2
II.	Southward passage mainly through E. Kenya (also eastern in spring)														
	Acrocephalus griseldis						2	1							
	Cercotrichas galactotes (T)						2							1	1
	Irania gutturalis (T)						2						1	1	
	Luscinia luscinia				1	2	3					1	2	2	2.
	L. megarhynchos (T)					1	2						1	1	1
	Monticola saxatilis (T)				1	1	1	2					1	1	1
	Oenanthe oenanthe (T)		1		2	2	2	1					1		
	*Acrocephalus palustris**		1		1	2	3						1	2	2
	*Locustella fluviatilis**					1	3							2	
	*Hippolais olivetorum**						2							1	1
	Sylvia communis					2	3					1	2	3	1
	*S. nisoria**(T)		1			1	2						2	2	
	Oriolus oriolus	1		1	1	1	2	2	1			1	1	1	3
	Lanius isabellinus (T)		1	1	1	1	2						2	2	
III.	Other species														
	Delichon urbica		1	1	3	2	2					2	1		
	Hirundo rustica	3	3	3	3	3	3	3	3	3	3	3	3	3	3
	Acrocephalus arundinaceus											2	2	2	1
	Muscicapa striata	2	2	2	2	1	2	1	2	2	2	2	1	1	2

[a]A. Lake Edward; B. Kampala; C. central Nyanza; D. Nakuru (rift valley highlands); E. Nairobi; F. Ngulia (Tsavo); G. Mombasa. [b]Abundance catagories: *3*, abundant (up to 100+ per day); *2*, common (up to 10+ per day); *1*, frequent (a few per year up to a few per day).
* Includes west European populations; *T*, wintering commonly south only to central Tanzania.

highlands at above 1800 m, the house martin (*Delichon urbica*), the willow warbler, the yellow wagtail and the tree pipit (*Anthus trivialis*), are also common here as wintering species.

Four of the species which winter in southern Africa are almost above from southward movements through Kenya and Uganda. These are the great reed warbler (*Acrocephalus arundinaceus*), the icterine warbler, the collared flycatcher and the lesser grey shrike. Icterine warblers and collared flycatchers pass regularly through Rwanda (Vande weghe 1979, *in litt.*), and lesser grey shrikes are evidently more common during October-November in Zaire (Dowsett 1972). These species thus seem to migrate mainly to the west of Uganda. The great reed warbler is common in September in central and eastern Sudan (Nikolaus 1983, and *in litt.*), and arrives in southern Zaire (De Roo and Deheegher 1969) and elsewhere in southern Africa (Pearson 1975) from November onwards. It probably overflies large areas of central Africa, but again perhaps uses a route west of Uganda. Sedge warblers occur on passage in Uganda and in west and central Kenya, but only locally and sporadically. This may be another species which largely overflies East Africa on the way south.

Species such as the willow warbler, the garden warbler, the spotted flycatcher and the yellow wagtail reach southern Africa on a broad front, but others have well-defined routes. Some seem to use remarkably narrow corridors through central and southeastern Kenya, even though large numbers of birds from a wide breeding range may be involved. November-December records of sprosser, marsh warbler and river warbler, plotted in Fig. 2, illustrate this phenomenon well. They are practically confined to a belt about 150 km wide east of Lake Turkana and the highlands. These species are seldom recorded west of Nairobi, or in northeast Kenya, the lower Tana valley or the coastal strip; their numbers in south Somalia are very low (Pearson 1989).

2.2 Timing

Peak passerine migration through the Middle East and immigration into coastal Ethiopia (Eritrea) is during late August and September (Smith 1957; Passburg 1959; Safriel 1968; Watson 1971; H. Biebach, *in litt.*). Precise passage times have recently been ascertained on the Sudan coast (Nikolaus and Pearson 1982; Nikolaus 1983, and *in litt.*). Species which later move on to equatorial and southern Africa cross this coast, some 400 km from the Ethiopian highlands, between mid-August and early October (Table 2), with peak migration about 5–10 September. They do not, however, become common in Kenya and Uganda, some 1600 km further south, until many weeks later.

A few migrants do penetrate quickly to the East African equator. Thus roosts of swallows build up locally in August, and willow warblers, yellow wagtails, whinchats (*Saxicola rubetra*) and sand martins appear in Uganda from mid- or late September. The main Uganda arrival is over 1 month later, however, and passage peaks there between late October and early December (Table 2). The so-called short rains are usually well established in Uganda by September, but in central and east Kenya conditions during September-October are much harsher at the end of a long dry period, and this presumably explains a slightly later migrant arrival. A few hirundines, wheatears and yellow wagtails appear at Nairobi in September, but the first arrivals of most species are typically late in October, and most passage occurs here from early November to mid-December. Elsewhere along the edge of the highlands passage also peaks during mid- to late November. Further east, the bushlands of Meru, Kitui and Tsavo receive rain

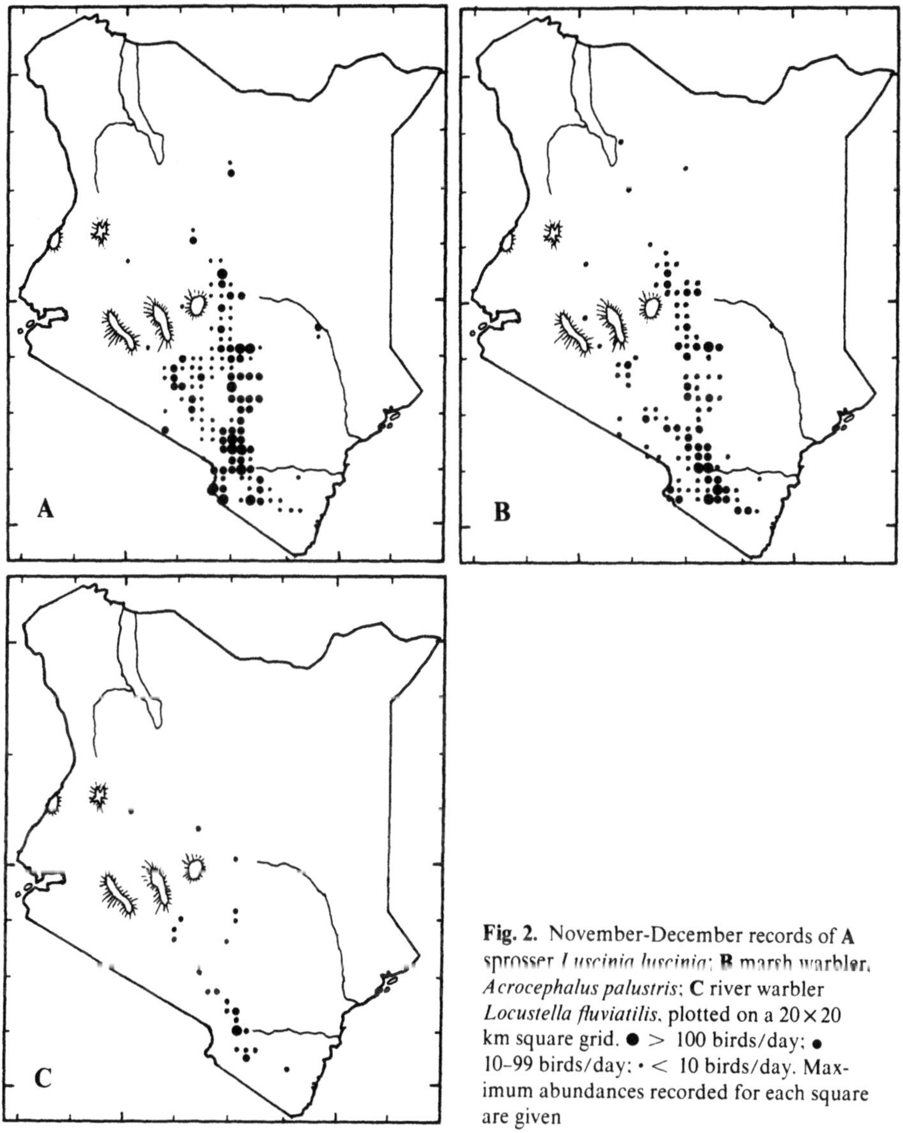

Fig. 2. November-December records of **A** sprosser *Luscinia luscinia*; **B** marsh warbler, *Acrocephalus palustris*; **C** river warbler *Locustella fluviatilis*, plotted on a 20×20 km square grid. ● > 100 birds/day; • 10–99 birds/day; · < 10 birds/day. Maximum abundances recorded for each square are given

later and leaf out only in November, and passage migrants remain abundant throughout December and into early January. The timing of migration across Tsavo (Table 2, Fig. 3) has been ascertained from capture rates in mist at night at Ngulia Lodge (Backhurst and Pearson 1984). Movement builds up quickly in early November at a time when the surrounding bush is usually still leafless. It peaks during late November and tails off during late December and early January. The spotted flycatcher, the rufous bush chat and the red-backed shrike practically complete their passage in November, and the sprosser by mid-December. In the whitethroat, the marsh warbler, the river warbler, the Basra reed

Table 2. Times of southward passage of various species in coastal Sudan (Nikolaus 1983, and *in litt.*), southern Uganda (Kampala) and southeast Kenya (Ngulia)[a]

	Sudan coast	S Uganda	SE Kenya
Hirundo rustica	m Aug – l Oct	l Aug – Nov	
Riparia riparia	l Aug – e Oct	m Sep – Nov	
Oriolus oriolus	l Aug – e Oct	l Oct – l Nov	e Nov – e Dec
Luscinia luscinia	m Aug – l Sep		e Nov – m Dec
Acrocephalus griseldis	m Aug – m Oct		m Nov – e Jan
A. palustris	e Aug – l Sep		m Nov – e Jan
Hippolais olivetorum	l Aug – e Sep		e Nov – e Dec
Locustella fluviatilis	m Aug – l Sep		m Nov – e Jan
Phylloscopus trochilus	l Aug – l Sep	l Sep – Nov	e Nov – e Jan
Sylvia borin	l Aug – e Oct	m Oct – m Dec	m Nov – m Jan
S. communis	l Aug – l Sep		e Nov – e Jan
S. nisoria	m Aug – l Sep		e Nov – m Jan
Muscicapa striata	l Aug – m Oct	e Oct – l Nov	l Oct – m Nov
Anthus trivialis	m Sep – m Oct		e Nov – e Dec
Motacilla flava	e Sep – l Oct	l Sep – Nov	
Lanius collurio	m Aug – e Oct	l Oct – l Nov	e Nov – e Dec
L. minor	l Aug – m Sep	m Oct – m Nov	

[a] e, m, and l indicate early, mid, and late

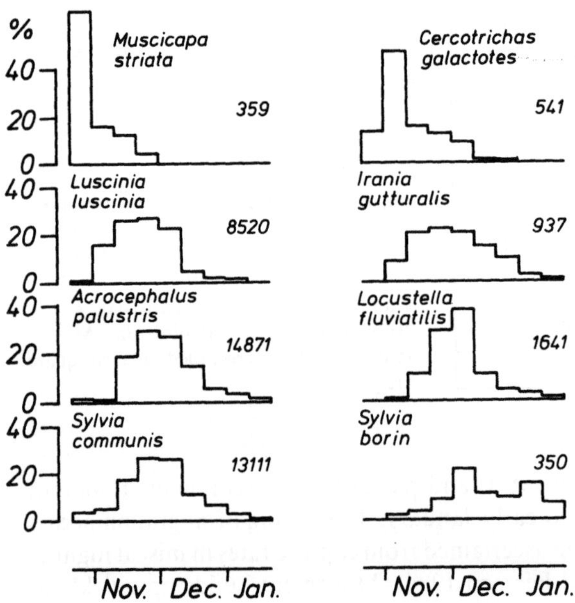

Fig. 3. The timing of southward passage of selected species at Ngulia, Kenya. Indices of occurrence for ⅓-monthly periods, based on night-catching rates (see Backhurst and Pearson 1984) are expressed as a percentage of the total late October to mid-January occurrence. *Numbers* in each species sample are given

warbler and the irania, on the other hand, passage is about equally divided between November and December; garden warblers and barred warblers are clearly still on the move well into January (Fig. 3).

In general then, passage through southeast Kenya is centred about 2 weeks later than through Uganda, and about 12 weeks later than across the Sudan coast. Species such as the marsh warbler and the river warbler cross the equator more than 3 months after reaching tropical northeast Africa. Most migrant passerines which overwinter in East Africa also arrive late. Ringing has shown that some warblers reach territories in south Uganda by early November (Pearson 1972), but most birds take up wintering sites in the bushlands of southeast Kenya in December or January. Most wintering *Acrocephalus* warblers delay arrival until after moult, appearing only during December, January or early February (Pearson 1972, 1973, 1982). One species whose wintering arrival seems to be earlier is the yellow wagtail. Many birds ringed at Nairobi in October have been retrapped during subsequent months (D.J. Pearson, G.C. Backhurst and D.E.G. Backhurst, unpublished work).

2.3 Behaviour and Migration Strategy

Many passage birds pause in transit in Kenya and Uganda in areas with green cover and good feeding potential. They frequently occupy the same thicket for days at a time. Song is common in sprossers, garden warblers and willow warblers, and also frequent in southeast Kenya in marsh warblers and river warblers. Ringed birds retrapped over periods of up to 2 to 3 weeks have usually been found to have put on weight. In a series of garden warblers retrapped at Kampala over late November – early December, most increased in weight by 5–10%, and one gained 25% (Pearson 1971). M.P.L. Fogden (*in litt.*) recorded similar passage fattening in this species in southwest Uganda.

Weights and visible fat ratings of birds grounded at night at Ngulia indicate that many carry fat loads of 15–30% over and above lean weight (see e.g. Pearson and Backhusrt 1976a). They may therefore be on long flights, bound directly for areas 7–800 km or more to the south. Some birds certainly do move through East Africa very quickly. A marsh warbler was recovered in south Malawi only 4 days after being ringed at Ngulia (Backhurst 1978). The fact that passage is already heavy across Ngulia earlier in November, with the bushland still dry and leafless for hundreds of kilometres to the north, suggests rapid migration at this time, perhaps directly from staging areas in Ethiopia.

3 Northward Passage

Migrants leave wintering areas in southern Africa only a few weeks before arrival back in the Palaearctic. The return passage through Kenya and Uganda between late March and early May is thus more urgent than the southward movement.

3.1 Species Distributions and Migration Routes

Migration appears generally less channelled in April than in November-December. It is pronounced in most parts of Uganda and west Kenya, and along the eastern edge of the highlands; also, in contrast to southward passage, through the Kenya rift, including the hot, low Lake Turkana area, and along the Kenya coast.

Concentrations of migrants occur locally at insect-rich sites which are less utilized at other times. These include the shores of the Kenya rift lakes and also two sites near Nairobi where some 4000 birds were caught and ringed in spring between 1971 and 1984, a *Lantana* scrub area adjacent to Kariobangi sewage works on the eastern edge of the city and an *Acacia* grassland area with ponds and ditches at Athi River, 25 km to the southeast.

By contrast with November-December, the bushland east and north of the Kenya highlands holds few migrants in April, even though large areas may already be in leaf with the beginning of the "long" rains. Falls sometimes occur during rain or mist, however, and these emphasize the volume of migration normally passing unseen across these low to medium altitude plateau areas. Falls are most frequent near the lights of isolated lodges or towns (e.g., Pearson 1981; Boothroyd 1987). At Ngulia, most April rain falls in afternoon or evening storms and showers, and strong winds tend to prevent mist formation at night. However, on the few occasions that attractive conditions have been encountered, birds were grounded here in hundreds or thousands, much as in November-December (Britton and Britton 1977; Pearson 1980).

Table 2 shows species abundances at various sites. Willow warblers tend to dominate northward movements from Uganda to east of the Kenya highlands. Sedge warblers are also common on a broad front from Lake Victoria to the coast. Lesser grey shrikes and red-backed shrikes occur widely, but the latter is much scarcer in Uganda and most abundant in east Kenya. Movements of yellow wagtails, tree pipits, garden warblers and reed warblers seem to extend east only to about central Kenya. By contrast, the whitethroat is seldom seen west of Nairobi, but predominates in the east Kenya bushland. The marsh warbler and sprosser are also confined to the east, including the coast. Species from wintering areas in northern and central Tanzania, such as the nightingale, the red-tailed shrike (*Lanius isabellinus*), the rock thrush (*Monticola saxatilis*) and the barred warbler, are also largely confined to an east Kenya route.

It is clear that many species use a more easterly route in spring than in autumn. Thus, movements of red-backed shrikes, sedge warblers and willow warblers are concentrated more in Uganda in autumn, but are heavy in central and eastern Kenya in spring. Lesser grey shrikes and great reed warblers appear to pass west of Uganda in autumn, yet are common in Kenya in spring. The species involved in the eastern arm of the autumn movement clearly return via a similar southeast Kenya route in April. However, the appearance of marsh warblers, sprossers, nightingales, shrikes and golden orioles along the coast is without parallel in autumn, and would suggest considerable northward passage across Somalia.

3.2 Timing

Movements of willow warblers, tree pipits and golden orioles begin about mid-March. Strong passage is established, however, only at the end of March and lasts until about 25 April. A few species continue to pass through until early May, and falls at a lighted site on the Upper Tana revealed small numbers of warblers up to the middle of May (Boothroyd 1987).

Figure 4 shows passage timing in the main species caught at Nairobi (Kariobangi). The sprosser occurs in late March and early April, while the nightingale, the irania, the red-tailed shrike and the rock thrush, all en route from Tanzanian winter quarters (and not featured in Fig. 4) show a similar, early timing. Willow warblers occur mainly from 25 March to 20 April and whitethroats from 5–25 April. Reed warblers and great reed warblers are slightly later, and sedge warblers decidedly so, passing mainly from 10 April to 10 May (Pearson et al. 1979). Overall, passage peaks about mid-April. This is about 2 weeks later than the main departure of wintering passerines from sites in Kenya and Uganda, which takes place between 20 March and 10 April. In the sedge warbler, the whitethroat and the garden warbler, in which the timing of passage and wintering exodus can be compared, the departure of the wintering birds by early April would suggest that they have a more southern breeding destination than the birds which pass through later from wintering areas further south. The "long" rains usually begin in mid- or late March. Central and east Kenya may remain dry, however, until early April. In such years most wintering migrants have departed and passage is well under way before the effects of rain are experienced.

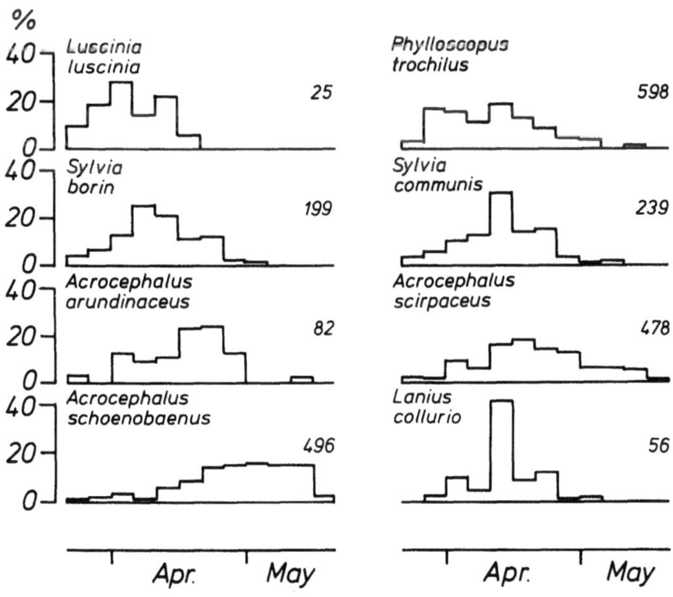

Fig. 4. The timing of northward passage of selected species at Kariobangi, Nairobi, Kenya. Mean daily capture rates for 5-day periods are expressed as a percentage of the sum of the mean rates for periods from late March to mid-May. *Numbers* in each species sample are given

3.3 Behaviour and Migration Strategy

Overall passage numbers appear to be lower in April than in November-December. This is especially so in years when the long rains are heavy. An abundance of new leaf and insect food might be expected to attract more passing migrants in wet years, but the converse is the case, Kenya passage is heavier in all areas when the rains are poor and savanna habitats remain relatively bare until well into April.

Considering the low migrant numbers and the short period within which passage is completed, it must be concluded that many birds overfly Kenya and Uganda on their way north, or pause only fleetingly. This is clearly the case with whitethroats, sprossers and marsh warblers, which seem to be confined to an east Kenya route, yet are seen only in small numbers in April unless grounded in local falls. Spring birds typically remain for only a day or two. They tend to occur in loose groups, and show less tendency to territorial behaviour or vocal activity than migrants in November-December. Certain species may however stay and fatten in suitable sites for up to 1–2 weeks, using these as staging points (e.g. Pearson et al. 1979).

Most spring passage birds caught for ringing in Kenya have been quite lean. Weights and visible fat ratings have indicated that less than one-third carried the 15–20% excess fat reserves necessary to support a direct flight to the Ethiopian highlands (see Nisbet et al. 1963; Pennycuick 1975). Data for Kariobangi are summarized in Table 3. It could be argued, however, that it would be mainly the less well provisioned birds which come to ground at Kenya ringing sites, those with adequate fuel loads tending to overfly undetected. An April fall at Ngulia in severe stormy weather involved birds which were nearly all fat (Britton and Britton 1977).

It has been estimated that passerine migrants with fat loads 60% or more above their lean weight might embark from Kenya and Uganda on direct 2500-km flights to the Middle East (Pearson 1971; Pearson et al. 1979). Spring

Table 3. Weights of the main spring passage migrants at Kariobangi, Nairobi. Percentages of fat birds (rated *3–4*; see Pearson and Backhurst 1976a) are also given. Lean weights (in parentheses) are means for samples of 17–100 birds with a fat rating of *1* caught in Kenya or Uganda during November – February

	n	Mean ±SD	Range	Fat 3–4 (%)
Luscinia luscinia (21.8)	23	23.4 ± 2.2	18.9–26.9	18
Acrocephalus schoenobaenus (10.1)	470	11.5 ± 1.5	9.1–16.5	30
A. scirpaceus (10.5)	479	11.1 ± 1.1	0.0–15.0	18
A. arundinaceus (26.8)	85	31.8 ± 4.9	24.6–45.7	32
Sylvia communis (13.2)	240	15.2 ± 1.9	11.3–22.3	42
S. nisoria (21.6)	17	27.2 ± 3.1	21.4–31.4	17
S. borin (17.4)	187	21.2 ± 2.6	16.1–28.7	58
Phylloscopus trochilus (7.9)	568	8.6 ± 1.1	6.5–14.6	17
Lanius collurio (24.9)	51	28.6 ± 3.3	24.0–39.8	15

weights indicative of such loads have been encountered occasionally in sedge warblers and great reed warblers, but seldom in other species. The majority of passage birds must be bound for staging areas in Ethiopia.

4 Discussion

4.1 The Two-Stage Southward Migration

Passerine migration from southern Africa back to breeding-grounds in eastern Europe and Asia takes only about 6 weeks. Species such as the red-backed shrike, the marsh warbler and the whitethroat, still common as far south as Botswana at the beginning of April (N. Hunter pers. comm.), pass through Kenya mainly in mid-April and Ethiopia about 10 days later (Ash 1980), reaching latitudes 50°-60°N late in May (Dement'ev and Gladkov 1954). By contrast, the corresponding southward journey, with its two-stage nature (Pearson and Backhurst 1976a; Curry-Lindahl 1981), can take up to over 4 months. After a rapid first stage, and arrival in Ethiopia and central Sudan in late August and September, most birds remain there for 1-3 months before completing migration, a further 800-4000 km to the south, during November-December.

The autumn stopover area is presumably just as important to these birds as the southern winter quarters. Most birds clearly interrupt their migration and spend long periods sedentary and territorial at latitudes north of about 10°N in Sudan and 7°N in Ethiopia (Pearson et al. 1988). Many undergo their annual wing moult there (Pearson 1973, 1975; Pearson and Backhurst 1976a). It is only when conditions begin to dry out during late October and November that birds resume migration. The first rain is then usually occurring, at least locally, in Kenya and Tanzania.

The first appearances of migrants in East Africa usually involve falls in rain or attraction to insects concentrated at the leading edge of the intertropical rain front (Sinclair 1978). Local weather conditions and thus patterns of migrant arrival tend to vary somewhat from year to year, especially in more arid environments. Appearances of grounded birds are not, however, always a good reflection of the strength of migration overhead, and it is unlikely that large-scale movements out of Ethiopia would be much influenced by local conditions many hundreds of kilometres to the south. Passage over Ngulia seems to be precisely timed, and builds up regularly in early November whether the surrounding bushlands are green or not. On the other hand, rain at Nairobi earlier in October does not produce the migrant falls that it would at the end of the month. The southward exodus from Ethiopia has presumably evolved so that its onset coincides with a time when suitable reception conditions could be found in Kenya, at least locally, in a normal year. Uganda is more equable, and receives rain in September, yet also experiences a delayed passage. This suggests that resources are so abundant in the northern tropics during September and early October that further southward migration would be generally disadvantageous at this time.

4.2 Flight Strategies During Intra-African Migration

It is well established that some passerines perform long, continuous migrations, flying by day as well as by night to cross ecological barriers such as oceans and deserts (e.g. Cooke 1905; Moreau 1961; Ward 1963). To what extent are long flights involved in migrations within sub-Saharan Africa? The limited occurrence of sprossers and marsh warblers during April would indicate that these species overfly Kenya on their way north, while in wet springs the low number of migrants generally suggests large-scale overflying by many species. In November, overflying might similarly account for the scarcity of sedge warblers and the absence of great reed warblers in Kenya and Uganda.

The April return passage from southern Africa is particularly rapid. Do birds proceed north mostly in short stages, flying only by night and replenishing their reserves each day? This strategy might prove hazardous, for it would depend on locating adequate food sources in a new environment each day. The substantial fat loads (15–30% above lean weight) found in many Palaearctic passerines in sub-Saharan Africa at migration times would indicate that longer distances, up to 1000 km or so, are often covered between opportunities for feeding. Do typical night migrant species continue flights over equatorial Africa by day? Or do they in any case come to ground and sit quietly, even if they have no need to feed, as do birds forced down on migration across the Sahara (Biebach 1988)? The scarcity of some passage species suggests a direct spring crossing of Kenya, involving daytime flying, but grounded birds, perched quietly, might be very unobtrusive. Considerable fat loads are common in November in birds attracted to Ngulia lights just before dawn, and it is perhaps unlikely that such individuals are about to land and search for food. Their spare fat may be merely an insurance, but it is difficult to explain weight excesses of 25–40% in this way.

4.3 Narrow Front and Loop Migration

Both passages through East Africa are characterized by a marked east-west segregation of species. This depends partly on breeding origin and on the preferred habitat in Africa, but is also clearly influenced by the topography of Kenya and north Tanzania, where areas over 3000 m probably represent real barriers to migration. Similar segregation is also a feature of southward migration through Sudan and Ethiopia (Pearson et al. 1988).

Some species migrate through Kenya on a front no more than 100–200 km wide. How do they navigate with such precision? How important are topographical clues, and how could these be utilized by inexperienced young birds? To refer again to one of the most intriguing of species examples, the marsh warbler on its southward journey through Africa must alter its migration direction at least twice, once during the stopover in Ethiopia and again on entering Tanzania (Fig. 5). The movements of specific populations of Palaearctic migrants in other parts of sub-Saharan Africa may well be generally more precise and at the same time more complex than has hitherto been recognized.

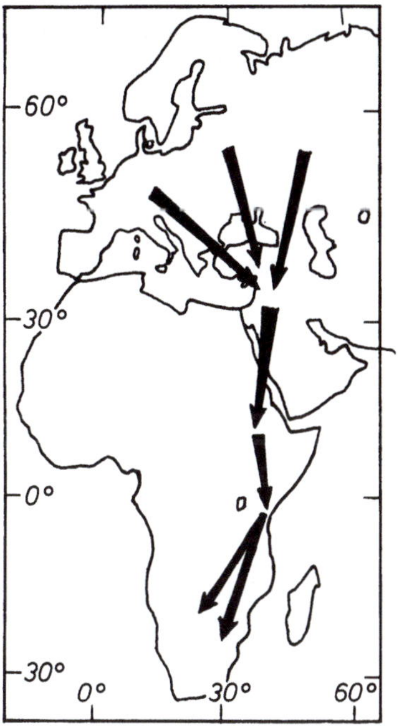

Fig. 5. Routes of southward migration of the marsh warbler (*Acrocephalus palustris*) from the Palaearctic to southern Africa

Several species evidently perform large loop migrations between the Middle East and southern Africa, with the more easterly spring route concentrated through Kenya and Ethiopia. The examples of the red-backed shrike and the lesser grey shrike have been documented (Moreau 1972; Dowsett 1972), but similar strategies probably apply in the sedge warbler, the great reed warbler and the willow warbler. The scarcity of these and other species in central Sudan in spring (Hogg et al. 1984; G. Nikolaus pers. comm.) is explained by the very arid conditions at this time. On the other hand, east Ethiopia, Somalia and the Arabian Gulf area are more likely to enjoy the effects of rainfall in April than in August-September.

Acknowledgments. This review utilizes ringing data kindly made available by G.C. Backhurst, D.E.G. Backhurst and P.L. Britton, to whom I am most grateful. I also thank many people for supplying unpublished records and observations, and especially B. Boothroyd, J.D. Cape, M.A.C. Coverdale, B.W. Finch, M.P.L. Fogden, P.C. Lack, J.M. Lock, B.S. Meadows, J.G. Rolfe, T. Stevenson, D.A. Turner and A.P. Zeigler.

References

Ash JS (1980) Migrational status of Palaearctic birds in Ethiopia. Proc IV Pan- Afr Ornithol Congr, pp 199–208

Backhurst GC (1978) East African bird ringing report 1974–77. J East Afr Nat Hist Soc Natl Mus 163:1–10

Backhurst GC, Pearson DJ (1977) Southward migration at Ngulia, Tsavo, Kenya, 1976/77. Scopus 1:12–17

Backhurst GC, Pearson DJ (1984) The timing of the southward night migration of Palaearctic birds at Ngulia, southeast Kenya. Proc V Pan-Afr Ornithol Congr, pp 361–369

Backhurst GC, Pearson DJ (1988) Ringing and migration at Ngulia, Tsavo, autumn 1986. Scopus 10:133–135

Biebach H (1988) Ecophysiology of resting willow warblers (*Phylloscopus trochilus*) crossing the Sahara. Proc 19th Int Ornithol Congr, pp 2162–2168

Boothroyd B (1987) The attraction of Palaearctic migrants to lights at Kiambere, Kenya. Scopus 11:38–40

Britton PL, Britton HA (1977) An April fall of Palaearctic migrants at Ngulia. Scopus 1:109–111

Chapin JP (1953) The birds of the Belgian Congo, Part 3. Bull Am Mus Nat Hist 75a:1–821

Cooke WW (1905) Routes of bird migration. Auk 22:1–11

Curry-Lindahl K (1981) Bird migration in Africa. Academic Press, London.

Dement'ev GP, Gladkov NA (eds) (1968) Birds of the Soviet Union, vol 6. Israel Program for Scientific Translations, Jerusalem

Dowsett RJ (1972) The lesser grey shrike in Africa. Ostrich 42:259–269

Dowsett RJ, Backhurst GC, Oatley TB (1988) Afrotropical ringing recoveries of Palaearctic migrants. I. Passerines (Turdidae to Oriolidae). Tauraco 1:29–63

Hogg P, Dave PJ, Rintoul JV (1984) Palaearctic migrants in the central Sahara. Ibis 126: 307–331

Moreau RE (1961) Problems of Meditteranean-Saharan migration. Ibis 103a:373–427, 580–623

Moreau RE (1972) The Palaearctic-African bird migration systems. Academic Press, London

Nikolaus G (1983) An important passerine ringing site near the Sudan Red Sea coast. Scopus 7:15–18

Nikolaus G, Pearson DJ (1982) Autumn passage of marsh warblers *Acrocephalus palustris* and sprossers *Luscinia luscinia* on the Sudan Red Sea coast. Scopus 6:17–19

Nisbet ICT, Drury WM, Baird J (1963) Weight loss during migration. Bird Banding 34:107–159

Passburg RE (1959) Bird notes from northern Iran. Ibis 101:153–169

Pearson DJ (1971) Weights of some Palaearctic migrants in southern Uganda. Ibis 113:173–184

Pearson DJ (1972) The wintering and migration of Palaearctic passerines at Kampala, southern Uganda. Ibis 114:43–60

Pearson DJ (1973) Moult of some Palaearctic warblers wintering in Uganda. Bird Study 20: 24–36

Pearson DJ (1975) The timing of complete moult in the great reed warbler *Acrocephalus arundinaceus*. Ibis 117:506–509

Pearson DJ (1980) Northward spring passage of Palaearctic passerines across Tsavo. Scopus 4:25–28

Pearson DJ (1981) Spring falls of Palaearctic passerines at Mtito Andei, Kenya. Scopus 5:80

Pearson DJ (1982) The migration and wintering of Palaearctic *Acrocephalus* warblers in Kenya and Uganda. Scopus 6:49–59

Pearson DJ (1989) Palaearctic migrants in the Middle and Lower Jubba valley, southern Somalia. Scopus 12:53–60

Pearson DJ, Backhurst GC (1976a) The southward migration of Palaearctic birds over Ngulia, Kenya. Ibis 118:78–105

Pearson DJ, Backhurst GC (1976b) Palaearctic passerine migration at Kariobangi, Nairobi. EANHS Bull 1976:23–28

Pearson DJ, Backhurst GC, Backhurst DEG (1979) Spring weights and passage of sedge warblers *Acrocephalus schoenobaenus* in central Kenya. Ibis 121:8–19

Pearson DJ, Nikolaus G, Ash JS (1988) The southward migration of Palaearctic passerines through northeast and east tropical Africa: a review. Proc VI Pan Afr Ornithol Congr pp 243–262

Pennycuick CJ (1975) Mechanics of flight. In: Farner DS, King JR (eds) Avian biology, vol 3. Academic Press, London

Roo A De, Deheegher J (1969) Ecology of the great reed warbler *Acrocephalus arundinaceus* wintering in the southern Congo savannah. Gerfaut 52:260–275

Safriel U (1968) Bird migration at Elat, Israel. Ibis 110:283–320

Sinclair ARE (1978) Factors affecting the food supply and breeding season of resident birds and movements of Palaearctic migrants in a tropical African savannah. Ibis 120:480–497
Smith KD (1957) An annotated check list of the birds of Eritrea. Ibis 99:1–26, 307–337
Vande weghe JP (1979) The wintering and migration of Palaearctic passerines in Rwanda. Gerfaut 69:29–43
Watson GE (1971) A serological and ectoparasite survey of migratory birds in northeast Africa. Smithsonian Institution, Washington
Ward P (1963) Lipid levels in birds preparing to cross the Sahara. Ibis 105:109–111
Zink G (1971) The migrations of European swallows *Hirundo rustica* to Africa from data obtained through ringing in Europe. Ostrich Suppl 8:211–222
Zink G (1973) Der Zug europäischer Singvögel. Part 1. Vogelzug, Radolfzell

Migration Across the Alps

B. BRUDERER and L. JENNI[1]

1 Environmental Conditions for Migratory Birds in the Alps

Climate and habitat in the Alps are governed by altitude. An altitudinal change of 200 m corresponds to a shift in temperature of 1° or to a N–S shift in the vegetational zones of about 200 km. Rich broad-leafed woods and wetlands are rare. A bird resting in the Alps has to endure low temperatures, accentuated by wind chill, and important changes in temperature. Frequency and amount of precipitation are increased at the windward side of the main Alpine chains. In the dry valleys between the chains limited surfaces of Mediterranean and steppe vegetation occur. For migrants on stop-over the chance of meeting harsh conditions and unsuitable habitats is increased compared with the lowlands. On the other hand, locally better conditions may be offered in the Alps by the delay in vegetation phenology and insect availability as well as by the spots of drier climate in some valleys. Some migrants preferring seeds of coniferous trees may find optimal conditions in the subalpine forests.

The weather conditions are also less predictable and less uniform in the Alps than in the lowlands. Geostrophic winds with limited directional deviation from the course of the mountain ridges are funnelled between the ridges, on the local scale and over the larger distance between the Alps and the Jura. Winds across the ridges are slowed down below the tops. The diurnal cycle of warming and cooling of the differently exposed slopes induces diurnally changing air currents along the slopes and valleys. Rising moist air leads to condensation and accumulation of clouds on the windward slopes, while on the downwind side clouds disappear. Intense insolation of rocky slopes may cause the accumulation of high clouds and local thunderstorms. The progress of frontal systems is slowed by friction along the Alps; high cloud barriers and prolonged precipitation may prevent any crossing of the Alps and sometimes cause important losses of migrants trapped in the Alpine valleys. In anticyclonic conditions inversions, inducing haze or fog layers, may build up in the valleys and basins (Eichenberger 1966; G.A. Gensler pers. comm.). For a migrant bird on the wing the most important weather features are probably (1) the funnelling of the winds between the Alps and the Jura; (2) the frequent wind changes from one slope to another; and (3) more clouds and precipitation in the mountain area.

[1] Schweizerische Vogelwarte, CH-6204 Sempach, Switzerland

E. Gwinner (Ed.) Bird Migration

2 Migratory Directions in Central Europe and Along the Alps

2.1 Questions and Data Base

The general directions of nocturnal migrants crossing central Europe during autumn aim mainly at the Iberian peninsula or southern France. For birds crossing the area covered by the radar stations in Fig. 1, the directions leading to these (intermediate) goal areas are between 210° and 250°, centered around 230°. A smaller proportion of birds, especially of long-distance migrants, might cross the Mediterranean Sea E of Iberia, thus heading more S. SE directions will be very rare, as most of the migratory divides are E or N of the area considered. For nocturnal long-distance migrants crossing Switzerland, ringing recoveries show mainly directions around 220° (Baumgartner and Bruderer 1985); the recruiting directions of a wide variety of species can be found in Figs. 6 and 7. Thus, according to the ringing recoveries, the main bulk of night migration would approach the Swiss Alps from the NE and meet them at a small angle.

The questions are, whether the actual approach directions towards the Swiss lowlands and the Alps correspond to those expected, and whether the birds proceed in the same direction or adjust their course to cross or avoid the Alps.

The data and methods used were tracking radar observations during autumn 1987 and 1988 at several sites in southern Germany and Switzerland (B. Bruderer unpublished work) together with older data on nocturnal migration (Steidinger 1968; Bruderer 1975, 1977, 1978, 1981, 1982; Rüsch and Bruderer 1981; Baumgartner and Bruderer 1985; Liechti and Bruderer 1986). We tried to separate topographical and meteorological factors by the comparison of simultaneous observations at different sites (Fig. 1). We observed the tracks (Figs. 1a and 3) in order to determine the resulting movements of the birds under the influence of winds. Our goal was to recognize the underlying behaviour of the birds by calculating the headings (Fig. 1b) or by selecting birds flying in low wind speeds, where wind drift would be compensated (Fig. 2). We checked whether birds above the surrounding relief (i.e. > 1500 m above radar) behave differently from those at low levels (< 1000 m above radar). The mean vectors and modes (if appropriate) of all the distributions (Table 1) supported the visual comparison of the diagrams.

To interprete the directional distributions it is important to be aware that the numbers of birds tracked at the different altitudes are not proportional to the density of migration, because we tried to track sufficient birds at all altitudes for a statistical comparison. The quantification of migration is not done by tracking single birds, but by fixed-beam counting or systematic scanning of the sky (Bruderer 1971, 1980). In order to allow a rough quantitative judgement of the directional behaviour at different levels, we recall that (according to Bruderer 1971) on average 50% of nocturnal migration is below 700 m above the Swiss lowlands, and 90% is below 2000 m. As the passes and lower ridges of the Alps are about 1500 m higher than the lowlands, most birds have to climb to cross the Alps. In fine weather with following winds or above fog layers the main mass of migration may be found above 1000 m above ground level (AGL), while in bad

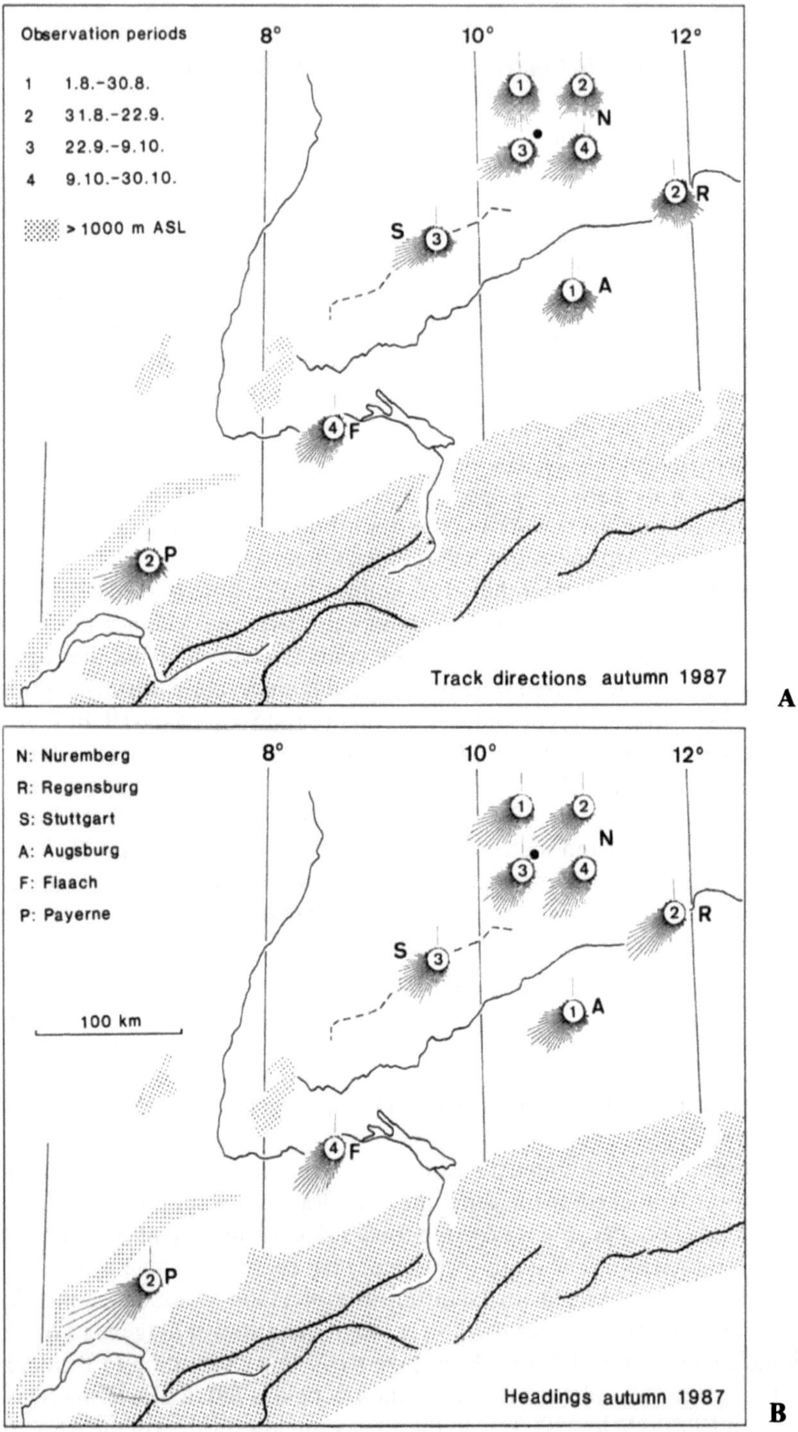

Fig. 1. A Distributions of track directions and **B** headings in southern Germany and in the Swiss lowlands during four periods of autumn 1987. At *N* the exact location of the radar is indicated by a point between the distributions from the four observation periods. In the other cases the *circle* marks the radar site. The letters *P* and *F* indicate the nearest village; *N, R, A*, and *S* are larger towns in the area. For statistics see Table 1

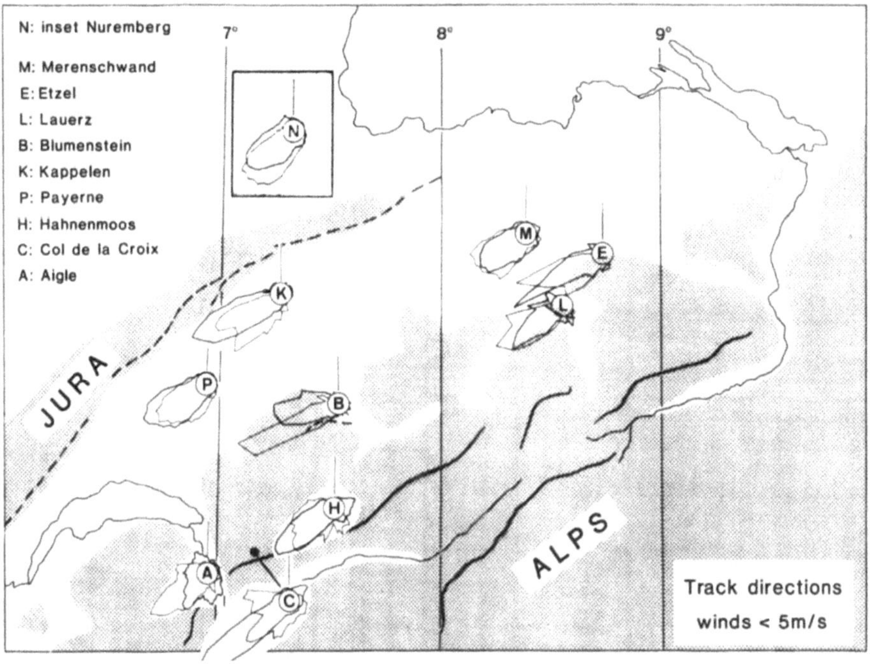

Fig. 2. Distribution of track directions at two flight levels (cf. Fig. 3) in weak winds (< 5 m/s). The distributions are expressed as percentages of flights towards each direction (smoothed by running means over three direction classes of 5°; the end points of the single vectors are connected); *E, L*, and *B* are based on 10°-classes without smoothing. It is supposed that under weak winds drift is completely compensated, at least below 1000 m AGL. For statistics see Table 1

weather and/or opposing winds nocturnal migration above the lowest few hundred meters may be very weak (see also Bruderer and Jenni 1988). Thus, in Fig. 1 high flying birds are overrepresented. In Figs. 2 and 3 we separated two flight levels, emphasizing the directions below 1000 m AGL. These so-called low flying birds are about 70% of all migrants in Fig. 3 (all winds) and about 50% in Fig. 2 (weak winds); the fractions of high flying birds (above 1500 m AGL) may represent about 20 and 30%, respectively.

2.2 Approach Directions in Southern Germany

In weak winds (< 5 m/s; Fig. 2 inset "N" and Table 1) the directions of nocturnal migration in southern Germany are around 230° to 235° below 1000 m AGL. The migrants passing north of the Danube valley will usually not reach the Alps, but pass N of the Jura. At higher levels the directions (headings and tracks) are usually more S, giving evidence of a southward aiming part of migration, which will approach the Alps. Birds heading SE are rare, however, and mainly confined to period 1. Leading lines met at small angles, as the Suebic Alb near "S" or the Rhine valley near "F", induce minor deviations of about 10°.

Table 1. Characteristic directions at the different locations. Mean vectors in the first row, maxima of distribution (the mode of the smoothed distributions according to Fig. 2) in the second row of each site. (n = number of birds per distribution)

Location	Period	All winds All heights Track	(n)	Head	All winds < 1000 m Track	(n)	Head	All winds > 1500 m Track	(n)	Head	Winds < 5 m/s < 1000 m Track	(n)	Head	Winds < 5 m/s > 1500 m Track	(n)	Head
Nuremberg	30.7.–30.10.87	209	(8969)	225	220	(4630)	228	196	(2287)	220	232	(1327)	233	213	(795)	216
		235		225	245		225	225		235	235		240	225		215
	30.7.–31.8.87	194	(2983)	227												
		200		240												
	31.8.–22.9.87	192	(1485)	231												
		?		240												
	22.9.–9.10.87	228	(1671)	221	242	(896)	221	209	(418)	220	233	(156)	231	215	(153)	217
		240		215	250		210	235		240	255		?	?		?
	9.10.–30.10.87	222	(2864)	220												
		240		225												
Augsburg	1.8.–30.8.87	189	(2335)	219	210	(903)	227	172	(822)	209	233	(366)	230	205	(198)	216
		?		240	240		240	170		255	235		245	230		225
Regensburg	31.8.–22.9.87	180	(2034)	229	195	(962)	233	175	(559)	226	229	(320)	233	224	(97)	232
		170		230	220		235	170		225	230		235	?		?
Stuttgart	22.9.–9.10.87	227	(1826)	225	241	(905)	226	220	(499)	226	248	(201)	245	225	(274)	228
		240		235	250		240	235		235	250		245	235		225
Flaach	9.10.–30.10.87	219	(2053)	216	225	(1008)	219	215	(554)	215	222	(573)	218	215	(242)	216
		225		215	230		225	195		205	225		225	?		225
Payerne	31.8.–26.9.87	222	(1537)	234												
		240		240												
Payerne	23.8.–25.9.86 + 1.9.–30.10.87	228	(4976)	230	236	(1681)	235	213	(1876)	224	234	(894)	236	223	(547)	227
		240		240	240		245	220		235	240		245	230		235
Merenschwand	2.8.–20.9.82	224	(2579)	223	227	(1257)	225	220	(651)	220	228	(964)	226	228	(431)	226
		240		230	235		230	250		235	235		230	250		225

Kappelen	20.8.–12.9.88	224	(1425)	232	234	(768)	238	204	(345)	222	239	(422)	240	224	(97)	225
		240		235	240		240	215		230	245		240	?		230
Etzel	9.8.–12.10.80	238	(4626)		223	(2432)		197	(1185)		237	(176)		234	(59)	
		240			250			210			240			250		
Lauerz	9.8.–15.10.80	214	(4948)		232	(2277)		190	(1502)		233	(126)		234	(115)	
		240			?			?			230			240		
Blumenstein	5.9.–24.10.83	230	(3324)		250	(2143)		211	(1181)		261	(293)		246	(190)	
		250			260			250			270			250		
Hahnenmoos	27.7.–11.9.75	217	(3172)	222	225	(1741)	228	211	(720)	218	227	(1213)	230	226	(303)	228
		230		230	230		235	230		230	230		235	220		230
Col de la Croix	1.8.–23.9.88	229	(4313)	232	233	(2711)	231	218	(803)	231	233	(1778)	232	229	(222)	232
		235		230	235		230	225		230	235		230	?		?
Monthey	10.8.–14.9.88	226	(2209)	232	223	(673)	231	227	(1046)	227	220	(560)	229	235	(484)	236
		235		230	215		225	235		230	215		225	235		235

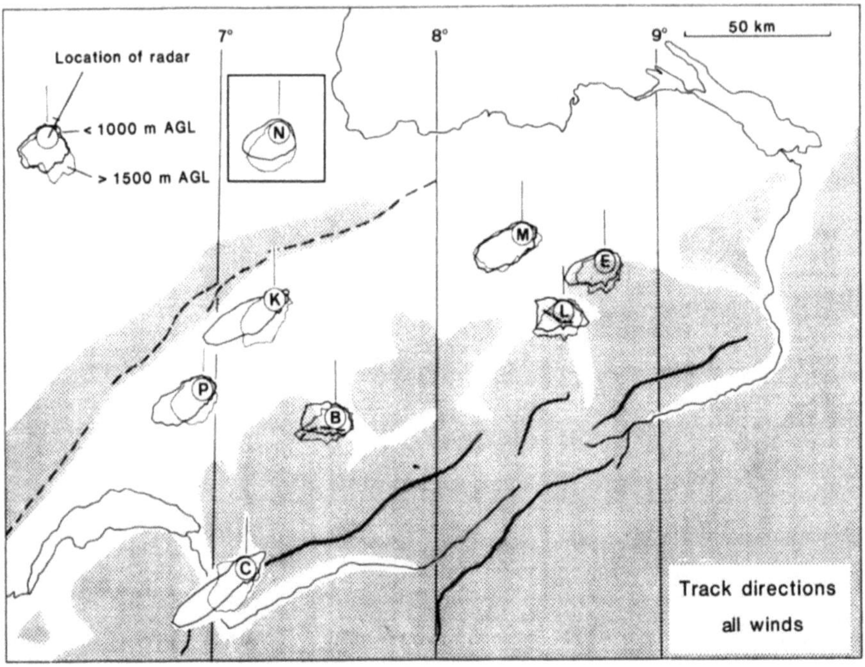

Fig. 3. Distribution of track directions including all wind conditions. Same representation as Fig. 2

The influence of wind results in drift, compensatory reactions, differential migratory activity (pseudodrift) or selection of flight levels in order to avoid unfavourable winds. The fact that the headings in Fig. 1b are much more uniform than the tracks in Fig. 1a indicates that there is more drift than compensation above southern Germany. Only relatively weak winds (up to 5 m/s) are fully compensated by birds flying lower than 1000 m AGL. At higher levels or with higher wind speeds only partial or no compensation takes place (F. Liechti unpublished work). The distributions of track directions in Fig. 1a differ more between different periods at the same site than at different sites during the same period (except "P", see below). A comparison of the wind situations (weak winds from all directions in period 3, rather strong westerly winds during the other periods) suggests that the influence of winds prevails over that of topography above the hilly landscape of southern Germany.

Under the frequent and often strong westerly and northwesterly winds most birds are drifted S towards the Alps (see track directions in Fig. 1a). The wind conditions at the moment of passage have a paramount effect on the flight paths of nocturnal migrants, inducing a large variation in directions from day to day. The proportion of SE heading birds (Fig. 1b) is very small and only relevant in period 1; it is small even at the easternmost point, at Regensburg. Thus, the tracks pointing S and SE are in most cases caused by drifted birds and not by birds heading in these directions;pseudodrift of populations wintering to the SE is negligible in the area considered, at least after period 1. Pseudodrift towards S may account for the limited fraction of migrants flying southward with southerly headings under all wind conditions.

2.3 *Directions in the Area of the Alps*

Birds at low levels and in weak winds (Fig. 2 and Table 1), approaching the Swiss lowlands from the area of the Danube valley and S of it, are able to proceed in the same directions unless deviated by high mountain ridges. Directions to the right of the main direction (230°) will lead towards the southern flank of the Jura; those to the left lead towards the border of the Alps or tangentially along it. Birds arriving at a ridge below its top may climb above or follow the ridge. The higher the ridge top relative to the flight level, and the less a ridge deviates from the bird's previous course, the more likely it is that the bird will follow the ridge: At "B" the low flying birds meet a high ridge (1500 m above radar) at an angle of about 30°; they are completely deviated (Fig. 2). At "L" they meet a ridge (1000 m above radar) at an angle of about 90°; some birds are deviated to the left or to the right, while most of them climb to cross the ridge (Fig. 2). At "A" the birds fly along the valleys joining at this point. At the Alpine passes "H" and "C" the axis of the passes is close to the preferred direction; accordingly the funnelling along the ridges is very distinct (including a notable fraction of reversed migration).It is even more pronounced when all winds are included (Fig. 3). Climbing and crossing of ridges is generally reduced in opposing winds and nearly stops when the mountain tops are in clouds (Fig. 3; Liechti and Bruderer 1986). The tendency of migrants to follow mountain ridges met at small angles relative to their primary direction (even under calm conditions) is certainly one of the major reasons for the funnelling of migration between the Jura and the Alps; this results in extremely constant directions around 224° under all weather conditions near Geneva (Baumgartner and Bruderer 1985). It is, however, not readily understood why birds even far off the mountain ridges maintain their track directions under westerly winds. In period 2 with strong westerly winds, the headings at "P" are extremely concentrated along the axis of the lowlands, indicating no compensation (Fig. 1b); the tracks (Fig. 1a) unlike those at "N" and "R" are scattered but not deflected in a particular direction by the strong westerly winds. The absence of cross winds due to the funnelling of winds between the Alps and the Jura is one of the reasons for this constancy of tracks and headings; however, the difference relative to the results from the flat country in Germany persists when birds flying under equally strong southwesterly winds are considered in both areas. The birds in Germany shift their headings slightly S of 230° and are drifted ESE. The birds at "P" maintain their headings and fly with increased air speeds, but with low or even negative ground speeds along the lowlands. In spite of the low migratory activity under such conditions, this example shows that there is a directing influence of the mountain ranges in addition to the funnelling of the winds (B. Bruderer unpublished work).

Birds approaching the area of the Alps at high altitudes (higher than 1500 m above the lowlands) show a wider variation of directions, often resulting in different cohorts. This is most conspicuous at "K", where one group of birds follows the Jura (which is below their flight level) in the same way as the low flying birds (245°); the other group has (probably) crossed the Jura with a mean track of about 220° (corresponding to the mean directions of high flying birds above southern Germany). At "M", most distant from the Alps and the Jura, the same tendencies are slightly indicated; the mean direction, however, equals that at low levels (see also Bruderer 1978, 1982). At the ridges bordering the Alps and at the main chains of the Alps, the birds flying above the ridges show two or even three directional tendencies: (1) maintaining the (supposed) primary direction; (2) following a leading line; (3) flying westward, possibly towards the lowest part of the horizon, the Lake of Geneva basin.

2.4 Conclusions and Discussion

Three (overlapping) groups of nocturnal migrants approach the area of the Swiss Alps from southern Germany: (1) The birds which under weak or following winds cross the area of the Danube valley and S of it, and proceed with directions close to their supposed primary directions around 235°, being deviated by the funnelling effect of the Alps and the Jura to reach Geneva with directions around 224°. (2) The birds with more southerly directions, originating further N, which seem to prefer higher flight levels and are prone to cross mountain ridges. (3) The huge mass of birds drifted towards the Alps and the Jura by the frequent westerly winds (including also individuals of groups 1 and 2). Considering group (3), those at low levels (a) either shift their directions first WSW and later SW (because of the protection against cross winds or additional leading effects of the mountain ranges) or get intercepted at single ridges, enter a valley, zigzag through the mountain ranges, and cause important concentrations of migrants at Alpine passes; a limited fraction of these birds may also pass through N-S directed valleys and cross the Alps towards Italy; group (3) birds at high levels (b) are either drifted across the ridges or concentrated along the highest Alpine chains.

The cohorts flying westward away from the Alpine chains at high altitudes, the extreme concentrations of high flying birds at "B", and those flying parallel to far off mountain ranges call for additional explanations. Birds released in dark nights often fly towards the brightest part of the surroundings (e.g. a valley, a white cloud, a white wall, the moon or another light source; B. Bruderer unpublished work; Bruderer and Neusser 1982). Free flying birds under hazy or misty conditions often change their flight paths to approach the light dome above a town (B. Bruderer unpublished work), or are attracted by lamps in the fog. These data suggest the following hypothesis: Nocturnal migrants might tend to avoid flying towards dark areas like the unlighted flank of a mountain; instead, they would prefer to fly towards lower, brighter parts of the horizon. Other interpretations might involve a preference to fly above a known habitat (even at night) resulting in avoidance of the mountain areas, or unknown effects such as detecting nearby mountain areas by infra-sound or other atmospheric features.

3 Which Birds Cross the Alps?

Almost all species migrating regularly through Switzerland in autumn have been observed in the Alps (Jenni and Naef-Daenzer 1986). However, several species seem to be underrepresented in the Alps, especially soaring birds and species dependent on wetlands (Bruderer and Winkler 1976; Sutter 1955; Thiollay 1966, 1967). In this section, we first examine whether flight capabilities and habitat requirements correlate with the proportion of birds crossing the Alps.

Radar studies suggest that birds crossing the Alps either fly southward at high altitudes or approach the Alps at low altitudes on tracks determined largely by wind drift, thus, deviated from their innate direction of migration; of course, intermediate behaviour between these two extremes exists. At the border of the Alps, the migrants show a great variation of behavioural reactions. It is possible that this variation as well as the altitudinal segregation are related to certain properties of the birds. In the second part of this section, we try to identify the high flying birds crossing the Alps and examine, both at the inter- and intraspecific level, whether there are correlations between crossing the Alps and certain characteristics of the birds, i.e. (1) origin and direction of migration and (2) energy reserves.

3.1 Flight Capabilities

Soaring raptors have been frequently observed to abandon crossing alpine passes under strong head-winds when raptor species migrating chiefly by flapping flight still pass (Thiollay 1967; H. Schmid pers. comm.). Therefore we might expect that crossing the Alps depends on flight capabilities.

Indeed, there is a negative correlation between the relative frequency of 11 raptor species in the Alps and their proportion of soaring during migration (Fig. 4). For *Buteo buteo,* one of the most distinct soaring species, the few individuals flying in the Alps occur early during the migration season. This is due to the fact that the climbing rate when soaring decreases by one-third over the migration season because of decreasing lift in thermals (Schmid et al. 1986).

Among passerines as well as among waders/waterbirds, flight techniques are similar and flight speed in isometric birds depends on size (Rayner 1988). Wing-beat pattern and frequency from radar-tracked birds allow us to distinguish waders/waterbirds (continuous wing beats) from passerines (wing beats alternating with pauses) and two separate size classes in each group. With tail winds, there are no differences in the proportions of large and small birds among passerines or waders/waterbirds at different flight altitudes in the Alps. With southwesterly winds, however, when birds are flying lower and are drifted towards the Alps, the proportion of large birds increases with altitude both in passerines and waders/waterbirds (Bloch et al. 1981). This indicates that large birds can cope better with opposing winds and are able to fly higher under adverse conditions. A comparison of the proportion of large and small passerines during the same period in late autumn between sites "K" and "C" shows that with

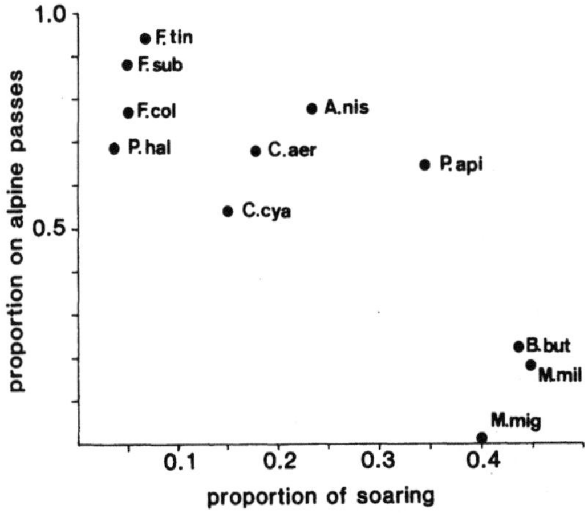

Fig. 4. Proportion of soaring in 11 raptor species during migration (determined by tracking radar) vs relative frequency in Alps vs lowlands. The latter is the number of birds observed at two alpine passes (Col de Bretolet and Wasserscheide) divided by the total observed at Col de Bretolet, Wasserscheide and Fort l'Ecluse, the extreme end of the lowlands near Geneva) (data from H. Schmid, pers. comm.). n = 11, least squares correlation coefficient r = –0.84, $P < 0.01$. A. nis = *Accipiter nisus*; B. but = *Buteo buteo*; C. aer = *Circus aeruginosus*; C. cya = *C. cyaneus*; F. col = *Falco columbarius*; F. sub = *F. subbuteo*; F. tin = *F. tinnunculus*; M. mig = *Milvus migrans*; M. mil = *M. milvus*; P. api = *Pernis apivorus*; P. hal = *Pandion haliaetus*

opposing winds, there are 40% large passerines over the Alps but only 30% over the lowlands (n = 1509, $chi^2 = 18.1$, $P < 0.001$); this confirms that large birds have a better capability to cross the Alps under opposing winds than do small birds.

3.2 Habitat Requirements

The number of captures per species at the alpine pass Col de Bretolet correlates with the abundance of the species in the breeding areas (estimations in five countries in the area of origin of Swiss migrants). The residuals of this relationship depend on the size of the bird, whether its migration is nocturnal or diurnal, and other factors (Jenni and Naef-Daenzer 1986 and below). This suggests that species of the same group (with respect to size, etc.) are caught roughly in proportion to their abundance at this alpine pass; here they normally do not rest.

An analysis of the residuals of this relationship with respect to habitat requirements reveals that species dependent on wetlands are underrepresented on Col de Bretolet (Jenni and Neaf-Daenzer 1986). However, this could be due to differences in flight behaviour between wetland and non-wetland species which in turn could influence catching rates.

Yet, there is circumstantial evidence that habitat requirements might play a role. When night migrants land in the early morning at the alpine pass Col de Bretolet, species dependent on the general type of habitat found at this site (bushes) are caught in higher proportions (compared with the number of night

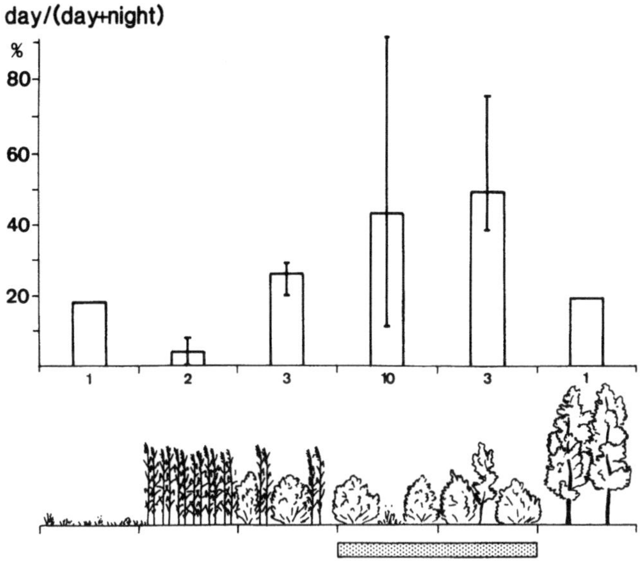

Fig. 5. Proportion of day captures (number of day captures divided by number of day and night captures) of night-migrating passerine species according to six habitat types. The number of species is indicated below the column, the range by a *line*. The types of habitats available at Col de Bretolet are marked by the *shaded bar*

captures) than species dependent on wetlands and forests (Fig. 5). This indicates that night migrants are prone to land in approximately appropriate habitats. The poor availability of wetlands in the Alps (the next large reed beds are 60 km further S of Bretolet) might therefore be a reason for wetland species to avoid the Alps or to cross the Alps non-stop with high energy reserves and at altitudes high enough to prevent capture.

3.3 Origin, Direction, and Altitude of Migration

Interspecific Level. An analysis of the residuals from the expected number of captures at Col de Bretolet (see Sect. 3.2) with respect to direction of migration within Europe reveals that species migrating S or SE (120–180°) are caught in very low numbers because the valley leading to the Col de Bretolet is oriented southwest and acts as a funnel only for southwestern migrants. Species showing a migratory divide within central Europe are also caught in lower numbers than expected (Jenni and Naef-Daenzer 1986). Moreover, for passerine species not dependent on wetlands with generally southwesterly directions of migration, and no migratory divide within Europe, the residuals from the expected number of birds correlate with the mean direction of migration within Europe (Fig. 6). Species with directions around 210° are caught in higher numbers than expected from their estimated number of breeding pairs than are species with more westerly directions (around 230°).

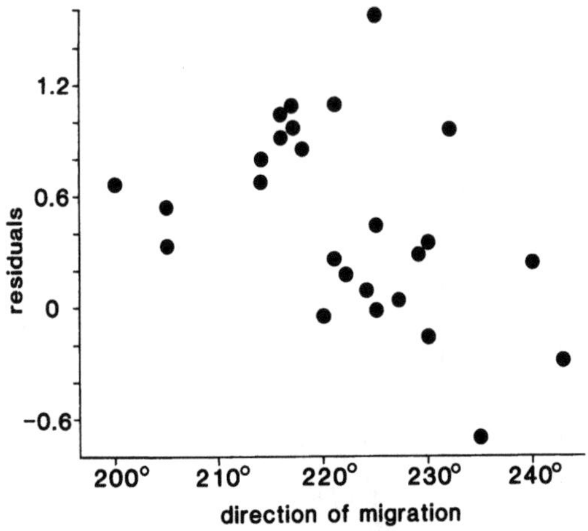

Fig. 6. Mean recruiting direction of migration versus residuals from the expected number of birds at Col de Bretolet (see Jenni and Naef-Daenzer 1986). The mean direction of birds migrating towards Switzerland is calculated from Swiss ringing recoveries (n> 10) or estimated from Zink (1973) for seven species. Included are 26 passerine species not dependent on wetlands, without a migratory divide within Europe, and which migrate generally to the southwest (> 180°); $r = -0.45$, $P < 0.03$

It is difficult to obtain data on the species-specific altitudinal distribution of migrants in order to test if a southerly direction implies a higher flight altitude for the species (but see Sect. 3.1). Wing-beat patterns do not allow us to recognize the species in passerines.

Intraspecific Level. At this level the origin of migrating birds can be evaluated by ringing recoveries and by the geographical variation of morphological characters. Both methods are applied here. Two species provide enough recoveries to compare the origin of birds occurring in the Swiss Alps and in the Swiss lowlands north of the Alps during autumn migration: In *Erithacus rubecula*, a short-distance night migrant, birds occurring in the Alps during autumn migration originate from more distant areas (Scandinavia) while birds occurring in the lowlands originate predominantly from Germany and Czechoslovakia. The latter have a wider spread of directions and a more westerly mean direction than birds occurring in the Alps (Jenni 1987). Also in *Fringilla coelebs*, a short-distance day migrant, birds occurring in the Alps originate from more distant areas and arrive with more southerly directions than birds from the lowlands (L. Jenni unpublished work). Other species consistently show the same tendency but the small number of recoveries available does not give significant differences.

Among the breeding populations of the long-distance night migrant *Sylvia borin* there is a clinal increase of wing-length over Europe towards the north-east (Klein et al. 1973; L. Jenni unpublished work). A detailed analysis of wing-length and feather-length (Berthold and Friedrich 1979; Jenni and Winkler 1989) at nine different ringing sites in Switzerland shows that longer-winged and therefore northern populations are caught at night in the Alps and at a stop-over site S of the Alps; this indicates that northern populations are more likely to cross the Alps (Jenni and Jenni-Eiermann 1987). In *Sylvia borin*, northern populations also migrate with more southerly directions over Europe than southern populations (Klein et al. 1973; Zink 1973; Jenni and Jenni-Eiermann 1987).

Do northern populations migrate at higher altitudes? On the alpine pass Col de Bretolet, birds are normally caught during the night in mist nets of 8 m height. On foggy nights lamps are used to attract the birds. It is assumed that, in the fog, birds normally flying higher than 8 m are caught in addition. In the three most common species caught during the night (*Erithacus rubecula, Ficedula hypoleuca, Sylvia borin*, n > 400 each), significantly longer wings occur in the captures of foggy nights than in those of clear nights (wing-length corrected for seasonal trends by regression analysis). Northern populations of these species have longer wings. Hence it can be inferred that they migrate at higher altitudes. Furthermore, in *Sylvia borin*, the mean wing-length of trapped birds increases with head-winds which induce lower flight altitudes and apparently force even the larger northern birds to fly low enough to be caught (Jenni and Jenni-Eiermann 1987).

At the inter- and intraspecific level, we conclude that passerines meeting the Alps at large angles are more likely to cross the Alps than birds approaching at small angles. At the intraspecific level, the former are normally northern populations (i.e. from Scandinavia) migrating within Europa with more southerly directions and higher above ground than southern populations (i.e. from Middle Europe).

3.4 Energy Reserves

It is possible that higher energy reserves balance the harsher environmental conditions and risks met by birds during migration and stop-over in the Alps.

Interspecific Level. For night-migrating passerines, we showed that species with high weight surplus over premigratory weight in the lowlands north of the Alps, indicating high fat reserves, are caught in higher numbers during night at Col de Bretolet than expected (Bruderer and Jenni 1988). Using new data (fat scores, ranging from 1 to 5) of a greater number of species (day and night migrants), this correlation can be confirmed; there is no clear difference between day and night migrants.

The mean fat scores in the lowlands north of the Alps during the migration season of the species (Jenni 1984), however, correlate with the mean recruiting directions of migration according to ringing recoveries (Fig. 7). This correlation prevents the variable fat score from contributing significantly to explain the variance in the number of birds caught on Col de Bretolet in a multiple regression analysis incorporating mean direction (see Sect. 3.3) and the measure of population size as other independent variables.

Thus, species arriving with more southerly directions in Switzerland have larger fat reserves and are more likely to cross the Alps than species migrating parallel to the Alps. For a given direction of migration, there are no clear

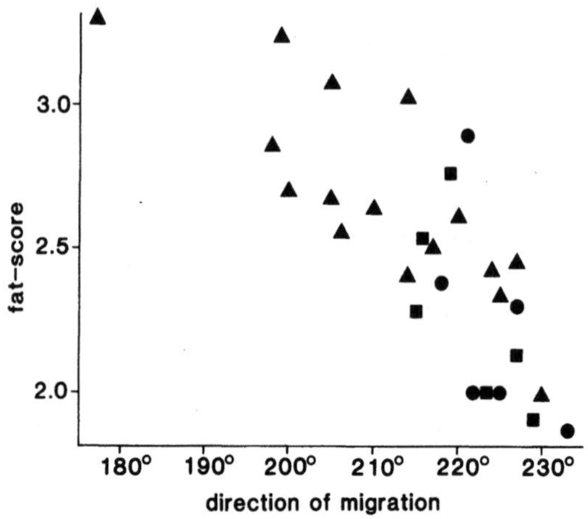

Fig. 7. Mean recruiting direction of migrants versus mean fat score in the lowlands north of the Alps during the migratory season of the species concerned. *Triangles* Species crossing the Mediterranean and the Sahara; *squares* crossing very often the Mediterranean; *dots* wintering predominantly on the continent; n = 28, r = -0.80, $P < 0.001$

differences in fat score between long-distance migrants crossing the Mediterranean and the Sahara, short-distance migrants crossing the Mediterranean regularly, and short-distance migrants wintering mainly on the continent; a migratory divide or specialized habitat requirements also do not have an effect.

This suggests, for the passerine species examined until now, that the larger energy reserves of species migrating S are not due to a fundamentally different destination or migratory strategy, but could be in response to the Alps. Further analyses are needed to test this hypothesis.

Intraspecific Level. In *Sylvia borin, S. atricapilla, Ficedula hypoleuca* and *Oenanthe oenanthe*, birds flying higher above the alpine pass Col de Bretolet (attracted by lamps, see Sect. 3.3) are significantly heavier than birds flying low over the pass on clear nights (weights corrected for seasonal trends, size and time of day by multiple regression analysis; Bruderer and Jenni 1988). Other night-migrating species show the same tendency.

3.5 Discussion

Species arriving in Switzerland with southerly directions cross the Alps in higher proportions and with higher energy reserves than species paralleling the Alps on migration. The former might correspond to the increased proportion of southward heading birds seen at high levels by radar. This general pattern might be influenced by differences in flight capability and habitat requirements in direct response to the difficulties imposed by crossing the Alps. Most birds are drifted towards the Alps with westerly winds, the most frequent winds in autumn. Therefore, we expect a differential reaction to weather and wind conditions according to species, amount of energy reserves or other yet unknown properties, either by refraining from migration under certain conditions or by different reactions at the border of the Alps.

Intraspecifically, the same general pattern shows up: Individuals arriving with southerly directions have more energy reserves and are more likely to cross the Alps. They originate from northern populations and apparently fly higher above ground. The analysis of *Sylvia borin* data from nine different ringing sites revealed that the endogenous differences in the population and age-specific migration program, found in the laboratory (Berthold et al. 1974), are not blurred by environmental factors. They are still recognizable in the field by the seasonal course, sequence, and energy reserves of the populations and age groups migrating through Switzerland (Jenni and Jenni-Eiermann 1987, unpublished work). Therefore, it is probable that crossing the Alps is facilitated by the endogenous migration program and the general migration strategy.

However, the question remains, whether higher energy reserves are built up proximately in response to the Alps or endogenously without exogenous stimulus. The findings at Falsterbo support the former: there, the northward reorientation of night and day migrants is thought to serve for fattening up as a direct response to the Baltic sea, a relatively small and flat obstacle of only 25–100 km (Alerstam 1978; Sandberg et al. 1988; Lindström and Alerstam 1986). It would of course be a much safer strategy to adjust the endogenous build up of fat reserves in direct response to perceived environmental barriers.

The ultimate causes of crossing or avoiding the Alps are difficult to assess. We merely mention some of the advantages and disadvantages. The disadvantages regarding climate, habitat and weather are listed in Section 1. They demand more energy and a better risk insurance. The advantages might be that the distance of migration is shorter for some destinations, the more so since the bird can tolerate drift by the frequent westerly winds. Energy expenditure would be smaller and migration more rapid. It might be an advantage to migrate more rapidly because of time pressure, food availability, competition and because first arriving birds might occupy better and nearer winter territories. We expect that the trade-off between disadvantages and advantages varies for different species and populations. Birds wintering S of the Sahara might be less dependent on the land mass of Iberia if they are able to put on fat as far north as central Europe or southern France and Italy; the distance from central Europe to western Africa (Switzerland – Lagos) is 10% longer over Gibraltar than directly. Many species moult completely in the winter quarters where they establish territories; first arriving birds might have an advantage. Therefore long-distance migrants should have more southerly directions (Fig. 7). Northern populations of long-distance migrants are especially under time pressure. Short-distance migrants migrating through Switzerland predominantly winter in southern France and Iberia and have no gain in distance by crossing the Alps. As seen by the time of migration (e.g. *Sylvia atricapilla*: Klein et al. 1973; Turrian and Jenni 1989), southern populations stay longer in the area of the breeding grounds before migration, while northern populations pass through; to arrive early on the wintering grounds does not seem to be important. Short-distance migrants depend more than long-distance migrants on the temperature gradient running NE-SW over Europe during winter. Therefore it is understandable that they take more westerly directions.

Acknowledgement. This work was supported by the Swiss National Foundation for Scientific Research, grant Nos. 3.171-0.85, 3.161-0.81.

References

Alerstam T (1978) Reoriented bird migration in coastal areas: dispersal to suitable resting grounds? Oikos 30:405–408

Baumgartner M, Bruderer B (1985) Radarbeobachtungen über die Richtungen des nächtlichen Vogelzuges am nördlichen Alpenrand. Ornithol Beob 82:207–230

Berthold P, Friedrich W (1979) Die Federlänge: Ein neues nützliches Flügelmass. Vogelwarte 30:11–21

Berthold P, Gwinner E, Querner U (1974) Vergleichende Untersuchung der Jugendentwicklung südfinnischer und südwestdeutscher Gartengrasmücken *Sylvia borin*. Ornis Fenn 51:146–154

Bloch R, Bruderer B, Steiner P (1981) Flugverhalten nächtlich ziehender Vögel – Radardaten über den Zug verschiedener Vogeltypen auf einem Alpenpass. Vogelwarte 31:119–149

Bruderer B (1971) Radarbeobachtungen über den Frühlingszug im Schweizerischen Mittelland. (Ein Beitrag zum Problem der Witterungsabhängigkeit des Vogelzuges). Ornithol Beob 68:89–158

Bruderer B (1975) Zeitliche und räumliche Unterschiede in der Richtung und Richtungsstreuung des Vogelzuges im Schweizerischen Mittelland. Ornithol Beob 72:169–179

Bruderer B (1977) Beitrag der Radarornithologie zu Fragen der Orientierung, der Zugphysiologie und der Umweltabhängigkeit des Vogelzuges. Vogelwarte 29 (Sonderheft):83–91

Bruderer B (1978) Effects of alpine topography and winds on migrating birds. In: Schmidt-Koenig K, Keeton WT (eds) Animal migration, navigation and homing. Springer, Berlin Heidelberg New York, pp 252–265

Bruderer B (1980) Vogelzugforschung unter Einsatz von Radargeräten. In: Koller S, Reichertz PL, Überla K (eds) Med Informatik Statistik 17:144–154

Bruderer B (1981) Stand und Ziele der Radar-Vogelzugforschung in der Schweiz. Rev Suisse Zool 88:855–864

Bruderer B (1982) Do migrating birds fly along straight lines? In: Papi F, Wallraff H (eds) Avian navigation. Springer, Berlin Heidelberg New York, pp 3–14

Bruderer B, Jenni L (1988) Strategies of bird migration in the area of the Alps. Proc 19th Congr Int Ornithol Ottawa, pp 2150–2161

Bruderer B, Neusser V (1982) Nächtliche Aktivität und "Heimfinden" beim Hausrotschwanz *Phoenicuros ochruros*. Ornithol Beob 79:145–157

Bruderer B, Winkler R (1976) Vogelzug in den Schweizer Alpen. Angew Ornithol 5:32–55

Eichenberger W (1966) Flugwetterkunde. Schweizer Verlagshaus AG, Zürich

Jenni L (1984) Herbstzugmuster von Vögeln auf dem Col de Bretolet unter besonderer Berücksichtigung nachbrutzeitlicher Bewegungen. Ornithol Beob 81:183–213

Jenni L (1987) Analysis of recoveries of same species subgroups with similar potential recovery rate. Acta Ornithol 23:129–132

Jenni L, Jenni-Eiermann S (1987) Der Herbstzug der Gartengrasmücke *Sylvia borin* in der Schweiz. Ornithol Beob 84:173–206

Jenni L, Naef-Daenzer B (1986) Vergleich der Fanghäufigkeiten von Zugvögeln auf dem Alpenpass Col de Bretolet mit Brutbeständen im Herkunftsgebiet. Ornithol Beob 83:95–110

Jenni L, Winkler R (1989) The feather-length of small passerines: a measurement for wing-length in live birds and museum skins. Bird Study 36:1–15

Klein H, Berthold P, Gwinner E (1973) Der Zug europäischer Garten- und Mönchsgrasmücken (*Sylvia borin* und *S. atricapilla*). Vogelwarte 27:73–134

Liechti F, Bruderer B (1986) Einfluss der lokalen Topographie auf nächtlich ziehende Vögel nach Radarstudien am Alpenrand. Ornithol Beob 83:35–66

Lindström Å, Alerstam T (1986) The adaptive significance of reoriented migration of chaffinches *Fringilla coelebs* and bramblings *F. montifringilla* during autumn in southern Sweden. Behav Ecol Sociobiol 19:417–424

Rayner JMV (1988) Form and function in avian flight. In: Johnston RF (ed) Current ornithology, vol 5. Plenum, New York, pp 1–66
Rüsch E, Bruderer B (1981) Einfluss der Topographie auf nächtlich ziehende Vögel. Rev Suisse Zool 88:865–874
Sandberg R, Pettersson J, Alerstam T (1988) Why do migrating robins, *Erithacus rubecula*, captured at two nearby stop-over sites orient differently? Anim Behav 36:865–876
Schmid H, Steuri T, Bruderer B (1986) Zugverhalten von Mäusebussard *Buteo buteo* und Sperber *Accipiter nisus* im Alpenraum. Ornithol Beob 83:111–134
Steidinger P (1968) Radarbeobachtungen über die Richtung und deren Streuung beim nächtlichen Vogelzug im Schweizerischen Mittelland. Ornithol Beob 65:197–226
Sutter E (1955) Vogelzug in den Schweizer Alpen. Acta XI Congr Int Orn Basel, pp 172–175
Thiollay JM (1966, 1967) La migration d'automne des rapaces diurnes aux cols de Cou et Bretolet. Nos Oiseaux 28:229–251, 29:69–97, 105–126
Turrian F, Jenni L (1989) Etude de trois espèces de fauvettes en période de migration postnuptiale à Verbois, Genève, phénologie du passage et utilisation du milieu. Alauda 57:133–154
Zink G (1973) Der Zug europäischer Singvögel, 1. Liefg. Vogelwarte Radolfzell, Möggingen

Timing of Bird Migration in Relation to Weather: Updated Review

W.J. RICHARDSON[1]

1 Introduction

Effects of daily changes in weather on numbers of birds taking off on migration have been investigated for many decades. This field was especially active in the 1960s and 1970s when many systematic radar and visual studies of migration were done. Also, multivariate and other statistical methods became widely used, providing a partial solution to the problem of quantifying and separating the effects of interrelated weather variables on migration. Work up to 1977 was reviewed by Richardson (1978a), hereafter cited as R78a.

Since 1977 some migration studies of types similar to those of the preceding period have been done. However, there have been fewer general "migration vs weather" surveys in favour of more focused analyses of methodological biases, species differences, site-specific effects, and effects on orientation. Of particular importance, Alerstam (1978–81) combined field data with a theoretical treatment of optimal migration strategies relative to weather. In many respects the change in study approach was helpful. The broad outlines of relationships between weather and numbers aloft as detected by various field methods were evident by 1977. However, there were some discrepancies among species, locations and observation methods. Recent work has helped resolve some of these questions.

This chapter summarizes recent evidence on effects of weather on numbers of migrants taking off each day. No attempt is made to review the larger pre-1978 literature. Thus this chapter is a supplement, not a replacement, for R78a. To save space, publications cited in R78a are marked "*" here and excluded from the References section. Studies cited by R78a as theses or "in press" and later published (Beason 1978; Richardson 1978b–80) are treated as pre-1978 work and are rarely mentioned. Winter and seabird movements are excluded (see Elkins 1983; Blomqvist and Peterz 1984), as are the strong relationships between reverse migration and weather (see R78a, 1982a). R78a also included more details about methodological biases, selection pressures that may shape migration-weather correlations, and weather effects on waterfowl, shorebird and hawk migration. Recent data on these three groups are referenced below under appropriate headings, but are not collated by bird group. I deal almost exclusively with north temperate latitudes, given the meagre data from elsewhere. Also, this review is

[1]LGL Ltd., Environmental Research Associates, POB 280, King City, Ontario, L0G 1K0, Canada

E. Gwinner (Ed.) Bird Migration

restricted to effects of short-term changes in weather on migration timing, and excludes long-term climatological effects (cf. Gauthreaux 1980).

2 Why Should We Care?

The large day-to-day differences in numbers of migrants aloft and their strong correlations with weather are of interest for several reasons. These correlations provide interesting examples of the fine attunement of behaviour to the physical environment. Variations in weather can have strong effects on the energy costs of flight, the ease with which a preferred course can be maintained, and the probability of direct mortality en route (R78a, 1990; Alerstam 1981). The ways in which migration strategies of different species are adapted to weather provide insight into selection pressures.

Migration timing relative to weather is important in understanding some aspects of orientation. The finding that many birds seem to orient well when flying above or below opaque cloud provided some of the impetus to search for non-visual orientation cues (Emlen 1975*; Able 1982b, 1985).

Many migration studies since the 1960s were done to develop methods of reducing numbers of bird-aircraft collisions, e.g. by developing multivariate or other procedures to forecast migration amount a few hours in advance, based largely on weather forecasts (Hild 1969; Blokpoel 1973; Geil et al. 1974*; Richardson 1974*). However, even strong correlations may provide only a limited forecasting ability (Blockpoel and Richardson 1978*, 1979).

Weather effects also need to be considered when developing population monitoring systems based on migration counts. Statistical methods can be used to help separate long-term trends in numbers from the confounding effects of weather on daily and annual counts (Ulfstrand 1958*; Hussell 1981, 1985).

3 Methodology and Biases

3.1 Field

Ideally, effects of weather on propensity to migrate should be studied by measuring the proportion of the birds present that take off each day. In theory this can be determined from direct observations of takeoff (e.g. Hebrard 1971*; Bolshakov and Rezvyi 1982) or turnover of marked birds (Rabøl 1978; Rabøl and Hansen 1978; Evans 1979; Mehlum 1983). However, this has been practical in only a few studies.

In practice, we usually measure numbers of migrants seen or detected by radar. These measures of number aloft are correlated with the proportion taking off, but are also affected by number of birds present in the takeoff area, which varies from day to day. Furthermore, numbers detected aloft can be affected by concentrating effects of topography and by the increased rate of passage (independent of numbers taking off) when wind is following vs opposing. Also, all common observation methods have biases, i.e. detectability varies with species, flight altitude, ground speed, etc. (R78a). Thus, migration counts are an imprecise reflection of numbers or proportion taking off, and are not ideal for analyses of weather effects on propensity to migrate. Observation biases were discussed in R78a, and only a few key points and recent developments are mentioned here.

The altitude bias is one of the most severe problems. Visual observers tend to detect low-flying birds, whereas most radars detect primarily birds at moderate and high altitude. Techniques involving a "cone of observation", e.g. vertical telescopic watches, moon watching, ceilometer, also tend to undersample low-flying birds because the cone is narrowest at low altitudes. Greatly compounding these problems is the fact that migrants tend to fly lower with headwinds than tailwinds. Thus, visual observers commonly overestimate the amount of migration with headwinds; radar and "cone of observation" methods are known (radar) or suspected to underestimate headwind migration.

With careful procedures some biases can be greatly reduced. Bird counts from long-range surveillance radars should be done for areas close to the radar, minimizing the number of birds below the radar horizon. When using "cone of observation" methods, the altitude bias can be largely corrected if the distance (height) of each migrant is measured, as in Bruderer (1971*) and Gauthreaux (1984). The biases of visual observation methods are especially severe along coasts, ridges or other linear features that concentrate migrants (Evans 1966b*; Alerstam 1978–81). Alerstam (1978) concluded that at Falsterbo, a complex coastal site in Sweden, it was impossible to determine from visual counts – even with the most elaborate analyses feasible – whether or not most species tended to take off with following winds. Visual observations in relatively flat inland areas can greatly reduce such problems (Gruys-Casimir 1965*; Laske and Immelmann 1981; Helbig and Laske 1986; Helbig et al. 1986, 1987). However, there is still a natural tendency to study visible migration at places where visible migrants concentrate, which are usually not the places where numbers of passing migrants most accurately reflect numbers taking off.

The biasing effects of wind on ground speed and thus the "Migration Traffic Rate" (MTR) can, in theory, be overcome if ground speeds are known. The migration density can be estimated as MTR/(ground speed), e.g. Bruderer (1977), Bol'shakov (1981). However, allowance should ideally be made for species and wind effects on air speed (e.g. Bloch and Bruderer 1982). Also, even with no correction, the bias caused by using MTR rather than migration density as the measure of migration amount is small when compared with the extreme (often 100 or even 1000 x) day-to-day differences in numbers of migrants aloft.

By combining two or more observation methods, it may be possible to overcome some of these problems. Besides several earlier radar vs visual comparisons (R78a), three teams recently compared migration data acquired simultaneously by different methods (Lindgren and Nilsson 1975*; Dolnik 1981; Buurma et al. 1986). Observation biases are more severe for diurnal than for nocturnal migration (Bol'shakov 1981 and Lyuleeva et al. 1981 in Dolnik 1981).

Although "traditional" observation methods can, if carefully used, reveal much about weather effects on migration, other methods with fewer (or different) biases deserve more use, namely aerial transects (Bellrose 1971*), radio-telemetry (Sladen et al. 1970*; Coon et al. 1976*; Beske 1982; Wege and Raveling 1983; Cochran and Kjos 1985), turnover of marked birds at staging sites (Rabøl 1978; Rabøl and Hansen 1978; Evans 1979); release of birds aloft (Demong and Emlen 1978), and cage studies of weather effects on *Zugunruhe* (Muller 1976*; Viehmann 1982), ideally under controlled conditions.

3.2 Analysis

Types of statistical procedures that can be used to assess weather effects on migration timing were discussed by Richardson (1974*, 1978a). Some continuing problems and recent developments merit comment here.

Migration counts are an imperfect measure of propensity to migrate; the latter is often the variable of more direct interest (Sect. 3.1). One confounding factor is date. With the same weather conditions, more migration is expected in the middle of the migration season than early or late in the season. Thus, date effects on numbers of migrants aloft need to be taken into account before evaluating weather effects, e.g. by truncating the earliest and latest days in the migration season and then subtracting the typical value for the date in question. The latter can be a "moving average" (Darby 1985), quadratic fit (many studies), or polynomial fit (Alerstam 1978; Hussell 1981, 1985).

Days of delay since the last day when weather was favourable for migration is another confounding factor. Many authors have suggested that, as the interval since the previous migratory flight increases and fat reserves are regained, the "migration threshold" diminishes and migrants become more likely to depart under marginal weather conditions (see Blokpoel 1973; Berthold 1978; Bruun 1986). Alerstam (1978) and Hilgerloh (1981) provided evidence of increased migration after periods with little or no migration, although it is difficult to discount natural cycles in weather as an alternative explanation. In caged garden warblers, Schindler et al. (1981) found no evidence for increased *Zugunruhe* after nights when it was suppressed by simulated bad weather. Few authors have incorporated allowance for delay in their analyses, partly because of difficulties in knowing how to measure it and in obtaining the necessary uninterrupted sequences of daily observations. If delay affects probability of takeoff, then delay needs to be taken into account to develop optimum models for predicting numbers aloft.

Weather variables whose typical values change greatly over the season, e.g. temperature, also need to be converted into deviations from seasonal norms. If this is not done, temperature and date effects cannot be separated. This is most important when analyzing data collected over many weeks, as during a radar or other study of many species combined.

Migration data (normally daily counts) must be independent in order to perform even approximate statistical tests. Even with daily or twice-daily counts there are minor problems with independence (Nisbet and Drury 1968*; Blokpoel and Richardson 1978*). Significance levels are unreliable if based on one case per day at each of several adjacent study areas, as in Beason (1978) or, worse, one case per hour, as in Titus and Mosher (1982).

Also, data must be collected daily over several seasons if one is to apply multivariate analyses meaningfully to 15–30 weather variables, as is often done. Some workers have reported such analyses on very small data sets (e.g. Hawkings 1982; McCrary et al. 1983; Millsap and Zook 1983). The results may not represent weather-migration relationships with much precision.

Commonly used multivariate methods make assumptions about the distributions of data (Richardson 1974a*, 1978a). Analysis of residuals (Draper and Smith 1981) is still underused as a way of checking that assumptions are met. Transformations are often essential to meet assumptions. Sometimes no reasonable transformation satisfies the assumptions, in which case a less restrictive type of analysis is required. Also, caution must be exercised in interpreting analyses based on transformed data; the results apply to the transformed but not always the original data (Blokpoel and Richardson 1978*).

The multivariate methods most commonly used in weather-migration studies have been multiple regression and multiple discriminant analyses, sometimes preceded by factor analysis to reduce the number of variables (R78a; Bruderer 1977; Gauthreaux 1977; Hilgerloh 1977, 1981; Alerstam 1978; Blokpoel and Richardson 1978*, 1979; Rabøl and Hansen 1978; Richardson 1979, 1982a; Wege 1979; Evans 1980; Hussell 1981, 1985; Zalakevicius 1984; Darby 1985; Helbig et al. 1987). Methods that make less stringent assumptions about the data have been applied rarely, if at all, e.g. log-linear analysis of multiway contingency tables (Perdeck and Speek 1984) and multidimensional scaling. Other untried methods that may be useful are canonical correlation for interpreting simultaneous counts of many species in relation to weather, and repeated measure procedures (as applied, for example, to hourly counts within a day). It is now practical to perform even the most elaborate of these analyses with a personal computer and readily available software.

Despite the increasing sophistication of analytical procedures, it is virtually impossible to determine from field observations whether birds respond directly to strongly interrelated variables like pressure, temperature and humidity. Multivariate analysis cannot reliably separate causal from coincidental relationships when there are many intercorrelated variables. One cannot assume that the few seemingly significant variables according to multivariate analysis have direct effects on migration amount; they may be included in multivariate models simply as representatives of groups of interrelated variables. Other things being equal, the most precisely measured weather variable will be included in a multivariate model even if its effects on migration are no more direct than those of other less precisely measurable weather variables. In most situations, all that can be assumed is that birds respond to some aspect of the weather that is interrelated with the seemingly important variable(s). Controlled experimentation on caged or released birds has its own biases, but is needed to separate such effects.

4 Synoptic Weather Features

The atmosphere at north temperate latitudes is organized into high and low pressure systems moving approximately eastward. Winds blow clockwise around highs and counterclockwise around lows (Fig. 1). Thus, as a high moves away to the east and/or a low approaches from the west, there are warm southerly winds and falling barometric pressure. Conversely, as a low moves away and/or a high approaches, there are cool north winds and rising pressure. Surface winds are light near the centre of a high. Fronts separate air masses in these systems, e.g. warm fronts where advancing warm air is replacing cold air, usually E or SE of an approaching low; and cold fronts where cold air is replacing warmer air, usually S or SW of a low. Precipitation and thick cloud are most common near lows and fronts (Fig. 1; see also Elkins 1983).

Migrants presumably react to locally measurable weather variables (e.g. wind, temperature, rain) rather than to synoptic features. However, given the synoptic patterns in weather plus the likely importance of wind direction and fair weather for migration, migration timing must be adapted to large-scale atmospheric features even if birds sense only their local manifestations (R78a). Correlations between specific weather variables and migration amount may reflect "a well-adapted response to weather by the birds, canalizing migration into windows

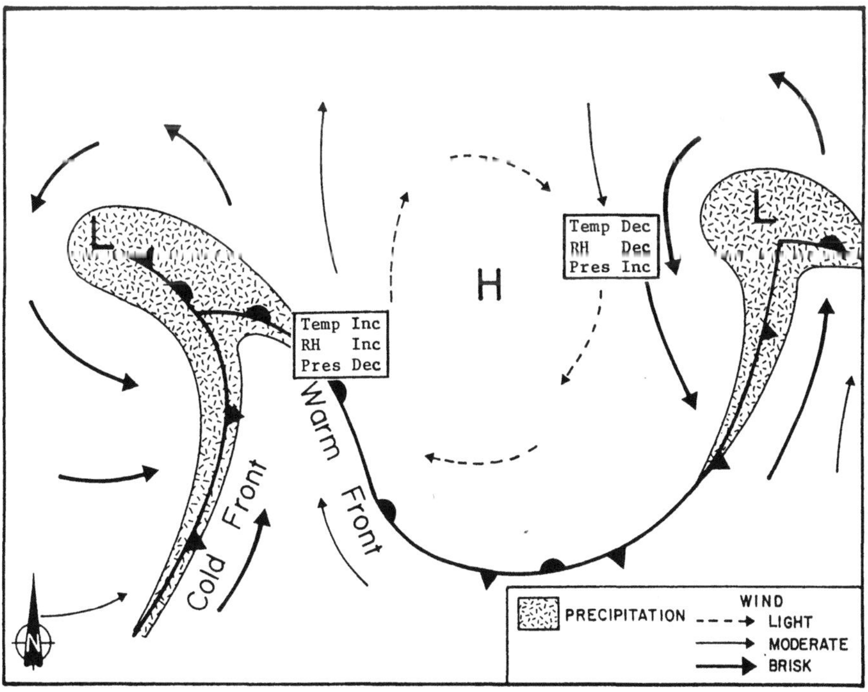

Fig. 1. Typical configurations of high (*H*) and low (*L*) pressure systems, fronts, precipitation and winds at north temperate latitudes. Also shown is the normal direction of change (increasing or decreasing) of temperature, relative humidity and barometric pressure in two synoptic situations

of relatively favourable weather in a highly complex and rapidly varying weather environment" (Alerstam 1978; see also Nisbet and Drury 1968*). Most studies of migration vs synoptic weather use daily data from a single site. However, relationships and potential selective advantages are best understood if migration is monitored over areas as large as pressure systems (e.g. Lowery and Newman 1966*; Williams et al. 1977*, 1978).

Pre-1978 studies showed that peak northward migration tends to occur with warm southerly tailwinds as a high moves away to the east, a low approaches from the west, or both. Conversely, peak fall migration occurs with cool northerly tailwinds as a low moves away to the east, a high approaches from the west, or both (R78a). In both seasons, much migration occurs with light winds near the centre of a high. Recent studies that have obtained similar results include, for spring, Gauthreaux (1977), Blokpoel and Gauthier (1980) and Richardson (1985); and for fall Alerstam (1978), Richardson (1979, 1980, 1982b, 1985), Evans (1980), Weir et al. (1980) and Hilgerloh (1981). Most data came from Europe or North America, but patterns seem similar in China (Hemmingsen 1951*; Ying 1985; Williams 1986). Most radar, visual, and moon-watch studies reveal the same patterns, indicating that the apparent trends are not strongly affected by observational biases. The few radio-telemetry and "release-aloft" results are also consistent (Coon et al. 1976*; Demong and Emlen 1978; Wege and Raveling 1983).

Pre-1978 results indicated that the most favoured synoptic situations differed between birds flying NW vs NE in spring, or SE vs SW in fall. Many groups tend to take off in peak numbers when winds are following relative to their own flight directions (R78a), i.e. with a low to the birds' left, a high to their right, or both. For example, birds flying W or NW in spring often do so with E or SE tailwinds north of a low (R78a; Swem 1985; Alerstam et al. 1986), whereas those flying NE often do so with SW tailwinds south of a low or north of a high (R78a). Similarly, southeastward migration often peaks with NW tailwinds S-W of a low or N-E of a high, whereas southwestward migration more often peaks with NE tailwinds W-N of a low or E-S of a high (R78a, 1985).

Not all birds follow these "rules of thumb" for peak migration. Some migrants are aloft under almost any conditions. Furthermore, when migrants are concentrated by coasts, ridges or valleys, numbers passing a given point may be quite different than suggested above. Several recent studies whose results differ from the above trends (van Dijk and Wassink 1980; Dyrcz 1981; Hawkings 1982; McCrary et al. 1983) involved counts in mountain passes or along coasts where visible migration is often not representative of total migration. Also, a reduced connection between synoptic weather and migration amount is likely when long-distance migrants are observed far from their takeoff points (Richardson 1976*; Alerstam et al. 1986; Williams and Williams 1988). At south temperate latitudes, different (but related) trends are expected because winds there blow the opposite directions around lows and highs.

Pre-1978 studies provided abundant data on migration timing relative to cold fronts, and some data on warm fronts. In spring, strong northward migration can either precede or follow warm fronts. In fall, strong southward migration commonly follows cold fronts, often with peak SE migration occurring closer behind the cold front than peak SW migration (R78a). Recent studies have corroborated many of these points, but have provided little new insight. One point of interest is that cold fronts are associated with onset of autumn migration as far south as the Gulf of Mexico (Able 1972*; Buskirk 1980). Also, birds flying south over the western Atlantic behind cold fronts sometimes seem to become disoriented when they catch up with the front; they also reduce their air speeds, probably to maximize endurance (Larkin et al. 1979). Elkins (1979, 1988) speculated that some of these birds are among the Nearctic birds carried to Europe by strong SW winds. Millsap and Zook (1983) suggested that the greater numbers of migrating accipiters seen after cold front passage in fall were attributable to increased sightability, not to increased numbers aloft. However, their evidence was inconclusive and inconsistent with much previous work (R78a).

5 Wind Direction and Speed

Pre-1978 data showed that most birds migrate in peak numbers with following and calm winds, in smaller numbers with side or light opposing winds, and in smallest numbers with strong opposing winds (R78a). Thus, birds with different preferred directions, e.g. SE vs SW, move in peak numbers with different wind directions (R78a, 1985; Crawford 1980; see also Sect. 4). This trend is understandable because wind can greatly affect the energetic cost of migration. If birds maintain a constant air speed, wind affects ground speed and thus flight duration. If birds adjust their air speed to maintain a specific ground speed, wind affects the necessary power output (Pennycuick 1978). The normal situation appears to be intermediate (e.g. Bloch and Bruderer 1982). In any case, following wind minimizes the energetic cost of migration, which is important to some if not all migrants (see Sect. 2 in R78a).

Wind direction and speed usually vary with altitude, sometimes only slightly but at other times drastically (Elkins 1983). At least some migrants adjust their altitudes to select the most favourable (or least unfavourable) wind within the usual range of flight altitudes (e.g. Sect. 4 in R78a; Alerstam 1978, 1981; Hilgerloh 1981; Jacoby 1983; Buurma et al. 1986). Consequently, many workers have considered both surface winds and winds aloft in evaluating numbers aloft relative to weather.

Following winds are expected to be especially important for birds that migrate long distances non-stop over inhospitable habitats. Energetic analyses indicate that some species could not complete their normal flights across oceans or deserts without following winds (e.g. Wood 1982; Stoddard et al. 1983). In areas where such flights have been studied, birds depart selectively with following winds (e.g. R78a, 1979, 1980; Kloeckner et al. 1982; Alerstam et al. 1986), but often encounter less favourable winds en route (e.g. Richardson 1976*; Larkin et al. 1979; Wood 1982; Williams 1985).

Relative numbers aloft with various winds are difficult to measure because wind affects conspicuousness as well as numbers aloft. However, pre-1978 data showed that the association with following winds was strong for some species and seemingly absent for others. As noted by Rabøl (1974*), energetic considerations are not the only factors affecting the relative advantages of flight under various wind conditions, especially for short-distance migrants. Some recent field and theoretical studies have provided useful data about species differences in wind selectivity, and factors that sometimes cause higher counts when winds are not following.

Results from different observation methods are reasonably consistent in showing that numbers aloft at night are strongly correlated with following or light winds. This trend has been shown with radar (R78a; Bruderer 1977; Gauthreaux 1977; Richardson 1979, 1980, 1985; Zalakevicius 1984), moon-watching or ceilometer (R78a; Richardson 1982b), radiotelemetry (Coon et al. 1976*; Wege and Raveling 1983), aerial transects (Bellrose 1971*), releases aloft (Demong and Emlen 1978), net captures of migrating owls (Evans 1980; Slack et al. 1987), and studies using more than one method (Lindgren and Nilsson 1975*; Bol'shakov 1981). Some studies of grounded nocturnal migrants also show more migration with tailwinds (e.g. Hilgerloh 1977; Rabøl 1983; Sykes 1986) even though such counts are a very imperfect reflection of the amount of migration during preceding hours (R78a).

Many studies of diurnal migration also show more birds aloft with following, light or side winds than with opposing winds, e.g. radar studies (R78a, 1979, 1985; Blokpoel and Gauthier 1980; Hilgerloh 1981; Kerlinger 1982; Perdeck and Speek 1984; Zalakevicius 1984; Kerlinger and Gauthreaux 1985; Buurma 1987); visual studies (R78a; Alerstam 1978, some species; Dunne 1978, one area; Gatter and Penski 1978; Wege 1979; Brueggeman 1980; Laske and Immelmann 1981; Hawkings 1982, autumn; Titus and Mosher 1982, two of three species; Robinson 1987; Willimont et al. 1988); and telemetry (Beske 1982; Wege and Raveling 1983).

However, unlike the consistent nocturnal results, many diurnal studies show an apparent preference for opposing winds. Virtually all of these studies were conducted along leading lines and using visual methods, which detect mainly the low altitude migrants (e.g. R78a; Alerstam 1978, some species; Lindskog and Roos 1980; van Dijk and Wassink 1980; Dyrcz 1981; Hawkings 1982, spring; Camphuysen and van Dijk 1983). However, some visual studies in flat inland areas also show much migration with headwinds (e.g. Buurma et al. 1986).

There may be several reasons for occurrence of upwind visible migration, especially along leading lines. Because wind speed is generally lower near the ground than aloft, birds flying upwind can reduce energy costs by flying low, where they are more easily seen than those flying at higher altitudes on days with tailwinds. Also, numbers of diurnal migrants migrating along a coast can be more strongly influenced by concentrating factors encountered en route than by the numbers that took off that morning (Evans 1966b*). Alerstam (1978) proposed a theory concerning wind effects on numbers of diurnal migrants visible along a coast, and he evaluated the theory using counts of 20 groups of migrants at Falsterbo. Factors believed to affect relative numbers seen with tail- and headwinds included degree of selectivity for following winds, flight speeds, sightability at low vs high altitude, and propensity for coasting. Selectivity for tailwinds was expected to be higher when winds are variable, the migration is comparatively short but extended in duration, and when it is divided into a small number of long flights (Alerstam 1979a). Coasting was expected to be maximal with side- and headwinds, and least with tailwinds (Alerstam and Pettersson 1977*; Alerstam 1978, 1979b).

According to Alerstam's (1978) theory, even if all birds preferred tailwinds, some groups would be seen in larger numbers with headwinds because of greater tendencies to coast and to fly low with headwinds. Field data from Falsterbo confirmed this for several species, mainly of small birds whose detectability is low when they fly high. Some of the more conspicuous species were seen in larger numbers with following winds. It was difficult to determine the degree of preference for following winds based on observations at a coastal site. However, Alerstam (1979a) concluded that "When allowance is made for the effects of wind on the migrants' flight altitude, propensity for coasting and travelling speed, records of visible migration are also compatible with a general positive correlation between [following wind] and migratory activity."

There are probably at least two reasons for the more consistent correlation with following winds at night than by day. (1) Migrants concentrate along topographic leading lines to a much lesser extent at night (Richardson 1978b; Able 1985). [Indeed, one of the few exceptions to the trend for maximum nocturnal migration with following winds is Weir et al. (1980), whose results probably were affected by coastal effects at night.] (2) Most night studies are based on radar, moon-watch or ceilometer methods, which all survey primarily moderate and high altitudes. The undersampling of low altitudes might cause overestimation of selectivity for following winds if nocturnal migrants tend to fly low with side or opposing winds, as is likely. However, two arguments suggest that this bias is not severe at night. First, proportionally fewer migrants may fly very low at night because of the danger of hitting obstacles. Second, studies using techniques that could detect nocturnal migration at low altitude, namely radio-telemetry, netting of owls, and to some extent counts of grounded birds, also show strong association with following, partially following, or light winds, and much reduced migration with opposing winds. Based on combined results of several methods, Bol'shakov (1981) concluded that numbers aloft at night are highest with following wind, although the trend may not be as strong as sometimes suggested.

Alerstam (1976a*, b*, 1978) pointed out that ground speed (or ground speed/air speed ratio) is a direct measure of wind effects on the energy cost of migration. He proposed using that ratio rather than various measures of wind direction and speed in analyses of weather effects on numbers aloft. When the ratio has been used, strong correlations have been found between it and numbers aloft (Alerstam 1976a*; Blokpoel and Gauthier 1980; Perdeck and Speek 1984). However, the ratio is closely related to the following wind component, and it is not clear which variable is better related to migration density (Blokpoel and Richardson 1978*; Richardson 1979).

Headwinds can affect the timing or duration of flights as well as numbers aloft. Birds sometimes take off later than normal in the day or evening when winds are unfavourable or improving; they may also descend or land earlier than normal when winds are unfavourable or deteriorating (R78a; Bloch et al. 1981; Bol'shakov 1981; Wege and Raveling 1983; Kerlinger and Gauthreaux 1985; Helbig and Laske 1986).

Soaring birds, e.g. many hawks, cranes and storks, are a special case. Updraughts strongly reduce the energetic cost of migration in these groups (see R78a and Sect. 7). Hawks often fly with side or light opposing winds if updraughts are present (R78a; Titus and Mosher 1982; Currie et al. 1985; Hoffman 1985; Kerlinger et al. 1985; Smith 1985; Welch 1987). Light winds promote thermal formation accounting in part for the often-reduced numbers of migrating hawks with strong winds (e.g. R78a; Hussell 1985; Welch 1987).

Swallows and swifts are another special case. During both migration and summer "weather movements", peak numbers are seen in many areas with opposing winds, sometimes strong (Sect. 7.2 in R78a; Alerstam 1978; Laske and Immelmann 1981; Elkins 1983).

Pre-1978 data indicated that, for most types of migrants, numbers aloft are negatively correlated with wind speed only when winds are opposing (R78a). Several recent studies have confirmed the suppression of migration with strong headwinds (e.g. Alerstam 1978; Bergman 1978; Lindskog and Roos 1980; Hilgerloh 1981; Currie et al. 1985; Buurma 1987; McIntosh and Barnhurst 1988). Significant negative correlations often exist between wind speed and migration intensity even when all wind directions are combined (Gauthreaux 1977; Alerstam 1978; Evans 1980; Bolshakov and Rezvyi 1982; Maran 1983; Rabøl 1983; Zalakevicius 1984). In a few cases, numbers seen have been positively related to the speed of crosswinds (Alerstam 1978 for swift; Willimont et al. 1988 for hummingbirds). Strong tailwinds may also be preferred by certain birds (Hubbard and Flock 1974).

In at least two waterfowl, year-to-year differences in the "earliness or lateness" of migration are related to the proportion of days with following winds during the normal spring migration period: Bewick's swans (Rees 1982); lesser snow geese (Ball 1983). Also, in greater snow geese, migration timing in both spring and autumn seems related to the latitude of the tracks of most low pressure areas (Smith and Hayden 1984). In seasons when storm tracks are farther north than average, spring migration tends to be early and autumn migration late. This may be a result of more frequent south winds, warmer average temperatures, or both, when storm tracks are father north.

In summary, most types of birds are more likely to fly with following than neutral or especially opposing winds. However, wind selectivity varies widely among species. Also, detection biases strongly influence apparent correlations between wind and numbers aloft. Methods most sensitive to birds at high altitudes, including radar, often overestimate the association with following wind; counts along leading lines often underestimate it.

6 Pressure, Temperature, and Humidity

These weather variables, and especially their short-term trends, are strongly correlated with one another, with synoptic situation, and with wind. South winds favouring spring migration tend to occur with falling pressure, rising temperature and rising humidity as a high moves away to the east, a low approaches from the west, or both. Conversely, north winds favouring fall migration tend to occur with rising pressure, falling temperature and falling humidity as a low moves away to the east, a high approaches from the west, or both (R78a; Elkins 1983).

Migration amount often is correlated with pressure, temperature and (less often) humidity. Directions of correlation usually are as suggested above (R78a). These are several reasons why it might be selectively advantageous for birds to respond directly to one or more of these three variables. Direct responses might occur because these variables can be indicators of future food supply (e.g. insect numbers), habitat suitability (e.g. freeze/thaw cycles), or travelling conditions en route (Lack 1960*; Nisbet and Drury 1968*; R78a). However, from field observations alone, even if analyzed by multivariate methods, it is virtually impossible to identify the specific variable(s), among a group of interrelated ones, to which birds respond.

6.1 Barometric Pressure

Pre-1978 data showed that in spring, migration amount is not consistently correlated with pressure per se. However, migration amount tends to be high when pressure is decreasing, especially for groups of birds that fly NE-ENE (R78a). A few recent studies provide generally similar results (Wege and Raveling 1983; Ying 1985; Williams 1986). Several other recent workers found no correlation, although sample sizes often were small. In contrast, Hawkings (1982) saw more geese travelling NW in S Alaska with high than low pressure, partly because days with low pressure were cloudy and unfavourable for seeing migrants. In another exception to the general trend, two of four groups of spring migrants studied in Lithuania tended to migrate with high or rising pressure, other factors being equal (Zalakevicius 1984).

Pre-1978 data showed that in fall, migration amount is often correlated with pressure per se as well as pressure trend. Groups flying SSW-W tend to travel with high and rising pressure; those going SE often travel with low but rising pressure. The difference is probably related to the fact that birds flying SE tend to fly in peak numbers in the NW winds close behind a departing low and/or cold front; birds flying SW tend to move in peak numbers a day or so later when the low is farther east and/or a high is closer (R78a).

Most recent studies have confirmed the association of fall migration with rising pressure, e.g. Alerstam (1978) for 13 of 20 groups of visible migrants, Wege (1979) and Wege and Raveling (1983) for geese, Zalakevicius (1984) for two of four groups, Evans (1980) for owls, and Ying (1985) for total fall migration. In some cases – Alerstam, Evans, Zalakevicius, some pre-1978 studies (R78a) – the association with rising pressure was significant after allowance for other variables. Recent work has confirmed that relationships of autumn migration volume to pressure per se are not as consistent as those to pressure trend. However, numbers often are higher with high than low pressure (e.g. Hilgerloh 1977; Evans 1980). Among visible migrants travelling SW at Falsterbo, Sweden, 12 of 20 groups exhibited significant positive correlations with pressure per se, and none exhibited significant negative correlations. Few exceptions to these trends have been reported recently. However, Zuckerbrot et al. (1980) found a strong negative correlation between pressure and numbers of *Coturnix* quail caught in Sinai in fall. Perhaps this was related to the tendency of southeastward migrants to fly with low pressure.

Some differences between studies probably occurred because pressure trend has been measured over intervals ranging from 4 or 6 h to 24 h. The latter may be too long to reflect current weather trends. If birds can sense small changes in pressure (Kreithen and Keeton 1974*), they probably do so over intervals < 24 h.

Correlations between barometric pressure and migration are doubtless partly indirect effects of the reactions of birds to other interrelated weather variables. However, controlled studies on caged birds suggest that some birds may react to pressure directly (Stolt 1969*, 1977*). A recent study under partly controlled conditions is consistent with this view (Viehmann 1982), although effects of temperature and pressure were confounded. Partial correlations sometimes persist between pressure trend and migration amount after other variables have been taken into account. This does not prove that birds react directly to pressure, but it is consistent with that hypothesis. Further controlled experiments are needed.

6.2 Temperature

The seasonal progression of temperature strongly affects freeze/thaw cycles, energy needs, food availability, and vegetation development. There are obvious adaptive advantages for migrants to adjust their schedules in response to year-to-year temperature variations. Lack (1960*), R(78a) and Gauthreaux (1982) cite many reports indicating that timing of spring migration is related to "earliness or lateness" of the season as measured by average temperature. This has been further corroborated for many species, e.g. various waterfowl (Flickinger 1981; Rees 1982; Hammond and Johnson 1984), red kite (Zimmerman and Sutter 1962), European cuckoo (von Haartman and Söderholm-Tana 1983) and brambling (Mikkonen 1981); see also Harmata (1980). Based on a novel type of analysis of snow goose counts, Smith and Hayden (1984) suggested that such relationships may be spurious side-effects of variables other than temperature. However, their analysis is inconclusive regarding causal factors.

Pre-1978 studies indicated almost unanimously that peak spring and fall migration tend to occur on warm and cold days, respectively. This behaviour probably has direct selective advantages (R78a), so some birds probably respond directly to temperature when determining when to fly. However, the correlation between wind direction and temperature is doubtless partly responsible for that between temperature and number of birds aloft.

Many recent studies confirm the extensive earlier evidence that peak spring migration tends to occur on days of high and/or rising temperatures: waterfowl (Evans 1979; Blokpoel and Gauthier 1980; Wege and Raveling 1983), some hawks (Hussell 1985), European woodcock (Creutz 1983), passerines (Rabøl and Hanzen 1978; Hussell 1981; Perdeck and Speek 1984), and birds in general (Gauthreaux 1977; Zalakevicius 1984; Ying 1985). Two recent spring studies that showed negative correlations between numbers and temperature employed biased observation methods (Hawkings 1982; Taylor and Kershner 1986).

Most pre-1978 studies showed that migrants travelling SSW-WSW in autumn tend to fly on days with low and/or decreasing temperatures. Birds travelling SE seemed to show less correlation with temperature, although more data were needed (R78a). Few new data are available on this point; however, southeastward departures of Canada geese were strongly associated with falling temperature (Wege 1979; Wege and Raveling 1983). Most recent studies of southwestward migration have confirmed a tendency to travel on days with cool or falling temperatures: waterfowl (Bergman 1978; Evans 1979; Leito and Renno 1983), owls (Evans 1980), wood pigeons (Alerstam 1978; Gatter and Penski 1978), woodcock (Creutz 1983), and many passerines (Hilgerloh 1977, 1981; Lindskog and Roos 1980). Hawks that soar are a special case, since warm days sometimes provide favourable travelling conditions. For hawks, consistent correlations with low or decreasing temperature were not evident in either pre-1978 autumn data (R78a) or recently (Alerstam 1978; Titus and Mosher 1982; Hoffman 1985).

In his extensive analysis of visible autumn migration at Falsterbo, Alerstam (1978) found that correlations between numbers seen and the trend in daily minimum temperature were significantly negative for 9 of 20 groups of migrants, and not significantly positive for any group. However, correlations with other measures of temperature were less consistent. Indeed, counts of hawks and some other groups tended to be higher on autumn days with high temperatures. The extent to which leading line effects and detection biases affected this result is unknown. However, the radar data of Zalakevicius (1984) also revealed exceptions to the trend for most autumn migration to occur in cool weather.

Pre-1978 data suggested that some correlations between numbers aloft and temperature represent direct reactions of birds to temperature (R78a). More recent evidence of the same types includes a study of *Zugunruhe* (Viehmann 1982) and multivariate analyses showing that partial correlations with temperature sometimes persist after other variables are taken into account (Gauthreaux 1977; Hilgerloh 1977, 1981; Alerstam 1978; Rabøl and Hansen 1978; Evans 1980; Hussell 1981, 1985; Perdeck and Speek 1984; Zalakevicius 1984). However, this evidence is weak. Most studies of temperature vs. *Zugunruhe* did not fully control other weather variables, and multivariate analyses of migration counts are necessarily inconclusive about causal links.

6.3 Relative Humidity

There is no proof that migrants respond directly to humidity, although it might sometimes be advantageous to do so (R78a). Pre-1978 radar studies suggested that, in both spring and fall, peak migration tends to be either unrelated to humidity or associated with low and decreasing humidity. Nisbet and Drury (1968*) suggested that, in the NE USA, a direct reaction of spring migrants to low humidity may be important to avoid flight when an approaching low is about to cause poor conditions en route and at the destination. However, some pre-1978 studies of visible migration and caged migrants (Muller 1976*) indicated an association with high, not low, humidity.

Of the few recent spring studies considering humidity, most found no association (Gauthreaux 1977; Wege 1979; Blokpoel and Gauthier 1980; Flickinger 1981; Hilgerloh 1981). Zalakevicius (1984) found no correlation for three of four groups of migrants; one group tended to fly with high but decreasing humidity. Wege and Raveling (1983) found that geese tended to fly with increasing humidity.

In autumn, few strong correlations of migration amount with humidity have been found recently (Alerstam 1978; Hilgerloh 1981; Zalakevicius 1984; Darby 1985). Hilgerloh (1977) found a negative correlation of thrush numbers with humidity, consistent with many earlier results. However, positive trends were found for owls (Evans 1980) and geese (Wege and Raveling 1983).

Overall, recent results suggest that humidity is not a key factor affecting migration timing. The weak negative trend evident up to 1977 (R78a) seems even less consistent than then thought.

7 Stability and Updraughts

There is little evidence that atmospheric stability affects migration timing in non-soaring birds. Pre-1978 evidence on this point was meagre and inconsistent (R78a), and few new data are available. Kerlinger (1982) found that common loons migrating overland travelled in stable air above the convective layer. Coordinated high-resolution observations of birds and atmospheric structure (e.g. Larkin 1982) are needed to assess effects of small-scale atmospheric features on migrants. Entomologists are well ahead of ornithologists in this area (e.g. Drake and Farrow 1988).

The importance of the many types of updraughts to soaring birds is well known (e.g. Haugh 1974*, 1975*; R78a; Harwood 1985). Atmospheric instability and other conditions promoting thermals, slope lift, etc. (see Wallington 1977; Elkins 1983) doubtlessly favour the migrations of soaring birds. However, true correlations between weather and numbers of hawks aloft are difficult to determine because weather also affects local routes and detectability. The occasional positive correlations of autumn hawk counts with temperature (Sect. 6.2), contrary to the trend for most birds, is probably a side-effect of an association between warm, fair weather and favourable conditions for updraught formation.

At Falsterbo, counts of two species of hawks were positively related to the presence of convective clouds, indicative of thermals (Alerstam 1978). In general, although many data are available on effects of updraughts on flight routes and behaviour, few studies have examined specific relationships between updraught presence and numbers of soaring birds aloft. [See Kerlinger (1989) for many related data.]

8 Precipitation, Cloud, and Visibility

Pre-1978 data showed that cloud and precipitation reduce numbers of birds aloft, other factors being equal, but rarely suppress migration totally. Birds often avoid entering fog, but except for soaring birds there is little evidence that fog alone reduces numbers aloft (Lack 1960*; R78a). Recent evidence supports these conclusions, and provides additional details.

8.1 Precipitation

Most recent visual and radar studies of migration in progress have shown significantly fewer birds aloft with rain or snow (e.g. Gauthreaux 1977; Alerstam 1978; Zalakevicius 1984; Hussell 1985). Other methods have given similar results, namely radio-telemetry (Wege and Raveling 1983), turnover of migrants on islands (Hilgerloh 1977; Rabøl and Hansen 1978; Mehlum 1983), and release of birds aloft at night (Demong and Emlen 1978). If precipitation stops after the usual departure time, migrants sometimes take off at a later hour than usual (R78a; Kerlinger and Gauthreaux 1985). Birds that encounter rain en route sometimes divert around it (Bruderer 1971*; R78a; Cochran and Kjos 1985). Migrants apparently land earlier on rainy nights (Bol'shakov 1981), although his counts of birds landing in a lit area may have been biased by attraction to lights during rain. Daily counts of grounded migrants are often positively correlated with rain (e.g. Riddiford 1985), since rain tends to interrupt migration. Alerstam (1981) suggested that wind and precipitation were the two weather factors most affecting migration intensity.

Even so, some migration often continues in light rain (e.g. Lack 1960*; R78a; van Dijk and Wasink 1980; Currie et al. 1985). Rain suppresses migration more strongly in some species than others (Alerstam 1978). In eastern Canada, rain suppresses autumn migration more strongly at night than by day (Richardson 1985). In late autumn, snowstorms can interrupt migration (R78a; Wege and Raveling 1983), but "forced departures" can occur near snowstorms (R78a; Gatter and Penski 1978; Camphuysen and van Dijk 1983; Leito and Renno 1983). Conversely, arrival dates of northern species are often related to snowmelt (Mikkonen 1981; von Haartman and Söderholm-Tana 1983).

8.2 Cloud

Cloud cover is correlated with precipitation, but for many species and groups overcast alone is sufficient to reduce the numbers of birds aloft (R78a; Hilgerloh 1977; Rabøl and Hansen 1978; Alerstam 1978; Titus and Mosher 1982; Zalakevicius 1984; Hussell 1985; Helbig et al. 1987). As a corollary, overcast is associated with increased counts of grounded migrants (Darby 1985; Riddiford 1985). However, many studies indicate that some birds migrate under solid overcast, usually with little or no detectable disorientation (e.g. Emlen and Demong 1980; Able 1982a,b, 1985; Bolshakov and Rezvyi 1982; Cochran and Kjos 1985). Most of these migrants fly either below or above cloud; the few flying in cloud often are less well oriented. Overall, overcast appears unfavourable for migration, no doubt partly because it reduces the number of cues available for orientation. However, many birds can maintain more or less "normal" courses under overcast.

Few migration studies have distinguished different cloud types. Zalakevicius (1984) found strong correlations of migration volume to cloud type. Alerstam (1978) confirmed that counts of some soaring birds are higher with convective clouds, which indicate the presence of updraughts (and increase sightability of high-flying birds).

8.3 Visibility and Fog

Horizontal visibility is negatively related to cloud cover, and their effects on migration are difficult to distinguish. For many types of birds, no correlation between migration volume and visibility is evident; for others there is more migration with good visibility (R78a; Gatter and Penski 1978; Lindskog and Roos 1980; Bolshakov and Rezvyi 1982). In some cases, a positive partial correlation with visibility persists after allowing for cloud and/or precipitation (R78a; Alerstam 1978; Titus and Mosher 1982; Zalakevicius 1984). This suggests but does not prove that birds react directly to visibility.

Although good visibility usually is a positive or neutral factor concerning the onset of migration, some workers have found negative correlations (R78a; van Dijk and Wasink 1980; Hilgerloh 1981). Reasons for this probably vary. In Hilgerloh's study, the negative correlation probably was spurious, reflecting the common occurrence of poor visibility during stable, fair-weather conditions favoured for migration. Increased counts of grounded migrants on islands and coasts often occur with poor visibility (Hussell 1981; Rabøl 1983; Riddiford 1985), again showing that conditions favouring ground concentrations often are contrary to those when large numbers are aloft. Darby (1985) also found correlations between visibility and ground counts of some passerine species but did not show the direction of correlation.

Effects of fog on numbers aloft are difficult to study systematically because of visual and radar limitations plus the often-local nature of fog. Migrants frequently fly above ground fog (R78a; also Alerstam and Ulfstrand 1974*; Jellmann 1979; Lyuleeva et al. 1981). However, some turn back when they encounter fog (R78a). Soaring birds rarely fly with fog, since fog is indicative of stable air without updraughts (but see Currie et al. 1985).

9 Conclusions

Recent studies of weather-migration relationships have corroborated many hypotheses developed from pre-1978 studies. Recent work has also helped clarify effects of detection biases on apparent weather-migration correlations. Among-species variations in wind selectivity are also better understood now than in 1978, although hypothesized reasons for these species differences (Alerstam 1978) require further testing.

For most types of migrants, it is clear that peak numbers migrate with roughly following winds relative to the preferred direction. Given the intercorrelations among weather variables, it is not surprising that numbers aloft are also correlated with several other aspects of weather. Although it is difficult to distinguish causal from coincidental relationships, even with multivariate methods, there is a distinct tendency to avoid migrating in inclement weather (overcast, rain, poor visibility). Direct reactions to certain other variables that seem to lack selective importance may occur if those variables are useful predictors of important phenomena that cannot be sensed directly, e.g. winds aloft en route or conditions at the takeoff point or destination during future hours or days (Nisbet and Drury 1968*; R78a).

In 1978 I listed 11 topics deserving more study in order to understand weather-migration relationships (Sect. 16 in R78a). These 11 topics can be divided into three groups.

1. Considerable progress has been made on 4 of 11 topics:
   a) evaluating methodological biases, combining the use of two or more complementary methods, and conducting some visual studies in featureless inland areas;
   b) evaluating species and other differences in wind selectivity;
   c) determining whether migration-weather correlations for particular species and areas are consistent with general patterns evident among other birds;
   d) tailoring observation and analytical methods to specific groups under study (e.g. Blokpoel and Richardson 1978*; 1979; Evans 1980; Lindskog and Roos 1980; Alerstam 1985; Kerlinger and Gauthreaux 1985; Helbig et al. 1987).
2. For 5 of 11 topics only a few isolated studies have been done since 1978:
   a) measuring local weather features such as turbulence and updraughts while collecting fine-scale data on migratory behaviour (Larkin 1982);
   b) using "unusual" methods that may be less biased or may allow experimental control, e.g. telemetry, caged birds, and releases aloft;
   c) reanalyzing existing data with modern statistical methods (Alerstam 1978);

d) using novel analytical methods (Perdeck and Speek 1984) and simulation techniques (Klein 1980; Stoddard et al. 1983);
e) analyzing possible modifying influences of non-meteorological variables like moon phase, magnetic disturbance, and tide phase.

3. Two of the 11 topics have not received attention since 1978 insofar as I am aware:
   a) evaluating the adaptive value of different potential responses to weather by analyzing spatial and temporal interrelationships of weather;
   b) studying weather effects on migration at south temperate latitudes.

Further work on these topics, especially those in groups (2) and (3), would be valuable.

Another area warranting more effort is the interaction between physiological readiness to migrate and responses to external cues like weather. The propensity of a migrant to travel on a given day is undoubtedly related to the degree of fat depletion during the previous flight, ecological conditions at the stopover point, time interval and weight gain since the last flight, and whether the bird is ahead of or behind its "normal" migration schedule. Several important studies have been conducted on these topics recently (e.g. this Vol.). However, the interrelationships of migratory readiness and numbers taking off have rarely been studied in much detail, partly because of differences in methodology (studies of grounded or caged birds vs counts of passing migrants). As Berthold (1980) has pointed out in a related context, it will be important to bring together experimental and observational methods if we are to better understand weather-migration relationships and numbers flying each day, e.g. for use in predicting daily variation in bird hazards to aircraft.

Summary

This paper reviews recent evidence about effects of short-term variations in weather on numbers of birds migrating each day or night, supplementing a review of pre-1978 literature on the same topic (Richardson 1978). For many groups of birds, maximum numbers fly when winds are roughly following relative to the preferred flight direction. However, at least a few migrants fly with all but the strongest headwinds. During visual studies, the preference for tailwinds is often obscured by the greater detectability of birds flying low against headwinds. Conversely, radar and other methods that fail to detect many low-altitude migrants often overestimate the association with tailwinds, especially for diurnal migration. Recent work has helped clarify effects of these detection biases on apparent correlations between weather and migration. It has also shown much among-species variation in selectivity for tailwinds. Present hypotheses about reasons for this variable wind selectivity require further testing.

Peak counts of most types of migrants are related to many weather variables besides wind, e.g. falling temperature and rising pressure in autumn, rising temperature and falling pressure in spring, at least partly clear skies, and absence of rain or snow. However, because weather variables are intercorrelated, causative and coincidental relationships are difficult to distinguish based on uncontrolled field studies, even with multivariate methods. Experimental work on this topic is still rare.

Acknowledgements. I thank H. Achtman, A. Shenitzer, and the Canadian Dept. of State (via H. Blokpoel) for translating Russian literature. Dorothy Richardson assisted in locating and organizing references, and D.H.T. Hussell provided helpful comments on the manuscript. I also thank the organizers and sponsors of this symposium for encouraging preparation of the review.

References[1]

Able KP (1982a) Field studies of avian nocturnal migratory orientation I. Interaction of sun, wind and stars as directional cues. Anim Behav 30:761–767

Able KP (1982b) The effects of overcast skies on the orientation of free-flying nocturnal migrants. In: Papi F, Wallraff HG (eds) Avian navigation. Springer, Berlin Heidelberg New York, pp 38–49

Able KP (1985) The contribution of radar and visual techniques to orientation research. Acta 18th Congr Int Ornithol Moscow 1982:293–299

Alerstam T (1978) Analysis and a theory of visible bird migration. Oikos 30:273–349

Alerstam T (1979a) Wind as selective agent in bird migration. Ornis Scand 10:76–93

Alerstam T (1979b) Optimal use of wind by migrating birds: combined drift and overcompensation. J Theor Biol 79:341–353

Alerstam T (1981) The course and timing of bird migration. In: Aidley DJ (ed) Animal migration. Cambridge University Press, Cambridge, pp 9–54

Alerstam T (1985) Strategies of migratory flight, illustrated by arctic and common terns, *Sterna paradisaea* and *Sterna hirundo*. Contrib Mar Sci (Suppl) 27:580–603

Alerstam T, Hjort C, Högstedt G, Jönsson PE, Karlsson J, Larsson B (1986) Spring migration of birds across the Greenland inlandice. Medd Grønl Biosci 21, 38 pp

Ball T (1983) The migration of geese as an indicator of climate change in the southern Hudson Bay region between 1715 and 1851. Clim Change 5:85–93

Beason RC (1978) The influences of weather and topography on water bird migration in the southwestern United States. Oecologia (Berl) 32:153–169

Bergman G (1978) Effects of wind conditions on the autumn migration of waterfowl between the White Sea area and the Baltic region. Oikos 30:393–397

Berthold P (1978) Das Zusammenwirken von endogenen Zugzeit-Programmen und Umweltfaktoren beim Zugablauf bei Grasmücken: eine Hypothese. Vogelwarte 29:153–159

Berthold P (1980) Experimental methods used in the study of bird migration: a survey and evaluation. Acta Ornithol 17:11–23

Beske AE (1982) Local and migratory movements of radio-tagged juvenile harriers. Raptor Res 16:39–53

Bloch R, Bruderer B (1982) The air speed of migrating birds and its relationship to the wind. Behav Ecol Sociobiol 11:19–24

Bloch R, Bruderer B, Steiner P (1981) Flugverhalten nächtlich ziehender Vögel–Radardaten über den Zug verschiedener Vogeltypen auf einem Alpenpass. Vogelwarte 31:119–149

Blokpoel H (1973) Bird migration forecasts for military air operations. Can Wildl Serv Occas Pap 16, 17 pp

[1]References marked "*" in the text were also cited in an earlier review (Richardson 1978a) and are excluded from this section to save space.

Blokpoel H. Gauthier MC (1980) Weather and the migration of Canada geese across southeastern Ontario in spring 1975. Can Field-Nat 94:293-299
Blokpoel H. Richardson WJ (1979) On the predictability of spring migration of snow geese across southern Manitoba. Canada. Proc 14th Mtg Bird Strike Comm Europe. The Hague. WP9
Blomqvist S. Peterz M (1984) Cyclones and pelagic seabird movements. Mar Ecol Prog Ser 20:85-92
Bolshakov KV (1981) Reconstruction of the total picture of nocturnal bird migration and the comparative effectiveness of different methods of detection. Proc Zool Inst Acad Sci USSR 104:95-123 (Transl TR-RUS-437. Can Secr State. Ottawa. 1984)
Bolshakov KV. Rezvyi SP (1982) An analysis of prestart and start activity of the redbreast robin (*Erithacus rubecula* L.) during spring night migration. In: Gavrilov VM. Potapov RL (eds) Ornithological studies in the USSR. vol 2. Zool Inst. USSR Acad Sci. Moscow. pp 285-305
Bruderer B (1977) Weather-dependence of height. density and direction of migration in Switzerland. Proc 12th Mtg Bird Strike Comm Europe. Paris. WP25
Brueggeman JJ (1980) Coastal occurrence of birds at Point Barrow. Alaska. in spring. Murrelet 61:31-34
Bruun B (1986) "Urgency" in the migration of European raptors with special emphasis on the Middle East. Gerfaut 76:271-279
Buskirk WH (1980) Influence of meteorological patterns and trans-Gulf migration on the calendars of latitudinal migrants. In: Keast A. Morton ES (eds) Migrant birds in the Neotropics. Smithsonian Inst. Washington. pp 485-491
Buurma LS (1987) [Patterns of high bird migration over the North Sea area in October.] Limosa 60:63-74 (in Dutch)
Buurma LS. Lensink R. Linnartz LG (1986) [Altitude of diurnal broad front migration over Twente; a comparison of radar and visual observations in October 1984.] Limosa 59:169-182 (in Dutch)
Camphuysen CJ. Dijk J van (1983) [Seabirds and estuary birds along the Netherlands coast. 1974-79.] Limosa 56:87-230 (in Dutch)
Cochran WW. Kjos CG (1985) Wind drift and migration of thrushes: a telemetry study. Ill Nat Hist Surv Bull 33:297-330
Crawford RL (1980) Wind direction and the species composition of autumn TV tower kills in northwest Florida. Auk 97:892-895
Creutz G (1983) Die Wetterabhängigkeit des Zugablaufes bei der Waldschnepfe (*Scolopax rusticola* L.) in Mitteleuropa. Beitr Vogelkd (Jena) 29:107-117
Currie NW. Hopkins DA. Harwood M (1985) A review of broad-winged hawk migrations. In: Proc Hawk Migr Conf IV. Hawk Migr Assoc N Am. pp 167-191
Darby KV (1985) Migration counts and local weather at British bird observatories – an examination by linear discriminant analysis. In: Morgan BJT. North PM (eds) Lecture notes in statistics. vol 29. Springer. Berlin Heidelberg New York. pp 37-64
Demong NJ. Emlen ST (1978) Radar tracking of experimentally released migrant birds. Bird-Banding 49:342-359
Dijk J van. Wassink A (1980) [The course of the visible spring migration of some waders along the coast of Noord-Holland and Zuid-Holland.] Limosa 53:109-120 (in Dutch)
Dolnik VR (1981) [Methods of bird migration discovery and estimation.] Proc Zool Inst Acad Sci USSR 104 (in Russian)
Drake VA. Farrow RA (1988) The influence of atmospheric structure and motions on insect migration. Annu Rev Entomol 33:183-210
Draper NR. Smith H (1981) Applied regression analysis. Wiley. New York
Dunne PJ (1978) Spring hawk movement along Racoon Ridge. N.J. – 1977. New Jersey Aud 4:39-48
Dyrcz A (1981) [Autumn migration of birds at Szrenica Pass. Karkonosze Mts.. Poland.] Acta Zool Cracov 25(3):33-67 (in Polish)
Elkins N (1979) Nearctic landbirds in Britain and Ireland: a meteorological analysis. Br Birds 72:417-433
Elkins N (1983) Weather and bird behaviour. Poyser. Calton. UK
Elkins N (1988) Recent transatlantic vagrancy of landbirds and waders. Br Birds 81:484-491
Emlen ST. Demong NJ (1978) Orientation strategies used by free-flying bird migrants: a radar tracking study. In: Schmidt-Koenig K. Keeton WT (eds) Animal migration. navigation. and homing. Springer. Berlin Heidelberg New York. pp 283-293

Evans DL (1980) Multivariate analyses of weather and fall migration of saw-whet owls at Duluth, Minnesota. MS thesis, North Dakota State Univ, Fargo, ND

Evans ME (1979) The effects of weather on the wintering of Bewick's swans *Cygnus columbianus bewickii* at Slimbridge, England. Ornis Scand 10:124–132

Flickinger EL (1981) Weather conditions associated with beginning of northward migration departures of snow geese. J Wildl Manage 45:516–520

Gatter W, Penski K (1978) Der Wegzug der Ringeltaube *Columba palumbus* nach Planbeobachtungen am Randecker Maar (Schwäbische Alb). Vogelwarte 29:191–220

Gauthreaux SA Jr (1977) The influence of weather variables on the density of nocturnal migration in spring. Proc 12th Mtg Bird Strike Comm Europe, Paris, WP-28

Gauthreaux SA Jr (1980) The influence of global climatological factors on the evolution of bird migratory pathways. Acta 17th Congr Int Ornithol, Berlin, pp 517–525

Gauthreaux SA Jr (1982) The ecology and evolution of avian migration systems. In: Avian biology, vol 6. Academic Press, London, pp 93–168

Gauthreaux SA Jr (1984) The use of small mobile radars to detect, monitor, and quantify bird movements. In: DOT/FAA/AAS/84-1, Fed Aviat Admin, Washington, DC, pp 121–131

Haartman L von, Söderholm-Tana L (1983) Bird arrival times revisited. Ornis Fenn Suppl 3:7–13

Hammond MC, Johnson DH (1984) Effects of weather on breeding ducks in North Dakota. US Dep Inter Fish Wildl Tech Rep 1, 17 pp

Harmata W (1980) Phenological dynamics of arrivals and departures on [sic] migratory birds in Cracov, Poland, and the environments in the 19th and 20th century. Int J Biometeorol 24:137–140

Harwood M (ed) (1985) Proceedings of hawk migration conference IV. Hawk Migr Assoc N Am, 393 pp

Hawkings JS (1982) Migration and habitat relationships of geese on the eastern Copper River Delta, Alaska. MS thesis, Univ Alaska, Fairbanks

Helbig A, Laske V (1986) Zeitlicher Verlauf und Zugrichtungen beim Wegzug des Stars (*Sturnus vulgaris*) im nordwestdeutschen Binnenland. Vogelwarte 33:169–191

Helbig AJ, Wiltschko W, Laske V (1986) Optimal use of the wind by Mediterranean migrants. Ric Biol Selvaggina 10(Suppl):169–187

Helbig AJ, Orth G, Laske V, Wiltschko W (1987) Migratory orientation and activity of the meadow pipit (*Anthus pratensis*): a comparative observational and experimental field study. Behaviour 103:276–293

Hild J (1969) Meteorologische Aspekte zum Problem Vogelschlag im Luftverkehr. Wetter Leben 21:1–11

Hilgerloh G (1977) Der Einfluss einzelner Wetterfaktoren auf den Herbstzug der Singdrossel (*Turdus philomelos*) über der Deutschen Bucht. J Ornithol 118:416–435

Hilgerloh G (1981) Die Wetterabhängigkeit von Zugintensität, Zughöhe und Richtungsstreuung bei tagziehenden Vögeln im Schweizerischen Mittelland. Ornithol Beob 78:245–263

Hoffman SW (1985) Raptor movements in inland western North America: a synthesis. In: Proc Hawk Migr Conf IV, Hawk Migr Assoc N Am, pp 325–338

Hubbard JD, Flock WL (1974) Radar observation of migratory waterfowl at Cold Bay, Alaska. AFWL-DE-TN-74-008. Rep for US Air Force Weapons Lab, Kirtland AFB, NM. 17 pp

Hussell DJT (1981) The use of migration counts for monitoring bird population levels. Stud Avian Biol 6:92–102

Hussell DJT (1985) Analysis of hawk migration counts for monitoring population levels. In: Proc Hawk Migr Conf IV, Hawk Migr Assoc N Am, pp 243–254

Jacoby VE (1983) Radar and visual observations of spring mass migration of sea ducks on the western coast of Estonia. Ornis Fenn Suppl 3:44–45

Jellmann J (1979) Flughöhen ziehender Vögel in Nordwestdeutschland nach Radarmessungen. Vogelwarte 30:118–134

Kerlinger P (1982) The migration of common loons through eastern New York. Condor 84:97–100

Kerlinger P (1989) Flight strategies of migrating hawks. Univ Chicago Press, Chicago

Kerlinger P, Gauthreaux SA Jr (1985) Seasonal timing, geographic distribution, and flight behavior of broad-winged hawks during spring migration in south Texas: a radar and visual study. Auk 102:735–743

Kerlinger P, Bingman VP, Able KP (1985) Comparative flight behaviour of migrating hawks studied with tracking radar during autumn in central New York. Can J Zool 63:755–761
Klein H (1980) Modifying influences of environmental factors on a time-distance-program in bird migration. Acta 17th Congr Int Ornithol, Berlin, pp 529–534
Kloeckner PD, Williams JM, Williams TC (1982) Radar and visual observations of transpacific migrants. 'Elepaio 42:77–80
Larkin RP (1982) Spatial distribution of migrating birds and small-scale atmospheric motion. In: Papi F, Wallraff HG (eds) Avian navigation. Springer, Berlin Heidelberg New York, pp 28–37
Larkin RP, Griffin DR, Torre-Bueno JR, Teal J (1979) Radar observations of bird migration over the western North Atlantic Ocean. Behav Ecol Sociobiol 4:225–264
Laske V, Immelmann K (1981) Eine Beziehung zwischen Handflügelindex und Windempfindlichkeit bei Zugvögeln. Vogelwarte 31:174–177
Leito A, Renno O (1983) Über die Zugökologie der an der Barentssee heimischen Population der Weisswangengans (*Branta leucopsis*) in Estland. Vogelwarte 32:89–102
Lindskog H, Roos G (1980) [The influence of weather on the autumn migration of tits, especially the blue tit *Parus caeruleus*, at Falsterbo, south Sweden.] Anser 19:1–10 (in Swedish)
Lyuleeva DS, Zalacjavicius MM, Shumakov ME, Yablonkevich ML, Bol'shakov KV (1981) Comparison of the results obtained by five different methods of counting birds daily migrating past at the Kurische Nehrung during the fall of 1977. Proc Zool Inst Acad Sci USSR 104:57–70 (Transl ENV TR-2474, Environm Can, Ottawa, 1984)
Maran J (1983) On the moult migration and moult of diving ducks on the North Estonian coast. Ornis Fenn Suppl 3:46–47
McCrary MD, McKernan RL, Landry RE, Wagner WD, Schreiber RW (1983) Nocturnal avian migration assessment of the San Gorgonio wind resource study area, spring 1982. Rep 83-RD-108 from Nat Hist Mus Found, Los Angeles, for S Calif Edison Co, Rosemead, CA. 121 pp
McIntosh MW, Barnhurst RJ (1988) Hawk migration in the Montreal area. Newsl Hawk Migr Assoc N Am 13:5–9
Mehlum F (1983) Resting time in migrating robins *Erithacus rubecula* at Store Færder, Outer Oslofjord, Norway. Fauna Norv Ser C, Cinclus 6:62–72
Mikkonen A (1981) Factors influencing the spring arrival of the brambling *Fringilla montifringilla* in northern Finland. Ornis Fenn 58:78–82
Millsap BA, Zook JR (1983) Effects of weather on accipiter migration in southern Nevada. Raptor Res 17:43–56
Pennycuick CJ (1978) Fifteen testable predictions about bird flight. Oikos 30:165–176
Perdeck AC, Speek G (1984) A radar study on the influence of expected ground speed, cloudiness and temperature on diurnal migration intensity. Ardea 72:189–198
Rabøl J (1978) A field method of estimating the migratory readiness of birds. Oikos 30:398–400
Rabøl J (1983) Nocturnal migrants at Christiansø bird station. Ornis Fenn Suppl 3:37–38
Rabøl J, Hansen K (1978) Environmental variables and the nightly emigration ratio of the robin (*Erithacus rubecola* [sic]) on the Island Hjelm, Denmark. In: Schmidt-Koenig K, Keeton WT (eds) Animal migration, navigation, and homing. Springer, Berlin Heidelberg New York, pp 294–301
Rees EC (1982) The effect of photoperiod on the timing of spring migration in the Bewick's swan. Wildfowl 33:119–132
Richardson WJ (1978a) Timing and amount of bird migration in relation to weather: a review. Oikos 30:224–272
Richardson WJ (1978b) Reorientation of nocturnal landbird migrants over the Atlantic Ocean near Nova Scotia in autumn. Auk 95:717–732
Richardson WJ (1979) Southeastward shorebird migration over Nova Scotia and New Brunswick in autumn: a radar study. Can J Zool 57:107–124
Richardson WJ (1980) Autumn landbird migration over the western Atlantic Ocean as evident from radar. Acta 17th Congr Int Ornithol, Berlin, pp 501–506
Richardson WJ (1982a) Northeastward reverse migration of birds over Nova Scotia, Canada, in autumn. Behav Ecol Sociobiol 10:193–206
Richardson WJ (1982b) Nocturnal landbird migration over southern Ontario, Canada: orientation vs. wind in autumn. In: Papi F, Wallraff HG (eds) Avian navigation. Springer, Berlin Heidelberg New York, pp 15–27

Richardson WJ (1985) The influence of weather on orientation and numbers of avian migrants over eastern Canada: a review. Contrib Mar Sci (Suppl) 27:604–617

Richardson WJ (1990) Wind and orientation of migrating birds: a review. Experientia (in press)

Riddiford N (1985) Grounded migrants versus radar: a case-study. Bird Study 32:116–121

Robinson LJ (1987) Wachusett Mountain: eight years and 100,000+ broad-wings later. Bird Obs E Mass 15:122–126

Schindler J, Berthold P, Bairlein F (1981) Über den Einfluss simulierter Wetterbedingungen auf das endogene Zugzeitprogramm der Gartengrasmücke *Sylvia borin*. Vogelwarte 31:14–32

Slack RS, Slack CB, Roberts RN, Emord DE (1987) Spring migration of long-eared owls and northern saw-whet owls at Nine Mile Point, New York. Wilson Bull 99:480–485

Smith NG (1985) Thermals, cloud streets, trade winds, and tropical storms: how migrating raptors make the most of atmospheric energy in Central America. In: Proc Hawk Migr Conf IV, Hawk Migr Assoc N Am, pp 51–65

Smith TJ III, Hayden BP (1984) Snow goose migration phenology is related to extratropical storm climate. Int J Biometeorol 28:225–233

Stoddard PK, Marsden JE, Williams TC (1983) Computer simulation of autumnal bird migration over the western North Atlantic. Anim Behav 31:173–180

Swem T (1985) Sitkagi Beach raptor migration study, spring 1982. In: Proc Hawk Migr Conf IV, Hawk Migr Assoc N Am, pp 13–26

Sykes PW Jr (1986) Autumn land-bird migration on the barrier islands of northeastern North Carolina. Occ Pap N Carol Biol Surv 1986-11. 49 pp

Taylor WK, Kershner MA (1986) Migrant birds killed at the Vehicle Assembly Building (VAB), John F. Kennedy Space Center. J Field Ornithol 57:142–154

Titus K, Mosher JA (1982) The influence of seasonality and selected weather variables on autumn migration of three species of hawks through the central Appalachians. Wilson Bull 94:176–184

Viehmann W (1982) Orientierungsverhalten von Mönchsgrasmücken (*Sylvia atricapilla*) im Frühjahr in Abhängigkeit der Wetterlage. Vogelwarte 31:452–457

Wallington CE (1977) Meteorology for glider pilots, 3rd ed. Murray, London

Wege ML (1979) Migration behavior of giant Canada geese. PhD thesis, University of California, Davis

Wege ML, Raveling DG (1983) Factors influencing the timing, distance, and path of migrations of Canada geese. Wilson Bull 95:209–221

Weir RD, Cooke F, Edwards MH, Stewart RB (1980) Fall migration of saw-whet owls at Prince Edward Point, Ontario. Wilson Bull 92:475–488

Welch B (1987) Hawks at my wingtip. North Country, Thorndike, ME

Williams MD (ed) (1986) Report on the Cambridge Ornithological Expedition to China 1985. mimeo

Williams TC (1985) Autumnal bird migration over the windward Caribbean islands. Auk 102:163–167

Williams TC, Williams JM (1988) Radar and visual observations of autumnal (southward) shorebird migration on Guam. Auk 105:460–466

Williams TC, Williams JM, Ireland LC, Teal JM (1978) Estimated flight time for transatlantic autumnal migrants. Am Birds 32:275–280

Willimont LA, Senner SE, Goodrich LJ (1988) Fall migration of ruby-throated hummingbirds in the northeastern United States. Wilson Bull 100:482–488

Wood B (1982) The trans-Saharan spring migration of yellow wagtails (*Motacilla flava*). J Zool (Lond) 197:267–283

Ying Mao (1985) Wader migration in Haizhou Bay, eastern China. Interwader 6:10–11

Zalakevicius MM (1984) [The role of weather in determining the magnitude of seasonal bird migrations, based on radar studies in Lithuania.] In: Il'ichev VD (ed) [Preventing bird damage to materials and technical structures.] Nauka, Moscow, pp 214–219 (in Russian; Engl transl available from WJR)

Zimmermann P, Sutter E (1962) Über das Zugverhalten des Rotmilans, *Milvus milvus*, in der Schweiz. Ornithol Beob 59:33–53

Zuckerbrot YD, Safriel UN, Paz U (1980) Autumn migration of quail *Coturnix coturnix* at the north coast of the Sinai Peninsula. Ibis 122:1–14

Aspects of the Molt Migration

J.R. JEHL JR.[1]

1 Introduction

In late spring or shortly after the breeding season, many temperate zone ducks and geese migrate to traditional assembly points, where they congregate in large numbers and undergo a simultaneous molt of their wing feathers, becoming flightless for several weeks. While knowledge of these assemblages of flightless birds was doubtless exploited by primitive hunters for millenia, the phenomenon of the molt migration was largely unappreciated by ornithologists until Salomonsen (1968) documented it in waterfowl and a few other groups. He considered that in its "best developed form" it involved birds moving from the breeding grounds to a special molting area, where they could rapidly replace their flight feathers at low risk from predation before resuming their flight to the winter quarters. He viewed this movement as distinct from the start of fall migration because: (1) the direction to the molting area usually differed from – and might even be opposite to – that to the wintering grounds; (2) all age and sex classes might not participate, (3) molting sites were highly localized; and (4) the numbers and densities of birds involved were often higher than at any other time.

The expression of molt migration is highly variable, as illustrated by the following examples. In many diving and dabbling ducks, for example, males take no role in parental care and leave the breeding grounds soon after mating to join in unisex molting assemblages. In geese, because both parents remain with the young until they have fledged and into the fall, molt migration is precluded for adults, which replace their remiges on the breeding grounds. Molt migrations are, however, undertaken by nonbreeders or birds that have failed to nest successfully in at least 9 of the 14 goose species (Ogilvie 1978). The best-known example of molt migration is that of the common shelduck (*Tadorna tadorna*), in which an estimated 200 000 birds, comprising over 90% of the population over 1 year of age from northwestern Europe, gather at the German Wadden Sea. This migration involves nonbreeders in late spring and adult males and females later in summer. In the common goldeneye (*Bucephala clangula*), molt occurs first in juvenile males, followed by adult males, juvenile females, and, finally, adult females (Jepsen and Jensen 1973).

[1]Sea World Research Institute, Hubbs Marine Research Center, 1700 South Shores Road, San Diego, CA 92109, USA

E. Gwinner (Ed.) Bird Migration

Salomonsen's comprehensive review – now required reading for waterfowl biologists (Ratti et al. 1982) – has stimulated much further investigation, to the extent that the routes to and locations of molting sites, as well as the timing of the movements, have been established for many species (e.g., Reed 1986; Geldenhuys 1981; Zicus 1981; Abraham 1980; Kumari 1980; van der Wal and Zomerdijk 1979). It has also been confirmed that in some species some apparently "eligible" birds do not participate (Jehl 1988; Reed 1986), that males and females may use different molting areas (Gaston and Cooch 1986), and that there may be population differences in movement patterns and behavior. For example, Lock (1986) found that common eider (*Somateria mollisima*) males in Labrador, "did not gather in large flocks, but dispersed in smaller flocks deeper in the bays and closer inshore," whereas "in northern Europe male eiders . . . gather in large flocks at the outermost edge of archipelagos." Also, there seems to be no definitive pattern to the distance of movement; in some geese, for example, they range from a few tens to many hundreds of kilometers (Ogilvie 1978).

Recent descriptive studies of molt migrations in waterfowl have largely confirmed and extended Salomonsen's general conclusions. The results have been discussed by many authors and are widely available (e.g., Palmer 1976; Cramp and Simmons 1979; Ogilvie 1978; Mead 1984). The emphasis in this chapter will be to (1) examine the evidence for molt migrations in taxa other than waterfowl, (2) consider aspects of the biology and physiology of birds during the molt; (3) discuss some comparative studies; (4) examine practical applications of our knowledge; and (5) comment on lines of future research.

2 Molt Migration in Taxa Other than Waterfowl

Because molt migrations represent a unique response to specific environmental conditions, each being constrained by the morphology and natural history requirements of individual species, variations are inevitable. Yet, in groups other than waterfowl these movements have been largely unappreciated, probably because our perceptions have been biased by the "classic" examples developed for waterfowl. Although molt migrations are not restricted to groups with synchronous wing molt (e.g., it occurs in shorebirds), the search for examples may logically begin there (Table 1).

Gaviidae. All species of loons undergo simultaneous loss of the remiges, but only the red-throated (*Gavia stellata*) does so in fall (September-December). In other species it is delayed until they reach the wintering grounds (February-April; Cramp and Simmons 1977). The difference seems related to size and nesting habitat (Woolfenden 1967). Because loons must be able to fly until after the young have fledged, only the small red-throated, which nests coastally, may be able to complete wing molt before migrating. The larger species, which take longer, mostly nest on freshwater lakes, and wing molt in fall would leave them susceptible to freeze-ups.

Table 1. Avian taxa having simultaneous molt of remiges[a]

Taxon	Comments
Gaviidae	All species
Podicipedidae	All species except *Rollandia microptera* (flightless)
Pelecanoididae	*Pelecanoides urinatrix* (all species?)
Anseriformes	All except *Anseranas*
Gruidae	All except *Balearica* and *Anthropoides virgo*
Rallidae	Most species
Heliornithidae	*Heliornis* only
Anhingidae	*Anhinga* (both species)
Phoenicopteridae	All as a rule
Jacanidae	All except *Jacana* and *Microparra*
Alcidae	All except *Aethia pusilla* and *Ae. pygmaea*
Bucerotidae	Females (only) in all species

[a] After Stresemann and Stresemann (1966).

No molt migrations have been described for loons and, given the above considerations, they might be expected only in the red-throated. Restricting the concept to the postbreeding season, however, may be unimaginative. In late winter flocks of up to several 1000 Arctic loons (*Gavia arctica*) in wing molt are found near productive fishing areas on the west coast of Baja California. If these assemblages are not present throughout the winter, but form specifically in late January through early April, they could represent a molt migration to a staging area for the northward, rather than southward, migration.

Podicipedidae. Recent studies have confirmed molt migrations in several grebes. Examples include movements of great crested grebes (*Podiceps cristatus*) to the IJsselmeer (Piersma 1987, 1988a,b) and of eared grebes (*Podiceps nigricollis*) to saline lakes in western North America (Storer and Jehl 1985). Indeed, the migration of hundreds of thousands of eared grebes to Mono Lake, California, is the largest yet described for any bird.

Phoenicopteridae, Gruidae. Molt migration has been reported in the African (*Phoenicopterus ruber roseus*), American (*P. r. ruber*), and Lesser (*Phoeniconaias minor*) flamingos (Cramp and Simmons 1977; Scott 1975; Brown 1975; Palmer 1962), though not yet in the South American forms. It is also thought to occur in most cranes (Cramp and Simmons 1979). Documentation of movements and molting sites in both groups is scanty. Evidently the phenomenon is irregular, perhaps because the timing of the wing molt in large birds is highly variable (cf. Snyder et al. 1987). In flamingos, it may occur "before, during, or after breeding, or outside of the breeding season altogether" (Brown 1975), and may not be an annual event (Stresemann and Stresemann 1966). Molt pattern is also variable and may be descending, simultaneous, or irregular (Cramp and Simmons 1979). In the cranes, similar variations have been described for *Grus canadensis* (Lewis 1979).

The scarcity of examples suggests that molt migrations in both groups are opportunistic and restricted to nonbreeders. Because food resources are typically superabundant at highly saline lakes used by flamingos, molt migrations are unnecessary, except during droughts. Adult cranes become flightless when attending young, so that molt migrations (Stresemann and Stresemann 1967) are restricted to nonbreeders.

Alcidae, Pelecanoididae. Alcids and diving petrels fly in air and water. The size of the wing and the molt pattern (but not the wing muscles; Raikow et al. 1988) represent a compromise between the requirements of these two modes of propulsion. Most alcids (wings longer than 107–115 mm) molt remiges simultaneously, whereas the two smallest (*Aethia pusilla* and *Ae. pygmaea*) molt sequentially, perhaps because further reduction in wing area would impair swimming (Storer 1958). In diving petrels Stresemann and Stresemann (1966) reported simultaneous molt in *Pelecanoides urinatrix* (wing 121 mm). This is likely the case in other species, with the possible exception of the smallest species (*P. georgicus*); its wing length (113 mm) is similar to that of the smaller alcids, which molt sequentially.

Although documenting molt migrations in pelagic birds will be difficult, I suspect that they are not uncommon. Bradstreet (1982) reported a near-textbook case in the little auk or dovekie (*Alle alle*). Nonbreeding adults and subadults leave Greenland colonies in late July or August, in advance of breeding adults, and move northward to molt in rich feeding areas along the edge of the pack ice. The only difference between this movement and that exhibited by many waterfowl is that the molting locality varies each year according to ice conditions. Similar migrations seem probable in auks, auklets, puffins, and guillemots (*Cepphus*), in which parental care ceases about the time the young fledge. In murres (*Uria*) the period of parental care is extended. Even so, at about the time the chicks fly from the nest site, "the breeding adults undergo a post-nuptial molt. Both the chick and the flightless adult accompanying it then swim northwards against the current. Murres from Newfoundland breeding colonies . . . move north along the Labrador coast, remaining in the more productive cold waters until they are able to fly" (Tuck 1960, p. 38). This movement is analogous to the postbreeding movements of Arctic geese that walk their young long distances toward coastal feeding grounds. Shall such movements during the flightless period be considered postbreeding dispersal or a kind of swimming (or walking) molt migration? The lines blur.

Rallidae, Anhingidae, Heliornithidae, Jacanidae. Simultaneous molt is widespread but not invariable in these families, except in the Anhingidae; reasons for the variability remain to be determined. Except for a few rails and anhingas, members of these families are typically solitary, nonmigratory residents of permanent tropical wetlands, and doubtlessly find food and solitude sufficient to make molt migrations unnecessary. Salomonsen (1968) reported possible molt migrations in coots (*Fulica americana, F. atra*) and perhaps rails. I know of no supplementary information.

Bucerotidae. Simultaneous wing molt occurs in female hornbills during their "imprisonment" in the nest chamber; molt migration is impossible.

Cinclidae. Dippers do not have simultaneous wing molt, but do lose three to four primaries at one time, so that they become flightless and can be captured by hand (Sullivan 1965) and incur high risks of predation. Because dippers use their stubby (90 mm) wings underwater, one might expect them to molt sequentially, like small alcids. That they do not indicates the wing's usefulness as a paddle may be impaired if some minimum number of primaries is not retained.

Dippers become flightless for 5–14 days shortly after the juveniles have dispersed. Then the adults disappear briefly from their territories and apparently move upstream (Price and Bock 1983). Unlike molt migrations, these movements are local (probably confined to the same drainage area) and do not result in aggregations in favored sites.

Scolopacidae, Charadriidae. Waders do not have synchronous wing molt, but at least four species: wood sandpiper (*Tringa glareola*), green sandpiper (*T. ocrophus*), lapwing (*Vanellus vanellus*), and Wilson's phalarope (*Phalaropus lobatus*), engage in postbreeding movements that combine elements of molt migration and staging at a single locality (Cramp and Simmons 1983; Hoffmann 1957). The best known occurs in Wilson's phalarope (Jehl 1987, 1988), in which adults replace body plumage and a few primaries at saline lakes in western North America before migrating to wintering areas. As in waterfowl, there are distinct peaks in migration, in this case with females preceding males by several weeks. Great Salt Lake, Utah (up to 400 000 birds in some years) is often the largest concentration point, but other lakes are used as well. In this species the sex ratio may vary greatly among molting areas.

3 Behavior, Ecology, and Physiology of Molting Birds

Interest in molt migration has recently broadened from documenting locations to studying the biology of the migrants. A bird's major challenge during this highly vulnerable period is to keep from getting killed. This can be met by selecting safe havens (dense marshes, open sea, etc; Salomonsen 1968), and also by minimizing the length of the molt period and regaining flight as quickly as possible. Accordingly, remiges are molted within a few days after arrival on the molting grounds. Regrowth takes 3–4 weeks in most waterfowl, with some species regaining flight when the primaries are 75% grown (Sjöberg 1988).

Weight gain prior to molting, a common pattern, is an adaptation to provide energy for the molt (e.g., Ankney 1979). Subsequent weight loss during the flightless period has also been considered adaptive, because light birds can resume flight earlier than heavy ones. The generality of this conclusion, however, is questionable, because many species maintain or gain weight during molt (e.g., Sjöberg 1988; Ankney 1979; Jehl 1988). Moreover, molting at low weights was not adaptive for the young geese studied by Owen and Ogilvie (1979), but resulted from harassment by adults. Furthermore, for species that become secretive and feed little during flightlessness, weight loss represents an unavoidable consequence, not a goal, of the behavior.

Although reduced activity will conserve energy that might be devoted to replacing the remiges, it can also delay the resumption of flight. Many species show a 15–20% loss of pectoral muscle mass during flightlessness, which must be rebuilt before the birds can fly again (Jehl 1988; Piersma 1988b; Ankney 1979). The reduction, too, has sometimes been deemed an adaptation, with the muscles inferred to provide nutrients for the rapid replacement of plumage. That interpretation is also questionable, for Ankney (1979) has shown that molting lesser snow geese are not nutritionally stressed and, moreover, that most of their breast muscle is lost before molt begins. The same seems true of eared grebes (Jehl 1988). That nutrients are shunted from breast to feathers is further negated by observations that leg muscles of waterfowl hypertrophy as breast muscles shrink (Ankney 1979; Piersma 1988b).

The opposite shifts in the size of breast and leg muscle mass have been considered "compensatory", a concept that again implies an adaptation for maintaining or regulating total body protein. It seems more likely, however, that the changes correspond to differences in work load. When geese become flightless (and breast muscles atrophy), they must rely on increased walking (and leg muscles hypertrophy; Ankney 1979). This effect of varied work load, as well as the inconstancy of total body protein, has been shown convincingly in great crested grebes (Piersma 1988b). This species relies entirely on its legs for aquatic propulsion. As its work load (swimming) during flightlessness is not much different from that at other seasons, leg muscle mass remains essentially constant, but the breast mass drops sharply. In my view, the loss of breast muscle mass has yet to be shown to be an adaptation for anything. Why, then, do birds allow it to occur, especially those that molt where food is superabundant (e.g., eared grebe; Jehl 1988)?

4 Comparative Studies

Comparative studies of molting birds have barely begun. Madsen and Mortensen (1987) studied pink-footed (*Anser brachyrhynchus*) and barnacle (*Branta leucopsis*) geese in East Greenland. Typically, these molt in somewhat different habitats but fed on similar plants, and foraging for each takes up ca. 45% of the day. When together, however, pink-foots are dominant, so that the barnacles shift to less nutritious food and increase foraging time by one-third.

Both Wilson's phalaropes and eared grebes use Mono Lake, California as a common molting station to exploit the superabundance of invertebrate prey in summer and autumn (Jehl 1988). Phalarope migration involves only post-breeding adults, with females arriving by mid-June and males 2 weeks later. Individuals remain for up to 6 weeks. In the first several weeks, in one of the most compressed molts yet described, the birds renew nearly all the body plumage, the tail, and a few primaries. At this time they spend much time resting in shoreline roosts, feeding for only a few hours each day; weight gains are modest. As the body plumage is renewed, they begin to increase the amount of time spent feeding. Before departing on a non-stop flight (4800 km) to South America in late July they will have doubled their arrival weight. Usually three to four primaries are molted prior to migration. The inner two are lost simultaneously and are almost fully regrown before other primaries are lost. Subsequent primaries are replaced singly, insuring that the wing will be essentially intact whenever migration begins. Phalaropes remain able to fly at all times and show no atrophy of breast muscles; indeed, total muscle mass increases as hypertrophy as wing loading increases and the birds prepare to migrate (Ellis and Jehl, unpub.). The Mono Lake flock is largely composed of adult females (65–70%) and adult males (ca. 30%); only a few juveniles 1–2% occur there.

Postbreeding eared grebes arrive at Mono Lake between late July and early October. Wing molt commences shortly after a bird's arrival, but apparently not before it has laid on 30–40 g of fat, which represents a 10–15% increase over arrival weight. Full regrowth of the remiges takes about 35 days. The birds fatten slowly during wing molt but rapidly afterward, even though body molt is still in progress, and by October many adults have more than doubled arrival weights. The same pattern occurs in nonbreeders, which arrive at the lake in late spring and may complete wing molt by late June.

From the time of their arrival until their ultimate departure in late fall, which occurs when food supplies ultimately fail, the grebes do not fly; for some this is a duration of 7 months. Their breast muscles begin to atrophy after only a few days at the lake and may eventually shrink to 50% of flying size. This reduction is not merely a contraction of the cells due to loss of water and fat but involves protein loss as well (Jehl unpubl.). About 2 weeks before departure, exercise flapping becomes conspicuous, the breast muscles hypertrophy, and the birds begin to regain flight.

The eared grebe's molt migration differs, in turn, from that of the great crested grebe at Lake IJsselmeer (Piersma 1988a), where the molting population consists of postbreeding adults (Table 2). The wing molt occurs in August-September and requires only 17 days, or apparently only half that of the eared grebe (see below). Great cresteds do not gain weight before or during molt and reduce diving during wing molt, perhaps to reduce the risk of breakage to growing feathers (Piersma 1988b). Eared grebes, by contrast, gain weight before and during molt, but do not decrease foraging, probably because prey are abundant and can usually be plucked from the lake surface.

The degree of breast muscle atrophy in great crested grebes is similar to that of some waterfowl but half that of eared grebes. Great cresteds begin exercise flapping shortly before primaries are regrown and depart shortly afterward. Eared grebes remain to stage at Mono Lake and do not rebuild their breast muscles until food runs out, when they have no other option. This may be several months after molt is completed.

Overall, the pattern of the fish-eating great crested grebe is more similar to that of some waterfowl than to that of the invertebrate-eating eared grebe. The

Table 2. A comparison of the molt migration of the great crested grebe at Lake IJsselmeer, The Netherlands, with that of the eared grebe at Mono Lake, California, USA

	Great Crested grebe[a]	Eared grebe[b]
Composition of population	Postbreeding adults	Adults and subadults
Size of population	40 000	500 000–750 000
Period of wing molt	Early-Aug–late Sept.	Late May–October
Duration of wing molt in individuals	17 days	35 Days[c]
Duration of flightlessness	Ca. 20 days	3–7 Months
Molt in period of maximum food	Yes	No
Weight gain before molt	No	Yes
Weight gain during molt	Yes	Yes
Molt rate maximized	Yes	Evidently not
Change in behavior	Yes, reduced foraging	Not apparent
Breast muscle atrophy	20–25%	Up to 50%
Breast muscle hypertrophy	Several days before primary growth completed	Ca. 2 weeks before migration; not related to end of molt
Exercise flapping	Yes	Yes
Diet	Fish	Invertebrates

[a] From Piersma (1987, 1988a,b).
[b] From Storer and Jehl (1985) and Jehl (1987, 1988).
[c] See text.

differences can be ascribed to the eared grebes' ability to exploit invertebrate prey in highly saline lakes for long periods.

The reported difference in the duration of wing replacement in the two grebes, and specifically the apparently protracted period in the eared grebe, requires further comment. Most workers determine the completion of wing molt by comparing the wing length of molting birds with standards established for the species (cf. Sjöberg 1988, p. 170, Piersma 1988a). By contrast, Storer and Jehl (1985) determined the replacement time in eared grebes by studying individual captives. They found that full regrowth required about 35 days, which is extremely long for a bird of that size. However, they also found that nearly all of this growth is completed in about 20 days. In field studies, such birds would be classified as having finished molt. Thus, the replacement period of the eared grebe appears to differ little from that of similarly sized waterfowl or the great crested grebe. In addition, Piersma (1988a) found that the maximum growth rate of the primaries in the two grebes was similar. These examples point out the need for standardizing definitions and procedures used in determining the flightless period.

5 Management Applications

For species like the paradise shelduck (*Tadorna variegata*), in which virtually the entire population assembles at a few molting stations year after year, molt migrations can provide a method for monitoring population changes (Williams 1979:388). Few species, however, meet this requirement. Indeed, for most species, the contention that the same birds occur at specific molting areas each year – while likely – is really an untested assumption. Furthermore, the location of molting areas is not necessarily constant. Some develop quickly, such as the establishment of a major molting area for over 100 000 common eiders in the German Wadden Sea since the 1970s, due in part to a shift of the molting areas of Danish birds (Nehls 1988; see also Madsen and Mortensen 1987); others can be abandoned almost overnight (Erskine and Smith 1986). And the age and sex composition of birds at different staging areas is not necessarily constant or even similar (see above, Wilson's phalarope).

In any case, analyzing data from molting areas is not straightforward and requires knowledge of a species' biology as well as of conditions on the breeding grounds and at other molting areas. Piersma (1987), for example, showed that because of turnover the actual number of great crested grebes at the IJsselmeer was double the peak counts. For eared grebes at Mono Lake, however, turnover is nil (Jehl 1988).

Mono Lake would seem ideal for monitoring eared grebes because the flock includes birds of all age and sex classes and of different breeding status. Between 1981 and 1987 peak counts ranged from 600 000 to 900 000. The annual differences seem large, but are difficult to interpret because of possible errors in estimation. Moreover, it appears that in some years significant numbers molt elsewhere, before moving to the staging area. Evidently the species is highly opportunistic and its use of seemingly "traditional" assembly points is less obligate that one might have assumed. I suspect that this conclusion applies to many other species, as well.

6 Future Studies

Discoveries of molt migrations open the door to future studies, some of which transcend molt migration per se. Because birds of different age and sex classes and breeding status may molt at different times in the same area, opportunities exist for studying the effects of the internal vs the external environment on the initiation and suppression of molt (e.g., Oring 1969; Brown 1975).

The ruddy duck (*Oxyura jamaicensis*) and other stiff-tailed ducks are unusual in undergoing two molts of the remiges annually (Siegfried 1973). Molt migration is known to occur in postbreeding adults (Palmer 1976). Is it possible that a second molt migration occurs prior to the spring molt, as well?

Comparative studies of species on a common molting area are few, and those comparing the same species in different localities have barely been attempted. To what degree can studies be extrapolated from one species or area to another?

The demonstration of rapid atrophy and hypertrophy of breast muscles calls for studies at the cellular level (e.g., Rosser and George 1985) to determine how a 25–50% change can be tolerated and accomplished so rapidly. To what degree is the hypertrophy due to exercise, or how much is mediated by endocrine changes?

What controls the direction and destination of molt migrations, which are often northward? Proposed benefits include longer days for feeding, reduced predation, or avoidance of competition with breeders, but the major advantage in geese is likely to be the exploitation of newly sprouted, nutritious vegetation (Owen and Ogilvie 1979; Madsen and Mortensen 1987).

While generalizations about the direction of movement may apply to high latitude grazers, they are unlikely to be valid for waterfowl in less seasonal environments (e.g., Africa; Curry-Lindahl 1981) or for species with different diets (e.g., eared grebe; Jehl 1988). If patterns exist, they remain to be determined. Truly, as Salomonsen (1968) reported, "in most species [there is] much to be done before [the molt migration] is properly understood."

7 Summary

Molt migrations exist in many taxa other than waterfowl. These have not been well documented because in some groups they are irregular and unpredictable, involve only nonbreeding birds, or fail to conform to the typical patterns envisioned for waterfowl. Molt migrations have limited utility for monitoring population changes, but knowledge of these movements provides opportunities for detailed studies of the comparative biology and physiology of birds at this vulnerable phase of the annual cycle.

Acknowledgments. My studies at Mono Lake were supported by the Los Angeles Department of Water and Power and Sea World Research Institute. I thank K.C. Parkes and T. Piersma for comments on the manuscript and G. Nehls for unpublished information.

References

Abraham KF (1980) Moult migration of lesser snow geese. Wildfowl 31:89–93

Ankney CD (1979) Does the wing molt cause nutritional stress in lesser snow geese? Auk 96:68–72

Bradstreet MSW (1982) Pelagic feeding ecology of dovekies, *Alle alle*, in Lancaster Sound and western Baffin Bay. Arctic 35:126–140

Brown LH (1975) East Africa. In: Kear J, Duplaix-Hall N (eds) Flamingos. Poyster, Berkhamsted, England, pp 38–48

Cramp S, Simmons KEL (1977) Handbook of the birds of Europe, the Middle East and North Africa, vol 1. Oxford University Press, London

Cramp S, Simmons KEL (1979) Handbook of the birds of Europe, the Middle East and North Africa, vol 2. Oxford University Press, London

Cramp S, Simmons KEL (1983) Handbook of the birds of Europe, the Middle East and North Africa, vol 3. Oxford University Press, London

Curry-Lindahl K (1981) Bird migration in Africa, vol 2. Academic Press, London

Erskine AJ, Smith AD (1986) Status and movements of common eiders in the Maritime Provinces. Can Wildl Serv Rep Ser 47:20–29

Gaston AJ, Cooch FG (1986) Observations of common eiders in Hudson Strait: aerial surveys in 1980–83. Can Wildl Serv Rep Ser 47:51–54

Geldenhuys JN (1981) Moults and moult location of the South African shelduck. Ostrich 52:129–134

Hoffmann L (1957) Le passage d'automne de chevalier sylvain (*Tringa glareola*) de France Mediterraneenne. Alauda 25:30–42

Jehl JR Jr (1987) Moult and moult migration in a transequatorially migrating shorebird: Wilson's phalarope. Ornis Scand 18:173–178

Jehl JR Jr (1988) Biology of the eared grebe and Wilson's phalarope in the nonbreeding season; a study of adaptations to saline lakes. Stud Avian Biol No. 12, 14 pp

Jepsen PU, Jensen AH (1973) The distribution and numbers of goldeneye (*Bucephala clangula*) moulting in Denmark. Dan Rev Game Biol 8:1–8

Kumari E (1980) Moult and moult migration of waterfowl in the Baltic Sea. Acta Ornithol (Warsaw) 17:33–44

Lewis JC (1979) Molt of the remiges of *Grus canadensis*. Proceedings of 1978 Crane Workshop, Rockport, Texas, Dec 1978. Colorado State University Printing Service, Fort Collins, Colorado

Lock AR (1986) A census of common eiders breeding in Labrador and the Maritime Provinces. Can Wildl Serv Rep Ser 47:30–39

Madsen J, Mortensen CE (1987) Habitat exploitation and interspecific competition of moulting geese in East Greenland. Ibis 129:25–44

Mead C (1984) Bird migration. Facts on file publications. Poyster, Berkhamsted, England

Nehls G (1988) Occurrence and food consumption of the common eider, *Somateria mollissima* in the Wadden Sea of Schleswig-Holstein. Proc 6th Int Wadden Sea Symp, pp 1–4, Nov 1988

Ogilvie MA (1978) Wild Geese. Poyster, Berkhamsted, England

Oring LW (1969) Summer biology of the gadwall at Delta, Manitoba. Wilson Bull 81:44–54

Owen M, Ogilvie MA (1979) Wing molt and weights of barnacle geese in Spitsbergen. Condor 81:42–52

Palmer RS (1962) Handbook of North American birds, vol 1. Yale University Press, New Haven, Connecticut

Palmer RS (1976) Handbook of North American birds, vol 3 Waterfowl (Part 2). Yale University Press, New Haven, CT, USA

Piersma T (1987) Population turnover in groups of wing-moulting waterbirds: the use of a natural marker in great crested grebes. Wildfowl 38:37–45

Piersma T (1988a) The annual moult cycle of great crested grebes. Ardea 76:82–95

Piersma T (1988b) Breast muscle atrophy and constraints on foraging during the flightless period of wing moulting great crested grebes. Ardea 76:96–106

Price F, Bock C (1983) Population ecology of the dipper (*Cinclus mexicanus*) in the Front Range of Colorado. Stud Avian Biol No 7, 84 pp

Raikow RJ, Bicanovsky L, Bledsoe AH (1988) Forelimb mobility and the evolution of wing-propelled diving in birds. Auk 105:446–451

Ratti J, Flake WD, Wentz WA (1982) Waterfowl ecology and management selected readings. The Wildfowl Society, Bethesda, MD

Reed A (ed) (1986) Eider ducks in Canada. Can Wildl Serv Rep Ser No 47

Rosser BWC, George CJ (1985) Effect of flightlessness during moult on the iron content in the pectoralis muscle of the giant Canada goose (*Branta canadensis maxima*). Can J Zool 63:480–483

Salomonsen F (1968) The moult migration. Wildfowl 19:5–24

Scott DA (1975) Iran. In: Kear J, Duplaix-Hall N (eds) Flamingos. Poyster, Berkhasted, England, pp 28–32

Siegfried WR (1973) Wing moult of ruddy ducks in Manitoba. Bull Br Ornithol Club 93:98–99

Sjöberg K (1988) The flightless period of free-living male teal *Anas crecca* in northern Sweden. Ibis 130:164–171

Snyder NFR, Johnson EV, Clendenen DA (1987) Primary molt of California condors. Condor 89:468–487

Storer RW (1958) Evolution in the diving birds. Proc 12th Int Ornithol Congr, pp 694–707

Storer RW, Jehl JR Jr (1985) Moult patterns and moult migration in the black-necked grebe *Podiceps nigricollis*. Ornis Scand 16:253–260

Stresemann E, Stresemann V (1966) Die Mauser der Vögel. J Ornithol 107:1–477
Stresemann E, Stresemann V (1967) Ein Sammelplatz mausernder Kraniche (*Grus grus*). J Ornithol 108:81–82
Sullivan J (1965) "Flightlessness in the dipper." Condor 67:535–536
Tuck LM (1960) The murres. Queen's Printer, Ottawa
Wal RJ van der, Zomerdijk PJ (1979) The moulting of tufted duck and pochard on the IJsselmeer in relation to molt concentration in Europe. Wildfowl 30:99–108
Williams M (1979) The moult gatherings of the paradise shellduck in the Gisborne-East coast district. Notornis 26:369–390
Woolfenden GE (1967) Selection for a delayed simultaneous wing molt in loons (Gaviidae). Wilson Bull 79:416–420
Zicus ML (1981) Molt migration of Canada geese from Crex Meadows, Wisconsin. J Wildl Manage 45:54–63

Chapter 2 Ecological and Behavioral Aspects of Migration

Site Attachment and Site Fidelity in Migratory Birds: Experimental Evidence from the Field and Analogies from Neurobiology

E.D. KETTERSON and V. NOLAN JR.[1]

1 Introduction

In this chapter we review briefly what field experiments have revealed about the formation of site bonds by avian migrants and, in so doing, consider whether it is fruitful to continue the practice of attributing site attachment to an imprinting-like process. We then refer briefly to recent findings in the neurobiology of recognition of objects and sites in three areas of investigation: filial imprinting, relocation of food by food-caching birds, and homing by pigeons. We suggest that analogies between these findings and site recognition in migratory birds may indicate new directions for research. These suggestions are speculative because current knowledge is slim; we hope the speculations may have heuristic value.

2 Site Attachment and Site Fidelity

2.1 Natural History of Site Attachment

Understanding of the development of site attachment in the migratory bird is incomplete and derives primarily from banding studies, supplemented by limited experimental work (see Baker 1978 for summary; Sokolov 1976, 1982, 1984, 1986, 1988). Newly independent birds are thought to disperse locally, to explore their surroundings, and perhaps to select potential breeding sites for the following spring. The distance dispersed prior to this choice, regardless of whether the choice is made in the bird's first autumn or in spring, is thought by many (e.g., Shields 1982; Greenwood 1987) to be determined by natural selection so that it balances the disadvantages of mating with close relatives and the advantages of mating with individuals whose genetic makeup has been molded by similar selective pressures. In general, the distance is greater for female birds than for males (e.g., Gauthreaux 1978; Greenwood 1980).

The individual then makes its first autumn migration, settles in a suitable winter site, becomes familiar with its surroundings, and, in many species, forms a bond that will cause it to return in succeeding winters (e.g., Ralph and Mewaldt 1975; Ketterson and Nolan 1982; Terrill, this Vol.). From this time on the

[1] Department of Biology, Jordan Hall, Indiana University, Bloomington, Indiana 47405, USA

E. Gwinner (Ed.) Bird Migration

migrations of site-faithful individuals are between known breeding and non-breeding areas. Additionally, in some species individuals are caught at the same stopover locations in successive migrations, indicating familiarity with migratory pathways and fidelity to sites along the way (Nolan and Ketterson, in preparation). The details of the development of bonds to seasonally occupied locations surely vary widely among migratory species, just as life history traits and ecology vary widely.

2.2 *Terminology*

We use the term site attachment for the processes leading to formation of a bird's preference for a location. These processes, which involve learning and memory (e.g., Löhrl 1959; Serventy 1967; Wiltschko and Wiltschko 1978; Berndt and Winkel 1979), result in the tendency to confine activities to a restricted location, and they occur in both sedentary and migratory species. It seems useful, therefore, to use a separate term for the expression of the preference by the free-living migrant, and we define site fidelity as the act of a migrant in returning to a location occupied in an earlier season or year.

2.3 *Components of Site Attachment/Site Fidelity*

Site fidelity implies not only that the migrant has learned the attributes of the site; it also requires the motivation and the ability to return, including the abilities to orient and to recognize the site upon arrival. These abilities may exist in sedentary species as well (Wiltschko and Wiltschko 1978; Nolan et al. 1986), but rarely have cause for expression. In any case, in migrants, recognition of the site to which there is a bond apparently interacts with the migratory physiological state and leads to the termination of that state (Ketterson and Nolan 1987a).

While the orientational component has received much productive attention, almost nothing is known of the recognition stage, i.e., what the relevant learned attributes of the site may be or how perception of them influences physiology. Some have proposed that animals develop an integrative cognitive map and possess the capacity to compare current environmental input to that internal representation (O'Keefe and Nadel 1978; Bingman 1990). In that view, recognition of the site upon arrival there might consist of matching the perception of spatial distribution of the landmarks with the memorized cognitive map. Alternatively, the animal might simply respond to one or two key stimuli as indicators of home. What the landmarks or stimuli might be is not known, although they are often tacitly assumed to be visual and there are some experimental data to support that view (Schmidt-Koenig and Walcott 1978).

2.4 Field and Laboratory Experiments Investigating Site Attachment and Site Fidelity

We are aware of two experimental approaches to the investigation of site attachment and thus site fidelity. The first, which we term staged release, is to displace birds at different stages of development (ages) and determine the correlation between their age or experience when displaced and the site to which they return, either in the year of displacement or later (e.g., Löhrl 1959; Ralph and Mewaldt 1975, 1976; Benvenuti and Ioalé 1980; Sokolov 1984). This approach has so far contributed little to identifying the cues necessary for recognition but has been quite useful for identifying the time or age at which attachment takes place.

The second method is to expose birds to a location that was their destination in a previous migration and to observe how this exposure affects their current season's migratory state (Stimmelmayr 1932; Gwinner and Czeschlik 1978; Ketterson and Nolan 1983, 1987a,b, 1988; Nolan and Ketterson, in press). If birds released or held at their migratory destination fail to migrate when it is time to do so, or to fatten and show Zugunruhe, this can be taken as evidence that they recognized the site. This approach, undertaken on caged birds, might ultimately prove amenable to cue manipulations. As yet, however, we know of only one attempt (Gwinner and Czeschlik 1978) to alter systematically the cues available during the phase in which migrants learn a site (acquisition stage), and we are aware of no manipulations during the recognition stage. The results of the lone experiment were negative, quite possibly becaused caged birds cannot learn the cues necessary for site attachment and recognition (Löhrl 1959; Berndt and Winkel 1979; Sokolov 1984; Nolan and Ketterson, in press).

A series of experiments in which, prior to autumn, we exposed both caged and free-living (released) dark-eyed juncos (*Junco hyemalis*) to their autumn migratory destination has been reviewed by us (Ketterson and Nolan 1988). Therefore, we present here only the results of a more recent study on indigo buntings (*Passerina cyanea*). Sniegowski et al. (1988) and Sniegowski (unpublished data) caught male buntings on their breeding territories, held them until the following spring, and then released them on their territories at a date when migration was just beginning among free-living conspecifics wintering far to the south. Controls were transported and released 1000 km to the south. If the birds released on territory recognized the site, and if recognition is sufficient to terminate migration, the experimentals would be expected to delete migration from that year's cycle, whereas the controls would be expected to migrate home. If the buntings released on territory did not remain there, then it might be concluded either that they did not recognize the site or that recognition could not override the seasonal physiology of migration. The results (Table 1) are equivocal but suggestive. Some of the buntings released on their territories remained (cf. the similar result in an early experiment by Stimmelmayr 1932, with one bluethroat, *Erithacus sveccica*, and one chiffchaff, *Phylloscopus collybita*). Furthermore, some released to the south returned. However, most individuals in both treatment

Table 1. Behavior of indigo buntings released prior to initiation of migration, at and away from their migratory destinations[a]

Release site	Number released	Remained at home	Migrated home	Unaccounted for
Home-Indiana	20	7	–	13
Away-Florida	20	–	8	12

[a] Combined data for 1985, 1986, from Sniegowski et al. (1988) and unpublished data by Sniegowski.

groups were unaccounted for, and it is possible that those released on territory may have migrated northward.

We conclude from results of the second experimental approach that, when migrants are exposed to their perennial migratory destination at the beginning of the migration season, at least some individuals recognize the site and this recognition overrides migration. However, further progress probably will require a system that can be brought under laboratory control, although we have just acknowledged the difficulty of controlling in the laboratory a phenomenon expected only in free-living birds [but see the laboratory studies of Glück (1984) and Roberts and Wiegl (1984) on habitat preference and of Hess (1973) on environmental imprinting].

3 Timing of Site Attachment and the Imprinting Model

Site attachment appears in certain ways to be analogous to imprinting, and imprinting is sometimes invoked in accounting for site attachment (Löhrl 1959; Sauer 1967; Serventy 1967; Ralph and Mewaldt 1975; Berndt and Winkel 1979; Sokolov 1984). Imprinting is usually described (e.g., Bateson 1979) as a special kind of learning in which (1) the information that is acquired restricts future preference and (2) acquisition occurs rapidly during a critical or sensitive period or phase. Some definitions require that the learning be accomplished without reinforcement and that the preference be irreversible or be retained for a long time without intervening practice.

Most research on site attachment that has drawn inspiration from imprinting models has focused on the timing of attachment, with at least an implicit expectation of finding a sensitive period. Existence of such a period is traditionally demonstrated with data graphed as an inverted U- or V-shaped curve (Fig. 1A). Subjects are exposed to a stimulus over a range of times and later tested to determine whether they established a preference for the stimulus. If exposure is before or after the sensitive period, no preference is exhibited. Only exposure during the sensitive period establishes a preference, which is graphed as the peak of the curve.

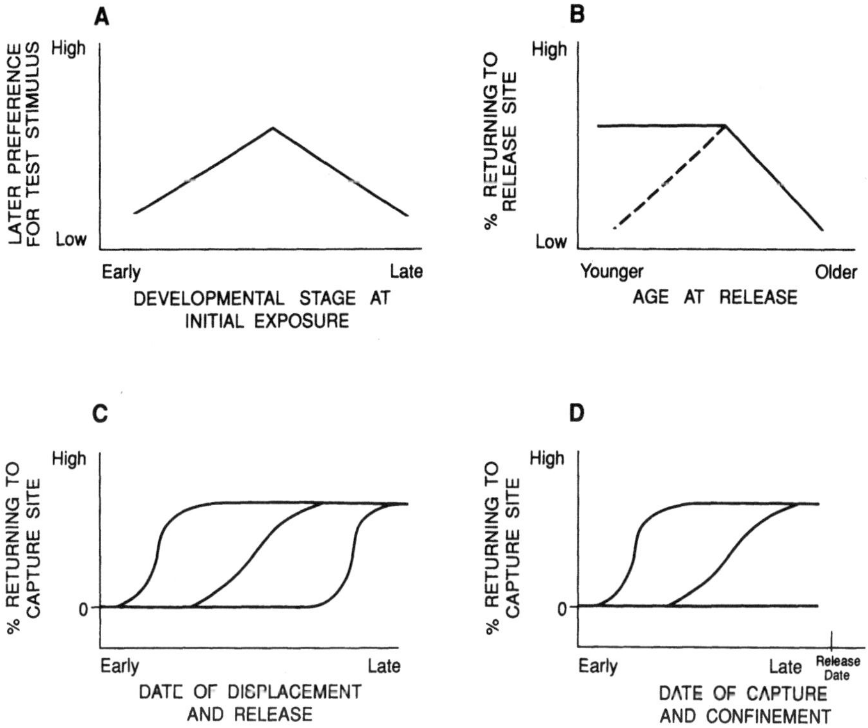

Fig. 1A-D. Expected results of staged-release experiments under various hypotheses and experimental designs to investigate timing of learning. **A** Classical evidence for a sensitive phase in imprinting. **B** Expected results in a staged-release experiment such as Löhrl's. *Solid line* shows observable results; *dashed line* the results necessary to demonstrate a sensitive phase. **C** Expected returns to capture site when displacements are at various dates and subjects are released immediately. In *left-hand curve*, a brief imprinting-like process occurs early in the season; in *right-hand curve* it occurs late in the season. In *central curve*, learning is by a gradual, individually varying process or possibly by an imprinting-like process whose date varies among individuals. See text. **D** Expected returns to capture site in a staged-delayed-release experiment. Captures are at various dates, and subjects are displaced and held until late in the season, then released simultaneously at the displacement site. *Left-hand rising curve* shows returns to the capture site if a brief imprinting-like process occurs early in the season. *Bottom horizontal curve* shows returns to the capture site if a brief imprinting-like process occurs late in the season. The *right-hand rising curve* shows returns to the capture site if learning is gradual and date of its completion varies among individuals

3.1 Attachment to the Natal Site

In field experiments to determine the timing of attachment to the natal site, individual birds are released at various ages and their later tendency to return is monitored. The problem with this method, however, is that the expected result is not an inverted U. The reason is that birds released prior to the start of the putative sensitive period will, if they do not disperse, still be present at the beginning of the sensitive period and therefore should return at the same rate as birds released when the period begins. Only individuals released after the sensitive period is complete would be expected to form no preference (Fig. 1B). Thus, this sort of experiment cannot reveal the existence of a sensitive period, although it can produce data consistent with the concept. In a field study, to produce the inverted U required for a rigorous demonstration of imprinting, birds would have to be released at a first site at a range of developmental stages, be recaptured and held after a period in which they lived free, and finally displaced and released at a second site so that those with a preference for the first site could be observed to express it. This experiment is so difficult that we do not expect it to be made.

The curve in Fig. 1B is very similar to a curve of the results in the classic experiment of Löhrl (1959), whose young collared flycatchers (*Ficedula albicollis*) were released at different ages at a location 90 km distant from their rearing site (which was not the natal site). Individuals released prior to or early in post-juvenal molt returned to the displacement site, whereas those that were released late in molt or after molt was completed (the stage at which autumn migration begins) did not return (Fig. 2). The few similar studies since Löhrl's (e.g., Serventy 1967; Berndt and Winkel 1979) have confirmed his finding that attachment to the natal and/or rearing site involves learning: birds transported when sufficiently young return to the displacement site. But with the possible exception of Serventy's work (1967), we are aware of no further demonstration of the termination of a phase in which learning occurs. We therefore suggest that there is little evidence to justify the general conclusion that natal site attachment is an imprinting-like process. The data seem to point just as consistently to a process of gradual learning, including exploration and expansion of a familiar area (Baker 1978), that culminates in attachment.

3.2 *Attachment to the Winter Site*

Investigators have also considered whether an imprinting-like process may account for attachment to the wintering site (Ralph and Mewaldt 1975, 1976). In these staged-release experiments marked birds are displaced at various dates in

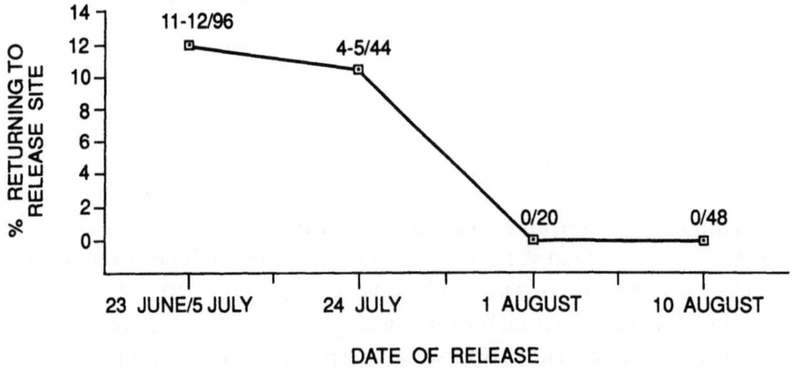

Fig. 2. Returns of collared flycatchers hand-reared away from natal site, displaced at various stages of development, and released at a distance from rearing site (After Löhrl 1959; cf. Fig. 1B)

their first winter and the percentages returning next winter to the home site are analyzed in relation to date of displacement. Ideally for such a study, all birds would remain at the release site until migrating northward in spring (i.e., would neither home to the capture site nor settle in some unknown place), and the survivors would indicate by their choice of the following winter's site whether they had become attached to the capture-home site or the displacement site. Figure 1C depicts the expected results, first, under the hypothesis that site attachment occurs rapidly during a sensitive period either early in winter (left-hand curve, Benvenuti and Ioalé 1980) or late in winter (right-hand curve, Schwartz 1963) and, second, under the alternate hypothesis that learning is gradual and varies with individual experience (middle curve). In all three possibilities, birds displaced prior to attachment fail to return to the capture-home site and those displaced after attachment return to that site. We emphasize, however, that the middle curve in Fig. 1C could also be generated by an imprinting-like process. If learning is rapid and precisely timed during a sensitive period in each individual, but the date of learning varies among individuals, the behavior of the population might be best represented by a normal distribution with the peak in mid-winter; the curve would look the same as if learning were gradual and variable. This kind of individual variation could be produced if members of the population differed in age by, say, 2 months according to whether they hatched during the preceding breeding season in early broods, mid-season broods, or late broods. One point of Fig. 1C is that these different possibilities are difficult to distinguish experimentally.

An experiment like that depicted in Fig. 1C was carried out by Ralph and Mewaldt (1975, 1976), who transported crowned sparrows (*Zonotrichia* spp.) various distances from their winter sites. The rate of return next year by young sparrows to the capture site varied according to date of displacement; those displaced prior to mid-January were less likely to return, and site attachment was believed to have occurred by mid-January. As the authors pointed out, birds displaced early in winter must survive for a longer period before their return can be observed next year; that is, the probability of mortality in the interval between displacement and the following winter presumably would be positively correlated with the length of that interval.

The experiment depicted in Fig. 1D is a variation on Fig. 1C and was designed to overcome the facts that transported birds do not necessarily remain at the release site until they migrate and that variation in length of the interval between release and return may affect the numbers that can be expected to return. In this staged-delayed-release design, birds are captured over a range of dates, displaced to and held at a second site, and released simultaneously at the displacement site late in winter, in time to migrate. This treatment prevents homing and also eliminates variable mortality after capture, whatever the date of capture. Next winter, return is monitored at both the capture site and the displacement site. If attachment occurs rapidly either early in winter or late in winter (Schwartz 1963), all returns should be to one site, the capture site or the displacement-release site, respectively. If learning occurs gradually, the percentage of birds returning to the capture site should be greatest among individuals captured latest, and some individuals might be expected to return to the displacement-release site.

Results of such an experiment appear in Fig. 3 (Nolan and Ketterson, in press). Dark-eyed juncos caught throughout December and January were displaced to an outdoor aviary surrounded by suitable junco habitat; they were released there in late winter. Intensive efforts in the following autumn and winter, both at the original capture site and at the aviary release site, yielded few returns of birds caught at any date, but there were two notable results. First, juncos caught late in winter tended to return at a greater rate than those caught earlier. When the data are analyzed with the capture period divided into thirds, chi square = 4.72, df = 2, $0.10 > p > 0.05$. Second, a few individuals caught during all thirds returned to the displacement site, indicating that the timing of site attachment was highly variable. Thus, whereas an individual caught as early as 12 December exhibited fidelity to the capture site, another caught as late as 24 January returned to the displacement-release site. (A young junco not included in this experiment was captured on 5 February, displaced to the aviary, and released there on 21 February; it returned next winter to the aviary.)

To summarize, it seems appropriate to ask whether the tradition of using the language of imprinting to describe site attachment should be continued. To be sure, information stored during site attachment can be retrieved after a long interval of no practice, but it seems to us that the evidence for a sensitive period – the most important criterion for imprinting – is slim at best. In order to study further the question of the relationship between imprinting and site attachment we need a study system with greater resolving power.

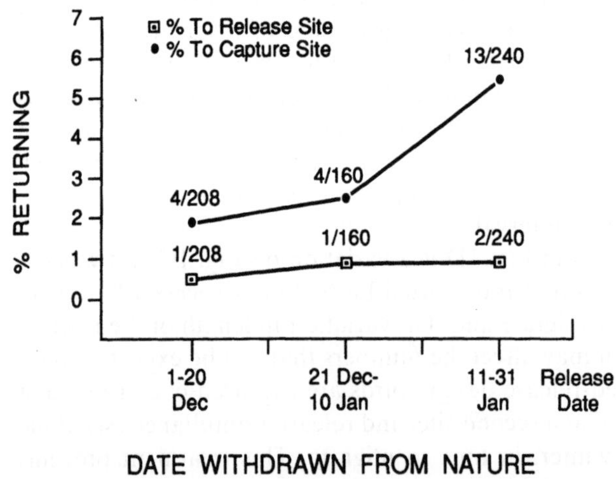

Fig. 3. Returns of young juncos in a winter staged-delayed-release experiment. Birds were caught at various dates in winter, displaced and held until late winter, then released simultaneously at the displacement site (cf. Fig. 1D)

4 Neural Bases of Behaviors Somewhat Analogous to Site Fidelity

Recent advances in understanding of the neural bases of three kinds of avian behavior should interest students of site fidelity: filial imprinting (summarized in Horn 1985), food-caching (Sherry and Vaccarino 1989; Krebs et al. 1989), and pigeon homing (summarized in Bingman et al. 1984, 1985, 1988, 1990). Each behavior involves recognition and thus has a component that is at least superficially analogous to a component of site attachment and site fidelity. Further, thus far at least, the behaviors have proved amenable to study. Space limitation permits only brief reference to this work and necessarily requires that we oversimplify, but we point out that two themes emerge. First, particular regions of the brain have come to be associated with a bird's ability to learn the attributes of, and later to recognize, significant places or objects. Second, this neurally localized ability to recognize can be uncoupled from other abilities, including the ability to learn tasks that require the very behaviors (motor patterns, visual discriminations) ordinarily associated with recognition.

4.1 Filial Imprinting

Filial imprinting occurs when young birds (e.g. *Gallus gallus*) follow their mother or some artificial substitute, with the well-known result that an attachment is formed to the object followed (this subsection is drawn from Horn 1985). Imprinting is demonstrated when the chick subsequently shows a preference for the followed object over some other potential imprinting stimulus. This preference requires that the chick recognize the stimulus on which it trained.

During imprinting a neural representation of the imprinting stimulus is stored in the brain, and autoradiography and lesioning implicate the intermediate and medial part of the hyperstriatum ventrale (IMHV) of the forebrain as the region in which storage takes place. When the IMHV is missing, a chick cannot acquire a preference. Further, when this region is lesioned in chicks that already have acquired a preference through imprinting, under some circumstances they can no longer recognize the imprinted stimulus.

One of Horn and his associates' most interesting findings is that it is possible to dissociate (uncouple) recognition of a stimulus from associative learning of that same stimulus. Chicks respond to potential imprinting stimuli as reinforcers. A chick placed in a cage with two foot pedals, only one of which will activate a view of an imprinting stimulus, will quickly learn to press the pedal that activates the view if it finds the view rewarding. The capacity to learn this task is unaffected by lesions to the IMHV. However, when these same lesioned chicks are compared to controls in their *preference for* the very stimulus they just worked to see, only the controls exhibit the preference. In other words, the IMHV is not necessary to learning a task that permits a chick to look at a stimulus it finds rewarding, but it is necessary to the chick's later ability to recognize that thing. This dissociability of recognition from task learning is the kind of observation that is leading neurobiologists to invoke the existence of multiple memory systems (Sherry and Schacter 1987).

4.2 Caching

In the last decade many studies have documented the memory capabilities of food-caching birds, some of which can retrieve their caches after several months (e.g., Kamil and Balda 1985; Balda et al. 1987). Two recent investigations have linked these capacities to a specific region of the brain, the avian hippocampus (HP). In comparisons within families, species with the greater tendency to store and recover food have larger hippocampal regions (Sherry et al. 1988; Krebs et al. 1989), and investigation of the impact of HP aspiration on cache recovery in chickadees (*Parus atricapillus*) shows that subjects lose their capacity to remember (recognize visually) the location of cache sites after aspiration (Sherry and Vaccarino 1988). These chickadees are still motivated to search and they retain all the motor and visual discriminatory abilities to do so after aspiration, but they apparently lack the ability to recall the spatial relationship between fixed feeding sites.

Whether these findings will prove relevant to site attachment is problematic. On the one hand, the task facing a bird attempting to recover stored food might be seen as involving components similar to those in the task confronting a bird about to migrate to a previously learned site: motivation to return, orientation, and recognition of the site upon arrival. On the other hand, at least two differences between food retrieval and site fidelity suggest underlying differences in the neural processes involved. In using food-storing memory, the bird makes only one visit to the cache and may learn only a single cue by which to recognize it, whereas site attachment is almost surely not based on one brief episode and, correspondingly, the memory is probably a composite of cues learned during various experiences. Further, for efficient food retrieval cache sites must either be forgotten once a seed has been retrieved or, if memory continues, the cache must be avoided unless or until it is used again (Sherry and Schacter 1987). Site fidelity is unlikely to involve such "programmed forgetting."

4.3 Orientation and Homing

Results of studies by Bingmam and associates (1984 summarized in Bingman et al. 1990) are consistent with the view that the avian hippocampus plays a necessary role in spatial recognition by homing pigeons (*Columba livea*). When HP-ablated pigeons are released 50 km from home, their ability to orient as well as controls depends upon the mode of orientation that they use, and they exhibit markedly reduced ability to reach home. When they have previously been familiar with landmarks at the release site and are forced to rely on visual recognition as their homing mechanism, homing is much impaired. Furthermore, when released nearer home, they evidently fail to recognize their home loft when in its vicinity. After postoperative retraining, however, many of the normal orientational and recognition abilities are restored.

According to Bingman (personal communication), a speculative application of these results to a migratory, potentially site-faithful bird would be as follows:

An individual that was caught in winter, subjected to hippocampal ablation, and released would have the opportunity to relearn its winter site after release but not to relearn its breeding site. If it migrated and succeeded in reaching its breeding site (and the pigeon results suggest that it might succeed), it probably would not recognize the site and so might not pass out of the migratory state. However, the following autumn it would be expected to recognize its winter site, again assuming that it reached that destination.

5 Summary and Further Speculation

There seems little reason to expect a priori that the relevant characteristics of a breeding or winter site would be stored for the long term in the same location in the brain as the quickly learned representation of a followed object. Rather, it is intuitively more appealing to describe the process of recognizing home after migration as one of matching current environmental input to a stored cognitive map. The implication of the hippocampus in recalling or recognizing food locations is provocative, but we have pointed out apparent dissimilarities between what is involved in the process of remembering the locations of a food cache and a home range. Further, the indication that the hippocampus is not necessary to the reacquisition of many spatial abilities in pigeons considerably obscures the importance of that structure. Whatever the relevance of these neural studies for site fidelity, it seems clear that we need to find a tractable study system that will permit us to dismantle the neural and behavioral components of this fascinating animal ability, preferably first in the laboratory and then in the field.

Acknowledgments. For kind assistance in the preparation of the manuscript we thank S.B. Robbins and L. Wolf. For generously sharing their work prior to publication we thank D.F. Sherry, V.P. Bingman, and P.D. Sniegowski. For stimulating discussion and helpful suggestions we thank G. Ball, W. Timberlake, A. Kamil, R. Wiltschko and R. Suthers. Aspects of our own research reported herein were supported by NSF DEB-81-10457 and BNS-83-15348 and by Indiana University's Office of Research and Graduate Development.

References

Baker RR (1978) The evolutionary ecology of bird migration. Holmes & Meier, New York

Balda RP, Bunch KG, Kamil AC, Sherry DF, Tomback DF (1987) Cache site memory in birds. In: Kamil AC, Krebs JR, Pulliam HR (eds) Foraging behavior. Plenum, New York, p 645

Bateson PPG (1979) How do sensitive periods arise and what are they for? Anim Behav 27:470–486

Benvenuti S, Ioalé P (1980) Homing experiments with birds displaced from their wintering ground. J Ornithol 121:281–286

Berndt R, Winkel W (1979) Verfrachtungs-Experimente zur Frage der Gerburtsortsprägung beim Trauerschnäpper (*Ficedula hypoleuca*). J Ornithol 120:41–53

Bingman VP (1990) Spatial organization in birds. In: Kesner R, Olton D (eds) Neurobiology of comparative cognition. Lawrence Erlbaum Assoc, Hillsdale NY (in press)

Bingman VP, Bagnoli P, Ioalé P, Casini G (1984) Homing behavior of pigeons after telencephalic ablations. Brain, Behav, Evol 24:94–108

Bingman VP, Ioalé P, Casini G, Bagnoli P (1985) Dorsomedial forebrain ablations and home loft association behavior in homing pigeons. Brain, Behav, Evol 26:1–9
Bingman VP, Ioalé P, Casini G, Bagnoli P (1988) Unimpaired acquisition of spatial reference memory, but impaired homing performance in hippocampal-ablated pigeons. Behav Brain Res 27:179–187
Bingman VP, Bagnoli P, Ioalé P, Casini G (in press) Behavioral and anatomical studies of the avian hippocampus. In: Chan-Palay V, Kohler C (eds) The hippocampus, New Vistas. Liss, New York
Gauthreaux SA (1978) The ecological consequences of behavioral dominance. In: Bateson PPG, Klopfer PH (eds) Perspectives in ethology, vol 3. Plenum, New York, p 17
Glück E (1984) Habitat selection in birds and the role of early experience. Z Tierpsych 66:45–54
Greenwood PJ (1980) Mating systems, philopatry and dispersal in birds and mammals. Anim Behav 28:1140–1162
Greenwood PJ (1987) Inbreeding, philopatry and optimal outbreeding in birds. In: Cooke F, Buckley PA (eds) Avian genetics. Academic Press, London, p 207
Gwinner E, Czeschlik D (1978) On the significance of spring migratory restlessness in caged birds. Oikos 30:364–372
Hess E (1973) Imprinting. Van Nostrand, New York
Horn G (1985) Memory, imprinting, and the brain. Clarendon Press, Oxford
Kamil AC, Balda RP (1985) Cache recovery and spatial memory in Clark's nutcrackers (*Nucifraga columbiana*). J Exp Psychol Anim Behav Processes 11:95–111
Ketterson ED, Nolan V Jr (1982) The role of migration and winter mortality in the life history of a temperate-zone migrant, the dark-eyed junco, as determined from demographic analyses of winter populations. Auk 99:243–259
Ketterson ED, Nolan V Jr (1983) Autumnal Zugunruhe and migratory fattening of dark-eyed juncos apparently suppressed by detention at the wintering site. Wilson Bull 95:628–635
Ketterson ED, Nolan V Jr (1987a) Spring and summer confinement of dark-eyed juncos at autumn migratory destination suppresses normal autumn behaviour. Anim Behav 35:1744–1753
Ketterson ED, Nolan V Jr (1987b) Suppression of autumnal migratory unrest in dark-eyed juncos held during summer on, near, or far from their previous wintering sites. Auk 104:303–310
Ketterson ED, Nolan V Jr (1988) A possible role for experience in the regulation of the timing of bird migration. In: Ouellet H (ed) Proc XIX Int Ornithol Congr, Univ Ottawa Press, Ottawa, pp 2169–2179
Krebs JR, Sherry DF, Healy SD, Terry H, Vaccarino AL (1989) Hippocampal specialization in food-storing birds. Proc Natl Acad Sci (US) 86:388–392
Löhrl H (1959) Zur Frage des Zeitpunktes einer Prägung auf die Heimatregion beim Halsbandschnäpper (*Ficedula albicollis*). J Ornithol 100:132–140
Nolan V Jr, Ketterson ED (in press) Experiments on winter-site attachment in young dark-eyed juncos. Ethology
Nolan V Jr, Ketterson ED, Wolf L (1986) Long-distance homing by nonmigratory dark-eyed juncos. Condor 88:539–542
O'Keefe J, Nadel L (1978) The hippocampus as a cognitive map. Clarendon, Oxford
Ralph CJ, Mewaldt LR (1975) Timing of site fixation upon the wintering grounds in sparrows. Auk 92:698–705
Ralph CJ, Mewaldt LR (1976) Homing success in wintering sparrows. Auk 93:1–14
Roberts EP, Wiegl PD (1984) Habitat preference in the dark-eyed junco (*Junco hyemalis*): the role of photoperiod and dominance. Anim Behav 32:709–714
Sauer EGF (1967) Migration habits of golden plovers. In: Sibley CG (ed) Proc XIII Int Ornithol Congr, Am Ornithol Union, Washington, DC, p 454
Schmidt-Koenig K, Walcott C (1978) Tracks of pigeons homing with frosted lenses. Anim Behav 26:480–486
Schwartz P (1963) Orientation experiments with northern waterthrushes wintering in Venezuela. In: Sibley CG (ed) Proc XIII Int Ornithol Congr, Am Ornithol Union, Washington DC, p 481
Serventy DL (1967) Aspects of the population ecology of the short-tailed shearwater *Puffinus tenuirostris*. In: Snow DW (ed) Proc XIV Int Ornithol Congr. Blackwell Sci, Oxford, p 165
Sherry DF, Schacter DL (1987) The evolution of multiple memory systems. Psychol Rev 94:439–454

Sherry DF, Vaccarino AL (1989) The hippocampus and memory for food caches in black-capped chickadees. Behav Neurosci 103:308–316

Sherry DF, Vaccarino AL, Buckenham K, Herz R (1988) Hippocampal size in food-storing birds. Soc Neurosci Abstr 14:235

Shields WM (1982) Philopatry, inbreeding, and the evolution of sex. State University of New York Press, Albany

Sniegowski PD, Ketterson ED, Nolan V Jr (1988) Can experience alter the avian annual cycle? Results of migration experiments with indigo buntings. Ethology 79:333–341

Sokolov LV (1976) Time of the formation of connections with the place of future nesting in some birds of passage on the Kurshsky Spit. Zool Zh 55:395–401

Sokolov LV (1982) Postbreeding movements and fidelity to birth regions of the chaffinches at the Kurische Nehrung. Tr Zool Inst Leningr 90:215–228

Sokolov LV (1984) The testing and the ability for imprinting and finding the site of future nesting in young chaffinches. Zool Zh 63:1671–1681

Sokolov LV (1986) Philopatry and dispersal of the chaffinch (*Fringilla coelebs*) in the Kurische Nehrung. Zool Zh 65:1544–1551

Sokolov LV (1988) Philopatry of migratory birds. Ornithol 23:11–25

Stimmelmayr A (1932) Neue Wege zur Erforschung des Vogelzuges. Verh Ornithol Ges Bayern 19:418–446

Wiltschko WW, Wiltschko R (1978) A theoretical model for migratory orientation and homing in birds. Oikos 30:177–187

Ecophysiological Aspects of Movements by Migrants in the Wintering Quarters

S.B. TERRILL[1]

1 Introduction

The annual cycle of migrants is typically divided into four fundamental periods: the breeding season, autumn migration, the wintering period, and vernal migration (Helms 1963). Each phase comprises distinct physiological and behavioral characteristics and each has unique properties with respect to the complex life histories of migrants. In general, more research emphasis has been placed on the breeding and migration periods compared to the winter period (e.g., Berthold 1988a). The overall winter distributions of many migrant species are well known, but detailed analyses of population structure, ecology, and physiology during this period are scarce.

Traditionally, the breeding and wintering areas have often been considered as relatively fixed sites, and migrants thought to traverse the distance separating these two regions as quickly as possible (although more rapidly in spring) given certain constraints (e.g., Curry-Lindahl 1981; Terrill and Ohmart 1984). Individuals that survive return to the same breeding and wintering sites annually. This is certainly not an incorrect picture, as we know that such a summary adequately characterizes the annual cycle for many migrants (see Ketterson and Nolan, this Vol.). It is, however, becoming increasingly clear that the concept that migrants generally show high degrees of winter site fidelity and tenacity is too simplistic, at least in some cases. One need only examine a recent winter issue of *American Birds*, a journal devoted to detecting and reporting population trends of North American avifaunna, to find numerous references to major distributional shifts, both within and between winters by temperate populations of annual migrants. Recent reviews of the wintering status of Palaearctic migrants in Africa have revealed a high degree of large-scale winter movements on that continent (e.g., Moreau 1972; Curry-Lindahl 1981; Berthold 1988a). As Berthold (1988a) asks: are these movements part of the normally occurring migration? Are they endogenously programmed (as are some aspects of autumn migratory behavior in at least several of the species involved)? Or, are they environmentally stimulated movements? As I intend to show, there is evidence to support affirmative answers to all three of these questions. These are important questions

[1]Siena College, Division of Science, Route 9, Loudonville, N.Y., 12211, USA

E. Gwinner (Ed.) Bird Migration

because they are equivalent to asking what forces, both proximate and ultimate, determine total migratory distance, a line of inquiry that has fundamental relevance both in terms of the regulation of migratory behavior in individuals and in terms of the dynamics and regulation of migrant populations.

The purpose of this review is to summarize what is known concerning movements within the winter quarters. After briefly reviewing the regulation of migratory distance in general, I will present an overall picture of what has been learned from field studies and from an experimental approach that has only very recently begun to explore the ecophysiological aspects of movements within the wintering quarters. Finally, I will present a general model of the regulation of migratory behavior that integrates this information and, hopefully, will stimulate further research in this area.

2 Determinants of Migratory Distance

Avian migratory behavior has ultimate roots in spatial and temporal dynamics of resource availability (Gauthreaux 1982). In breeding habitats where resource availability remains relatively constant and predictable throughout the year the majority of individuals are sedentary residents or perform only local movements. In regions where seasonal resource availability is somewhat less predictable, or changes in availability more frequent facultative migration occurs. By definition, whether a facultative migrant leaves, or remains on, the breeding (natal) grounds is a function of current (or impending) resource availability. Presumably, the total distance of facultative migration is a function of spatial characteristics of resource availability in a given year.

Facultative migration can involve entire populations or only portions of populations (partial migration) depending on the degree to which individual access to resources is affected. When only part of a facultative migrant population migrates, it is often members of classes considered socially subordinate that actually migrate, or migrate the farthest (Gauthreaux 1982). Thus, in these situations, it appears that dominant individuals monopolize access to resources on, or near, the breeding area and subordinates are forced to migrate to obtain access to resources necessary for overwinter survival. Although facultative migrants often comprise partial migrant populations, partial migrant populations do not necessarily indicate facultative migration, as the expression of migratory or sedentary behavior in annual partial migrant populations of some species has been demonstrated to have high heritability values (see Berthold 1988b for a detailed discussion of the genetic control of partial migration).

As seasonal changes in resource availability become more pronounced and frequent, the proportion of individuals (and species) that migrate increases (Gauthreaux 1982). Finally, the situation in which all individuals migrate every year (annual migrants) is reached when the probability of overwinter survival on the breeding grounds approaches zero for an entire population. Typically, annual migrants undergo autumn migration while food availability is still plentiful enough to enable them to make the energetically costly journey. Thus, seasonal changes in food availability are ultimately responsible for annual migration but not the proximate stimulus for migratory behavior. Rather, experimental evidence indicates that the underlying stimulus for annual migratory behavior, including the direction, duration, and distance of migration is endogenous in nature (see Gwinner 1986 and this Vol.; Berthold 1988b and this Vol, for recent reviews). In such cases, the migratory program is organized within the framework of an endogenous, circannual rhythm synchronized to environmental periodicity by changes in photoperiod (see Gwinner 1986, 1989 for recent reviews).

Substantial differences in the degree to which migratory behavior is regulated by endogenous versus exogenous factors have been typically recognized through interspecific, and to a lesser degree, interpopulational comparisons (Terrill 1989a). Evidence from field and laboratory studies, however, increasingly indicates that such variability exists even at the level of the individual. That is, the relative importance of exogenous and endogenous factors in regulating the overt expression of migratory behavior can change, in some predictable manner, with time and/or distance of migration (e.g., Perdeck 1964; Terrill and Ohmart 1984; Terrill 1987; Gwinner et al. 1988). In short, some annual migrants appear to change from a primarily endogenously stimulated migratory phase (obligate migration) to a period in which the fundamental stimulus for further migration is primarily environmental (Fig. 1). Theoretically, the obligate phase takes birds from, and across, regions where the probability of overwinter survival has been consistently low for generations. Individuals that attempted to overwinter in these regions and did not continue to migrate were eliminated. Presumably, the duration (and distance) of the obligatory phase is a function of the probability of overwinter survival along the migratory route. As this probability increases the birds switch to a facultative mode (in areas where there is relatively high spatio-temporal variance, for example, in food availability or climatic conditions) or to "winter residence" mode (in regions with relatively low variance in con-

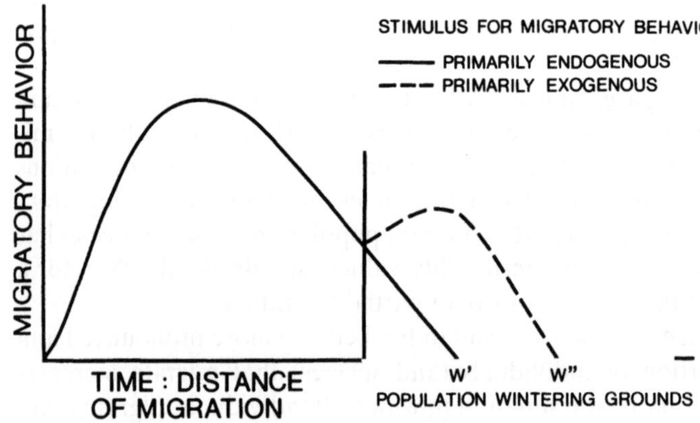

Fig. 1. Model of a change from a period in which the stimulus for migratory behavior is fundamentally endogenous to a period in which the stimulus for further migration is facultative. If environmental conditions are favorable, and relatively stable, no facultative migration occurs and birds terminate their autumn migration when the endogenous program is terminated

ditions within and between years). An assumption underlying this model is that selection is constantly acting to minimize the distance of migration while maximizing the probability of overwinter survival.

In summary then, annual migrants range from highly sedentary to very mobile within the wintering area. Theoretically, the degree to which sedentariness or mobility occurs is a function of spatial and temporal characteristics of resource availability (Terrill 1989b). In the following section, I will briefly summarize observational and experimental support for this proposition. Due to limited space, I will restrict my overview to passerines.

3 Movements Within the Wintering Grounds

By definition, the degree to which facultative migrants move from the breeding grounds in a given winter is a proximate function of resource availability. In years of severe food shortage, or exceptionally high conspecific density, or both, these birds appear far from the breeding grounds and more typical wintering areas. Such populations comprise classic "irruptive" migrants. Well-known examples include tits (*Parus*), nuthatches (*Sitta*), crossbills (*Loxia*), redpolls (*Acanthis*), and many other finches (see Gauthreaux 1982 for review). In terms of regulatory mechanisms, it is interesting to note that many, if not most, irruptive migrants migrate diurnally. Because members of these populations may or may not migrate in any given year they are not considered annual migrants (Terrill and Able 1988).

The notion that annual migrants can also show a great deal of movement within and between winters (and therefore variability in the total distance migrated between years) has been, until relatively recently, supported primarily by observations and studies involving short-distance, temperate migrants (e.g., Klein et al. 1973; Berthold 1975; Terrill 1988; Gwinner et al. 1988). Probably the first experimental evidence supporting the concept of a facultative phase following an obligate phase in an annual migrant comes from Perdeck's displacement experiments with starlings, *Sturnus vulgaris* (Perdeck 1964). Based on the results of these experiments, Perdeck concluded that juvenile starlings terminate their autumn migration if environmental conditions at a particular site are favorable for overwinter survival, or prolong migration if conditions are unfavorable, but only after a minimum period of internally motivated migration. Spaans (1977) found that migrant starlings do not show high degrees of winter site fidelity.

Evidence for similar phenomena exists from field studies of other temperate migrants including American tree sparrows, *Spizella arborea* (Niles et al. 1969), blackcaps, *Sylvia atricapilla* (Klein et al. 1973), chipping sparrows, *Spizella passerina* (Pulliam and Parker 1979), yellow-rumped warblers, *Dendroica coronata* (Terrill and Ohmart 1984; Terrill and Crawford 1988), snow buntings (Haila et al. 1986), and bramblings, *Fringilla montifringilla* (Jenni 1987). The general conclusions of these studies are that (1) after migrating some minimum distance in autumn these birds are performing facultative migratory movements in response to changing resource availability and (2) they are attempting to minimize overall migratory distance.

To present a somewhat more detailed example of such a system I will focus on the yellow-rumped warbler studies. The hypothesis that annual-migrant populations of this species were facultatively extending autumn migration was based on several lines of evidence. First, populations that winter in the Sonoran Desert (southwestern United States and northwestern Mexico) are highly restricted to insular patches of relatively lush habitats in an otherwise inhospitable, desert environment. Numbers of warblers wintering in these habitats vary greatly between years. Secondly, numbers of warblers decrease drastically in winters when unusually cold conditions occur. These decreases often involve large numbers of individuals within small areas. The initial hypothesis that the birds were dying was not supported because no dead birds were ever found following these disappearances. Local dispersal was also not a likely possibility because the surrounding habitats were generally less favorable than those the birds were evidently leaving.

The dynamic state of these populations is illustrated by numerical changes that occurred at a site at the northern edge of the Sonoran Desert wintering area (Phoenix, Arizona) and a site to the south in northern Mexico (Magdalena, Sonora). Decreases in numbers of warblers at the northerly sites corresponded to increases at southerly sites (Fig. 2). The region between these areas was generally unfavorable for overwinter survival (due to elevational increase) which suggests that the birds were making relatively long-distance movements from the Phoenix region. The magnitude of change in numbers of warblers correlated significantly with the availability of insects which, in turn, appeared to be a function of climatic conditions (Fig. 3). Analysis of stomach contents indicated that sweep samples

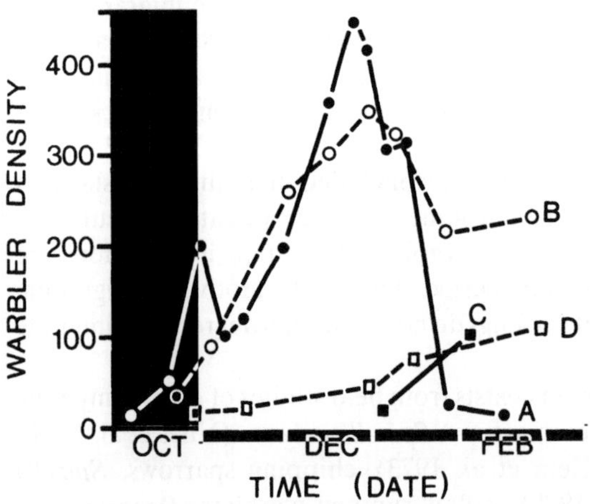

Fig. 2. Mean densities (individuals/40 ha) of yellow-rumped warblers in highly isolated riparian habitats near Phoenix, Arizona and to the south, near Magdelena, Sonora, Mexico. Note the dramatic decrease in numbers of birds in midwinter (1979–80) in the Phoenix area (**A**) in the wake of a major cold front, relative to a much lesser decrease the following winter during which no major cold fronts occurred (**B**). Decreases in the north corresponded to increases in the south (**C**: 1979–80; **D**: 1980–81)

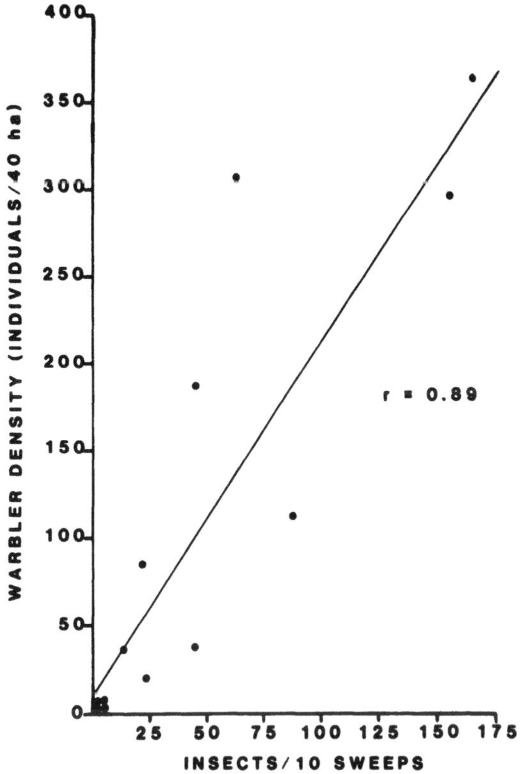

Fig. 3. Significant ($P < 0.001$) correlation between numbers of insects and numbers of yellow-rumped warblers at Arizona riparian sites from October through early February for 2 years (Terrill and Ohmart 1984)

used as a measure of insect availability accurately reflected the diet of the birds (Terrill and Ohmart 1984).

In terms of the regulation of these movements, it seems pertinent to ask whether midwinter movements are nocturnal as are fall (or "obligate phase") movements in this species. In other words, the obligate phase may reflect a fundamentally different behavior (nocturnal migration) than the facultative phase (for example, diurnal dispersal). Two lines of evidence indicate that facultative winter movements involve nocturnal migratory behavior: (1) Terrill and Ohmart (1984) found that yellow-rumped warblers showed significant southerly orientation in Emlen funnels (Emlen and Emlen 1966) when placed outdoors approximated 45 min after sunset in December and early January, and (2) nocturnal tower kills of migrating yellow-rumped warblers in the southeastern United States occur every month of the winter, but numbers vary greatly between winters (Terrill and Crawford 1988). With respect to the second line of evidence, the number of warblers killed in a given winter appears to be a function of climatic conditions, with many more individuals killed during, or just following, unusually cold conditions than during normal or mild periods. In an 8-year study of nocturnal tower kills of yellow-rumped warblers occurring between 15 December and 20 February in northern Florida, the greatest mortality occurred between 16 December and 29 December 1962, when some 200

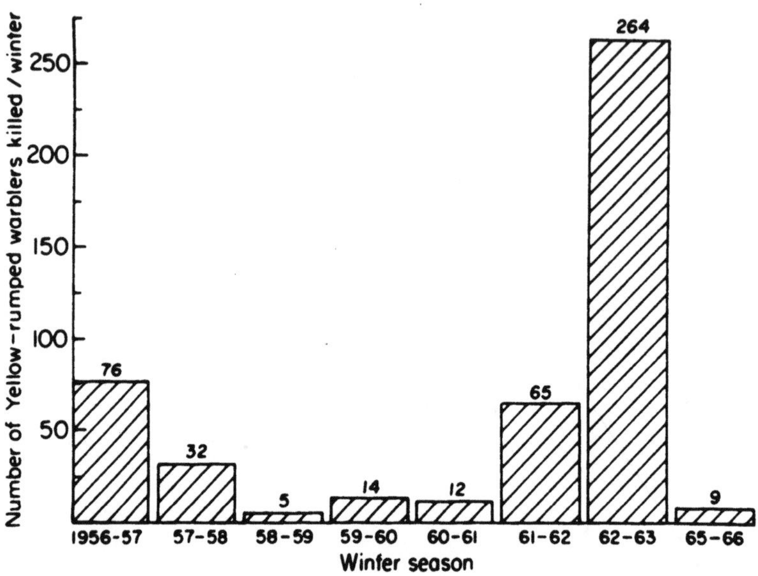

Fig. 4. Number of yellow-rumped warblers killed per winter (December 15 to February 20) at WCTV tower, Leon County, Florida (Terrill and Crawford 1988)

warblers were killed (Fig. 4). This major movement corresponded to the most severe, prolonged freeze of the 50 years preceding 1962 during which the entire southeastern United States was engulfed in subzero temperatures for nearly a week. Interestingly, Gochfeld (1985) found this species to be abundant on Jamaica during this winter, but virtually absent there the following winter which was much milder. In combination these results suggest that during a very cold winter large numbers of yellow-rumped warblers extended their migration to the Antilles, whereas in a mild winter they did not.

In summary, the association between midwinter migration, unusually cold weather, and changes in food availability support the concept that nocturnal migration by yellow-rumped warblers is highly facultative at this time of year. Results of this type are not confined to temperate, short-distance migrants. For example, there is evidence for large-scale winter movements by some North American migrants in the Neotropics (e.g., Morton 1980; Keast and Morton 1980), and in Africa it appears that wintering of many Palaearctic migrants is, to a large extent, dynamic (e.g., Curry-Lindhal 1981; Berthold 1988a). The observation that Eurasian migrants move around within their African winter quarters is not especially recent. For example, Elgood et al. (1966) noted that some species, including woodchat shrike (*Lanius senator*) and nightingale (*Luscinia megarhynches*), move slowly southward as the winter or dry season progresses.

It is largely unknown whether these extended winter movements are endogenously programmed or environmentally induced (Berthold 1988a) and, as I will show shortly, there is evidence supporting both possibilities. It is known, however, that midwinter movements of many Eurasian species that winter in

Africa are in fact nocturnal (e.g., Pearson and Backhurst 1976; Backhurst and Pearson 1977, 1984; Moreau 1972; for detailed discussion see Curry-Lindahl 1981). For example, in East Africa heavy southward nocturnal migration by many species including garden warbler (*Sylvia borin*), whitethroat (*Sylvia communis*), barred warbler (*Sylvia nisoria*), river warbler (*Lucustella fluviatilis*), willow warbler (*Phylloscopus trochilus*), and marsh warbler (*Acrocephalus palustris*) occurs regularly in December and January, and occasionally even in February.

The marsh warbler offers a rather striking example of a very protracted "fall migration". These birds begin leaving their European breeding grounds in July and reach Ethiopia around mid-August, but they do not arrive in Kenya until October or November (Dowsett-Lemaire and Dowsett 1987). Heavy migration has been noted in December and January at Amani, Tanzania (Moreau 1972) and at Ngulia, Kenya (Curry-Lindahl 1981). Finally, this species does not arrive on its southernmost wintering grounds in South Africa until late December or January, and spring departure there begins in March. Thus, the "fall" migratory period can last 6 or 7 months and the "wintering" (sedentary period may last for only 2 or 3 months. The hypothesis that this extended migration has an endogenous basis is supported by the result that captive marsh warblers show very extended autumnal *Zugunruhe*. In fact, autumnal migratory behavior continues up until approximately the time the winter molt begins (Berthold and Leisler 1980).

It may be that endogenous migratory programs in many species, as yet uninvestigated, are responsible for extended autumn migration. One might predict that extended (perhaps even stepwise) migration in winter would be endogenously programmed in the event that changes in winter resource availability have followed the same spatio-temporal patterns for many generations. Thus, it would be of considerable interest to examine resource availability, for example for marsh warblers, throughout the fall and winter periods.

On the other hand, evidence from both fieldwork and laboratory investigations on some Palaearctic migrants indicates that the impetus for extended migration within the wintering grounds is facultative (one would predict that this would be the case in regions were resource availability is relatively unpredictable between winters). One of the most dramatic examples of midwinter changes in populations of Palaearctic migrants wintering in Africa comes from work done on yellow wagtails, *Moticilla flava* (Wood 1979). Yellow wagtail populations are rather easily monitored and censused in the winter grounds because they congregate in large roosts. Wood focused on a large winter roost in Nigeria. Each year the roost begins assembling in October when the birds arrive from the European breeding grounds. The roost persists until March or April; however, Wood found a progressive decline in the numbers of wagtails at the roost from about 16 000 in November to 2000 or 3000 in March. As with the yellow-rumped warbler, food availability for yellow wagtails and their numbers at the roost declined in a parallel manner. Wood concluded that the birds were moving southwards as local conditions became increasingly severe and he presented evidence for such a shift based on ringing recoveries. Support for the proposition that these movements were facultative comes from the observation that the decline was substantially more marked for non-territorial birds, that feed in flocks, than for dominant, adult males who held, and remained on, feeding territories near the roost site throughout the entire winter.

There are indications that other species facultatively extend autumn migration in Africa. For example, Lack (1983) found close associations between food availability and winter movements of Palaearctic migrants in Tsavo East National Park, Kenya, and he suggested that these birds have adapted the strategy of remaining in areas with abundant resources, or moving on if resource availability becomes substantially reduced. Pearson (1978) states that whitethroat numbers remain higher in Kenya in years with prolonged winter rains than in dry winters.

4 An Experimental Approach

Nocturnal migrants express nocturnal migratory activity (*Zugunruhe*) in captivity. This attribute provides a convenient behavioral assay for experimental investigations of migratory behavior (e.g., see Gwinner, this Vol. and Berthold, this Vol.). Experimental investigations of the ecophysiological aspects of migrant winter movements are, for the most part, quite new. As was mentioned earlier, in one species (the marsh warbler) extended autumnal *Zugunruhe* indicates that winter movements in this species have an endogenous basis. In several other species, however, results indicate that stressful conditions can reactivate, or extend, nocturnal migratory activity in winter after termination of spontaneous autumnal migratory activity.

One such species is the dark-eyed junco (*Junco hyemalis*). These birds show spontaneous migratory activity from September into December, but after this period the behavior generally ceases (Ketterson and Nolan 1985). Terrill (1987) found that *Zugunruhe* was extended in winter if juncos were exposed to stressful environments. Female juncos were paired, or kept singly, in small activity cages (see Terrill 1987 for a detailed discussion of methods). The dominant member of each pair was determined. During the daytime, members of pairs were allowed to interact, and at night a partition divided each cage into two single cells so that the migratory activity of each bird could be measured. In general, juncos showed little or no nocturnal activity when they had access to high amounts of food during the winter months. During periods of food restriction in December and January (of 2 years), however, juncos, especially subordinates, showed increased (or induced) nocturnal activity (Fig. 5). Characteristics of this activity indicate that it was indeed migratory in nature (Terrill 1987, 1988). After late January, migratory activity could not be reactivated even when birds were subjected to restricted food treatments identical to those used earlier in the winter period. These results suggest that juncos have a migratory program that changes from (1) endogenously stimulated migratory behavior during the regular autumn migration period from September into January, to (2) facultatively stimulated migratory behavior in midwinter, and finally to (3) nonmigratory behavior in late winter.

A similar pattern has been found in the garden warbler. During the autumn migratory period (August through December) well-fed garden warblers show

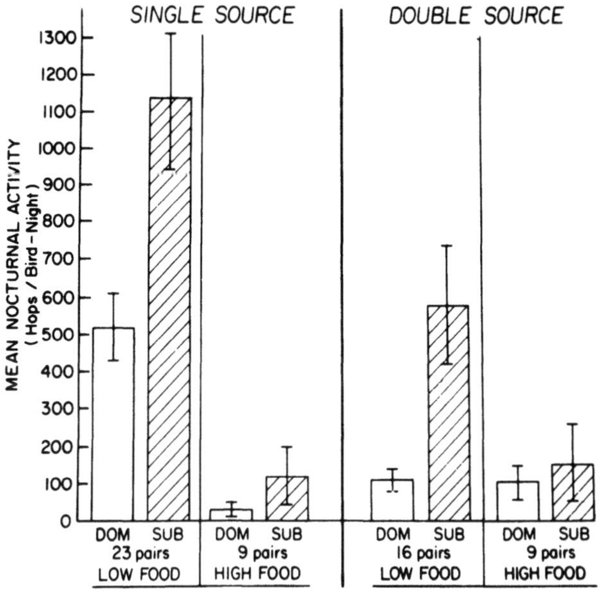

Fig. 5. Effects of social environment, food abundance, and distribution on *Zugunruhe* of dark-eyed juncos. Low food comprised 8 g of food per day for each pair, and high food comprised 14 g. *Single source* indicates that the food was placed in a single source at the center of each pair's cage and *double source* indicates that the food was equally divided into two sources placed at opposite ends of the cage. Subordinate members (*SUB*) of pairs with access to low amounts of food in a single source were significantly more active at night than were their dominant (*DOM*) partners (paired-sample $t = 3.67$, $P < 0.001$). Low food in a double source was also associated with significantly greater *Zugunruhe* in subordinates (paired-sample $t = 2.75$, $P < 0.005$). Subordinates with access to low food in a single source were significantly more active than subordinates with access to low food in a double source (two-sample $t = 2.17$, $P < 0.025$). Low food, single-source dominants showed significantly more *Zugunruhe* (two-sample $t = 3.34$, $P < 0.002$) than did their double-source contemporaries (Terrill 1987)

spontaneous *Zugunruhe* (Fig. 6). If, however, birds are deprived of food during this period, their migratory activity increases significantly (Gwinner et al. 1985, 1988; Terrill and Berthold, in prep.). When the birds again have access to food, *Zugunruhe* ceases until the birds have regained a substantial portion of their former weights [a pattern initially described in spotted flycatchers (*Muscicapa striata*); see Biebach 1985]. After termination of spontaneous migratory activity for the winter, migratory behavior was reactivated by subjecting the birds to intervals of food deprivation (Fig. 6). As was the case in autumn, migratory activity ceased as soon as food was returned and the birds began to regain weight. In contrast to autumn, however, the birds did not resume migratory activity once they had regained a substantial portion of their weights. As with the juncos, after mid- to late winter, food deprivation no longer stimulated migratory activity. In general, these patterns (at least during the autumn) of migratory behavior closely match the behavior of conspecifics in nature (e.g., Bairlein 1985; Beibach et al. 1986).

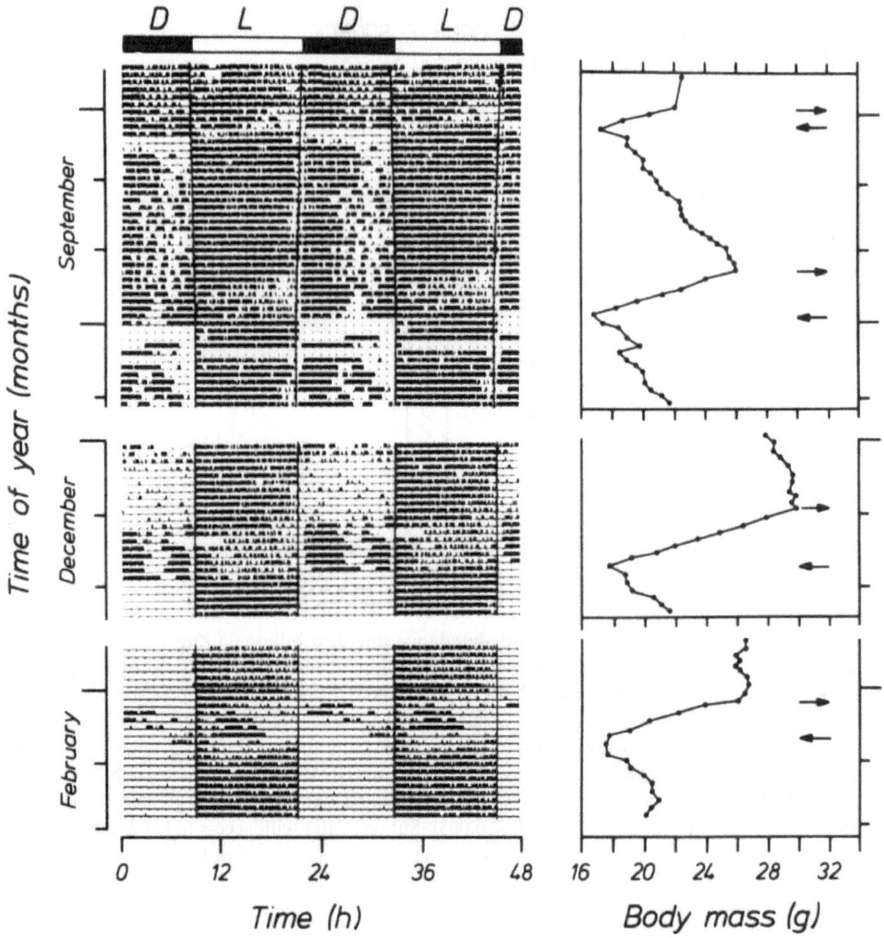

Fig. 6. *Left*: Record of locomotor activity obtained in September, December, and February from a garden warbler held under light (L)/dark(D) cycles indicated above. Each *horizontal line* from 0 to 24 h represents the activity recording of 1 day. Recordings from successive days have been mounted underneath each other in chronological sequence and the original record has been repeated on the right (24 to 48 h) to facilitate inspection of the data. *Right*: Changes in body mass; the *arrows* pointing to the *right* indicate food removal and the *arrows* pointing to the *left* indicate the point at which food was returned (Gwinner et al. 1988)

5 Synthesis

Migrants range from showing high degrees of winter-site fidelity and tenacity to being largely mobile throughout the winter period. Although there remains much empirical work to be done, theoretically this behavioral spectrum ultimately reflects an environmental continuum in resource availability that ranges from relatively stable, to predictable changes (e.g., within a winter from north to south) and finally to unpredictable changes (within and between winters).

It appears that the relative importance of endogenous and exogenous factors in regulating migratory behavior changes with time and distance of autumn migration in at least some annual migrants. Thus, during the initial stages of autumn migration these birds are obligate migrants, while later in the season they become, essentially, facultative migrants. This switch indicates a fundamental change in migratory behavior. The results outlined above suggest that during the obligate phase the endogenous impetus to migrate is (1) expressed when the birds are in an energetic state capable of meeting the metabolic demands of migration, (2) expressed, or enhanced, when birds are confronted with poor prospects for improving energy budgets, (3) inhibited when migrants are in poor condition energetically, but wind up in an area favorable for improving their condition; upon doing so, the birds return to state (1). The facultative period is expressed when state (1) is dropped ("switched off"), but states (2) and (3) remain operational. The remaining functional states are characteristics of facultative migration, and instead of using fat stores to fuel further migration birds now utilize their fat reserves to enhance overwinter survival at a favorable location. Within a species, some annual migrants known to change winter sites within and between years also demonstrate winter site fidelity and tenacity (e.g., Moreau 1972). Superficially, these observations may appear contradictory, but such variation can be interpreted as reflecting basic, facultative migratory decision rules: remain at, and return to, the same wintering site(s) if environmental conditions are, and remain, relatively favorable, but move on if conditions become unfavorable.

Finally, it is interesting to note that the results of the experiments on both the juncos and the garden warblers indicate that at some point in mid- to late winter facultative migratory behavior can no longer be stimulated. In addition to the possible importance of molt at this time of year (e.g., Gwinner et al. 1988; Berthold 1988a), this may have some significance with respect to the timing of winter-site fixation. This is approximately the time of year in which winter-site fixation has been shown to occur in migratory *Zonotrichia* sparrows (Ralph and Mewaldt 1975).

Because variation in the relative importance of exogenous and endogenous factors in regulating the expression of overt migratory behavior exists within individuals, it offers an optimal opportunity for experimental investigations of evolutionary and ecophysiological aspects of the regulation of a complex behavior that eliminates the confounding variables associated with between-species comparisons. On the other hand, further between-species comparisons will give us an idea of how universal and homologous these regulatory mechanisms are.

In conclusion, for the most part we know very little of the ecophysiological aspects of movements by migrants within the wintering grounds. As Curry-Lindahl (1981) points out (p. 500), "The discovery of midwinter migration . . . has shown how little we know about bird migration. . . . It has opened the door to a new field of ornithological research. It has widened our perspective on the complexity of bird migrations, and has given a new dimension to this phenomenon." Let us hope we can rise to meet the challenge.

Acknowledgements. I would like to thank the Alexander von Humboldt-Stiftung for support during the preparation of this review.

References

Backhurst GC, Pearson DJ (1977) Southwest migration at Ngulia, Tsavo, Kenya, 1976/77. Scopus 1:12–17

Backhurst GC, Pearson DJ (1984) The timing of southward night migration of Palaearctic birds over Ngulia, southeast Kenya. Proc V Pan-Afr Ornithol Congr, pp 361–369

Bairlein F (1985) Body weights and fat deposition of Palaearctic passerine migrants in the central Sahara. Oecologia (Berl) 66:141–146

Berthold P (1975) Migration: control and metabolic physiology. In: Farner DS, King JR (eds) Avian biology, vol 5. Academic Press, London, pp 77–128

Berthold P (1988a) The biology of the genus *Sylvia* – a model and challenge for Afro-European co-operation. Tauraco 1:3–28

Berthold P (1988b) The control of migration in European warblers. Proc XIX Intern Ornithol Congr, Ottawa, pp 215–249

Berthold P, Leisler B (1980) Migratory restlessness of the marsh warbler, *Acrocephalus palustris*. Naturwissenschaften 67:472

Biebach H (1985) Sahara stopover in migratory flycatchers: fat and food afffect the time program. Experientia (Basel) 41:695–697

Biebach H, Friedrich W, Heine G (1986) Interaction of body mass, fat, foraging and stopover period in trans-Sahara migrating passerine birds. Oecologia (Berl) 69:370–379

Curry-Lindahl K (1981) Bird migration in Africa. Academic Press, London

Dowsett-Lemaire F, Dowsett RJ (1987) European reed and marsh warblers in Africa: migration patterns, molt and habitat. Ostrich 58:65–85

Elgood JH, Sharland RE, Ward P (1966) Palaearctic migrants in Nigeria. Ibis 108:84–116

Emlen ST, Emlen JT (1966) A technique for recording migratory orientation of captive birds. Auk 103:225–227

Gauthreaux SA Jr (1982) The ecology and evolution of avian migration systems. In: Farner DS, King JR, Parkes KC (eds) Avian biology, vol 6. Academic Press, London, pp 93–168

Gochfeld M (1985) Numerical relationships between migrant and resident bird species in Jamaican woodlands. In: Buckley PA, Foster MS, Morton ES, Ridgely RS, Buckley FG (eds) Neotropical ornithology. Ornithol Mongr No 36:654–662

Gwinner E (1986) Circannual rhythms in the control of avian migrations. Adv Study Behav 16:191–228

Gwinner E (1989) Photoperiod as a modifying and limiting factor in the expression of avian circannual rhythms. J Biol Rhythms 4:237–250

Gwinner E, Biebach H, Kries I (1985) Food availability affects migratory restlessness in garden warblers (*Sylvia borin*). Naturwissenschaften 72:51–52

Gwinner E, Schwabl H, Schwabl-Benzinger I (1988) Effects of food deprivation on migratory restlessness and diurnal activity in the garden warbler *Sylvia borin*. Oecologia (Berl) 77:321–326

Haila Y, Tiainen J, Vepsäläinen K (1986) Delayed autumn migration as an adaptive strategy of birds in northern Europe: evidence from Finland. Ornis Fenn 63:1–9

Helms CW (1963) The annual cycle and *Zugunruhe* in birds. Proc XIII Intern Ornithol Congr, Ithaca, pp 925–939

Jenni L (1987) Mass concentrations of bramblings *Fringilla montifringilla* in Europe 1900–1983: their dependence upon beech mast and the effect of snow-cover. Ornis Scand 18:84–94

Keast A, Morton ES (1980) Migrant Birds in the Neotropics: Ecology, Behavior, Distribution and Conservation. Smithsonian Institution Press, Washington, DC

Ketterson ED, Nolan V Jr (1985) Intraspecific variation in avian migration: evolutionary and regulatory aspects. In: Rankin MA (ed) Migration: mechanisms and adaptive significance. Contributions in marine science, suppl vol 27, Univ of Texas Marine Science Institute, Port Aransas, Texas

Klein H, Berthold P, Gwinner E (1973) Der Zug europäischer Garten- und Monchsgrasmucken (*Sylvia borin* und *S. atricapilla*). Vogelwarte 27:73–134
Lack P (1983) The movements of Palaearctic landbird migrants in Tsavo East National Park, Kenya. J Anim Ecol 52:513–824
Moreau RE (1972) The Palaearctic-African bird migration systems. Academic Press, London
Morton ES (1980) Adaptations to seasonal changes by migrant land birds in the Panama Canal Zone. In: Keast AJ, Morton ES (eds) Migrant birds in the Neotropics: ecology, behavior, distribution and conservation. Smithsonian, Washington DC, pp 437–453
Niles DM, Rohwer SA, Jackson JA, Robbins JD (1969) An observation of midwinter nocturnal movement and tower mortality of tree sparrows. Bird-Banding 40:322–323
Pearson DJ (1978) The genus *Sylvia* in Kenya and Uganda. Scopus 2:63–71
Pearson DJ, Backhurst GC (1976) The southward migration of Palaearctic birds over Ngulia, Kenya. Ibis 118:78–105
Perdeck AC (1964) An experiment on the ending of autumn migration in starlings. Ardea 52:133–139
Pulliam HR, Parker TH III (1979) Population regulation of sparrows. Fortsch Zool 25:137–147
Ralph CJ, Mewaldt LR (1975) Timing of site fixation upon the wintering grounds in sparrows. Auk 92:698–705
Spaans AL (1977) Are starlings faithful to their individual winter quarters? Ardea 64:83–87
Terrill SB (1987) Social dominance and migratory restlessness in the dark-eyed junco (*Junco hyemalis*). Behav Ecol Sociobiol 21:1–11
Terrill SB (1988) The relative importance of ecological factors in bird migration. Proc XIX Intern Ornithol Congr, Ottawa, pp 2180–2190
Terrill SB (1989a) The regulation of migratory behavior: interactions between exogenous and endogenous factors. J Ornithol (in press)
Terrill SB (1989b) Food availability, migratory behavior and population dynamics of terrestrial birds during the nonreproductive season. In: Morrison ML, Ralph CJ, Verner J (eds) Food exploitation by terrestrial birds (in press)
Terrill SB, Able KP (1988) Bird migration terminology. Auk 105:205–206
Terrill SB, Crawford RL (1988) Additional evidence of nocturnal migration by yellow-rumped warblers in winter. Condor 90:261–263
Terrill SB, Ohmart RD (1984) Facultative extension of fall migration by yellow-rumped warblers (*Dendroica coronata*). Auk 101:427–438
Wood B (1979) Changes in numbers of over-wintering yellow wagtails, *Motacilla flava* and their food supplies in a west African savanna. Ibis 121:228–231

Control of Partial Migration and Autumnal Behavior

H. SCHWABL[1] and B. SILVERIN[2]

1 Introduction

Avian migration varies considerably both inter- and intraspecifically with respect to direction, timing and distance (Schüz 1971; Gauthreaux 1982). In populations of differential migrants all individuals migrate, but they differ in timing or distances (e.g., Gauthreaux 1982; Ketterson and Nolan 1983). Many species consist of breeding populations which reflect differential migratory behavior to the extent that some individuals migrate while others do not (Schüz 1971). This situation is referred to as partial migration (Schüz and Meise 1968). The definition should not be used for *species* that migrate in part of their breeding range and are sedentary elsewhere (Terrill and Able 1988).

Partial migration may be *obligate* or *facultative*. The first term refers to the situation in which the same individuals are either consistently migratory or sedentary across years regardless of fluctuations of ecological conditions, whereas the second term indicates that an individual may or may not migrate in a given year (Terrill and Able 1988). This distinction implies rather rigid endogenous control and independency from environmental conditions in *obligate* partial migrants and a strong response to environmental conditions, on the other hand, in *facultative* partial migrants. Obligate (endogenously) and facultatively (environmentally) controlled migration also applies to some long-distance migrants. In these birds migratory behavior can be reinduced by changes in environmental conditions after spontaneous ("obligate") migratory behavior terminates for the season (see Terrill, this Vol.; Terrill and Ohmart 1984; Gwinner et al. 1988).

2 Control Mechanisms

Control of partial migration can be discussed from two perspectives: First, *ultimate factors* such as intrasexual competition for breeding resources, differential tolerance to winter climate, differences in the ability to compete for resources during winter or differential mortality rates during migration led to the

[1]Max-Planck-Institut für Verhaltensphysiologie, Vogelwarte, D-8138 Andechs, FRG
[2]Department of Zoology, University of Gothenburg, Box 25059, S-400 31, Gothenburg, Sweden

E. Gwinner (Ed.) Bird Migration

evolution of partial migration (see reviews by Ketterson and Nolan 1983; Myers 1981; Gauthreaux 1978); and secondly, *proximate factors* regulate behavior in autumn. They will be the focus of this review.

The distinction between obligate and facultative partial migration reflects two hypotheses (*not* mutually exclusive) that have been proposed for the proximate control of partial migration:

1. The "genetic hypothesis' that was based on observations of Nice (1937) on song sparrows (*Melospiza melodia*), was proposed by Lack (1943, 1944).
2. The "behavioral/constitutional hypothesis" was first proposed by Kalela (1954) and Lack (1954) and later extended by Gauthreaux (1978, 1982), Baker (1978), Swingland (1984) and Lundberg (1987, 1988), among others.

The "genetic hypothesis" states that whether an individual is migratory or sedentary is largely dependent on genetic control and is relatively independent of environmental conditions. Substantial evidence supporting this concept is based on laboratory studies using a population of blackcaps (*Sylvia atricapilla*) from southern France (see Berthold, this Vol.), and European robins (*Erithacus rubecula*) from a south-west German population (Biebach 1983). Circumstantial evidence is available from a combined laboratory/field study for European blackbirds (*Turdus merula*; Schwabl 1981) and for a Belgium stonechat (*Saxicola torquata*) population (Dhondt 1983). Finally, Berthold (1984) reanalyzed Nice's field data (1937) on song sparrows and concluded that genetic factors were probably involved. Since the genetic aspect of the control of partial migration is discussed by Berthold (this Vol.) we will focus on the behavioral/constitutional hypothesis and proximate control mechanisms of autumnal behavior.

The "behavioral/constitutional" hypothesis assumes that the decision of whether to migrate or not depends on behavioral interactions in competition for resources prior to migration, or the ability of an animal to compete for them (e.g., Lundberg 1987). This approach implies that the complex behavioral and physiological bases of migratory behavior, for example direction of movement and ability to accumulate fat reserves prior to migration (see Wingfield et al., this Vol.) are present in all individuals, but are induced only in some of them by competition as proximate factor. Conversely, the genetic hypothesis implies that these mechanisms are only present in migrants (Berthold 1984), but that genes for migratory physiology are rapidly introduced into nonmigratory populations (Berthold 1988). However, the hypothesis that nonmigrant individuals will not show migratory activity in response to changing environmental conditions has yet to be tested.

3 Evidence for Proximate Control by Behavioral/Constitutional Factors

In contrast to control by genetic factors all the evidence for involvement of behavioral factors comes from field observations and is therefore correlational.

3.1 Autumnal Aggressive Behavior

In many partially migratory populations, aggressive activities occur in late summer and early autumn (Morley 1943) and birds compete for territories or establishment in a winter flock prior to the onset of autumnal migration by the migratory portion of the population (Kalela 1954). Mockingbirds (*Mimus polyglottus*), for example, are highly territorial in late summer and early autumn and in particular late born juveniles are apparently not able to establish a territory and then migrate (Michener and Michener 1935). Similar control is proposed for partial migration in British populations of European robins, a highly territorial species in autumn (Lack 1965) and in dippers (*Cinclus cinclus*; Lundberg et al. 1981). In European blackbirds, on the other hand, it is unlikely that competition for territories in autumn controls partial migration, since aggressive interactions are rather infrequent during this time (Schwabl 1988). In the willow tit (*Parus montanus*) only juveniles migrate and juvenile floaters will quickly occupy vacancies in territorial flocks during early autumn, but during late autumn no replacement occurs. There are no longer any floaters, possibly because they migrated (Ekman et al. 1981). Similar mechanisms are proposed for the control of partial migration in the goldcrest, *Regulus regulus* (Hilden 1982).

3.2 Social Status

As a rule, age and sex classes that are low in social status migrate while those high in social status are more likely to winter in the breeding area (Gauthreaux 1982). For example, adult male European blackbirds dominate females and young birds in competition for food during the nonbreeding season (Lundberg and Schwabl 1983). These dominance relationships are reflected in a higher proportion of females and young birds among the migratory fraction of the population (Fig. 1). The situation is similar in European robins (Lack 1965) and goldcrests (Hogstad 1984). In blue tits (*Parus coeruleus*) ringing recoveries also indicate that females are significantly more likely to migrate than males (Smith and Nilsson 1987). Although these correlations are suggestive, experimental evidence that social dominance proximately controls partial migration is still missing.

3.3 Behavioral Changes Between Years

Song sparrows, starlings (*Sturnus vulgaris*), and mockingbirds may migrate in one year and stay at the breeding area in another year (Nice 1937; Kessel 1953; Brackbill 1956). Migratory male European blackbirds are likely to become sedentary with increasing age, while females are more likely to stay migratory with increasing age (Fig. 1), suggesting that the distinction between facultative and obligate partial migration may even apply within populations. Changes from the sedentary to the migratory habit have not been proved so far, but the observation that those male European blackbirds which were present during one

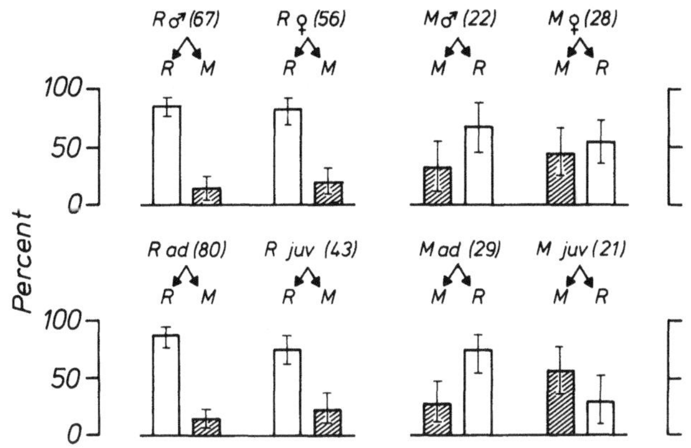

Fig. 1. Change of wintering status of European blackbirds in successive years in relation to sex (*above*) and age (*below*). The status of birds in a preceding winter is indicated above each pair of *bars* (*R* resident; *M* migrant). The *arrows* indicate which status (R or M) the birds had in the following winter. *Numbers in parentheses* indicate numbers of birds with known status in the successive year. Redrawn from Schwabl (1983)

winter and the following breeding season, were absent during the following winter but reappeared in the following breeding season, provides some evidence for this happening (Schwabl 1981). Whether these changes in behavior are induced by behavioral/constitutional factors needs to be investigated. Changes from the migratory to the sedentary habit occur in blackcaps under controlled laboratory conditions while the opposite never occurs (unpublished results). This may suggest genetic control (Berthold 1988), but habituation to captivity cannot be ruled out at this point.

3.4 Date of Hatching

Date of hatching may influence the tendency to migrate. European robins, hatching early during the breeding season, are likely to be recovered as migrants, while individuals born late are more likely to be recovered at the breeding area during the following winter (Adriaensen 1986), although local dispersal cannot be ruled out. For this species evidence for genetic control is also available (Biebach 1983), so that both genetic and environmental factors appear to play a role in the regulation of wintering behavior. Date of hatching also influences migration in blue tits: however, in this species early hatching individuals are represented in a higher fraction in residents than slightly later hatching birds (Smith and Nilsson 1987). In both species competition is assumed to be involved and the basis of the difference may be found in different ecology and social systems. Date of hatching may have its effect via prior residency and social dominance (Smith and Nilsson 1987).

3.5 Experimental Evidence

Solid experimental evidence for induction of migration by environmental factors or behavioral interactions is lacking. Such experiments are needed to evaluate the relative roles of genetic factors and behavioral interactions in control of partial migration of different species. Experimental studies similar to those conducted with differential migrants (see Terrill 1987 and this Vol.) are required. A host of supplementary environmental factors regulates migratory activity once Zugdisposition is developed in obligate migrants (see Wingfield et al., this Vol.). Whether individuals from partial migrant populations differ in their migratory responses to such factors as weather, temperature, food supply or population density needs to be tested.

4 Role of Hormones in Control of Partial Migration

The endocrine system is involved in a complex way in the induction of migratory state and modification of migratory behavior (see Wingfield et al., this Vol.). Control of partial migration may involve hormonal mechanisms of induction or inhibition of the migratory state.

4.1 Reproductive Hormones

Based on field observations that (1) the tendency to migrate is higher in females than in males, (2) it decreases with age and (3) there is territorial activity in early autumn, Lack (1954) proposed that the secretion of male sex hormones proximately controls migratory behavior in autumn: specifically, testosterone may inhibit migration. This hypothesis is attractive since testosterone induces territorial behavior in spring (Wingfield and Ramenofsky 1985) and apparently can inhibit migration (Wagner 1961; but see Schwabl and Farner 1989 and Wingfield et al., this Vol.). Runfeldt and Wingfield (1985) have shown that experimental prolongation of high levels of testosterone beyond the normal seasonal decline keeps male song sparrows in their territories well into autumn. Likewise, an experimental prolongation of spermatogenesis and high plasma levels of testosterone in a long-distant migrant, the pied flycatcher (*Ficedula hypoleuca*), prevents these males from migrating. Contrary to control males, which leave Sweden for Africa in late August/early September, experimental males stay and sing in potential territories as late as October (Silverin, unpublished data). Regulation of partial migration by differential testosterone secretion implies elevated testosterone levels in autumn prior to migration. Transient autumnal peaks of testosterone do occur in some species (Silverin 1984), but seasonal changes of levels of testosterone have been investigated with respect to the control of partial migration in only two species: the European blackbird and the willow tit.

4.1.1 European Blackbird

In the partially migratory population studied there is no autumnal surge in plasma levels of testosterone (Fig. 2). Levels are low both in adult and young males, but they are lower in females than in males (Schwabl et al. 1984) which is consistent with the observed higher tendency of females to migrate (Fig. 1). However, the comparison of levels of androgens in sedentary and migratory individuals within age and sex groups does not reveal higher levels in birds that do not migrate compared to those that apparently migrate (Schwabl et al. 1984). Further, those first-year male European blackbirds that develop autumnal migratory activity in captivity (*Zugunruhe*) do not have lower levels of androgens prior to and during migration than those that do not (Fig. 3). Levels of 5-α dihydrotestosterone are even higher in migrant than in resident first-year females, both in the fields (Schwabl et al. 1984) and in captivity (Fig. 4). Thus, neither the decline of migratory urge with age in male blackbirds nor sedentary behavior within age and sex classes are associated with elevated androgen levels that may inhibit migration.

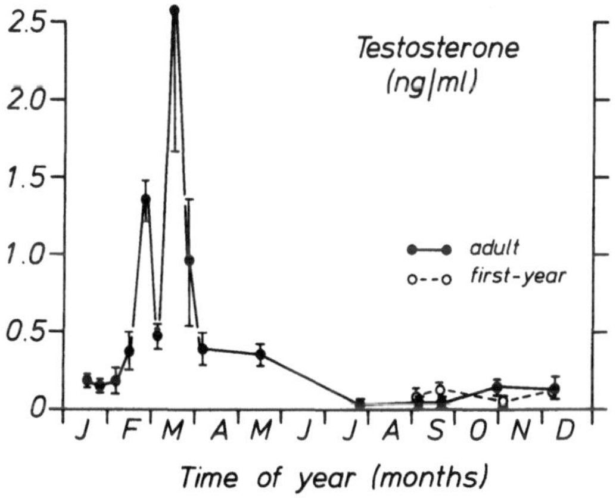

Fig. 2. Seasonal variation of plasma concentrations of testosterone in free-living male European blackbirds. Redrawn from Schwabl et al. (1980)

4.1.2 Willow Tit

In contrast to the European blackbird, male and female willow tits show a transitory autumnal rise in plasma concentrations of testosterone. This transitory increase in testosterone levels is restricted to juveniles (Fig. 5) and only about one-third of the juveniles have rather high levels (Fig. 6). At the time when this rise in testosterone occurs, juvenile willow tits are establishing themselves in winter groups that usually consist of four to six individuals, including an adult pair that previously bred in the area and genetically not related juveniles of both sexes (Silverin 1984; Hogstad 1987). However, not all juveniles succeed in becoming flock members (Ekman 1979). Part of the juveniles migrate (up to 200 km) and these individuals probably represent birds that could not establish themselves in

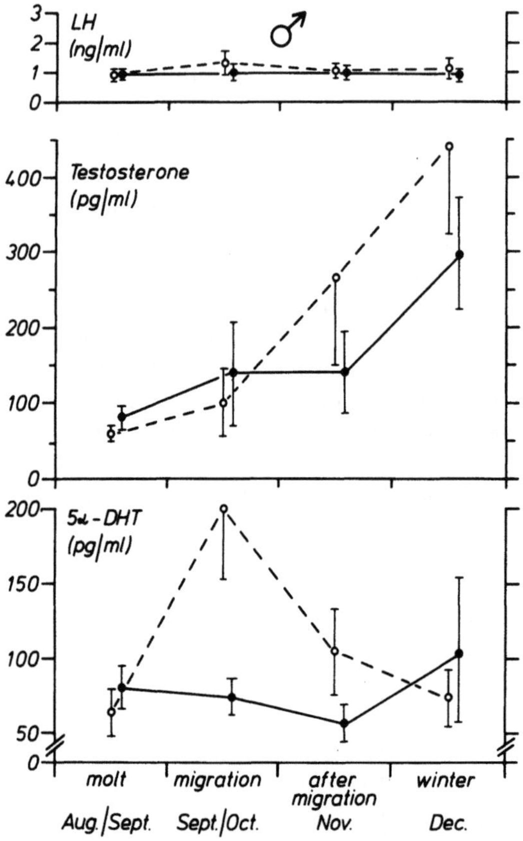

Fig. 3. Plasma levels of luteinizing hormone (*LH*) and androgens in first-year male European blackbirds that had developed migratory activity (*closed symbols*) or had failed to develop migratory activity (*open symbols*) during autumn in captivity. Redrawn from Schwabl et al. (1984)

winter flocks. These data suggest that the ability of juveniles to establish in a winter flock may depend on high testosterone levels which in turn may inhibit migration.

Levels of androgens, however, are not different between willow tits that migrate and those that are established in a wintering flock (Silverin et al. 1989). Further, implants of testosterone do not increase the probability of establishment in a winter flock or of sedentary behavior (Silverin et al. 1989). Thus, also in the willow tit, in which elevated levels of testosterone in late summer prior to migration occur, differential androgen secretion is not controlling partial migration. In summary, in the European blackbird and in the willow tit, two species with strikingly different social organization and physiology in autumn, testosterone secretion is apparently not involved in regulating partial migration.

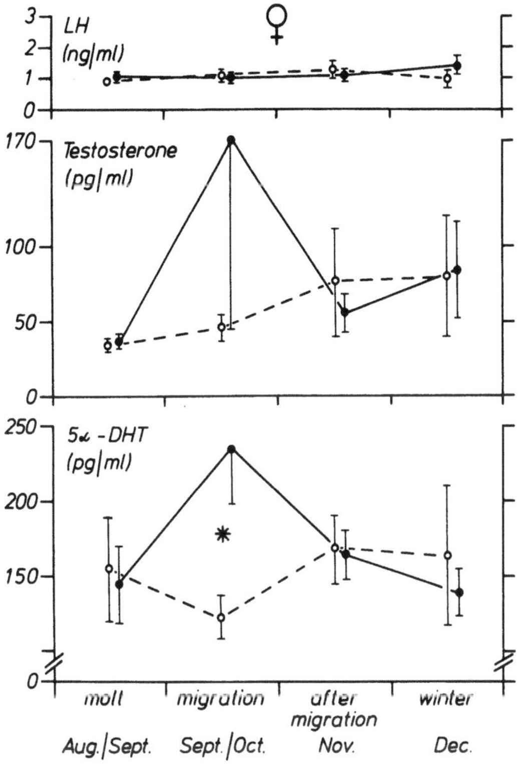

Fig. 4. Plasma levels of luteinizing hormone (*LH*) and androgens in captive first-year female European blackbirds with (*closed symbols*) or without (*open symbols*) migratory activity during autumn. Redrawn from Schwabl et al. (1984)

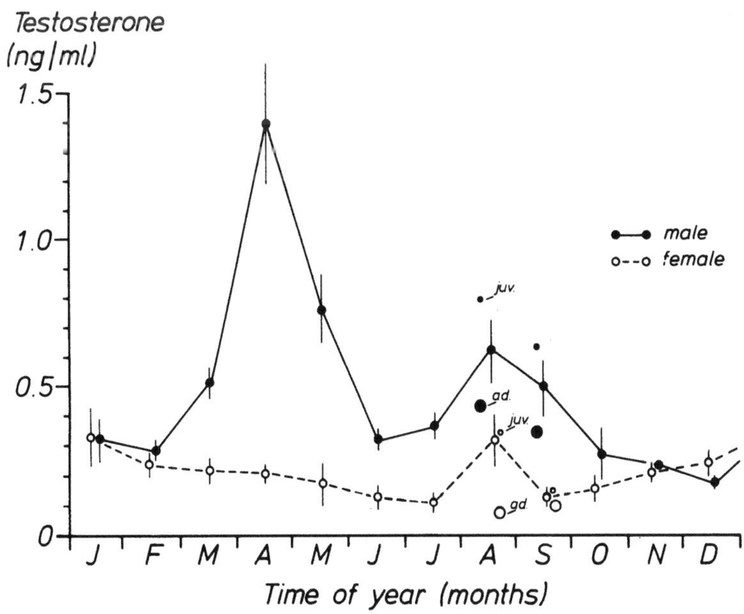

Fig. 5. Seasonal change of plasma levels of testosterone in free-living willow tits. *A d* Adult birds; *juv.* juvenile birds. Redrawn from Silverin et al. (1986)

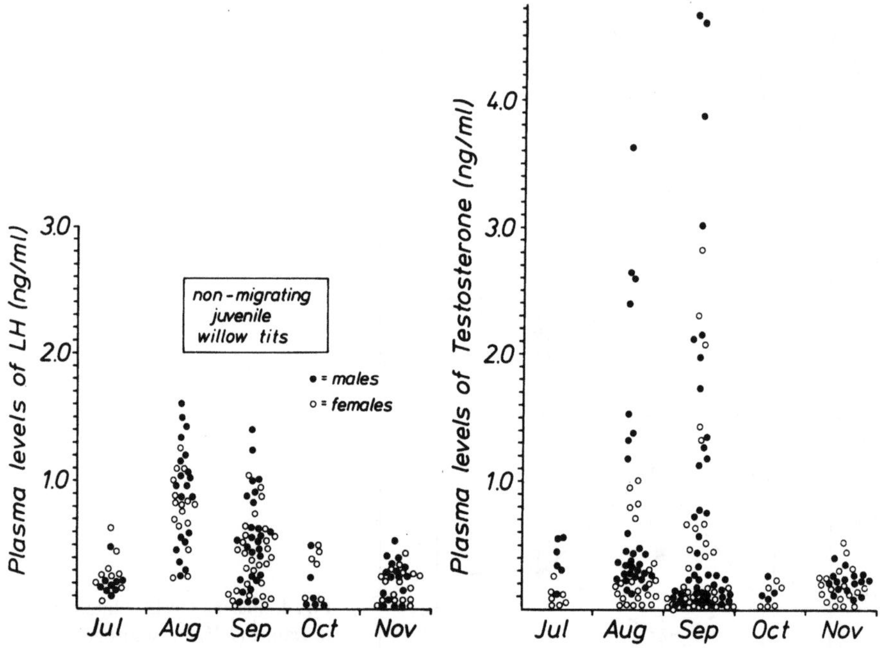

Fig. 6. Individual variability of plasma concentrations of luteinizing hormone (*LH*) and testosterone in juvenile willow tits during late summer. (Silverin et al. 1989)

4.2 Corticosterone

Corticosterone, the major avian corticosteroid, is not only involved in migratory physiology (see Wingfield et al., this Vol.), but is also associated with stress (Harvey et al. 1984), social status, and food availability (e.g., Schwabl et al. 1988). Thus, circulating levels of corticosterone may reflect differences in competitive potential or in stress, and measurements of this hormone may be useful in investigating the proximate control of partial migration. Corticosterone levels of female and first-year European blackbirds are, for example, elevated during periods of intense competition for food in winter, suggesting more stress in these groups than in adult males which have low levels of corticosterone (Schwabl et al. 1985). Levels, however, are not different between migratory and sedentary blackbirds in autumn, prior to the onset of migration (Schwabl et al. 1984). Therefore, it unlikely that stress (social or food) induces migration in blackbirds and that the pituitary-adrenal axis is proximately involved in the control of wintering strategy. On the other hand, migrating willow tits have higher baseline levels of corticosterone than those remaining in territorial flocks (Fig. 7). This may indicate stress, or lower social rank, in migrants compared to residents. However, caution is necessary here, since high levels of corticosterone may relate to actual demands of migratory flight (see Wingfield et al., this Vol.) rather than to differences in condition or social status. Curiously, migrating willow tits have rather low body mass and high levels of growth hormone, indicating poor body condition. Whether poor body condition or stress induces them to migrate needs to be investigated prior to the onset of migration (Silverin et al. 1989).

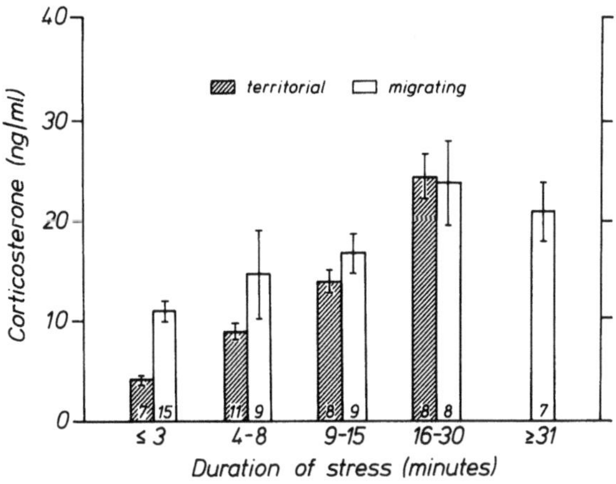

Fig. 7. Effect of duration of stress of capture on plasma concentrations of corticosterone in migrating and territorial willow tits during September. Sample sizes are indicated at the bottom of each column

5 Conclusion

Partial migration reveals a problem of great complexity. Genetic, environmental, and behavioral factors are apparently involved in its control and the relative roles of each still need to be established. Our knowledge of proximate control mechanisms is limited by lack of experimental studies both in the field and in captivity on behavioral and environmental induction of migratory behavior.

Studies concerning the genetic control of partial migration would greatly benefit from the consideration of environmental factors. The following aspects need to be evaluated: (1) do genetically programmed migrants and residents differ in their responses to environmental factors, and (2) is the ability to come into migratory state restricted to migrants, or can migratory state be induced in "genetic residents" under other conditions? Experiments performed with obligate migrants, which were previously thought to be strictly endogenously programmed, have revealed facultative responses (see Terrill, this Vol.; Gwinner et al. 1988). Most likely, such facultative mechanisms are also present in the initiation of migration of partial migrants, but the relative roles of endogenous and environmental control may vary according to species, sex, age, ecology, and social structure. Investigations of the hormonal control of partial migration may benefit from an approach that distinguishes between initially predictive, supplementary, and modifying factors.

References

Adriaensen F (1986) Differences in migratory behaviour between early and late brood nestlings of the robin *Erithacus rubecula*. Ibis 129:263–267

Baker RR (1978) The evolutionary ecology of animal migration. Hodder & Stoughton, London

Berthold P (1984) The control of partial migration in birds: a review. Ring 10:253–265

Berthold P (1988) Evolutionary aspects of migratory behavior in European warblers. J Evol Biol 1:195–209

Biebach H (1983) Genetic determination of partial migration in the European robin (*Erithacus rubecula*). Auk 100:601–606

Brackbill H (1956) Unstable migratory behavior in a mockingbird. Bird-Banding 27:128

Dhondt AA (1983) Variations in the number of overwintering stonechats possibly caused by natural selection. Ringing & Migr 4:155–158

Ekman J (1979) Non-territorial willow tits *Parus montanus* in late summer and autumn. Ornithol Scand 10:262–267

Ekman J, Cederholm G, Askenmo C (1981) Spacing and survival in winter groups of willow tit *Parus montanus* and crested tit *P. cristatus* – a removal study. J Anim Ecol 50:109

Gauthreaux SA (1978) The ecological significance of behavioral dominance. In: Bateson PPG, Klopfer P (eds) Perspectives in ethology. Plenum, New York, pp 17–54

Gauthreaux SA Jr (1982) The ecology and evolution of avian migration systems. In: Farner DS, King JR, Parkes KC (eds) Avian biology, vol 6. Academic Press, London

Gwinner E, Schwabl H, Schwabl-Benzinger I (1988) Effects of food deprivation on migratory restlessness and diurnal activity in the garden warbler *Sylvia borin*. Oecologia (Berl) 77:321–326

Harvey JG, Rees PA, Hall TR (1984) Stress and adrenal function. J Exp Zool 232:633–645

Hilden O (1982) Winter ecology and partial migration of the goldcrest *Regulus regulus* in Finland. Ornithol Fenn 59:99–122

Hogstad O (1984) Variation in numbers, territoriality, and flock size of a goldcrest *Regulus regulus* population in winter. Ibis 126:296–306

Hogstad O (1987) Social rank in winter flocks of willow tits *Parus montanus*. Ibis 129:1–9

Kalela O (1954) Populationsökologische Gesichtspunkte zur Entstehung des Vogelzugs. Ann Zool Soc Zool Bot, Fenn Vanamo 16:1–31

Kessel B (1953) Distribution and migration of starlings. Condor 55:49–67

Ketterson ED, Nolan V Jr (1983) The evolution of differential bird migration. In: Johnston RF (ed) Current ornithology, vol 1. Plenum, New York, pp 357–401

Lack D (1943/1944) The problem of partial migration. Br Birds 37:122–130, 143–150

Lack D (1954) The natural regulation of animal numbers. Oxford University Press, London

Lack D (1965) The life of a robin. Witherby, London

Lundberg P (1987) Partial bird migration and evolutionarily stable strategies. J Theor Biol 125:351–360

Lundberg P (1988) The evolution of partial migration in birds. Tree 3:172–175

Lundberg P, Schwabl H (1983) Structure and dominance behaviour of a partial migratory population of blackbirds *Turdus merula* during the non-breeding season. Ornithol Fenn Suppl 3:20–21

Lundberg P, Bergman A, Olsson H (1981) On the ecology of wintering dippers (*Cinclus cinclus*) in northern Sweden. J Ornithol 122:163–172

Michener H, Michener JR (1935) Mocking birds, their territories and individualities. Condor 37:97–140

Morley A (1943) Sexual behaviour in British birds from October to January. Ibis 85:132–158

Myers JP (1981) A test of three hypotheses for latitudinal segregation of the sexes in wintering birds. Can J Zool 59:1527–1534

Nice M (1937) Studies in the life history of the song sparrow. Trans Linn Soc NY 4:61–247

Runfeldt S, Wingfield JC (1985) Experimentally prolonged sexual activity in female sparrows delays termination of reproductive activity in their untreated mates. Anim Behav 33:403–410

Silverin B (1984) Animal gonadotropin and testosterone cycles in free-living male birds. J Exp Zool 232:581–587

Silverin B, Viebke PA, Westin J (1986) Seasonal changes in plasma levels of LH and gonadal steroids in free-living willow tits *Parus montanus*. Ornithol Scand 17:230–236

Silverin B, Viebke PA, Westin J (1989) Hormonal correlates of migration and territorial behavior in juvenile willow tits during autumn. Gen Comp Endocrinol 75:148–156

Smith GS, Nilsson JA (1987) Intraspecific variation in migratory pattern of a partial migrant, the blue tit (*Parus caeruleus*): an evaluation of different hypotheses. Auk 104:109–115

Swingland IR (1984) Intraspecific differences in movement. In: Swingland IR, Greenwood PJ (eds) The ecology of animal movement. Clarendon, Oxford

Schüz E (1971) Grundriß der Vogelzugskunde. Parey, Berlin Hamburg

Schüz E, Meise W (1968) Zum Begriff des Teilziehers. Vogelwarte 24:213–217

Schwabl H (1981) Untersuchungen über das Zugverhalten einer südwestdeutschen Population der Amsel, *Turdus merula*: zur Physiologie und Ökologie des Teilzuges. PhD-Dissertation, Ludwig-Maximilian-Universität, München
Schwabl H (1983) Ausprägung und Bedeutung des Teilzugverhaltens einer südwestdeutschen Population der Amsel, *Turdus merula*. J Ornithol 124:101–115
Schwabl H (1988) Raum-zeitliches Verhalten der Amsel (*Turdus merula*) im Herbst und seine Beziehung zum Phänomen des Teilzuges. Vogelwarte 34:312 318
Schwabl H, Farner DS (1989) Endocrine and environmental control of vernal migration in male white-crowned sparrows (*Zonotrichia leucophrys gambelii*). Physiol Zool 62:1–10
Schwabl H, Wingfield JC, Farner DS (1980) Seasonal variation in plasma levels of luteinizing hormone and steroid hormones in the European blackbird *Turdus merula*. Vogelwarte 30:283–294
Schwabl H, Wingfield JC, Farner DS (1984) Endocrine correlates of autumnal behavior of sedentary and migratory individuals of a partially migratory population of the European blackbird *Turdus merula*. Auk 101:499–507
Schwabl H, Wingfield JC, Farner DS (1985) Influence of winter on endocrine state and behavior in European blackbirds *Turdus merula*. Z Tierpsychol 68:244–252
Schwabl H, Ramenofsky I, Schwabl-Benzinger I, Farner DS, Wingfield JC (1988) Social status, circulating levels of hormones, and competition for food in winter flocks of the white-throated sparrow. Behaviour 171:107–121
Terrill SB (1987) Social dominance and migratory restlessness in the dark-eyed junco (*Junco hyemalis*). Behav Ecol Sociobiol 21:1–11
Terrill SB, Able KP (1988) Bird migration terminology. Auk 105:205–206
Terrill SB, Ohmart RD (1984) Facultative extension of fall migration by yellow-rumped warblers (*Deudroica coronata*). Auk 101:427–438
Wagner HO (1961) Beziehungen swischen dem Keimdrüsenhormon Testosteron und dem Verhalten von Vögeln in Zugstimmung. Z Tierpsychol 18:302–319
Wingfield JC, Ramenofsky M (1985) Testosterone and aggressive behavior during the reproductive cycle of male birds. In: Gilles R, Balthazart J (eds) Neurobiology. Springer, Berlin Heidelberg New York

Selection and Use of Habitat of Wintering Migrants

B. LEISLER[1]

1 Introduction

The objectives of this chapter are to discuss which habitats migrants choose for the winter quarters, and how they use these habitats. This discussion will focus on small migrant land birds that move from high to tropical latitudes. Therefore, the term "migrants" refers to north temperate species that winter at tropical and subtropical latitudes, and the term "residents" will refer to local, tropical, resident species.

Traditionally, ornithologists have regarded long-distance migrants as "invaders" to the tropics, because the birds are not resident, nor do they breed there. However, this concept has been questioned over the past 10–15 years. It has been suggested that migrants should properly be considered as peripatetic (Lack 1986a), in that they have several homes and tropical niches the same as resident species.

Investigations of how birds are adapted for seasonal occupation of different habitats is a rather recent endeavour. It has been approached by examining the ecology of migrants at three different levels, i.e., looking at individual species, ecological guilds, and whole bird communities. Despite the youth of this field, an explosion of interest in tropical migrants has resulted in a tremendous body of relevant literature. Although some summaries and review papers on special aspects of the ecology of migratory birds have been published (e.g. Moreau 1972; Karr 1976; Keast and Morton 1980; Curry-Lindahl 1981; Rappole et al. 1983; Hutto 1985; Greenberg 1985, 1986) most of the information is still widely scattered. Unfortunately, these studies vary in depth and objectives, and the habitat ecology of migrants is so diverse across the tropical biomes that it is difficult to draw generalizations.

Nowadays, habitat selection of migrants is seen as a hierarchical decision-making process (Hutto 1985) determined both by external and internal factors (Bairlein et al. 1986). Competitors and food availability have always been regarded as important factors in habitat selection. Less attention has been paid to such internal factors as morphological constraints (Winkler and Leisler 1985) and behaviour, e.g., the motivation during the obligate and facultative phase of a migrant (Terrill, this Vol.), its neophobia (i.e., an aversion to unfamiliar situations) or innate responses to particular cues. The relative importance of phys-

[1]Max-Planck-Institut für Verhaltensphysiologie, Vogelwarte, D-7760 Radolfzell-Möggingen, FRG

E. Gwinner (Ed.) Bird Migration

iological requirements in determining site selection by a bird away from its nest, or roost, has rarely been investigated (Walsberg 1985), but data from Karr and Freemark (1983) suggest that both local residents and migrants in a tropical forest may track microclimatic optima for physiological reasons.

The relative importance of these factors will be discussed by reviewing the literature and by focussing on our recent studies of the habitat utilization of wintering, and local, chat-like trushes in Kenya (Leisler et al. 1983; Leisler et al., in press).

2 Case Study

A group of chat-like thrushes (Erithacinae, Turdinae) and other open country birds were chosen for such an investigation for the following reasons: (1) in recent years the breeding ecology of these species has been investigated in great detail; (2) in the East African winter quarters they are numerous, widespread, winter visitors that encounter ecologically similar African residents; (3) the open habitat structure enables one to follow these birds for detailed field observations. Year-round wheatears (*Oenanthe*) live under extreme ecological conditions where food (i.e., terrestrial insects) is scarce or not permanently available. Species have adapted to these conditions by becoming generalized feeders (Potapova and Panov 1977). All wheatear species are potential competitors, although differing amounts of spatial separation diminish the degree of ecological overlap (Cornwallis 1975). In addition, where different species occur side by side in the same habitat, overlap is reduced by interspecific aggression (Ivanitzky 1980a,b). In their East African winter quarters, three migrant wheatear species (isabelline wheatear, *Oenanthe isabellina*; northern wheatear, *O. oenanthe;* pied wheatear, *O. pleschanka*) and two other wintering thrushes (the rock thrush, *Monticola saxatilis* and the whinchat, *Saxicola rubetra*) form an ecological guild together with three local species, the ant chat (*Myrmecocichla aethiops*), capped wheatear (*O. pileata*), and Schalow's wheatear (*O. lugen schalowi*). As such, this guild only exists in winter when the resident species are united with Palaearctic migrants. All species occupy fairly vegetated areas wherever open places occur that offer suitable perches and accessability to bare ground, thus making typical foraging possible. Typical foraging comprises scanning the ground from a perch, dropping down for an item and then returning to a perch, or making quick runs along the ground after prey (Cornwallis l.c.). Both sexes of the wintering species held separate short-term territories which are defended intra- as well as inter-specifically, but typical winter dispersion is represented by larger and more stable individual territories.

3 Which Habitats Do Migrants Choose for Winter Quarters?

North temperate migrants in the tropics seem to select approximately the same habitat types and microhabitats they use in the north, and for many species the niche gestalt during the non-breeding season closely resembles that of the breeding season (e.g. Pearson 1978; Fitzpatrick 1980; Curry-Lindahl 1981; Brosset 1984; Greenberg 1984b). However, shifts in the habitats used from summer to winter are quite common among migratory landbirds (Lack 1971) in that some species show a higher degree of flexibility in their winter quarters by using dissimilar habitats, or several kinds of habitats and moving frequently between them (e.g. Keast 1980a for North American parulid warblers). Fitzpatrick (1980, 1982) and Hutto (1985) have pointed out that North American tyrannids and wood warblers, which winter in the western parts of the Mexican plateau, generally encounter the least substantial change between breeding and non-breeding habitats. The similarity, and contiguity of breeding and wintering habitats allow these birds to use the more xeric western habitats year-round. In contrast, eastern North American warblers breeding in coniferous forests and wintering in the West Indies are the most flexible because of constraints imposed by lack of coniferous vegetation in the Caribbean basin (Greenberg 1979). Similarly, many Palaearctic migrants wintering in the different latitudinal vegetation belts of Africa had to adapt to a more geographically and ecologically varied environments.

Migrants have responded to the fact that many tropical habitats are seasonal during the period of their stay with two different strategies: (1) defending an area that will provide sufficient resources during most years [e.g. mesic sites which would not dry out (Morton 1980) or that coincide with existing territories of a native species (Greenberg 1984a)] or, (2) abandoning a location and reappear elsewhere in response to changing food resources, i.e., tracking the resources (as many migrants do in Africa, e.g. Pearson and Backhurst 1976; Lack 1983, 1987; Jones 1985; Pearson, this Vol.; see also Terrill, this Vol.).

What habitats migrants use depends largely on what habitats there are, more forest in America and Asia less in Africa. There is general agreement that migrants prefer to winter more in semiarid savannas, dry tropical forests (India), grasslands (Africa), riverine habitats (Central America), marginal and disturbed habitats (brushy and forest edges, early forest stages) and habitats modified by man (second growth forest, farm- and parkland, pastures, gardens) than in extensive areas of mature forests especially of wet lowland forests (e.g. Morel 1973; Karr 1976; Chipley 1977; Ulfstrand and Alerstam 1977; Morel and Morel 1978; Hutto 1980; Orejuela et al. 1980; Brosset 1984; Hilty and Brown 1986; Martin 1985). In the Neotropics forest habitats are generally more important for many migrant species than previously thought (Terborgh 1980). Although migrant densities are often higher in tropical second growth forests than in moist lowland forests and though only few species actually seem to prefer undisturbed forests (e.g. Terborgh 1980), this pattern is not consistent throughout the Neotropics. Whereas northern warblers and flycatchers dominate a variety of Central American and Caribbean habitats during the winter months they do indeed become more peripheral in South America (Fitzpatrick 1982). Robinson et al. (1988) and Fitzpatrick and Robinson (in prep.) have recently shown that it is only under specific eco-geographical conditions that northern migrants, which normally shun primary forests in the Amazon basin, use this habitat type to some extent. That is, migrants in this region appear to be concentrated only along a narrow zone at the eastern foothills of the Andes where productivity is exceptionally high, and resident species diversity is lower than in the lowland forest due to a reduced number of microhabitats.

In Africa, two-thirds of the Palaearctic migrants remain in savannas north of the equator and only few enter tropical forests (Brown et al. 1982).

The differential use of disturbed habitats by residents and migrants in Africa can be illustrated by an example from Liberia (Gatter and Mattes 1987). Formerly, this area was almost entirely covered by forest and offered just a few open sites to members of the pipit/wagtail family (Motacillidae). These scattered, open habitats (creeks, riverbanks, rocky hill sides, littoral zone) were used by four African species (*Motacilla clara, M. aguimp, Anthus leucophrys, A. similis*) and small numbers of wintering yellow wagtails (*M. flava*). Extensive deforestation for farmland has resulted in an enormous increase in open land, which is now occupied by the Afrotropical species to a lesser extent than by the Palaearctic migrants. The local species (especially *A. leucophrys*) plus an African invader, the yellow-throated longclaw (*Macronyx croceus*), were able to expand into formerly forested areas only where these have been severely changed by man into vegetationally poor savannas. The mobile Palaearctic migrants (yellow wagtail plus tree and red-throated pipit, *A. trivialis, A. cervinus*) extensively use all types of open land, and are capable of finding even small farm lots within huge forested areas. They seem to be well-adapted for using mosaics of cutover and mature forest and have gained most from forest destruction. Examples of comparable species that flourish in disturbed areas in the Neotropics have been compiled by Leck (1985), but the number and diversity of those species that are restricted to human-modified habitats is low. On the other hand, a large number of forest-dwelling species wintering in the disappearing humid forests of Central America and returning to their breeding grounds in reduced numbers indicate a loss of their wintering habitat (e.g. Fitzpatrick 1982; Rappole and Morton 1985; Terborgh 1989). The discussion whether migrants prefer secondary vs primary habitats has become even more complicated by results from Rappole and Morton (l.c.). They found that habitat alterations from mature to various forms of disturbed forest and secondary growth were associated not only with drastic decreases in total numbers of forest-dwelling migrant and resident species and individuals but also with changes in the behaviour of some migrants from territoriality to nomadism. Nomads and "floaters" found in the degraded vegetation can create a false sense of the habitat's suitability for those species.

Martin (1985) has recently provided evidence that frugivorous migrants in Panama are associated more with young tropical second growth than with mature forests especially during their premigratory fattening period due to a better fruit supply in the former. There are significantly more bird-dispersed fruit species in younger than in older forest, and the average size of fruits in the younger habitats is smaller (a distinct advantage for the migrants which are generally "small-mouthed" as compared to resident tropical fruit-eaters; Leck 1987). Although the relative amount of available, useable food in primary forest vs edge is unknown for most species of migrants and residents, insects seem to be more abundant in edge and second-growth situations (e.g. Janzen 1973; Hutto 1980). The reason why resident insectivorous birds inhabiting lowland evergreen forests fail to exploit edge and second-growth situations is not fully understood. It has been speculated that this might be caused by an innate "mental conservatism" (Willis 1966; Terborgh and Weske 1969; Lack 1971) which may have evolved in response to high nest predation rates in edge situations (Gates and Gysel 1978), whereas migrants might be less worried about nest predation because they do not breed. Generally, migrants occur in a wider range of habitats than residents (e.g. Greenberg 1985). This notion is supported by our findings on turdid chats in Africa where the residents occurred in a smaller range of habitats than the migrants mainly due to peculiar nest-site requirements. Migrants may be better able to explore alternative habitats and to incorporate assessments of food availability into their settling response because they are not restricted by breeding requirements.

4 Competition

The role of competition in migrants in the non-breeding season has recently been reviewed by Greenberg (1986). It appears that competition results in spatial separation (winter displacement) and in niche shifts, but most of the evidence is indirect (Table 1). For example, age- or sex-related geographic separation of populations of a migrant species is sometimes considered evidence for the role of intraspecific competition. Two other patterns of spatial separation of winter ranges offer compelling, but not necessarily convincing, evidence for interspecific competition: Allopatric/parapatric winter ranges such as those found in the genera *Dendroica, Empidonax* and *Vireo* are interpreted to have evolved through migrant-migrant competition; and an inverse relationship between native and migrant species numbers, which has been shown by Slud (1976) to hold at a global level, may be the result of resident-migrant competition. Several studies indicate that competition (intra- as well as interspecific) is more intense on the wintering grounds due to lower food availability (e.g. Rappole 1988) but food demand is higher in the breeding grounds. In many migrant species, males as well as females defend territories and in some species floaters have been shown to replace

Table 1. Summary of possible effects of competition at intraspecific and interspecific levels

	Indirect	Direct
Intraspecific		
age – related	Winter displacement	Interactions
intersexual	Niche shift	
Interspecific		
migrant – migrant	Winter displacement Niche shift	Interactions
resident – migrant	Inverse species diversity Niche shift	Interactions

removed territory owners (e.g. Rappole l.c.). Long-term winter territoriality is quite common, primarily in understory and terrestrial species (e.g., in *Acrocephalus, Hippolais, Luscinia* spp., Brosset 1971; Kelsey 1989). Additionally, short-term site defense has been reported for passerines that forage on fruit, nectar and localized sources of insects (e.g. Greenberg 1985).

Few examples of intersexual habitat segregation (i.e., males and females inhabit different vegetation structures) have been documented for long-distance migrants (e.g. eastern great reed warbler *Acrocephalus arundinaceus orientalis*, Nisbet and Medway 1972; hooded warbler *Wilsonia citrina*, Rappole and Warner 1980; Lynch et al. 1985). These differences could be caused either by greater dominance of males or by intrinsic differences in habitat preferences of the sexes. Recently, Morton et al. (1987) demonstrated that the original habitat segregation between the sexes in hooded warblers was maintained after removal of some males. Rappole (1988), in reviewing intra- and intersexual competition in migratory passerines, suggested that the partially non-overlapping use of microhabitats by males and females on the wintering grounds results from slight structural differences between the sexes and greater inter- and intraspecific competition for resources relative to the breeding grounds.

With respect to interspecific competition, I will not discuss the wealth of indirect information which has led to the formulation of the two hypotheses that migrant-migrant and resident-migrant interactions have evolutionarily shaped the winter distribution of migrant birds. Instead, I will briefly discuss conclusions which have been drawn from ecological investigations of how residents and migrants might influence each other in habitat selection and resource use (niche shift). Again, most of the evidence is indirect (e.g., Hutto 1980 found negative correlations between the numbers of migrants and residents over 13 habitat types and Keast 1980a and Terborgh and Faaborg 1980 reported within-habitat avoidance).

A crucial point in the evaluation of competition and in the examination of the ecological impact of migrants is their relative proportion in the winter quarters, i.e., how much of the species composition, of the density, or of the bird biomass, migrants constitute. Brown et al. (1982) give some new estimates for Palaearctic birds in Africa and conclude that their role as possible food competitors for residents has often been overstressed. They found that on a continent-wide basis and considering the entire year, only one bird in 10–20 was a migrant. However, the local effect of migrants is much greater than this as they can regionally and seasonally outnumber local residents, especially in some non-passerine groups. The relative contribution of Palaearctic passerines to the overall bird biomass in Africa varies with habitat from 4% in papyrus swamps to 50% in secondary growth of *Lantana* (Britton 1974, 1978). Comparably high figures have been reported from Jamaica from riverine forests (Gochfeld 1985).

Lack (1971) considered 10% of the Palaearctic migrants in Africa to be potential competitors for other wintering species and Moreau (1972) considered 28% of migrants to be potential competitors for residents.

Some attempts have been made to analyze niche relationships in guilds of migrant birds or to track a single species from its breeding area to its wintering grounds. For example, Bennett (1980) analyzed the foraging behaviour of the American redstart (*Setophaga ruticilla*) on different breeding and wintering sites in comparison with local, potential competitors and found that the redstart's niche breadth varied inversely with the number of species in the guild.

Chipley (1977) and Rabøl (1987) investigated the ecology of foliage gleaning migrants (mainly parulid warblers, willow warbler *Phylloscopus trochilus*) in their winter quarters in semi-open oak woodland in Colombia and in *Acacia* savanna in Kenya respectively.

Both found that the resident insectivorous species ecologically closest to the wintering warblers showed a significant shift in foraging height preferences between the period of migrant presence and absence in that residents foraged lower when the migrants were present. Another investigation of possible overlap of residents and migrants in a dry tropical forest in Mexico by Waide (1981) supports the view that canopy species tend to overlap more in their foraging behaviour than understory species. (Birds in the canopy are likely to make a greater use of leaf upper surfaces; Thiollay 1988.) Changes in foraging behaviour of residents occurred (they foraged lower) but substantial reductions in niche overlap did not occur and the resident community as a whole seemed to be little affected by wintering migrants.

Greenberg (1986) has critically commented on conclusions which have been drawn from some of these studies and has pointed out that a problem common to all of these analyses is that niche breadth of a migrant may change in response to variation in vegetation and prey diversity and not to the number of bird species. It is often the case that the resource spectrum is so drastically changed that interseasonal comparisons are weak (capturing tropical arthropods may take greater foraging specialization than gleaning budworms in spruce forests irrespective of competitive pressure) and changes in foraging parameters in tropical species may be associated with changes in insect distributions in periods of migrants' absence and presence.

Overt interspecific interactions which can result in:
1. interspecific territoriality (e.g. Leck 1972a);
2. interspecific dominance hierarchies (e.g. Greenberg 1984a);
3. or in supplantings in mixed species flocks are poorly documented.

The most detailed analyses stem from our own study on turdid chats in Africa. Interspecific interactions are frequent in this group, both among the wintering Palaearctic chats and between African and Palaearctic birds, because the territories of different species can overlap a great deal. In addition to interactions between the eight members of the "narrow guild", we could also observe interspecific interactions with other open country birds (species with similar ecology such as larks and pipits, see Table 2). We divided the species into three status categories: (1) African residents, species always present in the study areas; (2) African visitors; and (3) Palearctic migrants.

A total of 101 territorial disputes between African and Palaearctic species and 401 among only Palaearctic species have been observed.

Table 2. Dominance in interspecific encounters between open country birds in East Africa

Species	Status	Percentage of attacks	
		African-Palaearctic	Palaearctic-Palaearctic
Ant chat, *Myrmecocichla aethiops*	Afr. Resident	100	
Capped wheatear, *Oenanthe pileata*	"	100	
Schalow's wheatear, *O. lugens schalowi*	"	95.3	
Rufous-naped lark, *Mirafra africana*	"	100	
Fiscal shrike, *Lanius collaris*	"	100	
African pipits, *Anthus* sp.	Afr. Visitor	0	
Red-capped lark, *Calandrella cinerea*	"	0	
Isabelline wheatear, *O. isabellina*	Palaearct. Migrant	3.8	94.9
Wheatear, *O. oenanthe*	"	0	47.2
Pied wheatear, *P. pleschanka*	"	3.1	13.4
Rock thrush, *Monticola saxatilis*	"	0	100
Whinchat, *Saxicola rubetra*	"	0	6.8
Yellow wagtail, *Motacilla flava*	"	0	0
Red-throated pipit, *Anthus cervinus*	"	0	–

In Table 2, relative specific aggression (i.e., the percent of attacks in interspecific clashes) is shown. Thus, 100% means a species was always the aggressor, 0 a species was always subordinate/attacked. The data clearly show that all the African residents were dominant over the Palaearctic migrants but that this was not the case with African visitors. Within the Palaearctic migrants, there was considerable variation in dominance with yellow wagtail, whinchat and pied wheatear being the most subordinate species.

Because there is a general tendency for larger species to be dominant in aggressive encounters, we compared body weight in relation to dominance status in interspecific encounters. The results of interactions between African residents and Palaearctic migrants are shown in Fig. 1. Most of the African species are larger but, interestingly, there is a tendency to be dominant irrespective of body size. In contrast to this observation, dominance in interspecific encounters among the Palaearctic migrants was strictly size-dependent (Fig. 1). So it seems that African residents do have a real "home advantage" in interspecific disputes.

Other studies have also shown that migrants are at least sometimes subordinate to residents and are forced to feed at peripheral or less productive foraging sites whenever residents are present (Willis 1966, 1980; Leck 1972b; Tramer and Kemp 1980; Des Granges and Grant 1980) but such findings are not universal. Chipley (1977), in a very limited number of observations, found that migrant warblers were neither dominant nor subordinate to their resident equivalents. Johnson (1980) found that migrants were subordinate in their social relations with residents in mixed species flocks, but residents did not agonistically exclude them from the flocks (Powell 1980); clearly, more investigations of different flock systems are needed to draw generalizations.

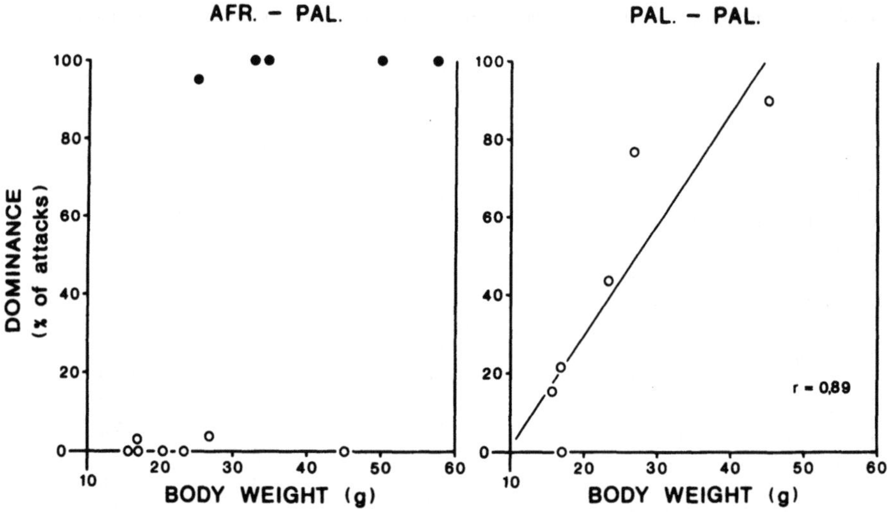

Fig. 1. Relationship between body weight and dominance in interspecific encounters between African residents and Palaearctic migrants and among Palaearctic migrants only. *Filled symbols* African residents; *open symbols* Palaearctic migrants

5 Food

Hutto (1985) has stressed the importance of food availability in habitat selection, especially for the actual settling reaction of migrants. During the winter months in the tropics migrants are faced with more diversified food sources, i.e., an increase of alternative/new food sources, of which many are sporadic, patchy or ephemeral. These sources include fruits, berries, nectar, army ant swarms, grass/bush fires and termite-emergence situations and other food flushes triggered by rain. Further, such seemingly stable resources as insects in tropical forests, may fluctuate considerably from year to year as well as seasonally (e.g. Fogden 1972; Willis 1976; Wolda 1978a,b, 1983; Leigh et al. 1982; Sabatier 1985). One hallmark of success of migrants is their ability to switch food resources opportunistically, i.e., to alternatively exploit resources as they become available (Willis 1966; Sick 1968; Morse 1971; Karr 1976). The turdid chats provide an impressive example of such an ability. All wintering species opportunistically use burnt grasslands where such prey as subterranean ants and termites have been more or less unaffected, and are easy to see. We compared the habitat use of wintering migrants in burnt areas to use in unburnt areas. In Fig. 2, the rate of movement of the three wintering *Oenanthe* species is shown. Time is plotted against distance travelled. The steeper the line, the faster the movement of a particular species. From the average number of stops and the mean time spent stationary, we constructed so-called saw-tooth curves (Cody 1968). It is apparent that all three wintering species were more active on burnt areas (more stops), they spent less time stationary and finally, their food-intake rates (not shown here) were higher. This indicates that migrants gain from switching to burnt grasslands.

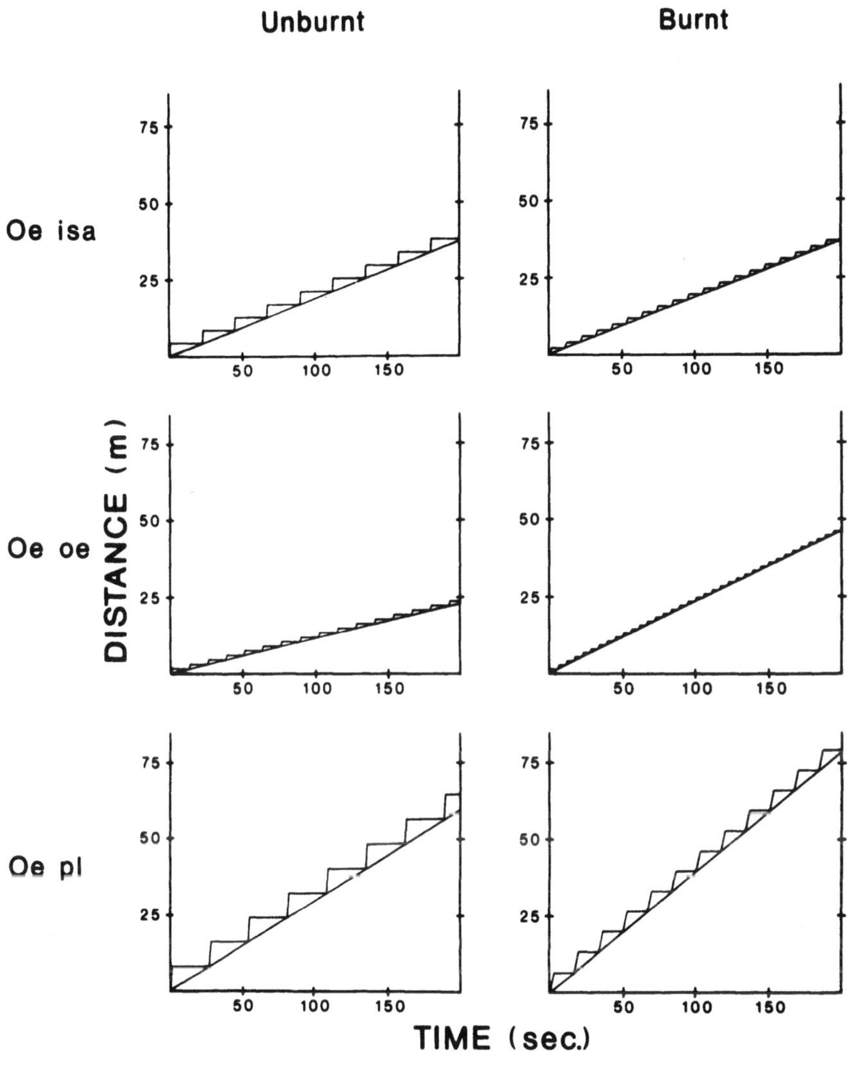

Fig. 2. Feeding curves of three wintering wheatear species in an East African savanna at unburned and burnt sites. Shown is the mean rate of movement and the mean number of stops. Oe isa, *Oenanthe isabellina*; Oe oe *O. oenanthe*; Oe pl *O. pleschanka*

The impact of migrants on fruiting trees and shrubs can be substantial in the Neotropics as well as in the Palaeotropics, e.g., up to 60% of feeding visits to fruiting trees can be by migrants (Leck 1972c).

Although most of the migrants remain more or less insectivorous, many become highly omnivorous. Much more attention has been paid to the mechanisms underlying this change in diet (e.g. Berthold 1976; Bairlein, this Vol.) than to ultimate questions on the evolutionary significance of this phenomenon. Highly omnivorous diets may be chosen by species which are unable to persist as strict insectivores during the non-breeding season (Greenberg 1985). One correlate of becoming omnivorous seems to be body size. Greenberg (l.c.) has shown that larger species (weighing > 15 g) of Neotropical migrants have a much greater chance of being omnivorous: of large migrant species 11 of 17 are omnivorous while migrants weighing 15 g or less (small) only 8 of 40 are omnivorous ($p < 0.05$). The same pattern holds true for Palaearctic migrants. Because of the lack of information on winter diet I have analyzed data from a SW-German ringing station (Brensing 1977) and found that more of the larger species become omnivorous (12 of 17 versus 4 of 17, $p < 0.02$). Larger species have less chance to subsist on insects in the tropics for two reasons: In the tropics larger insects are more seasonal than smaller ones (Lack 1986b; Gradwohl and Greenberg 1982) and large, active, tropical insects might be disproportionately more difficult to capture for the less specialized migrant species (Greenberg 1981, 1985).

Comparing ecological correlates of migrants and residents in a tropical African savanna, Lack (1986a) found that residents took a wider range of food types than migrants, contrary to the prediction by Herrera (1978) that migrants were food generalists and residents food specialists. Perhaps this is because the lean season in this study area is very harsh and even the wet seasons are rather unpredictable, thus forcing the residents to take a wide spectrum of foods. Also, Thiollay (1988) reports that the diversity of prey items was high in a guild of small foliage gleaners in a Neotropical forest. In a dry Neotropical forest Waide (1981) found that wintering migrants used food that occurred in both the resident's breeding and non-breeding seasons, but that this source was not utilized by the residents in normal years.

6 Behaviour

The breakdown of the monogamous territorial breeding system of most migrant passerines and an increase in plasticity of foraging behaviour in the non-breeding season (e.g. Rabenold 1980; Benett 1980) can determine the social behaviour, and hence habitat selection of a species during the winter. Correlations between foraging ecology and social behaviour are well documented but have been reviewed only for Neotropical migrants by Greenberg (1985). Most migrant species occur solitarily during the winter, but social behaviour can vary within one species in relation to food type and food distribution used, e.g. single Tennessee warblers (*Vermivora peregrina*) defend discrete productive flowers, but visit mass floral displays in large flocks (Tramer and Kemp 1979; Greenberg 1986). Similar changes have been reported by Pearson (1978) for wintering *Sylvia* warblers in Africa. In general, occurrence of single species flocks is rather uncommon, but has been reported for about a dozen Neotropical migrant species of tyrannids, vireos, thrushes, parulids, icterids and fringillids, which all are highly omnivorous. Intraspecifically, gregarious species are those which are the most omnivorous. Thus, in a species such as the eastern kingbird (*Tyrannus tyrannus*), which joins in nomadic flocks during the winter, fruiting trees on river and lake edges can determine the habitat use of the species. In a Mexican cloud forest Thiollay (1984) discovered that foraging rates of migrants were much higher than those of similar resident species. Comparing foraging success of tropical vs temperate foliage gleaning forest birds Thiollay (1988) found that foraging speed was significantly higher in temperate than in tropical species during their respective breeding seasons. Migrants may more often use their wings during habitat use according to Lack (1985), who found that in an African savanna, migrants took terrestrial arthropods mainly by pouncing on them, but the residents took them mainly by gleaning from the ground.

Also, in our turdid guild some wintering species, which are subordinate to residents, can cope with the dominant local species by making more flights and by frequently changing perches. This restless/fugitive strategy allows the migrants to: squeeze in, or even to superimpose their territories over that of others; to pounce to the ground; and to escape the attacks of the dominant species. Fitzpatrick (1980) reported that wintering eastern kingbirds escaped harassment by resident kingbirds, kiskadees and other large flycatchers in a similar way, i.e. by fleeing or retreating into the midst of the flock.

Several papers indicate that wintering migrants forage in a more generalized and flexible manner than their resident tropical counterparts (e.g. Tramer and Kemp 1980; Lack 1986a). A few migrants stand out as being extremely generalized (like the yellow-rumped warbler *Dendroica coronata*). At the opposite extreme, a few species are highly specialized on typical, tropical microhabitats. For example, worm-eating (*Helmitheros vermivora*) and the golden-winged warbler (*Vermivora chrysoptera*) forage on aerial leaf litter and the latter also forages in epiphytes (Greenberg 1986, 1987a). Greenberg (1987b) has shown that young worm-eating warblers have an innate preference for dead curled leaves which they can use substantially only during their stay in the tropics.

Whether a migrant forages more as a specialist or a generalist could be determined by the intensity of its neophobia. Experiments by Greenberg (1983, 1984c) revealed that generalized and plastic foragers (like bay-breasted warblers *Dendroica castanea*) were much less hesitant in feeding at novel microhabitats in the laboratory than specialists (chestnut-sided warblers *Dendroica pensylvanica*). In other experiments Greenberg (1984b) could show that an ontogenetic increase in neophobia restricts immature warblers to foraging at microhabitats most familiar and similar to those experienced as juveniles. It is tempting to suggest that differences in foraging flexibility between residents and migrants might also be caused by differences in neophobia and it would be most rewarding to search for physiological correlates of these differences.

7 Morphology

The following preliminary generalizations on differences in the morphology between small migrant and resident land birds have been made. Migrants are morphologically more uniform, residents more diverse, especially in their bill dimensions (Cox 1968). Migrants have shorter, and in the group of foliage gleaning birds broader bills as well, than their lowland tropical counterparts

(Greenberg 1981). Greenberg has speculated that these distinct differences in bill morphology might be determined by the type of insects that comprise the largest size classes of potential prey. These large insects are particularly important because they generally form the bulk of the nestling diet. Thus, long-distance migrants have summer-selected bill morphology, best adapted to feed on more sluggish caterpillars and grubs during summer (Thiollay 1988) and foraging on small conspicuous arthropods during winter, whereas in contrast fringillids appear to have winter-selected bills (Fretwell 1972). In tropical forest, mobile Orthoptera and Homoptera are the most abundant large, soft-bodied insects. They form an important resource for foliage gleaning birds during the rainy season (the breeding season of tropical species). Long bills are thought to close more rapidly than shorter, broader bills (Lederer 1975, 1980). Thus, tropical species having such long, fast bills seem to be adapted for efficient capture of active prey (Greenberg l.c.).

It has long been known that migratory species have more attenuated wings than residents (longer, narrower, more pointed). However, the exact ecological/behavioural correlates and the significance of these wing differences in habitat use have not yet been investigated in detail. Looking for meaningful correlations between morphology and behaviour in our turdid group we found that most migrants had lower wing loadings and this was negatively correlated both to the number of flights ($r = -0.76$, $p < 0.05$) and number of pounces to the ground ($r = -0.75$, $p < 0.05$). All migrant species had longer wing spans than the residents and this trait was positively correlated to the number of pounces ($r = 0.82$, $p < 0.05$). This means that differences in wing shape allow migrants to have a more aerial locomotion, and makes available certain modes of habitat use less practically utilized by residents.

8 Conclusions

Some of the general ecological differences in habitat utilization between residents and migrants in their winter quarters are summarized in Table 3:

1. Migrants resort more to drier, more open and peripheral habitats; they overlap strongly with tropical canopy and edge species.
2. Migrants usually occur in a wider range of habitats; they are more eurytopic than their tropical equivalents.
3. During competitive interactions, residents seem to be dominant over migrants, but further observations are needed to clarify this relationship. On the other hand, niche shifts of residents have been reported to be induced by migrants and Miller (1963) has speculated that the presence of insectivorous migrants may even influence the breeding season of residents in an equatorial cloud forest.
4. Higrants often use sporadically available food sources (sometimes in an opportunistic way) and food sources generally unutilized by residents, whereas residents rely on more stable resources. As a result of this difference,

Table 3. Summary of factors affecting differences in habitat use between resident and migrants

	Resident	Migrant
Habitat		Drier, more open, peripheral
Intra-, intersp. range of habitat types	Small	Wide, eurytopic
Competitive interactions	Dominant Niche shift	Subordinate Niche shift
Food	Stable resources	Sporadically available surplus "unused"
Range of food types	Wide?	Narrow?
of foliage – gleaners	Mobile	Less mobile
Behaviour		
Neophobia	?	?
Intrasp. repertoire of foraging tactics	Narrow	High
Feeding tactics	(Ground gleaning)	Using wing (pouncing)
Morphology	Diversified	Uniform
Body size	Larger	Smaller
Bill	Longer	Shorter
Bill length variability	High	Low
Wing	Shorter distal segments	Attenuated

residents may feed on a wider range of food types in seasonal tropical habitats.

5. Resident foliage gleaners in the New World are generally better adapted to feed on more mobile and more hidden prey (Thiollay 1988).
6. Differences in the intensity of neophobia between residents and migrants have not yet been investigated.
7. According to Herrera (1978), the repertoire of feeding tactics within one species is higher in migrants than in residents. There are also some indications that migrants may use the wings more often for habitat utilization than residents (pouncing versus ground gleaning).

The following morphological differences have been found:

1. The average migrant is smaller than equivalent residents (Tramer 1974; Karr 1976).
2. Residents are morphologically more diverse, migrants more uniform.
3. Residents have longer bills than migrants and bill length variability is higher in residents than in migrants.
4. Migrants have more attenuated wings, residents shorter distal wing segments.

Some differences between migrants and residents concerning several aspects of habitat selection have been portrayed in this review. Migrants seem to be better adapted to exploit environmental irregularities and have some of the basic characteristics of the island-colonizing species described by Ricklefs and Cox

(1972); Diamond (1975) and O'Connor (1981, this Vol.). This does not necessarily imply that migrants have to be regarded as opportunistic interlopers into the tropics. Especially in the Western Hemisphere the facts

- that nearly half of the 332 species of birds which breed in the Nearctic and winter in the Neotropics also have breeding populations there,
- that there are tropical resident congeners for 78% of the Nearctic migrant species,
- that essentially all Neotropical habitat types are used by the migrants, and
- that only few interactions between migrants and residents occur support the hypothesis that many species of migrants essentially are tropical species (Rappole et al. 1983).

Far more data from studies on well-defined guilds and communities will be necessary to make more conclusive assertions about similarities or differences between the Palaearctic-African and Nearctic-Neotropic bird migration systems (Keast 1980b) and for drawing general conclusions on factors determining habitat selection of migrants.

Acknowledgements. Dr. S. Terrill improved the English, Drs J. Haffer, R. Greenberg, J.H. Rappole, S.K. Robinson and J.M. Thiollay commented on an earlier version of the manuscript. The field study in East Africa was supported by the Deutsche Forschungsgemeinschaft.

References

Bairlein F, Leisler B, Winkler H (1986) Morphologische Aspekte der Habitatwahl von Zugvögeln in einem SW-deutschen Rastgebiet. J Ornithol 127:463–473

Bennett SE (1980) Interspecific competition and the niche of the American redstart (*Setophaga ruticilla*) in wintering and breeding communities. In: Keast A, Morton ES (eds) Migrant birds in the Neotropics. Smithsonian, Washington DC, pp 319–335

Berthold P (1976) Animalische und vegetabilische Ernährung omnivorer Singvogelarten: Nahrungsbevorzugung, Jahresperiodik der Nahrungswahl, physiologische und ökologische Bedeutung. J Ornithol 117:145–209

Brensing D (1977) Nahrungsökologische Untersuchungen an Zugvögeln in einem südwestdeutschen Durchzugsgebeit während des Wegzuges. Vogelwarte 29:44–56

Britton PL (1974) Relative biomass of Ethiopian and Palaeartic passerines in West Kenya habitats. Bull Br Ornithol Club 94:108–113

Britton PL (1978) Seasonality, density and diversity of birds of a papyrus swamp in western Kenya. Ibis 120:450–466

Brosset A (1971) Territorialisme et défense du territoire chez les migrateurs paléarctiques hivernant au Gabon. Alauda 39:127–131

Brosset A (1984) Oiseaux migrateurs européens hivernant dans la partie guinéenne du Mont Nimba. Alauda 52:81–101

Brown LH, Urban EK, Newmann K (1982) The birds of Africa, vol 1. Academic Press, London

Chipley RM (1977) The impact of wintering migrant wood warblers on resident insectivorous passerines in a subtropical Colombian oak forest. Living Bird 1976 15:119–141

Cody ML (1968) On the methods of resource division in grassland bird communities. Am Nat 102:107–147

Cornwallis L (1975) The comparative ecology of eleven species of wheatear (genus *Oenanthe*) in S.W. Iran. PhD Thesis, University Oxford

Cox GW (1968) The role of competition in the evolution of migration. Evolution 22:180–192

Curry-Lindahl K (1981) Bird migration in Africa, vol 2. Academic Press, London

Diamond JM (1975) Assembly of species communities. In: Cody ML, Diamond JM (eds) Ecology and evolution of communities. Harvard University Press, Cambridge, Mass, pp 342–344

Fitzpatrick JW (1980) Wintering of North American tyrant flycatchers in the neotropics. In: Keast A, Morton ES (eds) Migrant birds in the Neotropics. Smithsonian, Washington DC, pp 67–78

Fitzpatrick JW (1982) Northern birds at home in the tropics. Nat Hist 91:40–47

Fitzpatrick JW, Robinson SK (in prep) Distribution and abundance of North temperate migrant land birds in the Amazon basin and eastern Andes

Fogden MPL (1972) The seasonality and population dynamics of equatorial forest birds in Sarawak. Ibis 114:307–343

Fretwell S (1972) Populations in a seasonal environment. Princeton University Press, NJ (Monogr Pop Biol No 6)

Gates JE, Gysel LW (1978) Avian nest dispersion and fledging success in field-forest ecotones. Ecology 59:871–883

Gatter W, Mattes H (1987) Anpassungen von Schafstelze *Motacilla flava* und afrikanischen Motacilliden an die Waldzerstörung in Liberia (Westafrika). Verh Ornithol Ges Bayern 24:467–477

Gochfeld M (1985) Numerical relationships between migrant and resident bird species in Jamaican woodlands. In: Buckley PA, Foster MS, Morton ES, Buckley FG (eds) Neotropical ornithology. Ornithol Monogr 36:654–662

Gradwohl J, Greenberg R (1982) The breeding season of antwrens on Barro Colorado Island. In: Leigh E, Rand AS, Windsor D (eds) The ecology of a tropical forest: seasonal rhythms and long-term changes. Smithsonian, Washington DC

Granges JL des, Grant PR (1980) Migrant hummingbirds' accomodation into tropical communities. In: Keast A, Morton ES (eds) Migrant birds in the Neotropics. Smithsonian, Washington DC, pp 395–409

Greenberg R (1979) Body size, breeding habitat, and winter exploitation systems in *Dendroica*. Auk 96:756–766

Greenberg R (1981) Dissimilar bill shapes in New World tropical versus temperate forest foliage-gleaning birds. Oecologia (Berl) 49:143–147

Greenberg R (1983) The role of neophobia in determining the degree of foraging specialization in some migrant warblers. Am Nat 122:444–453

Greenberg R (1984a) The winter exploitation systems of bay-breasted and chestnut-sided warblers in Panama. Univ Calif Publ Zool 116:1–107

Greenberg R (1984b) Neophobia in the foraging-site selection of a neotropical migrant bird: an experimental study. Proc Natl Acad Sci USA 81:3778–3780

Greenberg R (1984c) Differences in feeding neophobia in the tropical migrant wood warblers *Dendroica castanea* and *D. pensylvanica*. J Comp Psychol 98:131–136

Greenberg R (1985) Social behavior and foraging ecology of neotropical migrants. Acta XVIII Congr Intern Ornithol Moscow, pp 648–653

Greenberg R (1986) Competition in migrant birds in the nonbreeding season. In: Johnston RF (ed) Curr Ornithol 3:281–307

Greenberg R (1987a) Seasonal foraging specializations in the worm-eating warbler. Condor 89:158–168

Greenberg R (1987b) Development of dead leaf foraging in a tropical migrant warbler. Ecology 68:130–141

Herrera CM (1978) Ecological correlates of residence and nonresidence in a Mediterranean passerine bird community. J Anim Ecol 47:871–890

Hilty SL, Brown WL (1986) A guide to the birds of Colombia. Princeton University Press

Hutto RL (1980) Winter habitat distribution of migratory land birds in western Mexico, with special reference to small foliage-gleaning insectivores. In: Keast A, Morton ES (eds) Migrant birds in the Neotropics. Smithsonian, Washington DC, pp 181–203

Hutto RL (1985) Habitat selection by nonbreeding migratory land birds. In: Cody ML (ed) Habitat selection in birds. Academic Press, London, pp 455–476

Ivanitzky VV (1980a) Interspecific relations of sympatric species of wheatears (*Oenanthe*, Turdinae, Passeriformes). I. Ecological prerequisites of competition and spatial structure of a mixed population. Zool Zh 59(4):587–597 (in Russian)

Ivanitzky VV (1980b) Interspecific relations of sympatric species of wheatears (*Oenanthe*, Turdinae, Aves). II. Behavioral aspects of coexistence in closely related species. Zool Zh 59(5):739–749 (in Russian)
Janzen DH (1973) Sweep samples of tropical foliage insects: effects of seasons, vegetation types, elevation, time of day, and insularity. Ecology 54:687–708
Johnson TB (1980) Resident and North American migrant bird interactions in the Santa Marta highlands, northern Colombia. In: Keast A, Morton ES (eds) Migrant birds in the Neotropics. Smithsonian, Washington DC, pp 239–247
Jones P (1985) The migration strategies of Palearctic passerines in West Africa. Rep XIV ICBP Conf 1983, pp 9–21
Karr J (1976) On the relative abundance of migrants from the north temperate zone in tropical habitats. Wilson Bull 88:433–458
Karr JR, Freemark KE (1983) Habitat selection and environmental gradients: dynamics in the "stable" tropics. Ecology 64:1481–1494
Keast A (1980a) Spatial relationships between migratory parulid warblers and their ecological counterparts in the Neotropics. In: Keast A, Morton ES (eds) Migrant birds in the Neotropics. Smithsonian, Washington DC, pp 109–130
Keast A (1980b) Synthesis: ecological basis and evolution of the Nearctic-Neotropical bird migration system. In: Keast A, Morton ES (eds) Migrant birds in the Neotropics. Smithsonian, Washington DC, pp 559–575
Keast A, Morton ES (eds) (1980) Migrant birds in the Neotropics. Ecology, behavior, distribution and conservation. Smithsonian, Washington DC
Kelsey MG (1989) A comparison of the song and territorial behaviour of a long-distance migrant, the Marsh Warbler *Acrocephalus palustris*, in summer and winter. Ibis 131:403–414
Lack D (1971) Ecological isolation in birds. Blackwell Sci, Oxford
Lack PC (1983) The movements of palaearctic landbird migrants in Tsavo East National Park, Kenya. J Anim Ecol 52:513–524
Lack PC (1985) The ecology of the landbirds in Tsavo East National Park, Kenya. Scopus 9:2–24, 57–96
Lack PC (1986a) Ecological correlates of migrants and residents in a tropical African savanna. Ardea 74:111–119
Lack PC (1986b) Diurnal and seasonal variation in biomass of arthropods in Tsavo East National Park, Kenya. Afr J Ecol 24:47–51
Lack PC (1987) The structure and seasonal dynamics of the bird community in Tsavo East National Park, Kenya. Ostrich 58:9–23
Leck CF (1972a) The impact of some North American migrants at fruiting trees in Panama. Auk 89:842–850
Leck CF (1972b) Seasonal changes in feeding pressures of fruit and nectar eating birds in the neotropics. Condor 74:54–60
Leck CF (1972c) Observations of birds at *Cecropia* trees in Puerto Rico. Wilson Bull 84:498–500
Leck CF (1985) The use of disturbed habitats by North American birds wintering in Mexico. Biotropica 17:263–264
Leck CF (1987) Habitat selection in migrant birds: seductive fruits. Trends Ecol Evol 2:33
Lederer R (1975) Bill size, food size and jaw forces of insectivorous birds. Auk 92:385–387
Lederer R (1980) Prey capture by flycatchers and the importance of morphology to behavior. Sociobiology 5:43–46
Leigh EG, Rand AS, Windsor DM (eds) (1982) The ecology of a tropical forest. Seasonal rhythms and long-term changes. Smithsonian, Washington DC
Leisler B, Heine G, Siebenrock KH (1983) Einnischung und interspezifische Territorialität überwinternder Steinschmätzer (*Oenanthe isabellina, O. oenanthe, O. pleschanka*) in Kenia. J Ornithol 124:393–413
Leisler B, Heine G, Drews C, Siebenrock KH (in press) Interspecific aggression and coexistence between overwintering and local chat-like trushes in Kenya. Proc Seventh Pan-African Ornithol Congr
Lynch JF, Morton ES, Voort M van der (1985) Habitat segregation between the sexes of wintering hooded warblers (*Wilsonia citrina*). Auk 102:714–721

Martin TE (1985) Selection of second-growth woodlands by frugivorous migrating birds in Panama: an effect of fruit size and plant diversity? J Trop Ecol 1:157–170

Miller AH (1963) Avifauna of an American equatorial cloud forest. Univ Calif Publ Zool 66:1–72

Moreau RE (1972) The Palaearctic-African bird migration systems. Academic Press, London

Morel GJ (1973) The Sahel Zone as an environment for palaearctic migrants. Ibis 115:413–417

Morel GJ, Morel MY (1978) Recherche écologique sur une savane sahelienne du Ferlo septentrional, Sénégal. Etude d'une communauté avienne. Cah ORSTOM (sér Biol) 13:3–34

Morse D (1971) The insectivorous bird as an adaptive strategy. Annu Rev Ecol Syst 2:177–200

Morton ES (1980) Adaptations to seasonal changes by migrant landbirds in the Panama Canal Zone. In: Keast A, Morton ES (eds) Migrant birds in the Neotropics. Smithsonian, Washington DC, pp 437–457

Morton ES, Lynch JF, Young K, Mehlkop P (1987) Do male hooded warblers exclude females from nonbreeding territories in tropical forests? Auk 104:133–134

Nisbet ICT, Medway Lord (1972) Dispersion, population, ecology and migration of eastern great reed warblers *Acrocephalus orientalis* in Malaysia. Ibis 114:451–494

O'Connor RJ (1981) Comparisons between migrant and non-migrant birds in Britain. In: Aidley DJ (ed) Animal migration. Soc Exp Biol Seminar Ser 13 Cambridge Univ Press, pp 167–195

Orejuela JE, Raitt RJ, Alvarez H (1980) Differential use by North American migrants of three types of Colombian forests. In: Keast A, Morton ES (eds) Migrant birds in the Neotropics. Smithsonian, Washington DC, pp 253–264

Pearson DJ (1978) The genus *Sylvia* in Kenya and Uganda. Scopus 2:63–71

Pearson DJ, Backhurst GC (1976) The southward migration of palaearctic birds over Ngulia, Kenya. Ibis 118:78–105

Potapova EG, Panov EN (1977) Structural patterns of jaw apparatus in wheatears of the genus *Oenanthe*, with special reference to the ways of decrease of food competition between closely related species. Zool Zh 56(5):743–752 (in Russian)

Powell GVN (1980) Migrant participation in neotropical mixed species flocks. In: Keast A, Morton ES (eds) Migrant birds in the Neotropics. Smithsonian, Washington DC, pp 477–483

Rabenold KN (1980) The black-throated green warbler in Panama: geographic and seasonal comparison of foraging. In: Keast A, Morton ES (eds) Migrant birds in the Neotropics. Smithsonian, Washington DC, pp 297–307

Rabøl J (1987) Coexistence and competition between overwintering willow warblers *Phylloscopus trochilus* and local warblers at Lake Naivasha, Kenya. Ornis Scand 18:101–121

Rappole JH (1988) Intra- and intersexual competition in migratory passerine birds during the nonbreeding season. Acta XIX Intern Ornithol Congr, Ottawa, pp 2308–2317

Rappole JH, Morton ES (1985) Effects of habitat alteration on a tropical avian forest community. In: Buckley PA, Foster MS, Morton ES, Buckley FG (eds) Neotropical ornithology. Ornithol Monogr 36:1013–1021

Rappole JH, Warner DW (1980) Ecological aspects of migrant bird behavior in Veracruz, Mexico. In: Keast A, Morton ES (eds) Migrant birds in the Neotropics. Smithsonian, Washington DC, pp 353–393

Rappole JH, Morton ES, Lovejoy TE III, Ruos JL (1983) Nearctic avian migrants in the Neotropics. Publ US Dept Interior, Fish and Wildlife Service

Ricklefs RE, Cox GW (1972) Taxon cycles in the West Indian avifauna. Am Nat 106:195–219

Robinson SK, Terborgh J, Fitzpatrick JW (1988) Habitat selection and relative abundance of migrants in south eastern Peru. Acta XIX Intern Ornithol Congr, Ottawa, pp 2298–2307

Sabatier D (1985) Saisonnalité et déterminisme du pic de fructification en forêt guyanaise. Rev Ecol (Terre Vie) 40:289–320

Sick H (1968) Vogelwanderungen im kontinentalen Südamerika. Vogelwarte 24:217–243

Slud P (1976) Geographic and climatic relationships of avifaunas with special reference to comparative distribution in the neotropics. Smithson Contrib Zool 212:1–149

Terborgh JW (1980) The conservation status of Neotropical migrants: present and future. In: Keast A, Morton ES (eds) Migrant birds in the Neotropics. Smithsonian, Washington DC, pp 21–30

Terborgh JW (1989) Where have all the birds gone? Princeton University Press, NJ

Terborgh JW, Faaborg JR (1980) Factors affecting the distribution and abundance of North

American migrants in the eastern Caribbean region. In: Keast A, Morton ES (eds) Migrant birds in the Neotropics. Smithsonian, Washington DC, pp 145–155

Terborgh JW, Weske JS (1969) Colonization of secondary habitats by Peruvian birds. Ecology 50:765–782

Thiollay JM (1984) Essai d'analyse du succès de la chasse chez les oiseaux insectivores arboricoles d'une foret neotropicale. Gerfaut 74:99–125

Thiollay JM (1988) Comparative foraging success of insectivorous birds in tropical and temperate forests: ecological implications. Oikos 53:17–30

Tramer EJ (1974) Proportions of wintering North American birds in disturbed and undisturbed dry tropical habitats. Condor 76:460–464

Tramer EJ, Kemp TR (1979) Diet-correlated variations in social behavior of wintering Tennessee warblers. Auk 96:186–187

Tramer EJ, Kemp TR (1980) Foraging ecology of migrant and resident warblers and vireos in the highland of Costa Rica. In: Keast A, Morton ES (eds) Migrant birds in the Neotropics. Smithsonian, Washington DC, pp 285–296

Ulfstrand S, Alerstam T (1977) Bird communities of *Brachystegia* and *Acacia* woodlands in Zambia. J Ornithol 118:156–174

Waide RB (1981) Interactions between resident and migrant birds in Campeche, Mexico. Trop Ecol 22:134–154

Walsberg GE (1985) Physiological consequences of microhabitat selection. In: Cody ML (ed) Habitat selection in birds. Academic Press, London, pp 389–413

Willis EO (1966) The role of migrant birds at swarms of army ants. Living Bird 5:187–231

Willis EO (1976) Seasonal changes in the invertebrate litter fauna on Barro Colorado Island, Panama. Rev Bras Biol 36:643–657

Willis EO (1980) Ecological roles of migratory and resident birds on Barro Colorado Island, Panama. In: Keast A, Morton ES (eds) Migrant birds in the Neotropics. Smithsonian, Washington DC, pp 205–225

Winkler H, Leisler B (1985) Morphological aspects of habitat selection in birds. In: Cody ML (ed) Habitat selection in birds. Academic Press, London, pp 415–434

Wolda H (1978a) Seasonal fluctuations in rainfall, food and abundance of tropical insects. J Anim Ecol 47:369–381

Wolda H (1978b) Fluctuations in abundance of tropical insects. Am Nat 112:1017–1045

Wolda H (1983) Spatial and temporal variation in abundance in tropical animals. In: Sutton SL, Whitmore TC, Chadwick AC (eds) Tropical rain forest: ecology and management. Blackwell Oxford

Some Ecological Aspects of Migrants and Residents

R.J. O'CONNOR[1]

1 Introduction

The ecology, and particularly the life history ecology, of migrancy and residency in birds has received surprisingly little attention. For some species, such as those migrating into harsh, arctic environments, the brief summer flush of productivity and the intolerable conditions at other times of year offer an obvious explanation superficially requiring little further study. For other species early work, by biologists resident in the temperate zone, asked merely how temperate zone species fitted into the tropical communities in which they wintered before returning to take their "proper" place in their breeding season community. Latterly this discussion has widened to entertain the notion that migrants breeding in the temperate zone may be species of tropical origin only temporarily visiting these breeding grounds (see review in Morse 1980). The importance of reproduction and survival rates and of migration costs in the evolution of migration has also received greater attention recently (Ketterson and Nolan 1976, 1983; Fretwell 1972, 1985; Greenberg 1980; O'Connor 1981, 1985).

Fretwell (1980) points out that three distinct questions as to the evolution of migrancy and residency need answering: (1) why are some species year-round residents; (2) why do some species leave areas where they can winter extremely successfully to breed elsewhere; and (3) why do others leave their breeding grounds to winter elsewhere?

2 Some New Data

To investigate these questions, the breeding biology of some 42 species breeding in Britain have been examined, either as migrants (17 species) or as residents (25 species). Partial migrants, in which some but not all individuals of the species migrate, were classified as residents here. Breeding data were obtained from records of the British Trust for Ornithology's nest record programme whilst population data were obtained from the BTO's Common Birds Census scheme. Procedures were essentially the same as used in a previous study (O'Connor 1985). The nest records data were used to estimate rates of nest and individual loss

[1]Department of Wildlife, University of Maine 240 Nutting Hall, Orono, Maine 04469, USA

E. Gwinner (Ed.) Bird Migration

at egg and chick stages of the nest cycle using the bias-free method of Mayfield (1975). Predation rates were similarly estimated from those cards for which this information was recorded, with appropriate allowance for reporting bias.

3 Hole-Nesting and Residency

Why do any species at all overwinter in harsh areas? Would a resident species perhaps support a larger population if it left, even at some cost to migration, to winter in a milder climate? What benefits accrue to remaining? Von Haartman (1968) postulated that some temperate zone species were crucially dependent on key breeding resources that were in short supply and that winter residency was forced on such species to guarantee them priority of access to those resources once the winter ameliorated. Fretwell (1969, 1980) similarly stressed the importance of site dominance – of being in possession of the resource from the outset for the advantages such possession seems to confer in subsequent ownership disputes – in switching the outcome of competition. Von Haartman (1968) developed his ideas in relation to resident hole-nesting species, for which the possession of a secure but scarce nest hole greatly reduced predation rates and the idea has spread that nest-holes are the crucial resource won by overwintering. Alerstam and Hogstedt (1980) confirmed that in Sweden sheltered nest sites, in tree holes, tunnels, crevices, buildings, and so on, were relatively more frequently used by resident species than by migrants.

In Britain resident and migrant species alike had significantly lower nest losses, particularly at the chick stage, if nesting in holes or crevices (Table 1). Hole-nesting residents also laid larger clutches. The incidence of hole and crevice nesting was also slightly higher among residents (36 versus 29%). On these points the data are in agreement with von Haartman's hypothesis. However, nest losses were markedly higher in the residents than in the migrants, even when compared across similar nest sites, and this is the opposite of the pattern expected on von Haartman's hypothesis.

Alerstam and Hogstedt (1980) found hole and crevice nesting in a number of the migrant species in Sweden. Many of these were either species with exposed foraging methods, e.g. aerial hunters such as the swallow *Hirundo rustica* or swift

Table 1. Clutch size and nest success in relation to migrancy and nest site

	Residents					Migrants					
	Hole or crevice (N = 9)		Other (N = 16)		Prob-ability level	Hole or crevice (N = 5)		Other (N = 12)		Prob-ability level	Migrants vs residents probability
	Mean	SD	Mean	SD		Mean	SD	Mean	SD		
Clutch size	6.11	1.70	4.48	0.96	$P<0.01$	5.26	1.07	4.96	0.59	n.s.	n.s.
Nests lost, %	42.33	13.10	56.77	11.72	$P<0.02$	30.14	9.99	42.30	9.26	$P<0.10$	0.01
Daily losses, %											
Egg stage	2.67	1.10	3.21	1.28	n.s.	1.18	0.62	2.14	1.01	$P<0.05$	0.01
Chick stage	1.20	0.72	2.32	1.27	$P<0.05$	0.70	0.40	1.63	0.57	$P<0.02$	n.s.

Apus apus, "sit and wait" predators such as the flycatchers and kingfishers, or exposed gleaners such as the wheatear *Oenanthe oenanthe* and the wagtails. These same groups provide the five migrant hole-nesters in Table 1. These species are particularly at risk whilst foraging for their nestlings, for they may be observed by a predator and followed back to the nest. A secure site confers obvious advantages in such circumstances and Alerstam and Hogstedt showed that these migrant hole-nesters arrived back earlier than did migrants using other habitats. Yet they arrive well after many resident species that use open nests. The availability for migrants of holes and crevices late in the season argues that hole nests cannot be a limiting factor to which winter residency is an adaptive response. This, together with the lower nest success experienced by the residents, argues against von Haartman's hypothesis of access to secure nest sites as a general explanation for winter residency.

4 Multiple Broods and Residency

An alternative explanation for winter residency is that the lower nesting success experienced by residents might be offset by packing more attempts into a breeding season and that the early start facilitates this. Mistle thrushes, *Turdus viscivorus*, for example, most frequently use habitats in which breeding can start early enough to allow a second nest cycle, and use other habitats with decreasing frequency as the chances of that second clutch decline (O'Connor 1986). If this strategy were general among the residents, then multiple brooding would be commoner among residents than among migrants. Earlier I found no evidence that the number of broods reared differed between the two groups (O'Connor 1981). In fact, the incidence of single brood species in the present sample is just the opposite to that required, with 40% of all residents being single brooded but only 17.6% of migrant species being single brooded. Hence it seems unlikely that increased opportunities for second broods can have led to winter residency.

5 Residency and Habitat Saturation

A further possible hypothesis recognizes that nesting success in a given habitat is usually a function of the density of conspecifics there (Lack 1966; Fretwell 1980; Rosenzweig 1985). Birds nesting in crowded habitats may do better to move into intrinsically poorer but less crowded alternatives, with average habitat quality – and therefore nest success – steadily worsening with continued population increase. This process, predicted on theoretical grounds by Brown (1969) and Fretwell and Lucas (1970), occurs widely among birds in Britain (reviews in O'Connor and Fuller 1985 and O'Connor 1985) and over a wide range of species there exists a strong positive correlation between population density and breadth of nesting habitat (O'Connor 1987). Hence it is worth considering whether nest success might vary seasonally in a way correlated with habitat saturation.

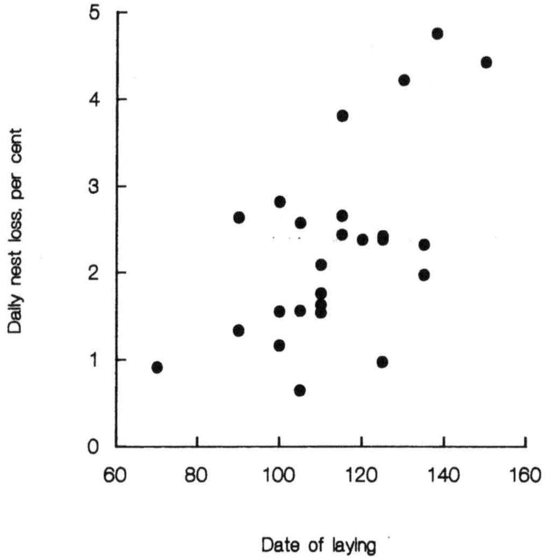

Fig. 1 Chick stage nest loss in relation to laying date in 25 resident species. Laying date is the modal date of laying for each species over the period 1962–80. Nest loss rates have been adjusted for the mean effect of hole-nesting documented in Table 1

Of the nest success components listed in Table 1, only chick stage nest losses among the residents varied markedly with date of laying (Fig. 1). Among migrants the generally lower nest losses (Table 1) hardly varied with season ($r = 0.007$, n.s.). Early nesting residents therefore experience fewer losses than do later nesting species. We can dismiss the possibility that the earliest residents win the safest habitats and the later species the poorest, since the migrants do better still, despite their late breeding (Table 1). When the residents are grouped separately into a group of early species nesting before the migrants start and a group of late residents breeding in parallel with the migrants, this becomes even clearer. Chick stage nest losses averaged 1.44 ± 1.45% per day within the early group but rose to 2.76 ± 1.37 within the late residents, against only 1.36 ± 0.68 among the migrants breeding at that time (Kruskal-Wallis test = 6.98, $P < 0.03$). The late residents' use of less safe habitats cannot, therefore, be merely the best option available to them: they could be expected, for example, to preempt those taken up by the migrants at that time.

In the present study nest losses proved correlated with nest habitat diversity in residents (Spearman's rho = 0.491, $P < 0.05$) but not in migrants (rho = 0.096, n.s.). Nest habitat diversity differed markedly between the early (nesting before migrants start) and late (breeding in parallel) residents, with diversity indices of 9.49 and 14.67 respectively (Mann-Whitney U-Test $P < 0.025$). For migrants mean nest habitat diversity was 8.88. The mean densities of those early and late residents for which census data were available were 5 and 9.5 pairs/km^2. Density estimates were not available for all the migrant species studied here, but the migrants were markedly less abundant than the residents: my earlier study, using data for a (slightly different) set of well-censused species (O'Connor 1981), obtained estimates of 2.84 and 6.38 pairs/km^2 respectively. Migrant densities were thus comparable to or slightly below those of the earliest residents. On

Table 2. Nest habitat diversity in resident and migrant hole-nesting and other species. For each species the diversity measure $D = 1/\Sigma p_i^2$ was computed from a sample of British Trust for Ornithology nest record cards and the results were averaged across species. Diversity varies significantly across the four groups (Kruskall Wallis test, $P = 0.008$)

	Mean SD (N)
Resident hole-nesters	8.21 ± 4.58 (9)
others	13.07 ± 5.39 (16)
Migrant hole-nesters	4.31 ± 1.50 (5)
others	10.78 ± 5.02 (12)

average, therefore, the late resident species are subject to twice the intensity of intraspecific competition and occupy a wider spectrum of nesting habitats than do either the early residents or the migrants, with an associated reduction in average nesting success. Moreover, these patterns were independent of hole-nesting status, for among hole-nesting species habitat diversity in migrants was just half that in residents and migrants using open nest sites were also more stenotopic than the equivalent residents (Table 2). Thus among hole- and open-nesters, the migrants used a narrower spectrum of nest habitats than did the residents.

6 Why Migrants Move to Breed

The results thus far imply that residents in Britain experience lower predation losses by breeding early. Given this, why should migrants, which necessarily return late, use the area at all? My explanation is that migrants are r-selected species taking opportunistic advantage of resources that shift spatially over time, an idea consistent with Ashmole's (1961, 1963) hypothesis that areas with greater seasonality in productivity would be more prone to invasion by migrants.

In a previous study (O'Connor 1981) I concluded that migrant species in Britain could well be classified as classic "r-selected" (in the sense of MacArthur and Wilson 1967) species whilst resident species were "K-selected". My evidence was four-fold. First, migrants were generally smaller in body size than were residents. Second, dimorphism promoted survival within resident species, probably by allowing ecological divergence during the harsh winter season, but was associated with reduced survivorship among migrants. Migration costs may be especially sensitive to wing shape and this cannot be optimized simultaneously in two wing-size morphs. Thirdly, the residents and migrants showed opposite dependencies on survival and reproductive rates: the most widespread and abundant resident species were those that displayed high adult survival but the most numerous migrants were those that laid most eggs in a season. Only during recovery from a population crash due to an extraordinarily severe winter did resident numbers prove at all a function of reproductive output. Finally, populations of migrants proved very labile, with large increases and decreases (but no net changes) around relatively low densities whilst resident populations were relatively better controlled to high density equilibrium levels.

As productivity increases after a seasonal low (e.g. each spring in the temperate zone), it can support many more birds than have survived the low. The survivors can immediately benefit from the increase in per capita resources through increased clutch sizes. However, no long-term advantage accrues to the survivors in defending the increased resources against migrant invaders since the flush will die away seasonally. Hence migrants should be relatively more frequent in more seasonal than in less seasonal areas. These ideas are supported by empirical studies. Herrera (1978) showed that the proportion of migrant species in bird communities increased from southern Europe northwards through Fennoscandinavia in a manner highly correlated with a measure of seasonality of climate. In addition, the expected increase in clutch size among resident species has been demonstrated by Ricklefs (1980) for bird communities in areas ranging from Costa Rica to Alaska. Again, if one compares the neighbouring islands of Britain and Ireland one finds the avifauna of the less seasonal Ireland is characterized both by a lower proportion of migrant species (Lack 1976; O'Connor, in prep.) and by higher average densities of the resident species (Wilson 1977; O'Connor, in prep.). All of these points are explicable within Ashmole's hypothesis. An important feature of the hypothesis is that it predicts these patterns as a function of seasonality and not of adversity. That is, either a winter "low" or a summer "high" in productivity will bring about the patterns described, a feature not present in winter adversity models.

Since early residents and migrant species have similar densities and habitat breadths, why do the early residents not adopt the migrant strategy? The answer appears to be that the early residents escape the diffuse competition afforded by the large number of populous resident species nesting in the later part of the season, whilst the migrants do not escape. Two pieces of evidence favour this hypothesis. First, the populations of most resident species in Britain vary from year to year, often declining following a severe winter. Whenever residents experience a severe winter, their collective numbers the following summer are reduced, however, the migrant population, their own numbers unaffected since they have wintered elsewhere, should be less constrained by diffuse competition from the resident species. They should be able to occupy better habitats in greater numbers and enjoy better breeding success as a result. This is then reflected in higher numbers of migrants returning the following summer. Low numbers of residents in any given season should then be correlated with higher numbers of migrants breeding the following summer, and conversely, at least as long as the total population is close to any equilibrium carrying capacity. In practice a strong negative correlation ($r = -0.921$, $P < 0.01$) of this type was found, providing strong support for the idea that residents at high numbers deny migrants access to habitats the latter could otherwise use. Data from the years immediately following the excessively severe (the worst for 300 years) winter of 1962–63 were omitted from the calculations because resident numbers were demonstrably depressed below equilibrium values in those years (O'Connor 1981). Nevertheless, this winter's effects on residents (whose numbers fell typically to 18–50% of their normal levels) yielded further confirmatory evidence. Five of the six migrant species for which data are available used a broader range of nesting habitat in the summer of 1963 than they did in 1962 whilst five of the six resident species so far examined showed the opposite trend. Thus migrant species experienced widespread ecological release as to nest habitat use when environmental conditions reduced intraspecific competition within the resident species. This picture is consistent with the findings of Cox (1968), who showed that migrant land birds in the tropics were primarily space and habitat segrators or specialists, with little tendency to divide the resources within a habitat by morphological adaptation. Hence when habitats were fully packed, as in the tropics during the breeding season, they moved elsewhere. But the same traits against morphological adaptation in the tropics continue to limit them in the temperate zone, allowing them to breed well in habitats to which they happen to be well adapted but extremely poorly outside these. Such characteristics imply marked susceptability to diffuse competition from other species, as described by O'Connor (1985).

In summary, nesting habitats at the peak of the breeding season are crowded and this reduces average breeding success there, principally through nest predation. Most residents are K-strategists adapted to such conditions but some escape the effects of this crowding by breeding early, when predation is lower. By doing so, however, they may be ecologically restricted to narrower habitat niches and thereby be limited in total population size. Migrants, in contrast, are r-strategists primarily returning to the temperate zone to a seasonal resource peak that residents surviving the local winter are too few to exploit fully. They are, however, especially sensitive to diffuse competition for habitat from the competitively superior residents and are therefore restricted to narrow habitat niches, again with corresponding limitations as to maximum population size.

References

Alerstam T, Hogstedt G (1980) Spring predictability and leap-frog migration. Ornis Scand 11:196–200

Ashmole NP (1961) The biology of certain terns. D Phil Thesis, Oxford

Ashmole NP (1963) The regulation of numbers of tropical oceanic birds. Ibis 103:458–473

Brown JL (1969) The buffer effect and productivity in tit populations. Am Nat 103:347–354

Cox GW (1968) The role of competition in the evolution of migration. Evolution 22:180–192

Fretwell SD (1969) Dominance behavior and winter habitat distribution in juncos (*Junco hyemalis*). Bird-Banding 40:401–25

Fretwell SD (1972) Populations in a seasonal environment. Princeton University Press, Princeton NJ

Fretwell SD (1980) Evolution of migration in relation to factors regulating bird numbers. In: Keast A, Morton ES (eds) Migrant birds in the Neotropics: ecology, behavior, distribution and conservation. Smithsonian, Washington DC, pp 517–527

Fretwell SD (1985) Why do birds migrate? Inter- and intraspecific competition in the evolution of bird migration contributions from population ecology. Proc Int Ornithol Congr 18:630–637

Fretwell SD, Lucas HL (1970) On territorial behaviour and other factors influencing habitat distribution in birds. I. Theoretical development. Acta Biotheor 19:16–36

Greenberg R (1980) Demographic aspects of long-distance migrations. In: Keast A, Morton ES (eds) Migrant birds in the Neotropics: ecology, behavior, distribution and conservation. Smithsonian, Washington DC, pp 493–504

Haartman L von (1968) The evolution of resident versus migratory habit in birds. Some considerations. Ornis Fenn 45:1–7
Herrera CM (1978) On the breeding distribution pattern of European migrant birds: MacArthur's theme reexamined. Auk 95:496–509
Ketterson ED, Nolan V (1976) Geographic variation and its climatic correlates in the sex ratio of eastern wintering dark-eyed juncos *Junco hyemalis hyemalis*. Ecology 57:679–693
Ketterson ED, Nolan V (1983) The evolution of differential bird migration. In: Johnston RF (ed) Current ornithology, vol 1. Plenum, New York, pp 357–402
Lack D (1966) Population studies of birds. Oxford University Press, Oxford
Lack D (1976) Island biology illustrated by the land birds of Jamaica. Blackwell Scientific, Oxford
MacArthur RH, Wilson EO (1967) The theory of island biogeography. Princeton University Press, Princeton NJ
Mayfield H (1975) Suggestions for calculating nest success. Wilson Bull 87:456–466
Morse DH (1980) Population limitation: breeding or wintering grounds? In: Keast A, Morton ES (eds) Migrant birds in the Neotropics: ecology, behavior, distribution and conservation. Smithsonian, Washington DC, pp 505–516
O'Connor RJ (1981) Comparisons between migrant and non-migrant birds in Britain. In: Aidley DJ (ed) Animal migration. Cambridge University Press, Cambridge, pp 167–195 (Society for Experimental Biology Seminar Series, Number 13)
O'Connor RJ (1985) Behavioural regulation of bird populations: a review of habitat use in relation to migration and residency. In: Sibly RM, Smith RH (eds) Behavioural ecology: ecological consequences of adaptive behaviour. Blackwell Scientific, Oxford, pp 105–142 (British Ecological Society Symposium Number 25)
O'Connor RJ (1986) Dynamical aspects of avian habitat use. In: Verner J, Morrison ML, Ralph CJ (eds) Wildlife 2000: Modeling habitat relationships of terrestrial vertebrates. University of Wisconsin Press, Madison, WI, pp 235–240
O'Connor RJ (1987) Organization of avian assemblages – the influence of intraspecific habitat dynamics. In: Gee J, Giller PR (eds) Organisation of communities – past and present. Blackwell Scientific, Oxford, pp 163–183 (British Ecological Society Symposium Number 26)
O'Connor RJ, Fuller RJ (1985) Bird population responses to habitat. In: Taylor K, Fuller RJ, Lack PC (eds) Bird census and atlas studies. British Trust for Ornithology, Tring, UK, pp 197–212 (Proceedings of the Eighth International Bird Census Conference)
Ricklefs RE (1980) Geographical variation in clutch size among passerine birds. Ashmole's hypothesis Auk 97:38–49
Rosenzweig ML (1985) Some theoretical aspects of habitat selection. In: Cody ML (ed) Habitat selection in birds. Academic Press, London, pp 517–540
Wilson J (1977) Some breeding bird communities of sessile oak woodlands in Ireland. Pol Ecol Stud 3:245–256

Chapter 3 Physiological Adaptations to Migration

The Visual Problems of Nocturnal Migration

G.R. Martin[1]

Amongst birds a completely nocturnal life-style is rare. Less that 1% of species habitually complete all aspects of their life-cycle outside of daylight hours. However, occasional nocturnal activity in otherwise diurnal birds is relatively common. The most widespread such behaviour is migration at night (Martin 1990). How diurnal birds cope with occasional nocturnal activity is not well understood, but in view of the restricted life-style of truly nocturnal birds (Martin 1986a), it is perhaps surprising to find that otherwise strictly diurnal birds are able to undertake their migratory flights at night.

1 Migration at Night

1.1 The Phenomenon

During most of the year diurnal bird species, especially passerines, roost as soon as light levels begin to fall to those experienced around dusk [but the exact relationship between light levels and roosting behaviour is not simple (Krantz and Gauthreaux 1975; Astrom 1976)]. Also when given a choice, passerine birds usually show a preference for daytime light levels for their activities. However, at the time of year when migration takes place many of these same birds may begin to show a preference for lower light levels and actually increase their level of activity after dusk (Gwinner 1975). Radar observations of the time of departure of birds from coastal watchpoints or through mountain valleys indicate that many normally diurnally active birds chose to depart on at least certain sections of their migratory flights at night and may land either during the night or the following morning.

1.2 The Species

Although there has been a wealth of research devoted to understanding the mechanisms whereby nocturnally migrating birds might determine the appropriate direction in which to fly (see for example, the reviews of Able 1980; Baker

[1] University of Birmingham, School of Biological Sciences and School of Continuing Studies, P.O. Box 363, Birmingham, B15 2TT, UK

E. Gwinner (Ed.) Bird Migration

1984; Moore 1987), this experimental work has concentrated on only a handful of species. At the same time, however, the reports of much of this work is prefaced by, or contains statements to the effect that, "Most songbirds (order Passeriformes) migrate almost exclusively at night" (Able 1982, p. 761). Despite such statements systematic lists of species which are known to make at least some of their migrations at night are not available. Even the question of whether bird species fall neatly into "nocturnal" or "diurnal" migrant categories does not seem to have been widely discussed. However, the question of whether nocturnal migratory behaviour is restricted to certain species is important since the presence of interspecific differences in the propensity to migrate at night could imply important differences in the sensory or perceptual capacities available to nocturnal and diurnally migrating birds.

Difficulty in deciding which species should be described as night migrants reflects of course the difficulty in identifying night-flying birds. However, information on the nocturnal migration of individual species or populations can be gained from the occurrence at dawn of presumed newly arrived migrants at migratory watch points and data on the birds attracted at night to artificially illuminated structures and bird-trapping facilities (Aldrich et al. 1966; Durman 1976; Verheijen 1981; Gauthreaux 1982).

It is known that the categories of night-time and day-time migrants are not necessarily mutually exclusive at either the species or population level. For example, within one species there is evidence that some individuals migrate by day and others at night even through the same migratory watch points, e.g. the European starling *Sturnus vulgaris* (Feare 1984). Also, species which by general concensus are regarded as daytime migrants, such as the swallow, *Hirundo rustica,* may occasionally arrive at migratory watch points during the night (Mead C, Personal communication, 1988). In addition, there are species in which the majority of a population regularly migrate continuously by both night and day, at least for part of their migratory journey. For example, the blackpoll warbler, *Dendroica striata*, and the wheatear, *Oenanthe oenanthe*, contain populations which apparently make non-stop journeys of more than 24 h across the Atlantic Ocean (Snow 1953; Dorst 1961; Nisbet 1970; Williams and Williams, this Vol.). Also, yellow wagtails, *Motacilla flava*, may make continuous night and day flights across the Sahara Desert (Wood 1982), but there is also evidence that these birds sometimes rest in the desert during the hottest part of the day (Biebach et al. 1986).

A recent survey of nocturnal-diurnal migratory behaviour of bird species passing through observatories positioned around the coast of the British Isles showed that the majority of species may migrate by night and/or day through these watch points and that for many species it was virtually impossible to allot them to a diurnal or nocturnal migratory category (Martin 1990). Of the 147 species (from 16 orders) surveyed, no species fell into an exclusively nocturnal category, but 75% of all species had been recorded as a night migrant at some time. In only one order (Gaviiformes) were there no species whose passage through these migratory watch points did not occur occasionally at night. Thus, while many birds do migrate at night it could not be concluded that any species always migrates at night through these particular migratory watch points and therefore, that certain species are totally dependent upon nocturnally available cues for their migratory orientation or are necessarily sensorily well equipped to cope with the problems of migration at night. However, it would also seem that a large majority of migrating birds will, at least occasionally, travel at night.

From the results of this survey it was proposed that migration at night by diurnally active birds may not be a general or preferred strategy for a particular species or population. Birds may perhaps travel by night or by day according to local factors. One such factor which could determine whether a bird migrates at night may result from the need to fly continuously across inhospitable terrain especially if that flight could not be readily completed in daylight hours. This may apply particularly to flights passing through coastal watch points around the British Isles. An individual bird may thus migrate by day or by night depending upon its geographical position along its migratory path. Also, it has recently been proposed by Kerlinger and Moore (1989) that certain bird species chose to travel by night due to the usually more favourable structure of the atmosphere for flight at this time. It seems likely that these two factors may account for at least some of the variability in reports of nocturnal-diurnal migration through particular watch points collated by Martin (1990).

2 The Use of Vision in Nocturnal Migration

2.1 The Sensory Cues Available for the Orientation of Migratory Birds

Much has been learnt from laboratory and field experiments of the sensory cues that night-migrating birds might use to orient themselves, at least at the beginning of their journeys. However, such studies deal with only a handful of passerine species drawn from the Sylviidae, Muscicapidae, Emberizidae and Turdidae. Cues identified so far include the position of the setting sun, the earth's magnetic field, the patterns of stars, the moon and topographical features (See Emlen 1975; Able 1980; Papi and Wallraff 1982; Baker 1984; Presti 1985; Moore 1987 for reviews). Thus, a number of cues may be available simultaneously to a migrating bird though which cues may be used by a given species is not clear. It may be misleading to assume that all of these cues are equally available to all birds when migrating at night.

There is evidence that whatever cues are available they are used in some kind of hierarchical or complementary fashion such that one cue can substitute for another in differing circumstances (Emlen 1975; Moore 1985). This means that one species or population of birds (or even age class within a species) in one set of weather conditions, or geographical locality, may rely primarily on one set of cues to determine the orientation of their nocturnal migratory journey, while another group of birds may use another set of cues (see the above mentioned reviews and, for example, the results of Able 1982, and Sandberg et al. 1988).

2.2 The Importance of Visual Cues

The importance of the earth's magnetic field in the migratory orientation of some bird species seems generally accepted (Presti 1985). However, it seems possible that an intimate relationship exists between the mechanisms of magnetic field detection and the visual system (Leask 1977; Semm et al. 1984; Semm and Demaine 1986). Furthermore, there is much evidence that cues based upon vision are of equal or primary importance to magnetic cues in the orientation of nocturnally migrating birds. The following experimental findings and field observations are pertinent to this point.

1. The majority of nocturnal migrations take place in weather which provides the "ideal" conditions of calm, light or following winds with little cloud cover and good visibility, both prior to the time of departure and during the actual flight (Richardson 1978; Elkins 1983; Kerlinger and Moore 1989).
2. Prior to their nocturnal migratory journey passerine birds usually cease their normal daytime activities and start to roost around dusk in the usual way (Palmgren 1944; Hebrard 1971). They then become active again and typically depart on their migratory flight between 30 and 45 min after sunset, usually after the time of civil twilight (Kerlinger and Moore 1989). It is thought that it is during this flightless period that migrant birds make the decision of whether to migrate and in which direction. Just what birds do at this time is unclear but the primary importance of visual cues associated with detecting the position of the setting sun or the pattern of polarized light in the sky (which is a direct correlate of the sun's position) has been demonstrated (Moore 1987). The extent to which a bird's migratory orientation is influenced at this time by the earth's magnetic field is unclear. If magnetic cues are employed they may be used to calibrate a compass based upon the stars rather than consulted directly but evidence is conflicting (Wiltschko and Wiltschko 1975a,b, 1976; Sandberg et al. 1988).
3. There is considerable evidence that visual cues are of primary importance not only in initiating the direction of orientation but also in maintaining it throughout the nocturnal journey. These visual cues may be associated with the stars and moon and also involve the use of topographical landmarks as guides or beacons.

That caged nocturnal passerine migrants (blackcap, garden warbler, lesser whitethroat and indigo bunting) may orient themselves using the pattern of the stars above has been clearly demonstrated (Sauer and Sauer 1955; Emlen 1967) However, the extent to which star patterns are used to orient birds when actually on their migratory flights is less clear. Evidence from both radar studies and direct visual observation have shown that although birds prefer to make their migratory flights under clear skies, those which do fly under overcast skies may be no less well oriented in their flight path than those which travel when stars are visible (Able 1974, 1982).

Nocturnally migrating birds seem to become disoriented in their flight paths only when both the ground and the stars are not visible to them. Thus disorientation among migrating passerines is thought to be rare (Bellrose and Graber 1963; Drury and Nisbet 1964; Bellrose 1967; Able 1974) and seems to occur only when birds are deprived of visual cues. So long as birds fly below the cloud ceiling, their flight relative to both the ground and the air mass is, "as straight, level and fast as comparable birds flying on clear nights" (Able 1982 p. 47). Cases of poor orientation are associated with low overcast, rain, fog or cloud cover often of several days duration (Emlen 1975), although there are reports that birds can still be well oriented even under these conditions (Griffin 1973). In addition, Able (1982) has argued that simple absence of access to sun and stars for several days need not result in disorientation of nocturnal migrants. "So long as birds were flying below the clouds, tracks were as straight as under clear skies and the distribution of mean headings of all birds was oriented" (Able 1982, p. 48).

The disorientation of birds in low cloud or fog does suggest that they cannot fly successfully guided by cues from the earth's magnetic field alone. Some kind of visual cues derived either from the stars and/or moon above or from the ground below would seem necessary for the correct orientation of birds at night. The fact that migrating birds "caught out" by bad weather conditions are attracted, often fatally, to illuminated structures suggests strongly that at night these birds may be dominated by visual cues from below them but that they are easily confused. Indeed, the attraction of birds to lighthouses may be due to their confusing the artificial light source for the moon which they would normally use for orientation of their flight (Baker 1984, p. 94). If this is the case it would certainly suggest that a bird's ability to conduct visual discriminations at night of even large, bright objects is relatively poor.

Furthermore, it seems that nocturnally migrating birds do not even need to be caught out by "bad weather" for them to be confused by artificial lights below them. D. Pearson (personal communication 1988) has regularly observed that passerine species migrating through East Africa are attracted at night to artificial lights but only when there is fog or mist at ground level in the area illuminated by the lights, and no moon. When these two conditions are fulfilled, birds apparently passing in the clear, starlit sky above the low cloud are attracted down through the mist to the lights. If a moon (of any phase) is present then these lights are ignored.

This observation is paralleled by data from Bardsey Island bird observatory (Wales) (Durman 1976). Here, the number of birds attracted to the lighthouse is correlated with the absence of the moon, even to the extent that birds were attracted temporarily to the light during an eclipse of the moon. On Bardsey, birds may be attracted to the light not only under overcast conditions but also when there is low level cloud or "smoke haze" and a clear starlit sky above.

It would seem that to attract birds down the light must be scattered by fog or mist to produce a large pool of diffuse light since flashing, as opposed to rotating or fixed, lights do not attract birds (C. Bibby, personal communication). The basis of this attraction remains unknown but it could be associated with the loss of a clear visual horizon rather than the absence of the moon itself (see section 3.2). These observations do however, suggest that the passerine bird's ability to make visual discriminations at night is rather limited and that the birds are influenced by visual cues from the ground beneath them even when there is a clear starlit sky above.

Thus it may be suggested that visual cues from below may dominate over both magnetic and star-based cues during actual migratory flight even though these latter cues may be used to determine direction of flight at the time of departure.

That day-flying migrating birds are typically influenced in their migratory routes by visual cues from the landscape below seems well accepted together with the idea that these birds frequently follow so-called leading lines. These are usually large topographical features such as rivers, hill and mountain ridges, coasts and valleys (Mead 1983; Baker 1984). It has also been shown that birds crossing extensive tracts of water can compensate for the drifting effects of wind by using the wave pattern below them to maintain a constant heading (Alerstam and Petterson 1976).

Evidence from radar studies suggests that nocturnally migrating birds may also be influenced by similar topographical features including river systems (Bingman et al. 1982), mountain ranges, valleys or individual peaks (Bruderer 1978, 1982) although Emlen and Demong (1978) have argued that the influence of topographical cues may be minimal at certain sites. The principle idea behind the use of landmarks in migrating birds is that the birds may determine their compass direction from some other cue but that this direction is then projected onto the landscape below. Topographical features which correspond with this direction can be noted and used for guidance. These may be in the form of leading lines, a single fixed beacon, or involve the alignment of two or more distant objects (which can produce a more direct course since it is then possible to make use of the effect of parallax (Bruderer 1982).

2.3 *Visual Fields*

The idea that a bird projects the compass direction for its migratory flight on to the landscape is very similar to what humans might do when orienting through unfamiliar terrain. However, the bird has two particular advantages over a person in this task. Not only can it gain altitude rapidly and therefore considerably broaden the range of topographical features brought into play, but the extent of its visual field will possibly allow the spatial relationships between different topographical features to be more easily appreciated.

By virtue of the optical structure of the eyes and of their position and movements in the skull, many birds gain complete coverage of the visual world both above and around them. Indeed in the mallard duck, *Anas platyrhynchos*, total panoramic vision, including complete visual coverage of the celestial hemisphere above the head, is achieved without eye movements (Martin 1986b), while in the European starling a similar degree of visual coverage can be achieved by virtue of eye movements (Martin 1986c). Thus while a bird with laterally placed eyes flies towards or over a given topographical feature, the position of these features can be constantly monitored with respect to other features both to the side and behind, and also to the complete celestial hemisphere above the bird. On a cloudless night this will mean that the star pattern of the whole celestial hemisphere is constantly available and can be related to the complete landscape surrounding the bird.

The general thrust of this section has been that during their nocturnal migratory flight visual cues of some kind play an important role in determining and sustaining the bird's migratory orientation. However, it was seen in the previous section that flight at night is not necessarily the most favoured migratory strategy of a species or population and that birds may frequently fly by day and/or night according to local circumstances (geographical and/or meteorological) along their migratory route. Use of visual cues has the advantage of providing continuity between day-time and night-time and this applies not only to topographical cues but also to star patterns and the sun. At the end of the day under clear skies the stars will become visible while the position of the sun can still be detected and vice versa at dawn. However, as remarked in the introduction these are birds which during most of their life cycle typically begin roosting as light levels approach those around sunset. It is worth considering therefore the visual problems which a diurnal bird might face on its nocturnal migrations and what its visual capacities to cope with these might be.

3 The Night Environment and the Visual Capacities of Birds

3.1 Night-time Light Levels

One obvious feature of night-time is that the light levels associated with it are far from uniform either within a single night or between nights. Light levels constantly change throughout the night as a function of the position of the sun, the position and phase of the moon and according to the presence of cloud cover. Night-time light levels also vary markedly with latitude and season. [See Martin (1990) for details of these influences on the light regime experienced at night.]

Figure 1 shows (1) the average luminance levels of the earth's surface when illuminated by various natural light sources and (2) the average luminance of the sky near the horizon. Luminance values are independent of viewing distance and Fig. 1 indicates the average luminance of the earth's surface when viewed from directly above. It can be seen that the light levels which may be experienced between sunset and sunrise can range over at least 6 log units (1 000 000 fold) under clear skies and 7 log units if cloud is taken into account (maximum cloud cover will reduce all of these ground luminance values by about 1 log unit). Strictly, these luminance values apply only to temperate latitudes during the summer. At lower latitudes natural light levels deviate little from those of Fig. 1 though the speed of transition between sunset and the various twilight levels will be more rapid. At higher latitudes not only do the actual levels of illumination vary but the period that can be called night-time also varies markedly. Thus during the summer at 60° latitude, night-time (between sunset and sunrise) lasts less than 5 h and light levels never fall below those of civil twilight.

The large variation in nocturnal light levels indicated in Fig. 1 raises serious problems for understanding the visual tasks which face nocturnal migrants and for interpreting reports of nocturnal migratory behaviour from different observatories especially the more northern ones. The range of light levels experienced at the earth's surface between that produced by maximum moonlight without

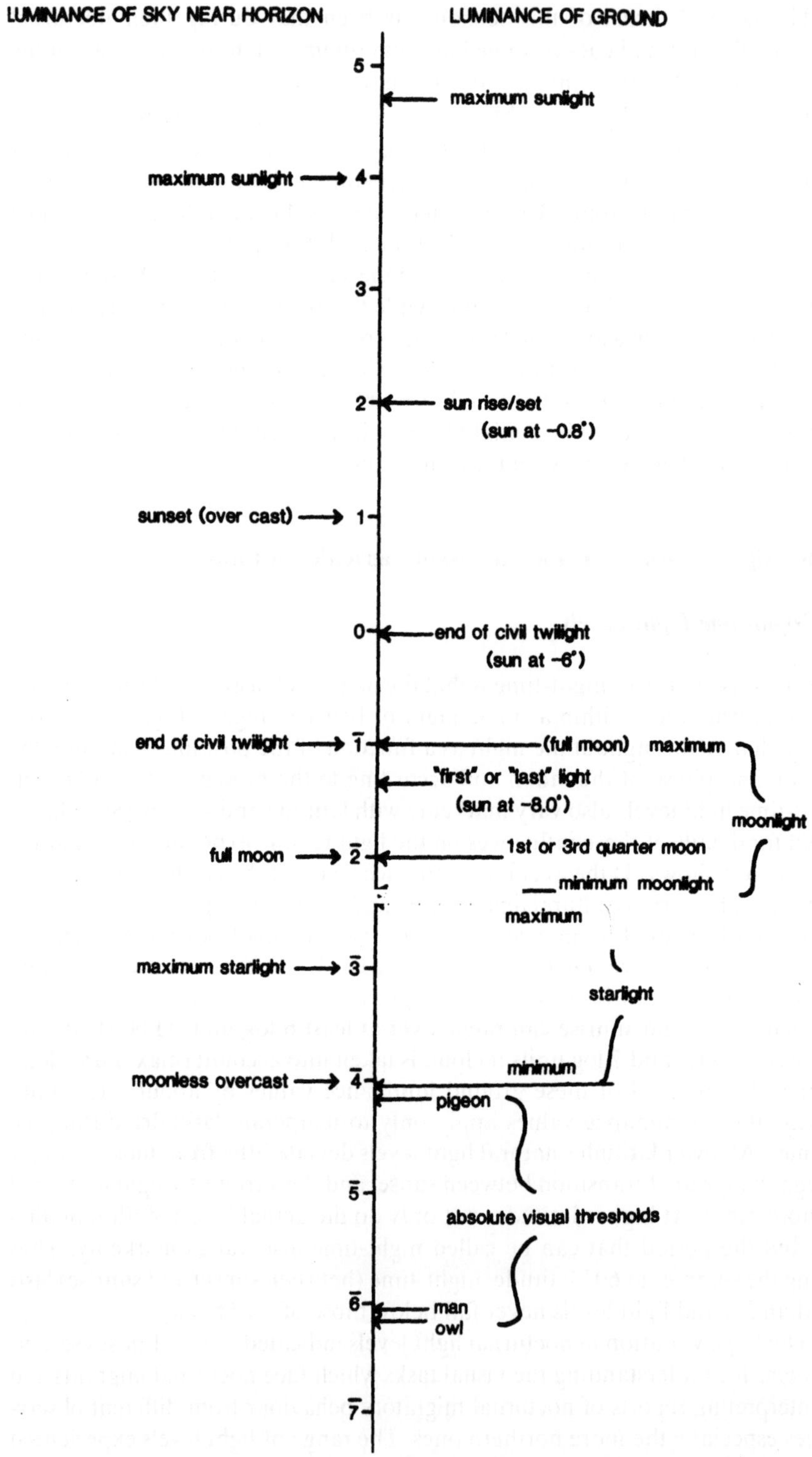
LUMINANCE OF SKY NEAR HORIZON
LUMINANCE OF GROUND
5
maximum sunlight
maximum sunlight
4
3
2
sun rise/set
(sun at −0.8°)
sunset (over cast)
1
0
end of civil twilight
(sun at −6°)
end of civil twilight
1̄
(full moon) maximum
"first" or "last" light
(sun at −8.0°)
moonlight
full moon
2̄
1st or 3rd quarter moon
minimum moonlight
maximum
maximum starlight
3̄
starlight
minimum
moonless overcast
4̄
pigeon
5̄
absolute visual thresholds
6̄
man
owl
7̄

cloud and minimum starlight with cloud is approximately 4 log units (10 000-fold) and clearly it may be that certain species or populations of birds will decide to travel or not on a nocturnal section of their journey depending on the ambient light levels as well as the weather conditions. It should be noted that weather conditions and light levels are not necessarily independent of each other.

Many shorebirds and wildfowl, and some passerine species, breed at very high latitudes, often between 70° and 80° N. At these high latitudes during the time of the year when such birds migrate towards or from temperate latitudes, night-time is virtually non-existant; light levels rarely fall below those of twilight. Hence, although these birds may fly for more than 24 h, at these high latitudes they would not necessarily entail flights at nocturnal light levels. Only when birds are at temperate latitudes and below can it be assumed that a continuous flight of 24 h necessarily implies migration at light levels below those of civil twilight.

3.2 Visual Capacities in Relation to Naturally Occurring Light Levels

In considering the vision of an animal in relation to the visual problems which the night environment poses, the simplest questions to ask concern (1) the amount of spatial detail which the animal can detect at a given light level, and (2) the minimum amount of light necessary to give some kind of visual response, the absolute visual threshold. Even in the absence of specific data for a particular species enough is now theoretically understood about the limits to visual sensitivity in vertebrate eyes in general (Snyder et al. 1977; Barlow 1981; Martin 1985) to discuss confidently the limits of visual performance and to determine what it is *not* possible for an animal to detect visually at a given light level. It is also possible to extrapolate with some caution from data on the visual performance of one species to that of another with similar eye structure.

Also shown in Fig. 1 are known absolute visual thresholds for man (Pirenne et al. 1957), the tawny owl (Martin 1977) and the pigeon, *Columba livia* (Blough

Fig. 1. The average luminance (cd/m^2) of the earth's surface when viewed from above and illuminated by various natural light sources. Based upon the data of Bond and Henderson (1963) for 50° latitude at summer solstice. The reflectance of the surface is taken to be 25% (0.6 log) which is similar to that of grass or leaf litter (Martin 1982), the reflectance of dark soil is approximately 10% (1.0 log). Civil twilight is defined according to the angle of elevation of the centre of the sun below the horizon between -0.8° (sunrise/set) and -6°. Night-time (the period between "last" and "first light") begins when the sun lies at 8° below the horizon. The illumination produced by the moon varies as a function both of the moon's elevation and of its phase. Shown are maximum luminance levels produced by the moon at three different phases and these define the approximate full range of luminance levels which may be considered as produced by moon light. Also shown is the range of light levels produced in the absence of the moon by starlight which varies according to time of year and geographical position. All of the surface luminance measures are for a clear sky. The maximum attenuation produced by cloud cover is approximately 90% (1 log unit) (Taylor and Kerr 1941); average cloud will reduce these luminance levels by approximately 0.3–0.5 log units. Also shown are the average luminance levels of the sky near the horizon (Middleton 1958) under various natural conditions, and the absolute visual thresholds of the pigeon (Blough 1955), man (Pirenne et al. 1957) and the tawny owl (Martin 1977)

1955). The last two are the only bird species for which absolute threshold data are available. The tawny owl is a strictly nocturnal species while the pigeon is strictly diurnal and as a rule goes to roost as soon as light levels approach those which occur at around dusk. While it is likely that other bird species will have absolute sensitivities which differ from those of the owl and the pigeon, consideration of both the optical structure of the eye and the structure of the retina in birds would suggest that no birds are likely to be found with greater sensitivity than the owl, and that the sensitivity of many diurnal birds will be similar to that of the pigeon (Martin 1985).

Figure 1 indicates that for all except the overcast starlit nights diurnal birds such as the pigeon are likely to have some kind of vision, but that there will be nights (minimum starlight, with or without overcast) when no vision is possible at all for many species (note that in open habitats men and owls are never in that position). Spatial resolution in birds at low light levels is unknown (except for the data on a single species of owl; Fite 1973). However, on theoretical grounds it is understood that a high degree of spatial resolution is not possible in any vertebrate eye at these low light levels whatever the level of contrast in the scene (Snyder et al. 1977). Thus it would seem safe to conclude that when migrating at night even under light levels in the maximum moonlight – maximum starlight range, birds are able to detect only the grossest detail and are unlikely to achieve the degree of spatial resolution that the human visual system is capable of under the same circumstances. Therefore on both theoretical grounds and by extrapolation from current data on bird vision, it seems safe to conclude that nocturnally migrating birds cannot be guided by fine detail.

Poor resolution of detail, however, need not necessarily be a handicap since migrating birds are flying in open air space well away from obstacles. Also, if birds are using distant topographical features to guide their flight, high visual acuity would not be of any particular value. Certainly under all clear sky conditions nocturnally migrating birds should be able to detect major topographical features on the ground (water surfaces will have higher reflectance, tree canopies a lower reflectance than the average value for grass used in the construction of Fig. 1) and in silhouette against the sky. The sky, and therefore the horizon, will always be visible except on moonless nights with overcast.

3.3 Comparisons with Man and Nocturnal Birds

It may be concluded that nocturnally migrating birds are unlikely to have any more visual information available to them (probably less) than a human being in the same circumstances. These birds' visual capacities at night are probably sufficient to detect only the gross details of topographical features and star patterns. On many occasions night migrants must be essentially flying blind compared with the kind of fine spatial detail which is assumed to guide their flight when completing the rest of their life-cycle. However, these birds are flying in the open air space and their extensive visual fields may facilitate detection of the overall star pattern and the spatial relationships between the major topographical features which encircle and lie below the bird.

It has been argued that in the strictly nocturnal birds, flight mobility must at times depend upon a detailed knowledge of the bird's environment (Martin 1986a). This knowledge facilitates the correct interpretation of those, often minimal, visual cues which are available. Thus the totally nocturnal life-style of these birds appears to rely upon cognitive as well as sensory capacities. The preceding discussion suggests that nocturnal migrants, although flying in spatially simple circumstances, may also at times be dependent upon what may be considered minimal visual cues which must be correctly interpreted. That these nocturnally available cues may sometimes be incorrectly interpreted is suggested by the attraction of birds to artificial sources of illumination under certain circumstances. An owl probably gains its ability correctly to interpret the minimal cues which are available to it through familiarity with its surroundings. For migrant birds such a behavioural strategy is not an option but clearly the ability to interpret minimal sensory information correctly is likely to increase with experience of both the general nature of the night environment and that of a specific migratory route. Limited evidence does indeed suggest that immature passerine birds are influenced more than adults by extraneous lighting cues during night-time tests of migratory orientation (Gauthreaux 1982).

4 Concluding Remarks

The problem of whether individual birds or species can be usefully considered "nocturnal migrants" as opposed to regarding migration at night as a local or temporal phenomenon is an important one which deserves to be reviewed in detail. Geographical location, meteorological conditions and light levels could all influence the likelihood that an individual bird will migrate at night on any particular part of its route. It seems likely that when birds do migrate at night they are primarily dependent upon visual cues for their guidance but that their visual capacities cannot cope adequately with all of the visual problems which the night environment can pose. There is a clear need for further data on the night-time visual capacities of birds.

Acknowledgements. I am most grateful to the wardens of British Bird Observatories who provided details from their records of the occurrence of nocturnal migration and to Chris Mead (British Trust for Ornithology), Robin Baker (University of Manchester), Colin Bibby (Royal Society for the Protection of Birds) and Peter Evans (University of Durham) for valuable correspondence and dicussion.

References

Able KP (1974) Environmental influences on the orientation of free-flying nocturnal bird migrants. Anim Behav 22:224–238

Able KP (1980) Mechanisms of orientation, navigation, and homing. In: Gauthreaux SA Jr (ed) Animal migration, orientation, and navigation. Academic Press, London, p 284

Able KP (1982) Field studies of avian nocturnal migratory orientation. I. Interaction of sun, wind and stars as directional cues. Anim Behav 30:761–767

Aldrich JW, Graber RR, Munro DA, Wallace GJ, West GC, Gahalane VH (1966) Mortality at ceilometers and towers. Auk 83:465–467

Alerstam T, Pettersson S-G (1976) Do birds use waves for orientation when migrating across the sea? Nature 259:205–207

Astrom G (1976) Environmental influence on daily song activity of the reed bunting *Emberiza schoeniclus* (L.). Zoon Suppl 2:1–82

Baker RR (1984) Bird navigation: the solution of a mystery? Hodder & Stoughton, London

Barlow HB (1981) Critical limiting factors in the design of the eye and visual cortex. Proc R Soc B 212:1–34

Bellrose FC (1967) Radar in orientation research. Proc XIV Int Ornithol Congr, pp 281–309

Bellrose FC, Graber RR (1963) A radar study of the flight direction of nocturnal migrants. Prox XIII Int Ornithol Congr, pp 363–389

Biebach H, Friedrich W, Heine G (1986) Interaction of body mass, fat, foraging and stopover period in trans-Sahara migrating passerine birds. Oecologia (Berl) 69:370–379

Bingman VP, Able KP, Kerlinger P (1982) Wind drift, compensation, and the use of landmarks by nocturnal bird migrants. Anim Behav 30:49–53

Blough DS (1955) Method for tracing dark adaptation in the pigeon. Science 121:703–704

Bond DS, Henderson FP (1963) The conquest of darkness, AD 346297, Defense Documentation Center, Alexandria, Va

Bruderer B (1978) Effects of alpine topography and winds on migrating birds. In: Schmidt-Koenig K, Keeton WT (eds) Animal migration, navigation, and homing. Springer, Berlin Heidelberg New York, p 252

Bruderer B (1982) Do migrating birds fly along straight lines? In: Papi F, Wallraff HG (eds) Avian navigation. Springer, Berlin Heidelberg New York, p 3

Dorst J (1961) The migrations of birds. Heinemann, London

Drury WH, Nisbet ICT (1964) Radar studies of orientation of songbird migrants in south-eastern New England. Bird Banding 35:69–119

Durman R (1976) Bardsey. In: Durman R (ed) Bird observatories in Britain and Ireland. Poyser, Calton, p 29

Elkins N (1983) Weather and bird behaviour. Poyser, Calton

Emlen ST (1967) Migratory orientation in the indigo bunting, *Passerina cyanea*, part 1: the evidence for celestial cues. Auk 84:309–342

Emlen ST (1975) Migration: orientation and navigation. In: Farner DS, King JR (eds) Avian biology, vol 5. Academic Press, London, p 129

Emlen ST, Demong NJ (1978) Orientation strategies used by free-flying bird migrants: a radar tracking study. In: Schmidt-Koening K, Keeton WT (eds) Animal migration, navigation and homing. Springer, Berlin, Heidelberg New York, p 283

Feare CJ (1984) The starling. Oxford University Press, Oxford

Fite KV (1973) Anatomical and behavioral correlates of visual acuity in the great horned owl. Vision Res 13:219–230

Gauthreaux SA Jr (1982) Age-dependent orientation in migratory birds. In: Papi F, Wallraff HG (eds) Avian navigation. Springer, Berlin, Heidelberg New York, p 68

Griffin DR (1973) Oriented bird migration in or between opaque cloud layers. Proc Am Philos Soc 117:117–141

Gwinner E (1975) Circadian and circannual rhythms in birds. In: Farner DS, King JR (eds) Avian biology, vol 5. Academic Press, London

Hebrard JJ (1971) Fall nocturnal migration during two successive overcast days. Condor 74:106–107

Kerlinger P, Moore FR (1989) Atmospheric structure and avian migration. In: Power D (ed) Current ornithology, vol 6. Plenum, New York, pp 109–142

Krantz PE, Gauthreaux SA Jr (1975) Solar radiation, light intensity, and roosting behaviour in birds. Wilson Bull 87:91–95

Leask MJM (1977) A physiochemical mechanism for magnetic field detection by migratory birds and homing pigeons. Nature 267:144–145

Martin GR (1977) Absolute visual threshold and scotopic spectral sensitivity in the tawny owl, *Strix aluco*. Nature 268:636–638

Martin GR (1982) An owl's eye: schematic optics and visual performance in *Strix aluco* L. J Comp Physiol 145:341–349

Martin GR (1985) Eye. In: King AS, McLelland J (eds) Form and function in birds, vol 3. Academic Press, London, p 311

Martin GR (1986a) Sensory capacities and the nocturnal habit in owls. Ibis 128:266–277

Martin GR (1986b) Total panoramic vision in the mallard duck *Anas platyrhynchos*. Vision Res 26:1303–1305

Martin GR (1986c) The eye of a passeriforme bird, the European starling (*Sturnus vulgaris*): eye movement amplitude, visual fields and schematic optics. J Comp Physiol 159:545–557

Martin GR (1990) Birds by night. Poyser, London

Mead C (1983) Bird migration. Country Life, Feltham

Middleton WEK (1958) Vision through the atmosphere. University Toronto Press, Toronto

Moore FR (1985) Integration of environmental stimuli in the migratory orientation of the savannah sparrow (*Passerculus sandwichensis*). Anim Behav 33:657–663

Moore FR (1987) Sunset and orientation behaviour of migrating birds. Biol Rev 62:65–86

Nisbet ICT (1970) Autumn migration of the blackpoll warbler: evidence for long flight provided by regional survey. Bird Banding 41:207–240

Palmgren P (1944) Studien über die Tagesrhythmik gekäfigter Zugvogel. Z Tierpsychol 6:44–86

Papi F, Wallraff HG (eds) (1982) Avian Navigation. Int Symp Avian Navigation (ISAN), Tirrenia (Pisa), Sept 11–14, 1981 (Proc Life Sci). Springer, Berlin Heidelberg New York, 380 pp

Pirenne MH, Marriott FHC, O'Doherty EF (1957) Individual differences in night vision efficiency. Med Res Counc GB Spec Rep Ser 294

Presti DE (1985) Avian navigation, geomagnetic field sensitivity, and biogenic magnetite. In: Kirschvink JL, Jones S, MacFadden BJ (eds) Magnetite biomineralization and magnetoreception in organisms. Plenum, New York, p 455

Richardson WJ (1978) Timing and amount of bird migration in relation to weather: a review. Oikos 30:224–272

Sandberg R, Pettersson J, Alerstam T (1988) Shifted magnetic fields lead to deflected and axial orientation of migrating robins, *Erithacus rubecula*, at sunset. Anim Behav 36:877–887

Sauer EGF, Sauer EM (1955) Zur Frage der nächtlichen Zugorientierung von Grasmücken. Rev Suisse Zool 62:250–259

Semm P, Demaine C (1986) Neurophysiological properties of magnetic cells in the pigeon's visual system. J Comp Physiol A Sens Neural Behav Physiol 159:619–625

Semm P, Nohr D, Demaine C, Wiltschko W (1984) Neural basis of the magnetic compass: interactions of visual, magnetic and vestibular inputs in the pigeon's brain. J Comp Physiol A Sens Neural Behav Physiol 155:283–288

Snow D (1953) The migration of the Greenland wheatear. Ibis 95:376–378

Snyder AW, Laughlin SB, Stavenga DG (1977) Information capacity of eyes. Vision Res 17:1163–1175

Taylor AH, Kerr GP (1941) Distribution of energy in daylight. J Opt Soc Am 31:3–8

Verheijen FJ (1981) Bird kills at lighted man-made structure: not on nights close to a full moon. Am Birds 35:251–254

Wiltschko W, Wiltschko R (1975a) The interaction of stars and magnetic field in the orientation system of night migrating birds. I. Autumn experiments with European warblers (Gen *Sylvia*). Z Tierpsychol 37:337–355

Wilstchko W, Wiltschko R (1975b) The interaction of stars and magnetic field in the orientation system of night migrating birds, part II. Spring experiments with European robins (*Erithacus rubecula*). Z Tierpsychol 39:265–282

Wiltschko W, Wiltschko R (1976) Interrelation of magnetic compass and star orientation in night migrating birds. J Comp Physiol 109:91–99

Wood B (1982) The trans-Saharan spring migration of yellow wagtails (*Motacilla flava*). J Zool (Lond) 197:267–283

Nutrition and Food Selection in Migratory Birds

F. Bairlein[1]

1 Introduction

The main fuel for migratory flights in many bird species is fat that is deposited in the bird's body in large amounts prior to, and during migration (Berthold 1975; Blem 1980; Dawson et al. 1983). The extent of migratory fat deposition correlates positively with distance (Berthold 1984). Thus, long-distance migrants exhibit the greatest changes in body mass, with gains of up to 100% above nonmigratory levels. In nature, fat deposition occurs primarily just before the crossing of major ecological barriers (e.g. seas and deserts), with maximum rates of net daily gain of more than 10% of lean body mass (e.g. Bairlein 1987b, 1988). Such a period of exceptional energetic demands requires appropriate nutritional strategies. Furthermore, free-living migrants must choose from a broad range of foods differing in morphology and nutrient contents. In this chapter, I will review nutritional adaptations and aspects of food selection in migrants which appear to correspond to their migratory requirements.

2 Food Intake and Assimilation

Among the mechanisms for migratory fat deposition hyperphagia ("overeating") is the most extensively studied (Blem 1980). In the mainly granivorous white-crowned sparrow *Zonotrichia leucophrys gambelli* (King 1961a,b), and in the bobolink *Dolichonyx oryzivorus* (Gifford and Odum 1965) a marked increase of daily food intake during spring migratory fattening occurred (Fig. 1). These studies found no evidence for changes in utilization efficiency. Other studies involving different species, however, concluded that hyperphagia alone did not entirely account for that increased mass (Merkel 1958; Fry et al. 1972; Bhatt and Chandola 1985). Recently, a long-term study under controlled, laboratory feeding conditions using the insectivorous garden warbler, *Sylvia borin*, demonstrated that migratory fattening in that species, both in autumn and spring, is associated with a pronounced increase in daily food-intake rates, as well as with a significant increase in the efficiency of food utilization. This enhanced efficiency

[1] Department of Zoology, Physiological Ecology Section, University of Köln, Weyertal 119, D-5000 Köln 41, FRG

E. Gwinner (Ed.) Bird Migration

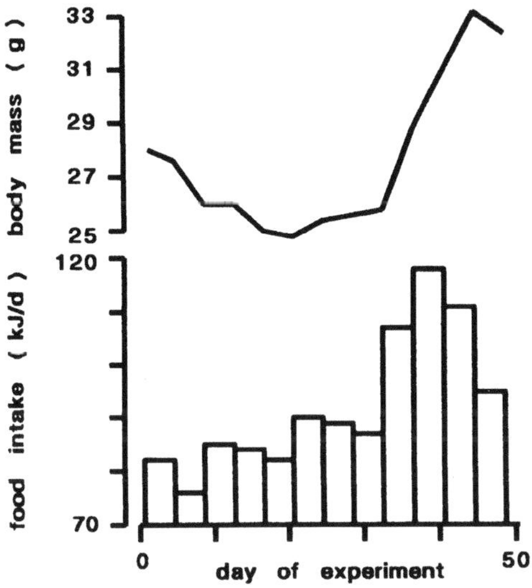

Fig. 1. Changes in body mass and food intake rates of an individual captive white-crowned sparrow during the spring migratory period (After King 1961a)

increases metabolizable energy and nutrient intake (Fig. 2; Bairlein 1985). These changes in assimilation efficiency accounted for, on average, almost one-third of the entire increase in food assimilated during migratory fattening compared to nonmigratory periods. Periods of spontaneous mass loss (though food was always available ad libitum) were characterized by decreasing both food intake and efficiency of food and nutrient assimilation. Most of the variation in assimilation efficiency was due to changes in the efficiency of fat utilization, but minor changes were also found in carbohydrate and protein utilization (Bairlein 1985). In free-living birds, such a strategy may reduce costs associated with foraging (Rayner 1982), and maximize energy intake and, in particular, the uptake of specific nutrients. These data, obtained on captive birds under controlled laboratory conditions, indicate that endogenous control mechanisms, corresponding to the seasonal changes in the birds' physiological states, are involved in the regulation of food intake and food and nutrient utilization, respectively. However, the mechanisms of such a regulation are poorly known. Variations between species (e.g. granivores versus insectivores), between seasons (fall versus spring), in relation to diet or environmental conditions, and, in particular, to the nutritional needs of the birds must be considered.

3 Seasonal Shifts in Food Selection

Traditionally, evidence for adaptive shifts in food selection during migratory fattening was thought to be weak (Blem 1976). Within the last decade, however, an increasing number of studies have demonstrated that many migrants show extreme seasonal shifts in dietary composition. Such seasonal changes in diet composition are perhaps most striking in passerine birds that switch from a diet

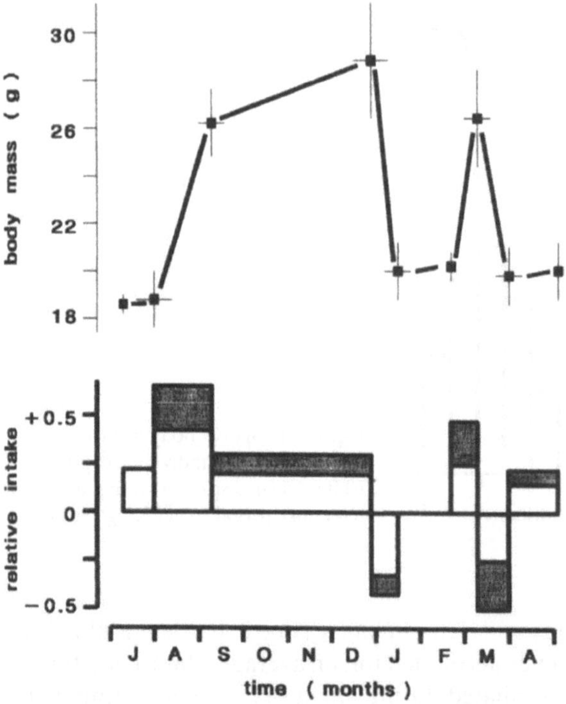

Fig. 2. Average net food intake of 10 captive garden warblers (LD 12:12, 20°C) relative to the low body mass phase in "winter" during various periods of their seasonal body mass cycle (*above*). The *open bars* show the changes in gross food intake (food eaten; dry matter), the *dashed bars* show additional changes in net food intake (food assimilated) due to changing assimilation efficiencies. For clarity, SD of food intake values are avoided. However, most of the variations are highly significant

consisting almost entirely of insects to one that comprises mostly fruits. In particular, just prior to, and during, autumn migration and on the wintering grounds, many migrants rely almost entirely on fruits which can comprise up to 100% of the diet (Nordberg 1942; Schmidt 1964; Morel 1968; Fry et al. 1970; Leck 1972; Moreau 1972; Ferns 1975; Brensing 1977; Howe and DeSteven 1979; Thomas 1979; Baird 1980; Greenberg 1981; Jordano 1981, 1985, 1987a; Debussche and Isenmann 1984, 1985; Herrera 1984a, 1987; Guitian 1987; Bairlein, unpublished work). Seasonal shifts in diet selection are also known from other bird taxa. For example, during migratory stopovers in inland Texas, feeding shorebirds ate seeds to an extent that seeds comprised up to 37% of the diet, and occurred in up to 100% of individuals in a given group (Baldassare and Fischer 1984). Furthermore, in many waders fruits comprise important dietary items during migration (Glutz von Blotzheim et al. 1975, 1977). Moreover, geese feed selectively on various plant species during migration, depending upon particular nutrient requirements (Fig. 3; e.g. Raveling 1979; Drent 1980; Drent et al. 1981; Boudewijn 1984). Excluding the geese, however, the role of seasonal changes in diet selection in migrants and the mechanisms involved in these changes are, in

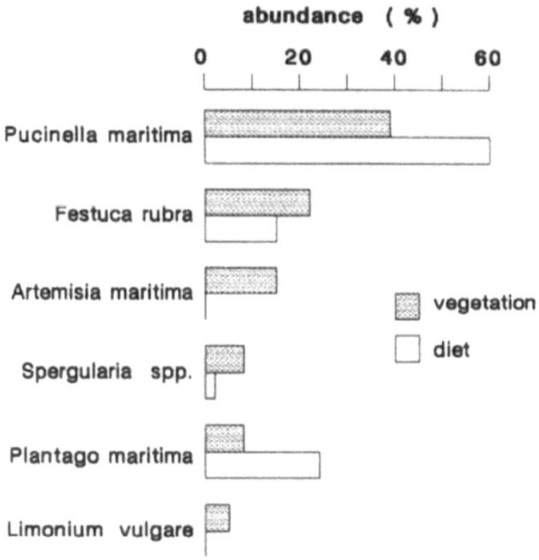

Fig. 3. Selective feeding of brent (*Branta bernicla*) on various plant species, indicated by differences in the relative occurrence of items in the vegetation and in droppings (After Dijkstra and Dijkstra de Vlieger 1977, cited in Bergmann 1987)

general, poorly known. In many regions, the time of maximum migration corresponds closely with the peak in fruit abundance (Stiles 1980; Willson et al. 1982; Herrera 1984a). Thus, seasonal frugivory in migrants may simply reflect seasonal changes in availability of various food items, both of insects and the alternate fruit diet, respectively. Insectivores may switch to the seasonal abundant fruits when insects became scarce.

It has been noted, however, that the degree of frugivory during migration, as well as species-specific preferences for specific fruits, often vary independently of local fruit abundance (e.g. Kroll 1972; Morton 1973; Thompson and Willson 1979; Baird 1980; Greenberg 1981; Sorensen 1981; Izhaki and Safriel 1985; Martin 1985a; Jordano 1987a). Furthermore, studies on food preference under controlled laboratory conditions and under unlimited availability of various food items showed that seasonal variation in food selection (and the extent of frugivory) can be independent of availability (Fig. 4; Berthold 1976; Wheelwright 1988; Bairlein, unpublished work). Thus, seasonal shifts in diet may also be associated with endogenous, annual changes in preference. These changes may then correspond to nutritional requirements of the individual bird and might vary with seasonal changes in the bird's physiological state. Specific nutritional needs during migratory fattening may be fulfilled better by some foods than others, and consequently, the nutritional quality of food items may significantly influence foraging decisions in migrants (Yarbrough and Johnston 1965; Raveling 1979; Bairlein 1985; King and Murphy 1985; Buchsbaum and Valiela 1987). Moreover, an endogenous component in the control of seasonal food selection may reduce risks involved when dealing with novel food resources (Greenberg 1983, 1987).

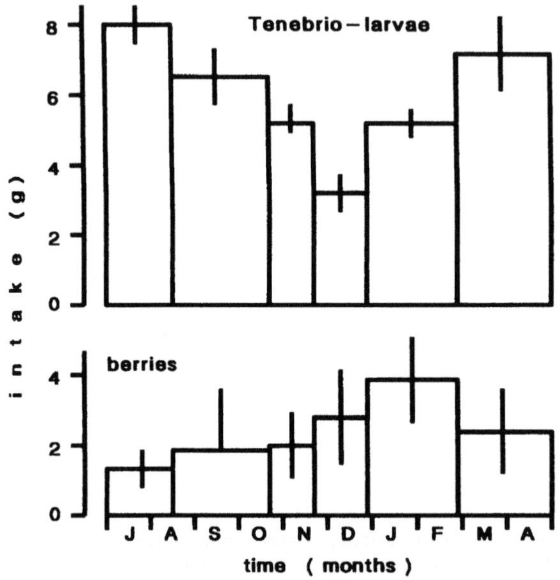

Fig. 4. Seasonal variation in intake of insect food (mealworms) and fruits, respectively, (means, SD) in garden warblers kept indoor in constant photoperiod and temperature. Both food items were always simultaneously available ad libitum (After Berthold 1976)

4 Selection for Energy and Nutrients

The factors governing food selection in wild animals are very complex, and yet little investigated (Robbins 1983; King and Murphy 1985). With respect to migrant frugivory, fruit selection and the rate of fruit intake are influenced by: (1) habitat factors such as spatial distribution and abundance of fruiting plants; (2) plant characteristics including conspicuousness and accessibility of fruits and crop size; (3) fruit characteristics, such as color, size, pulp mass, shape, taste, palatability, digestibility, energy and nutrient content, chemical composition and secondary plant metabolites; and (4) degree of ripeness or microbial infection (e.g. Turcek 1963; Snow 1971; Stiles 1976; Sorensen 1981, 1983; Herrera 1982, 1987; Willson and Thompson 1982; Willson and Melampy 1983; Morden-Moore and Willson 1982; Martin 1985a,b; Moermond and Denslow 1985; Davidar and Morton 1986; Willson 1986; Burger 1987; Borowicz 1988a,b). Furthermore, food choice and intake may be related to the birds' physiological needs as well as their morphological capacities (e.g. digestive capacity, intestine morphology, size of feeding apparatus, visual systems, etc.; Ricklefs 1977; Rybczynski and Riker 1981; Burkhardt 1982; Eriksson and Nummi 1982; Herrera 1984b; Wheelwright 1985; Jordano 1987b). In addition, behavioral aspects, like handling techniques of prey items, intra- and interspecific competition, or fear (Howe and Estabrook 1977; Howe 1979; Hughes 1979; Levey 1987) may interfere with diet selection and food intake, respectively.

Due to the complex interactions of these various factors, selective feeding, which in turn leads to migratory fattening is difficult to evaluate using field studies (Sorensen 1981; Herrera 1984a; Johnson et al. 1985; Willson 1986). However, such a period of exceptional energetic demands should select for appropriate strategies, i.e., choosing only the most rewarding items. Therefore, it should be advantageous for migrating birds to forage preferentially on high-fat and/or high-energy food (Snow 1971; Blem 1976; Stiles 1980). Only few studies, however, have established a relationship between diet quality and diet selection during migratory fat deposition.

Feeding on seeds by migrating shorebirds is considered to be related to the high carbohydrate content of seeds which should replenish body reserves faster than invertebrates (Baldassare and Fischer 1984). Furthermore, Thompson and Willson (1979) mentioned that fall passerine migrants seem to be able to seek the energy provided by fruits. Recently, the possibility that a migrant bird can sense the energy and nutrient level in diets was experimentally confirmed in captive garden warblers (Bairlein 1990), using a semisynthetic diet in which all food variables were kept constant and only nutritional composition was specifically varied (Bairlein 1986). Such experimental diets are a prerequisite for the analysis of nutritional relationships in animals (Taghon 1981; Murphy and King 1982; Bairlein 1986; Eggel and Ziswiler 1988).

Garden warblers were tested in replicated, dual-choice trials in which two diets differing in nutrient and/or caloric value were presented simultaneously. In trials in which only the protein content of the foods varied, the birds preference depended on the level of dietary protein (Fig. 5). Significantly different choices were made at low or moderate protein levels, but did not occur at high dietary protein levels. In trials in which lipid content and nonstructural carbohydrates, respectively, were varied, the birds always preferred food with higher fat and higher sugar content. However, with respect to a corresponding variation in energy content of the higher-nutrient foods, the observed variation in preference

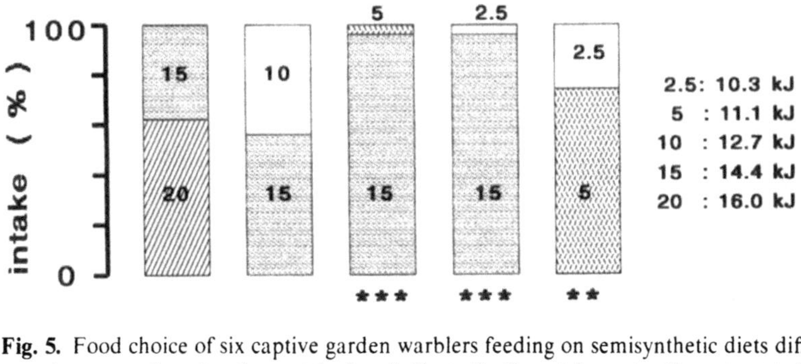

Fig. 5. Food choice of six captive garden warblers feeding on semisynthetic diets differing only in protein content (*numbers* in or at the *bars* indicate the % protein content in wet matter). In each trial two foods were presented simultaneously for 4 days. *Stars* below the bars indicate significantly different selection ($^{**}p < 0.01$; $^{***}p < 0.001$). The caloric values of the diets are shown on the *right* side of the figure (After Bairlein 1990)

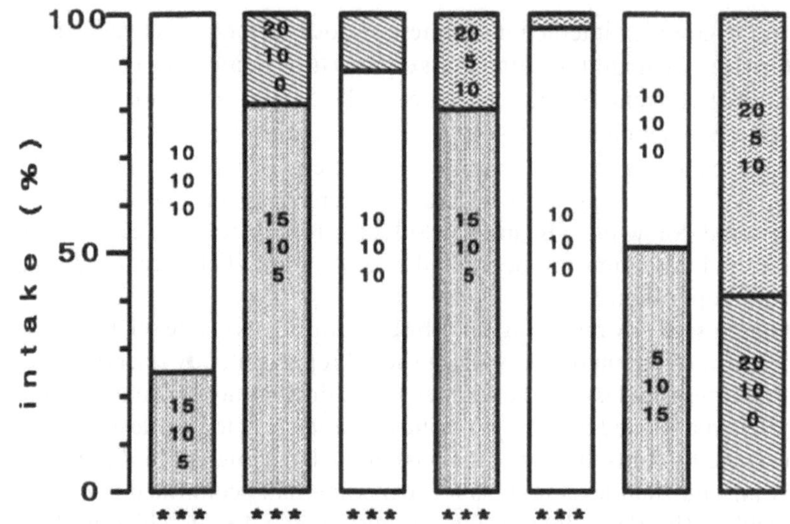

Fig. 6. Average food selection by six captive garden warblers feeding on isocaloric diets differing in nutritional composition. The *figures within the bars* show (*from top*) % protein, % fat and % nonstructural carbohydrates, respectively, in the wet diets (50% water content). *Stars* below the bars indicate statistically different selection (see Fig. 5) (After Bairlein 1990)

may be simply due to the higher energy content, though not necessarily as obtained on the high-protein levels. Trials with isocaloric diets, however, showed that garden warblers are indeed able to feed selectively on foods differing in dietary nutrients independent of the caloric value (Fig. 6). In general, the birds preferred the diets with the higher fat levels. Among foods similar in fat content, choice was closely related to the amount of nonstructural carbohydrates. Preference for fruits high in fat content has also been shown in northern mockingbirds (*Mimus polyglottos*) and gray catbirds (*Dumetella carolinensis*), but not in white-throated sparrows (*Zonotrichia albicollis*; Borowicz 1988b). Thus, migrants seem to be able to respond to variations in nutritional quality of foods by making selections appropriate for meeting their nutritional requirements most effectively, although species-specific differences may exist.

5 Nutritional Constraints of Fruit Diets

In contrast to insects and many seeds, fruits are, in general, relatively low in protein and lipids, but may be high in carbohydrates (Snow 1971; Moermond and Denslow 1985; Debussche et al. 1987; Herrera 1987). The extent of frugivory in omnivorous songbirds has been assumed to be limited by the paucity of proteins found in fruits and this may prevent the birds from relying on fruits for long periods of time (Morton 1973; Berthold 1976). However, low dietary protein levels are not necessarily disadvantageous for birds "whose immediate need is to lay down a reserve of fat in a short time" (Snow 1971), although protein content and protein quality may have a pronounced effect on fitness and consequently food selection during other stages of the birds life cycle (e.g. prior to and during reproduction or during molt; e.g. Friedman 1975; McLandress and Raveling 1981; Von der Decken 1983; Sedinger and Raveling 1984; Teunissen et al. 1985; Waterlow 1986; Bazely 1987; Buchsbaum and Valiela 1987; Murphy and King 1987; Ankney and Afton 1988).

Furthermore, it has been shown (Bairlein 1987a) that captive garden warblers are able to recover in body mass after an initial loss although they were maintained on low-protein diets for a couple of weeks (Fig. 7). The birds compensated for this loss by increasing their food intake, although this compensation occurred after a latency in response to very low protein levels (Fig. 8). Moreover, average daily protein requirements of garden warblers (2.7 mg g^{-1} d^{-1}) were much lower (only about 10%) than previously supposed (Berthold 1976),

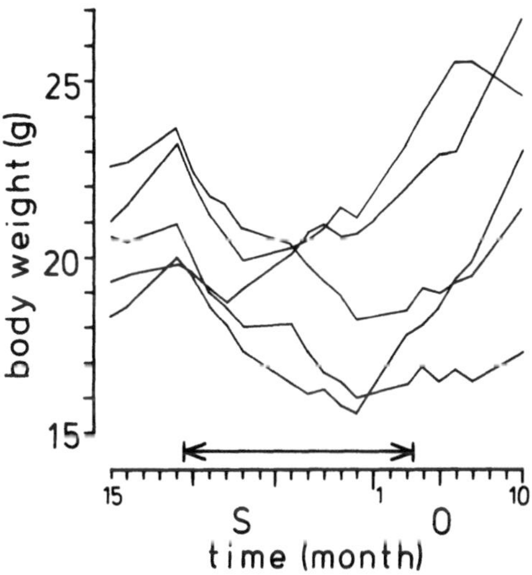

Fig. 7. Changes in body mass of five captive garden warblers fed on a low-protein diet (*arrow*; 2.4% protein vs 14.4% protein before and after) (After Bairlein 1987a)

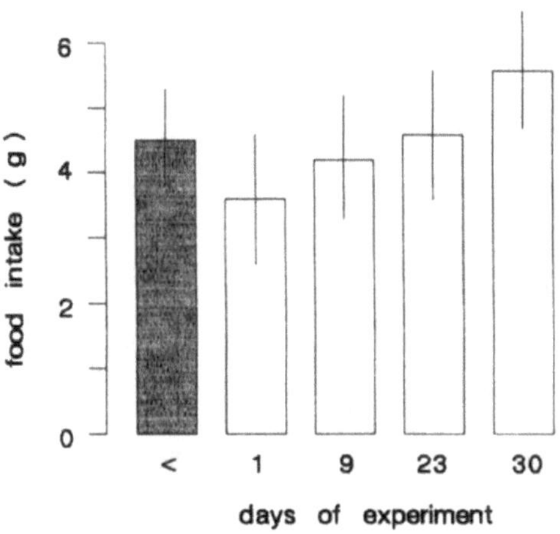

Fig. 8. Daily food intake of five garden warblers while fed on the low-protein diet (see Fig. 7; *open bars*). The *dashed bar* shows the average intake of the standard food (14.4% protein) (After Bairlein 1987a)

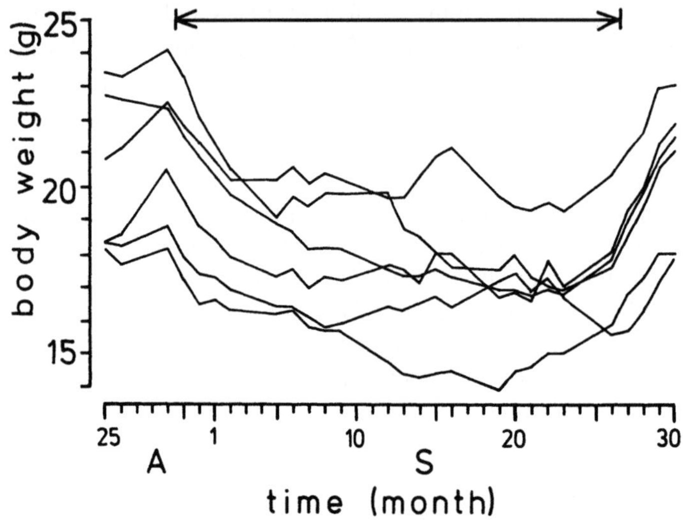

Fig. 9. Body mass changes of six garden warblers while being fed on a lipid-reduced diet (*arrow*; 5% vs 10% fat before and after; after Bairlein 1987a)

indicating that low protein levels of fruits appear to be much less critical for relying on fruits. Daily needs for lipids were much higher than for proteins (27 mg $g^{-1} d^{-1}$), and feeding the birds fat-reduced diets resulted in a much longer recovery in body mass (Fig. 9). In addition, compensatory food intake during low-fat feeding was more pronounced than at the protein restricted diets. Furthermore, low fat levels were also compensated for by a significant increase in fat and carbohydrate assimilation efficiency (Bairlein 1987a). Together, these data clearly indicate that these birds can compensate for nutritional deficiencies by effectively regulating their intake of specific food types and nutrients. The ability to detect specific nutrient deficiencies and to respond quickly and appropriately by selective feeding may be important in avoiding malnutrition, and in adapting to changing nutritional demands as well as changing food resources during migration (King and Murphy 1985; Murphy and King 1987).

6 Role of Frugivory

As in the case of the garden warbler, other migrants may also be equipped with the ability to make highly selective dietary choices and, as such, may necessitate a reevaluation of the role of frugivory in migratory fattening. The birds are able to feed selectively on the most rewarding food items, in particular on diets rich in fat. They respond to low dietary nutrient content by a compensatory increase in food intake and nutrient assimilation efficiency, which enables them to respond rapidly to these low levels. Furthermore, such migrants exhibit variations in their intestine morphology and physiology which seem to be adaptive related to their fruit diets (Walsberg 1975; Sorensen 1984; Jordano 1987b).

Relying on fruits during migration appears to be a significant adaptation. Fruits are often locally super-abundant, thus minimizing time spent searching for food; they are "easy prey" (Snow 1971); due to their generally low fiber content, fruits seem to require relatively little mechanical breakdown to render the nutritious cell content available (Moermond and Denslow 1985); and finally, they seem to be rich enough in nutrients to meet the specific physiological needs of feeding migrants (see also Catterall 1985 regarding the importance of fruits for winter survival).

Regarding the specific role of dietary lipids, it should be noted that many fruits in the Mediterranean region (where most of the Palaearctic passerine migrants fatten during fall migration) appear to be generally higher in lipids than in the temperate zone (Snow 1971; Herrera and Jordano 1981; Debussche et al. 1987; Herrera 1987), and that the occurrence of many frugivores coincides with a predominance of these lipid-rich fruits (Herrera 1984a). It should also be noted that many fruits also offer a rich carbohydrate source which may further support fat deposition.

In European robins (*Erithacus rubecula*) mass gain due to fat deposition in the Mediterranean showed a close relationship to the consumption of nutritive fruits, and these birds gained the most mass when feeding on fruits relatively rich in lipids (Herrera 1977, 1981). Similar relationships were found in autumn passage migrants in Portugal where frugivores increased body mass at about twice the rate of insectivores (Ferns 1975), and birds feeding on figs (*Ficus carica*) in Mediterranean stopover sites showed significantly higher average body mass than birds feeding elsewhere (Ferns 1975; Thomas 1979; Baccetti et al. 1985; Bairlein in prep.).

Similar results were obtained in a preliminary experiment on captive garden warblers (Fig. 10). Additional ad libitum feeding with black elder berries (*Sambucus nigra*) resulted in a rapid additional gain in body mass that was associated with a marked increase in total daily food intake. Interestingly, however, additional feeding with red elder berries (*Sambucus racemosa*) did not enhance body mass gain and food intake, but rather resulted in a loss of body mass, although both berries are relatively similar in their nutritional values (Berthold 1976), and standard food was simultaneously available ad libitum. These data indicate that various fruits (and foods in general) may have quite different effects on food intake and/or body mass. It may be further speculated that secondary plant chemicals, well known as feeding agents in herbivores (e.g. Moss et al. 1971; Bryant and Kuropat 1980; Buchsbaum et al. 1984; Buchsbaum and Valiela 1987; Linden 1984; Hudson and White 1985), play a significant role in migratory fattening, either by affecting diet selection, stimulating food intake, or affecting metabolism (Schmidt 1964; Ettinger and King 1981).

Another aspect, sometimes mentioned regarding frugivory in omnivorous songbirds, is the generally high water content of fruits which might induce migrants to consume fruits as a water source (Walsberg 1975; Herrera 1984). In blackcaps (Berthold 1976) and garden warblers (Bairlein, unpublished work), however, consumption of berries did not increase in trials without drinking water available at a constant laboratory ambient temperature of 20°C. At higher

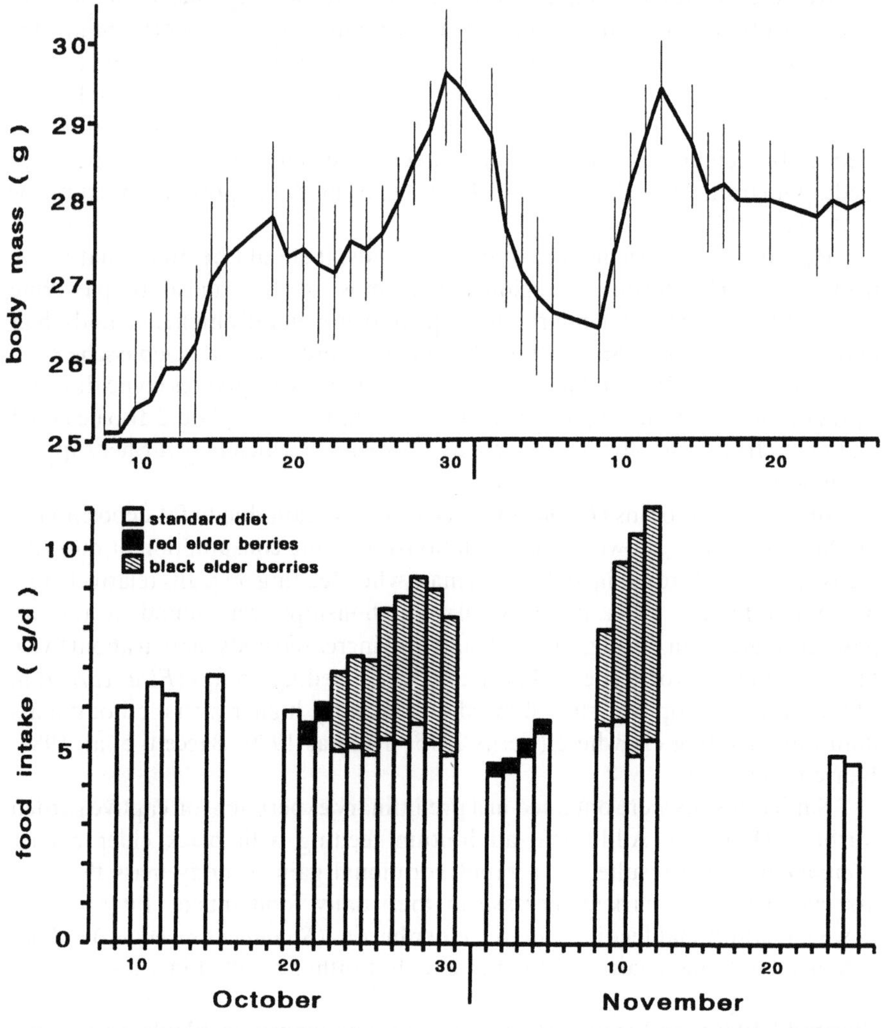

Fig. 10. Changes in average body mass (± SD; *above*) and changes in food intake (dry matter; *below*) of 8 captive garden warblers. The birds were continuously fed ad libitum on a standard lab diet. On some days, they additionally received fruits ad libitum

ambient temperatures, however, which result in increased water requirements (Bairlein, unpublished work), berries may provide an important water source, in particular when drinking water is less available in drier habitats (Walsberg 1975).

7 Conclusions

Most of the data available regarding nutrition and food selection in migrants are related to seasonal frugivory. Much less is known on the role of seasonal shifts in the invertebrate diets of migrants. For instance, differences in body mass of some wader species (little ringed plover, *Charadrius dubius*; wood sandpiper, *Tringa glareola*) between inland and coastal feeding sites (Glutz von Blotzheim et al. 1975, 1977) may indicate differences in prey availability and/or in nutritional quality of prey items. Furthermore, selective feeding of warblers on aphids during migratory stopover may be related to the high sugar content of aphids (Bibby and Green 1981; Glutz von Blotzheim 1986). However, although the data on nutrition and food selection are still scarce, they do provide evidence that the specific nutrient requirements of fattening migrants and, consequently, their specific appetites, may significantly influence their foraging decisions. Thus, for instance, foraging tactics (e.g. Drent 1980; McLandress and Raveling 1981; Moore and Simm 1985, 1986), and habitat selection (Bairlein 1981; Cody 1985; Martin 1985a,b) may be strongly related to both the bird's nutritional demands and the nutritional quality of food items. The nutritional requirements and the nutritional capacities of most wild birds, however, are poorly known (e.g. Robbins 1983; King and Murphy 1985), offering a worthwhile field of research.

Acknowledgements. J.R. King and M.E. Murphy made helpful comments on the manuscript. S.B. Terrill improved the English text. This work was supported by the Deutsche Forschungsgemeinschaft.

References

Ankney CD, Afton AD (1988) Bioenergetics of breeding northern shovelers: diet, nutrient reserves, clutch size, and incubation. Condor 90:459–472

Baccetti N, Frugis S, Mongini D, Piacentini D, Spina F (1985) Osservazioni sul peso di beccafichi, *Sylvia borin*, in transito in diverse aree Italiane. Riv Ital Ornithol 55:171–179

Baird JW (1980) The selection and use of fruit by birds in an eastern forest. Wilson Bull 92:63–73

Bairlein F (1981) Ökosystemanalyse der Rastplätze von Zugvögeln. Ecol Birds 3:7–137

Bairlein F (1985) Efficiency of food utilization during fat deposition in the long-distance migratory garden warbler, *Sylvia borin*. Oecologia (Berl) 68:118–125

Bairlein F (1986) Ein standardisiertes Futter für Ernährungsuntersuchungen an omnivoren Kleinvögeln. J Ornithol 127:338–340

Bairlein F (1987a) Nutritional requirements for maintenance of body weight and fat deposition in the long-distance migratory garden warbler, *Sylvia borin* (Boddaert). Comp Biochem Physiol A Comp Physiol 86:337–347

Bairlein F (1987b) The migratory strategy of the garden warbler: a survey of field and laboratory data. Ringing & Migr 8:59–72

Bairlein F (1988) Herbstlicher Durchzug, Körpergewichte und Fettdeposition von Zugvögeln in einem Rastgebiet in N-Algerien. Vogelwarte 34:237–248

Baldassarre GA, Fischer DH (1984) Food habits of fall migrant shorebirds on the Texas high plains. J Field Ornithol 55:220–229

Bazley DR (1987) Snow buntings feeding on leaves of salt-marsh grass during spring migration. Condor 89:190–192

Bergmann HH (1987) Die Biologie des Vogels. Aula, Wiesbaden

Berthold P (1975) Migration: control and metabolic physiology. In: Farner DS, King JR (eds) Avian biology, vol 5. Academic Press, London, pp 77–128

Berthold P (1976) Animalische und vegetabilische Ernährung omnivorer Singvogelarten: Nahrungsbevorzugung, Jahresperiodik der Nahrungswahl, physiologische und ökologische Bedeutung. J Ornithol 117:145–209

Berthold P (1984) The endogenous control of bird migration. Bird Study 31:19-27
Bhatt D, Chandola A (1985) Circannual rhythms of food intake in spotted munia and its phase relationship with fattening and reproductive cycles. J Comp Physiol A Sens Neural Behav Physiol 156:429-432
Bibby CJ, Green RE (1981) Autumn migration strategies of reed and sedge warblers. Ornis Scand 12:1-12
Blem CR (1976) Patterns of lipid storage and utilization in birds. Am Zool 16:671-684
Blem CR (1980) The energetics of migration. In: Gauthreaux SA (ed) Animal migration, orientation and navigation. Academic Press, London, pp 175-224
Borowicz VA (1988a) Do vertebrates reject decaying fruit? An experimental test with *Cornus amomum* fruits. Oikos 53:74-78
Borowicz VA (1988b) Fruit consumption by birds in relation to fat content of pulp. Am Midl Nat 119:121-127
Boudewijn T (1984) The role of digestibility in the selection of spring feeding sites by brent geese. Wildfowl 35:97-105
Brensing D (1977) Nahrungsökologische Untersuchungen an Zugvögeln in einem südwestdeutschen Durchzugsgebiet während des Wegzuges. Vogelwarte 29:44-56
Bryant JP, Kuropat PJ (1980) Selection of winter forage by subarctic browsing vertebrates: the role of plant chemistry. Annu Rev Ecol Syst 11:261-285
Buchsbaum R, Valiela I (1987) Variability in the chemistry of estuarine plants and its effect on feeding by Canada geese. Oecologia (Berl) 73:146-153
Buchsbaum R, Valiela I, Swain T (1984) The role of phenolic compounds and other plant constituents in feeding by Canada geese in a coastal marsh. Oecologia (Berl) 63:343-349
Burger AE (1987) Fruiting and frugivory of *Cornus canadensis* in boreal forests in Newfoundland. Oikos 49:3-10
Burkhardt D (1982) Birds, berries and UV. Naturwissenschaften 69:153-157
Catterall CP (1985) Winter energy deficits and the importance of fruit versus insects in a tropical island bird population. Aust J Ecol 10:265-279
Cody ML (1985) Habitat selection in birds. Academic Press, London
Davidar P, Morton ES (1986) The relationship between fruit crop sizes and fruit removal rates by birds. Ecology 67:262-265
Dawson W, Yacoe ME, Marsh RL (1983) Metabolic adjustments of small birds for migration and cold. Am J Physiol 245:R755-767
Debussche M, Isenmann P (1984) Origine et nomadisme des Fauvettes à tete noir (*Sylvia atricapilla*) hivernant en zone méditerranéenne francaise. L'Oiseau & RFO 54:101-107
Debussche M, Isenmann P (1985) Frugivory of transient and wintering European robins *Erithacus rubecula* in a Mediterranean region and its relationship with ornithochory. Holarct Ecol 8:157-163
Debussche M, Cortez J, Rimbault I (1987) Variation in fleshy fruit composition in the Mediterranean region: the importance of ripening season, life-form, fruit type and geographical distribution. Oikos 49:244-252
Decken A von der (1983) Experimental studies on the quality of food proteins. Comp Biochem Physiol B Comp Biochem 74:213-220
Drent R (1980) Goose flocks and food exploitation: how to have your cake and eat it. Acta XVII Congr Int Ornithol Berlin 1978:800-806
Drent R, Ebbinge B, Weijand B (1981) Balancing the energy budgets of arctic-breeding geese throughout the annual cycle: a progress report. Verh Ornithol Ges Bayern 23:239-264
Eggel W, Ziswiler V (1988) Ein standardisiertes Futter für ernährungsökologische Untersuchungen an fruchtfressenden Vögeln. J Ornithol 129:97-99
Eriksson K, Nummi H (1982) Alcohol accumulation from ingested berries and alcohol metabolism in passerine birds. Ornithol Fenn 60:2-9
Ettinger AO, King JR (1981) Consumption of green wheat enhances photostimulated ovarian growth in white-crowned sparrows. Auk 98:832-833
Ferns PN (1975) Feeding behaviour of autumn passage migrants in north east Portugal. Ringing & Migr 1:3-11
Friedman M (1975) Protein nutritional quality of foods and feeds. Dekker, New York

Fry CH, Ash JS, Ferguson-Lees IJ (1970) Spring weights of some palaearctic migrants at Lake Chad. Ibis 112:58–82
Fry CH, Ferguson-Lees IJ, Dowsett RJ (1972) Flight muscle hypertrophy and ecophysiological variation of yellow wagtail *Motacilla flava* races at Lake Chad. J Zool (Lond) 167:293–306
Gifford CE, Odum EP (1965) Bioenergetics of lipid deposition in the bobolink, a transequatorial migrant. Condor 67:383–403
Glutz von Blotzheim U (1986) Gelegenheitsbeobachtungen an Grasmücken der Gattung *Sylvia* (Aves). Ann Naturhist Mus Wien 88/89 B:15–23
Glutz von Blotzheim U, Bauer K, Bezzel E (1975, 1977) Handbuch der Vögel Mitteleuropas, vol 6 and 7. Akademische Verlagsgesellschaft, Wiesbaden
Greenberg R (1981) Frugivory in some migrant tropical forest wood warblers. Biotropica 13:215–223
Greenberg R (1983) The role of neophobia in determining the degree of foraging specialization in some migrant warblers. Am Nat 122:444–453
Greenberg R (1987) Seasonal foraging specialization in the worm-eating warbler. Condor 89:158–168
Guitian J (1987) *Hedera helix* y los pajaros dispersantes de sus semillas: tiempo de estancia en la planta y efficiencia de movilizacion. Ardeola 34:23–35
Herrera CM (1977) Ecologia alimenticia del Petirrojo (*Erithacus rubecula*) durante su invernada en encinares del sur de Espana. Donana Acta Vertebr 4:35–59
Herrera CM (1981) Fruit food of Robins wintering in southern Spanish Mediterranean scrubland. Bird Study 28:115–122
Herrera CM (1982) Seasonal variation in the quality of fruits and diffuse coevolution between plants and avian dispersers. Ecology 63:773–785
Herrera CM (1984a) A study of avian frugivores, bird-dispersed plants, and their interaction in Mediterranean scrublands. Ecol Monogr 54:1–23
Herrera CM (1984b) Adaptation to frugivory of mediterranean avian seed dispersers. Ecology 65:609–617
Herrera CM (1987) Vertebrate-dispersed plants of the Iberian Peninsula: a study of fruit characteristics. Ecol Monogr 57:305–331
Herrera CM, Jordano P (1981) *Prunus mahaleb* and birds: the high-efficiency seed dispersal system of a temperate fruiting tree. Ecol Monogr 51:203–218
Howe HF (1979) Fear and frugivory. Am Nat 114:925–931
Howe HF, Steven D De (1979) Fruit production, migrant bird visitation, and seed dispersal of *Guarea glabra* in Panama. Oecologia (Berl) 39:185–196
Howe HF, Estabrook (1977) On intraspecific competition for avian dispersers in tropical trees. Am Nat 111:817–832
Hudson RJ, White RG (1985) Bioenergetics of wild herbivores. CRC, Boca Raton
Hughes RN (1979) Optimal diets under the energy maximization premise: the effects of recognition time and learning. Am Nat 113:209–221
Izhaki I, Safriel UN (1985) Why do fleshy-fruit plants of the Mediterranean scrub intercept fall- but not spring-passage of seed-dispersing migratory birds? Oecologia (Berl) 67:40–43
Johnson RA, Willson MF, Thompson JN, Bertin RI (1985) Nutritional values of wild fruits and consumption by migrant frugivorous birds. Ecology 66:819–827
Jordano P (1981) Alimentacion y relaciones troficas entre los passeriformes en paso otonal por una localidad de Andalucia central. Donana Acta Vertebr 8:103–124
Jordano P (1985) El ciclo anual de los paseriformes frugivoros en el matorral mediterraneo del sur de Espana: Importancia de su invernada y variaciones interanuales. Ardeola 32:69–94
Jordano P (1987a) Notas sobre la dieta no-insectivora de algunos Muscicapidae. Ardeola 34:89–98
Jordano P (1987b) Frugivory, external morphology and digestive system in Mediterranean sylviid warblers, *Sylvia* spp. Ibis 129:175–189
King JR (1961a) On the regulation of vernal premigratory fattening in the white-crowned sparrow. Physiol Zool 34:145–157
King JR (1961b) The bioenergetics of vernal premigratory fat deposition in the white-crowned sparrow. Condor 63:128–142
King JR, Murphy ME (1985) Periods of nutritional stress in the annual cycles of endotherms: fact or fiction? Am Zool 25:955–964

Kroll H (1972) Zur Nahrungsökologie der Gartengrasmücke (*Sylvia borin*) beim Herbstzug 1969 auf Helgoland. Vogelwarte 26:280–285

Leck CF (1972) The impact of some North American migrants at fruiting trees in Panama. Auk 89:842–850

Levey DJ (1987) Seed size and fruit-handling techniques of avian frugivores. Am Nat 129:471–485

Linden H (1984) The role of energy and resin contents in the selective feeding of pine needles by the capercaillie. Ann Zool Fenn 21:435–439

Malmborg PK, Willson MF (1988) Foraging ecology of avian frugivores and some consequences for seed dispersal in an Illinois woodlot. Condor 90:173–186

Martin TE (1985a) Selection of second-growth woodlands by frugivorous migrating birds in Panama: an effect of fruit size and plant density? J Trop Ecol 1:157–170

Martin TE (1985b) Resource selection by tropical frugivorous birds: integrating multiple interactions. Oecologia (Berl) 66:563–573

McLandress MR, Raveling DG (1981) Changes in diet and body composition of canada geese before spring migration. Auk 98:65–79

Merkel FW (1958) Untersuchungen über tages- und jahreszeitliche Änderungen im Energiehaushalt gekäfigter Zugvögel. Z Vergl Physiol 41:154–178

Moermond TC, Denslow JS (1985) Neotropical avian frugivores: patterns of behavior, morphology, and nutrition, with consequences for fruit selection. In: Buckley PA, Foster MS, Morton ES, Ridgely RS, Buckley FG (eds) Neotropical ornithology. Ornithol Monogr 36:865–897

Moore FR, Simm PA (1985) Migratory disposition and choice of diet by the yellow-rumped warbler (*Dendroica coronata*). Auk 102:820–826

Moore FR, Simm PA (1986) Risk-sensitive foraging by a migrating bird (*Dendroica coronata*). Experientia (Basel) 42:1054–1056

Morden-Moore AL, Willson MF (1982) On the ecological significance of fruit color in *Prunus serotina* and *Rubus occidentalis*: field experiments. Can J Bot 60:1554–1560

Moreau RE (1972) The Palaearctic-African bird migration system. Academic Press, London

Morel G (1968) Contribution a la synécologie des oiseaux du Sahel Sénégalais. ORSTOM, Paris

Morton ES (1973) On the evolutionary advantages and disadvantages of fruit eating in tropical birds. Am Nat 107:8–22

Moss R, Watson A, Parr R, Glennie W (1971) Effects of dietary supplements of newly growing heather on the breeding of captive red grouse. Br J Nutr 25:135–143

Murphy ME, King JR (1982) Semi-synthetic diets as tool for nutritional ecology. Auk 99:165–167

Murphy ME, King JR (1987) Dietary discrimination by molting white-crowned sparrows given diets differing only in sulfur amino acid concentration. Physiol Zool 60:279–289

Nordberg S (1942) Über die Ernährungsweise der Vögel während der Zugzeit. Ornithol Fenn 19:85–88

Raveling DG (1979) The annual cycles of body composition of Canada geese with special reference to control of reproduction. Auk 96:234–252

Rayner JMV (1982) Avian flight energetics. Annu Rev Physiol 44:109–119

Ricklefs RE (1977) A discriminant function analysis of assemblages of fruit-eating birds in Central America. Condor 79:228–231

Robbins CT (1983) Wildlife feeding and nutrition. Academic Press, London

Rybczynski R, Riker DK (1981) A temperate species-rich assemblage of migrant frugivorous birds. Auk 98:176–179

Schmidt E (1964) Untersuchungen an einigen Holunder fressenden Singvögeln in Ungarn. Zool Abh Mus Tierkd (Dres) 27:11–28

Sedinger JS, Raveling DG (1984) Dietary selectivity in relation to availability and quality of food for goslings of cackling geese. Auk 101:295–306

Snow DW (1971) Evolutionary aspects of fruit-eating by birds. Ibis 113:194–202

Sorensen AE (1981) Interactions between birds and fruit in a temperate woodland. Oecologia (Berl) 50:242–249

Sorensen AE (1983) Taste aversion and frugivore preference. Oecologia (Berl) 56:117–120

Sorensen AE (1984) Nutrition, energy and passage time: experiments with fruit preference in European blackbirds (*Turdus merula*). J Anim Ecol 53:545–557

Stiles FG (1976) Taste preferences, color preferences and flavor choice in hummingbirds. Condor 78:10–26
Stiles EW (1980) Patterns of fruit presentation and seed dispersal in bird-disseminated woody plants in the eastern deciduous forest. Am Nat 116:670–688
Taghon GL (1981) Beyond selection: optimal ingestion rate as a function of food value. Am Nat 118:202–214
Teunissen W, Spaans B, Drent R (1985) Breeding success in brent in relation to individual feeding opportunities during spring staging in the Wadden Sea. Ardea 73:109–119
Thomas DK (1979) Figs as food source of migrating garden warblers in southern Portugal. Bird Study 26:187–191
Thompson JN, Willson MF (1979) Evolution of temperate fruit/bird interactions: phenological strategies. Evolution 33:973–982
Turcek FJ (1963) Color preferences in fruit- and seed-eating birds. Proc XIII Int Ornithol Congr, pp 285–292
Walsberg GE (1975) Digestive adaptations of *Phainopepla nitens* associated with the eating of mistletoe berries. Condor 77:169–174
Waterlow C (1986) Metabolic adaptation to low intakes of energy and protein. Annu Rev Nutr 6:495–526
Wheelwright NT (1985) Fruit size, gape width, and the diets of fruit-eating birds. Ecology 66:808–818
Wheelwright NT (1988) Seasonal changes in food preferences of American robins in captivity. Auk 105:374–378
Willson MF (1986) Avian frugivory and seed dispersal in eastern North America. Curr Ornithol 3:223–279
Willson MF, Melampy MN (1983) The effect of bicolored fruit displays on fruit removal rates by avian frugivores. Oikos 41:29–31
Willson MF, Thompson JN (1982) Phenology and ecology of color in bird-dispersed fruits, or why some fruits are red when they are "green". Can J Bot 60:701–713
Willson MF, Porter EA, Condit RS (1982) Avian frugivore activity in relation to forest light gaps. Caribb J Sci 18:1–6
Yarbrough CG, Johnston DW (1965) Lipid deposition in wintering and premigratory myrtle warblers. Wilson Bull 77:175–191

Fat Storage and Fat Metabolism in Relation to Migration

M. Ramenofsky[1]

1 Introduction

The pivotal role played by fat in suppyling energy for migration in birds has been appreciated for nearly 4 decades and reviewed extensively (Odum and Perkinson 1951; King and Farner 1956; Berthold 1976; Blem 1976, 1980; Dawson et al. 1983). The basic biochemistry, physiology, and bioenergetics of migratory fattening have been identified; yet, the regulatory mechanisms involved remain obscure. On the one hand, the behavioral and physiological functions associated with migration are thought by some to be an expression of an endogenous (genetic) program that relies little upon cues from the environment (Berthold 1985). Conversely, environmental information is regarded by others as playing an instrumental role in the regulation of such migratory functions as hyperphagia, fattening, and Zugunruhe (nocturnal restlessness) (Stetson and Erickson 1972; Schwabl and Farner 1989a,b; Wingfield, this Vol.). Recently, Wingfield et al. (this Vol.) proposed a scheme for classifying environmental and endogenous factors that regulate migratory phenomenon. Specifically, vernal (spring) fattening is triggered by initial predictive factors that include an increase in photoperiod and endogenous programs (Wingfield et al., this Vol.). However, the supplementary factors that guide the progression and termination of both vernal and autumnal fattening are not well understood. In an effort to describe the regulatory mechanisms involved in migratory fattening, the current information on avian lipid biochemistry is reviewed including the sources and sites of fat production and utilization in light of the two prominent physiological states of migration: (1) the premigratory phase during which fuel is absorbed and (2) the migratory or flight phase during which fuel is mobilized and utilized. These schemes are used to identify some of the regulatory enzymes involved in lipid metabolism that may serve as focal points for future research on the proximate factors regulating initiation, maintenance, and termination of fattening during migration.

2 Fat as a Source of Energy

In general, lipids are hydrophobic, organic compounds that can be extracted from cells and tissues with nonpolar solvents. Fatty acids (FA) combine with glycerol

[1]Department of Zoology, NJ-15, University of Washington, Seattle, Washington 98195, USA

E. Gwinner (Ed.) Bird Migration

1a. Formation of triacylglycerol from fatty acids and glycerol

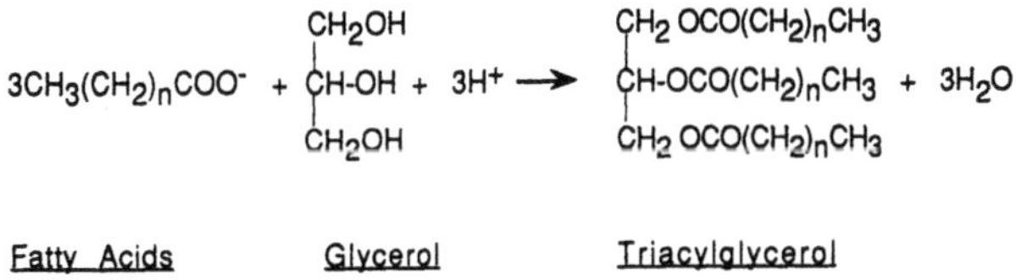

1b. Partial hydrolysis of triacylglycerol at positions 1 & 3 to produce 2-monoacylglycerol

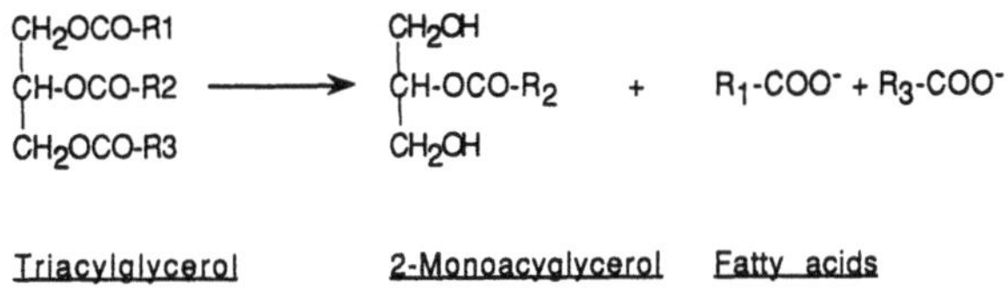

Fig. 1. **a** Fatty acids combine with glycerol to form triacylglycerol. **b** Hormone-sensitive lipase (*HSL*) hydrolyzes the ester bonds of the triacylglycerol molecule at positions 1 and 3 of the glycerol resulting in the production of 2-monoacylglycerol and 2 fatty acid molecules. R_n represent the hydrocarbon tails of fatty acids

to form triacyglycerol (TG) or fat that represent the most abundant form of lipid (Allen 1976) (Fig. 1a). Fats constitute the major chemical fuel utilized during migration by many organisms, including birds, insects, mammals (Odum and Perkinson 1951; Weis-Fogh 1952; King and Farner 1956; Hochachka et al. 1977; King 1961; Berthold 1976; Blem 1976, 1980; Dawson et al. 1983). Fats serve as an excellent source of energy for actively flying organisms. They are the most energy dense of all the stored fuels (Pond 1981; Newsholme and Leech 1983). When oxidized, they provide approximately twice the energy of carbohydrate or protein per unit weight (Hochachka et al. 1977; Schmidt-Nielson 1983). Hydrophobic and nonpolar in nature, fat is stored without water or protein. Thus, stores of fat account for less weight and bulk of an organism than do the other chemical fuels. Oxidation of fats for energy may result in a sparing of both glucose and muscle protein, thus preserving muscle strength for flight (Dawson et al. 1983). Most tissues in the body including kidney, liver, and muscle are able to oxidize FA. However, such tissues as the brain, red and white blood cells, kidney medulla, and eye, lack the aerobic capacity necessary for FA oxidation. In these tissues, carbohydrate or ketone bodies, a further reduced form of FA, provide the substrate for energy production (Hochachka et al. 1977; Newsholme and Leech 1983).

3 Fate of Dietary Fats

Figure 2 shows the sources, sites of production, deposition, mobilization, and utilization of fats in birds. Most of this information has been drawn from the literature on domesticated avian species and to a lesser extent from that on mammals, owing to the lack of information on free-living avian species.

3.1 Route of Ingested Fats

Initially, dietary fats are emulsified by conjugated bile salts in the duodenum. Exposure of the disrupted components to the hydrolytic activity of pancreatic lipase results in release of FA, glycerol (GLY), and monoacyglycerol (MG). FA and MG combine with fat-soluble vitamins, phospholipids, and sterols to form

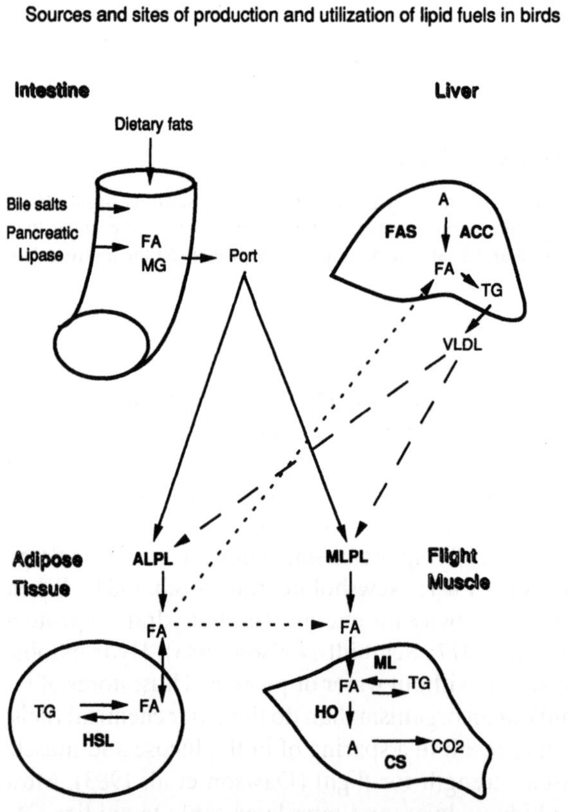

Fig. 2. Fate of fats ingested in the diet and synthesized in the liver in birds. Enzymes are indicated in *bold*, and *arrows* depict directional movement of metabolites. *Abbreviations*: *A* acetyl-CoA; *ACC* acetyl-Co carboxylase; *ALPL* adipose tissue lipoprotein lipase; *CS* citrate synthase; *FA* fatty acid; *FAS* fatty acid synthetase; *HO* β-hydroxyacyl-CoA dehydrogenase; *HSL* hormone-sensitive lipase; *MG* monoacylglycerol; *ML* muscle lipase; *MLPL* muscle lipoprotein lipase; *Port* portomicron; *TG* triacylglycerol; *VLDL* very-low density lipoprotein

micelles in which the nonpolar components face the center of the complex. The polar hydroxy groups are directed toward the micelle surface and the aqueous environment (Annison 1983; Newsholme and Leech 1983). Garrett and Young (1975) suggested that the mechanism by which the FA and MG of the micelle complexes are absorbed by the intestine in birds is comparable to that in mammals. In the jejunum, the major site of absorption, FA and MG dissociate from the micelle complexes and pass through the cell membrane of the intestine by simple diffusion when the micelle interacts, in some unknown fashion, with the brush border of the intestinal epithelium. FA and MG are reesterified to TG in the intestinal epithelium. Next, TG associates with protein, cholesterol, and such phospholipids as lecithin and sphingomyelin. The net result is production of a multimolecular complex described as lipoprotein. Many forms of lipoprotein exist, and they are classified on the basis of density, which is a function of the lipid/protein ratio. Portomicrons and very low density lipoproteins (VLDL) possess relatively more lipid than protein and represent the major species of lipoprotein that deliver TG to tissues. In birds, portomicrons are synthesized in the intestinal mucosa. They are then absorbed directly into the venous portal system that drains the small intestine (Bensadoun and Rothfield 1972). VLDL are produced in the liver and secreted directly into the circulation. Lipoproteins are water soluble and facilitate transport of TG to target tissues.

3.2 Lipogenesis

Lipogenesis is the de novo synthesis of long chain, unsaturated fatty acids from acetyl-CoA that arises from nonlipid sources such as lactate, alanine, or glucose (Goodridge 1985). In birds, lipogenesis occurs primarily in the liver (Leveille et al. 1968, 1975; Pearce 1980). The two major enzyme complexes involved in lipogenesis include acetyl-CoA carboxylase (ACC) and fatty acid synthetase (FAS). ACC is a biotin-dependent enzyme that catalyzes the conversion of acetyl-CoA to malonyl-CoA. FAS directs synthesis of long chain, saturated fatty acids from acetyl-CoA and malonyl-CoA. It consists of a complex of multifunctional polypeptides. The site of enzyme activity, however, is confined to one polypeptide chain that exists in various forms each expressing different catalytic activity (Qureshi et al. 1975). The process of lipogenesis is completed by enzymes of the mitochondria and endoplasmic reticulum that accomplish elongation and desaturation of FA, once they have been synthesized.

In general, ACC is considered the regulatory enzyme for lipogenesis (Donaldson 1979; Goodridge 1975, 1985). Such nutritional states as fasting or long-term subsistence on high-fat diets will decrease lipogenesis. These conditions lead to elevations of plasma FA, fatty acid or β-oxidation, and production of fatty acyl CoA. The depression of lipogenesis is due to the allosteric inhibition of the catalytic activity of ACC by FA and fatty acyl CoA (Goodridge 1972, 1975). During feeding after a period of fasting, lipogenesis increases rapidly in birds. Under these conditions, the regulatory enzyme for fatty acid synthesis appears to be FAS (Qureshi et al. 1975; Donaldson 1979). Donaldson (1979) has suggested that the action of both regulatory enzymes provides a rapid and continuous enhancement of the synthetic capacity of the liver to produce FA under various states of feeding and lipogenesis. FAS directs synthesis of FA when rates of lipogenesis are low, and ACC regulates synthesis when rates are high. Therefore, during migration, each enzyme system may play a specific role to enhance synthesis of FA. For example, FAS may be instrumental in increasing lipogenesis at the outset of premigratory fattening as well as during periods of refueling when fat deposits are depleted in the midst of migration.

Fatty acid synthesis requires the presence of NADPH, a coenzyme. NADPH is produced in the liver by reduction reactions catalyzed by glucose-6-phosphate dehydrogenase, 6-phosphogluconate dehydrogenase (both of which occur in the pentose phosphate shunt pathway), and malic enzyme or L-malate: $NADP^+$ oxidoreductase. In general, the activity of each enzyme correlates positively with the rate of fatty acid synthesis, and the utilization of NADPH. For this reason, none of these enzymes are considered to play a regulatory role in fatty acid synthesis (Goodridge 1985).

3.3 Metabolic Pathways for Fatty Acids Synthesized in Liver

Once synthesized in the liver, FA may undergo one of the following reactions depending upon the nutritional state of the organism. They may be (1) incorporated into TG and stored as fat, (2) broken down by β-oxidation to produce acetyl-CoA that, in turn, may be used to synthesize additional fat or may be further reduced for the production of ketone bodies (3-hydroxybutyrate), or (3) incorporated into diacylglycerol (DG) or TG that associates with VLDL and then is released into circulation and transported to target tissues.

3.4 Fat Deposition

Before FA can be deposited in or used by target tissues, hydrolysis of TG contained in the lipoproteins must take place. This hydrolytic activity is catalyzed by the lipoprotein lipase (LPL), which has also been described as a blood clearing-factor lipase (Hamosh and Hamosh 1983; Borensztajn 1987; Chan et al. 1988). In birds and mammals, LPL has been localized in capillaries serving adipose tissue, cardiac, skeletal, and smooth muscles, and liver (Husbands 1972; Gornall et al. 1972; Benson et al. 1975; Hamosh and Hamosh 1983). In adipose tissue, LPL is synthesized and released from the parenchymal cells. The enzyme is transferred through the vascular epithelium and binds to the luminal surfaces of the endothelial cells lining the capillaries that supply adipose tissue. Adipose tissue lipoprotein lipase (ALPL) binds specifically to the apolipoprotein C II site on the portomicron and VLDL. Once bound, ALPL hydrolyzes the ester bonds of the DG and TG components so that FA and GLY are released from lipoprotein. FA diffuse across the plasma membrane into the adipocyte and bind to GLY, MG, and DG to form TG for storage. GLY passes to the liver or kidney where it is incorporated into glucose. After TG have been removed from the VLDL and portomicrons by the hydrolytic actions of LPL, the remnant lipoproteins are removed from circulation and metabolized by the liver (Newsholme and Leech 1983). Avian and mammalian studies have identified a positive correlation between the catalytic activity of LPL and fatty acid uptake by mammary and adipose tissue, and by developing ovarian follicles (Garfinkel et al. 1967; Benson et al. 1975; Scow et al. 1976). Furthermore, injections of anti-lipoprotein lipase serum into domestic fowl inhibited LPL hydrolysis of TG in bloodborne lipoproteins and the resulting deposition of FA into ovarian follicles (Bensadoun and Kompiang 1979). Activity of ALPL is associated with fat deposition in developing and adult rats, during feeding after a period of fasting, and in mammals that hibernate (Gruen et al. 1978; Fried et al. 1983; Florant and Greenwood 1986; Eckel 1987).

In most mammalian species studied thus far activity of ALPL is regulated by the pancreatic hormone, insulin (Hamosh and Hamosh 1983; Eckel 1987). However, the hypoglycemic action of insulin in birds is far less pronounced (Hazelwood 1971). Therefore, the regulation of ALPL may differ from that in mammals. Research is needed to define the hormonal regulation of ALPL in birds, especially, migratory species.

3.5 Lipolysis

Within the adipocyte, hormone-sensitive lipase (HSL) plays a critical role in the mobilization of FA by regulating the rate of lipolysis, and the resulting release of FA from stored TG. HSL catalyzes the rate-limiting step, hydrolysis of the ester bonds at positions 1 and 3 of the TG molecule (Fig. 1b). Additional species of lipase in the adipocyte hydrolyze the ester bonds on the remaining MG molecules, thus allowing release of GLY and further FA from the adipocyte. In a number of avian species, lipolytic activity of HSL has been shown to be stimulated by epinephrine, norepinephrine, glucagon, growth hormone, oxytocin, and arginine vasotocin (Goodridge and Ball 1965, 1967; Gibson and Nalbandov 1966; Grande 1968, 1969; Prigg and Grande 1971; John and George 1973; Wingfield et al., this Vol.). Once released into the circulation, FA bind quickly to albumin because of their poor solubility in aqueous solutions (Annison 1983; Newsholme and Leech 1983). Next, bound FA may be transported either to target tissues directly for uptake and utilization or routed to the liver, where they are incorporated into TG, converted to VLDL, and released into the circulation to reach target tissues.

3.6 Flight Muscles: Site of Uptake and Utilization of Fatty Acids

The flight muscles (pectorales major and minor) of many birds, particularly migratory species, are especially adapted for utilizing FA for energy. These muscles are highly vascularized and well supplied with mitochondria and aerobic enzymes for complete oxidation of FA (George and Vallyathan 1964; Hochachka et al. 1977; Marsh 1981, 1984; Dawson et al. 1983; Lundgren and Kiessling 1988a). FA are delivered to the flight muscles through the circulation bound either to albumin or lipoproteins (portomicrons and VLDL). Release of FA from lipoprotein follows the catalytic actions of muscle lipoprotein lipase (MLPL). MLPL is bound to the endothelial cells lining the capillaries serving the red fibers of skeletal muscle (Borensztajn 1987). In muscle, as in adipose tissue, LPL binds to the apolipoprotein C II site on the lipoprotein and hydrolyzes the ester bonds of the DG and TG. Thus, FA are released into the capillary beds serving the flight muscles. The mechanism by which FA dissociate from albumin is not well understood. Nevertheless, FA diffuse from the capillary bed into the muscle fiber to be either stored or utilized. Once in muscle fiber, FA may be incorporated into TG and stored. Lipase contained within the muscle fiber (ML) hydrolyze the ester bonds of the stored TG, and thus release FA and GLY. FA may then undergo β-oxidation for the production of acetyl-CoA that then may enter the Krebs cycle and electron transport system for complete oxidation to CO_2, H_2O, and energy.

4 Premigratory Phase: Fuel Absorption (Fig. 3)

Preparation for vernal and autumnal migration in birds involves dramatic behavioral and physiological adaptations for establishing and sustaining extensive stores of fat throughout the migratory period (King et al. 1963; Berthold 1976; Blem 1976, 1980; King and Murphy 1985). The sequence of events includes an increase in appetite (hyperphagia) that leads to hyperlipogenesis, rapid deposition of fat in the subcutaneous and visceral depots, and elevation of body mass (King 1961; King and Farner 1956; Gifford and Odum 1965; Fry et al. 1970; Blem 1976, 1980; Pettersson and Hasselquist 1985). In captive golden- and white-crowned sparrows (*Zonotrichia atricapilla* and *Zonotrichia leucophrys gambelii*) premigratory hyperphagia involves an increase in the frequency of food intake without an alteration in meal size (Marsh 1965; Moldenhauer 1965). Bairlein (1985b and this Vol.) has shown in captive garden warblers, *Sylvia borin*, a long-distance migrant, that premigratory hyperphagia is associated with increased efficiency of digestion and assimilation of dietary fat, protein, and carbohydrate over the prefattening period. Taken together, the increased rate of feeding and heightened efficiency of nutrient utilization could provide an increase in the supply of substrate for production of portomicrons and VLDL that may contribute to hyperlipogenesis and result in premigratory fattening.

4.1 Liver Lipogenesis

In a number of avian species, lipid is deposited in the liver during the premigratory period (Odum and Perkinson 1951; King et al. 1963; de Filippo et al. 1988). The activities of enzymes involved in hepatic lipogenesis may also increase at this time. Shah et al. (1978) reported elevated activity of malic enzyme in the liver of the rosy pastor (*Sturnus roseus*) and white wagtail (*Motacilla alba*) during

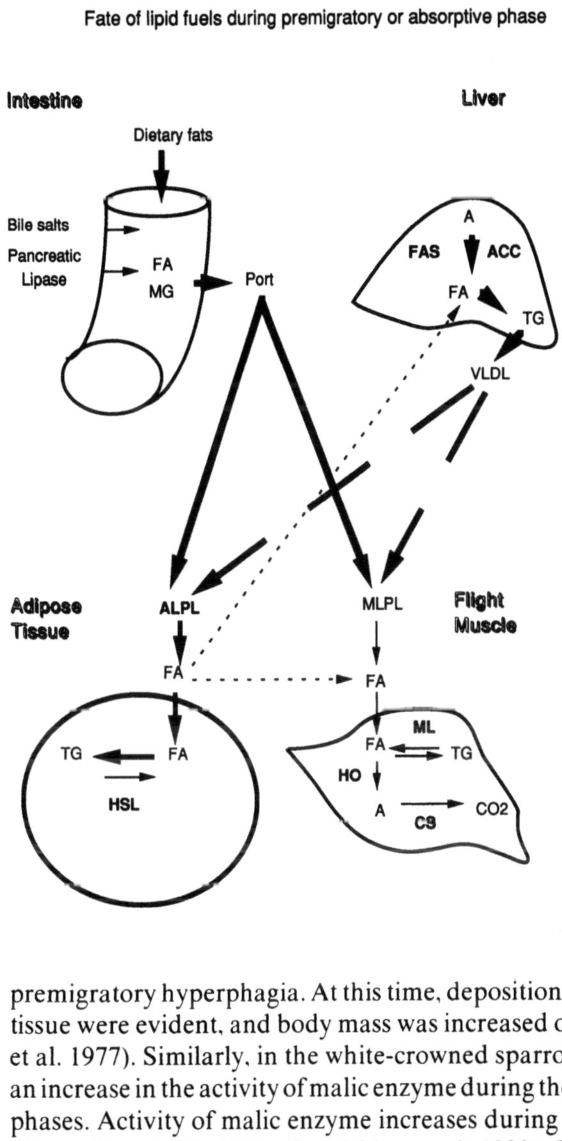

Fig. 3. Fate of fats ingested in the diet and synthesized in the liver in birds during the premigratory or absorptive phase of migration. Enzymes are indicated in *bold*, and *arrows* depict directional movement of metabolites. The thickness of the arrows indicates the relative flux of metabolites through that portion of the pathway. For *abbreviations*, see Fig. 2

premigratory hyperphagia. At this time, deposition of fat in the liver and adipose tissue were evident, and body mass was increased over prefattening levels (Patel et al. 1977). Similarly, in the white-crowned sparrow, DeGraw (1975) identified an increase in the activity of malic enzyme during the premigratory and migratory phases. Activity of malic enzyme increases during hyperphagia and lipogenesis (Goodridge 1975, 1985; Goodridge et al. 1989). However, it is not consider a regulatory enzyme for fatty acid synthesis (Goodridge 1985).

What is the evidence that hepatic ACC and FAS may be acting as regulatory enzymes in premigratory lipogenesis? To my knowledge, activity of hepatic ACC in free-living birds has not been investigated. In captive dark-eyed juncos (*Junco hyemalis*), a short-distance migrant, there is a vernal increase in the activity of hepatic FAS prior to the onset of premigratory hyperphagia, fat deposition, and elevation in body mass (M. Ramenofsky, R.S. Savard, M.R.C. Greenwood, in preparation). In domestic fowl, Donaldson (1979) has suggested that the nutritional state of the animal determines whether ACC or FAS regulates hepatic lipogenesis. Therefore, FAS may initiate the increase in the capacity for hepatic lipogenesis prior to premigratory hyperphagia and fattening. Once birds are hyperphagic and hyperlipogenic, ACC may regulate fatty acid synthesis. Both points deserve further study particularly to address the mechanisms underlying the initiation of premigratory hyperphagia, lipogenesis, and fat deposition.

During vernal migration, relationships between photostimulation, gonadal activity, and expression of premigratory hyperphagia and fattening have been identified in white-throated sparrows (*Zonotrichia albicollis*) and white-crowned sparrows (Weise 1967; Stetson and Erickson 1972; Mattocks 1976; and Schwabl and Farner 1989a,b). By contrast, autumnal hyperphagia and fattening are independent of gonadal function (Mattocks 1976; Moore et al. 1982). The elevation in activity of FAS preceded the onset of hyperphagia and fattening in juncos (M. Ramenofsky, R.S. Savard, M.R.C. Greenwood, in preparation). These results suggest that the vernal increase in activity of FAS may be triggered by the initial predictive factor, namely, the lengthening photoperiod and resulting endocrine activity, rather than a secondary response to changes in feeding behavior. Anterior pituitary, pancreatic, and steroid hormones are known to influence hepatic lipogenesis in birds (Goodridge and Ball 1966, 1967; Chandrabose and Bensadoun 1971; Wheeland et al. 1976; Ballantyne et al. 1988; Wingfield et al., this Vol.).

From these results, it is apparent that hyperlipogenesis during the premigratory phase does not depend entirely upon dietary contribution associated with hyperphagia, but is enhanced by an increased rate of hepatic synthesis of fatty acids.

4.2 Adipose Tissue

In birds, subcutaneous and visceral depots of adipose tissue are the major sites of fat storage (McGreal and Farner 1956; Pond and Mattacks 1985). Increased activity of hepatic lipogenesis, premigratory hyperphagia, and heightened efficiency of assimilation of nutrients result in an elevation of circulating levels of lipoproteins, which in turn leads to deposition of TG in adipose depots. Given this sequence of events, one would predict an increase in ALPL during the premigratory period. Preliminary results from studies on the metabolic changes occurring during premigratory fattening in the European quail (*Coturnix coturnix*), a long-distance migrant, confirm this speculation (M. Hall, M. Ramenofsky, unpublished work). Captive birds sampled during the autumnal premigratory period demonstrated elevations of food intake, activity of ALPL, and body mass. Measurements of ALPL activity in captive dark-eyed juncos sampled throughout the winter months and vernal migration, revealed no change in the level of enzyme activity despite dramatic increases in fat deposition and body mass (M. Ramenofsky, R.S. Savard, M.R.C. Greenwood, in preparation). These results are unexpected in light of the positive correlations of ALPL activity, fat deposition, and body mass in mammals (Gruen et al. 1978; Fried et al. 1983; Florant and Greenwood 1986; Eckel 1987). Additional measures of ALPL activity in birds are needed throughout the premigratory and migratory phases, when fat deposition is initiated and sustained. Considering the activity of ALPL in the European quail, one would expect a peak of enzyme activity during the period of hyperphagia and fattening. But why activity of ALPL, as measured in juncos, does not remain elevated once migratory fattening has been achieved remains unclear. Additional studies are needed to (1) clarify the regulatory mechanisms involved, and (2) resolve the apparent differences between birds and mammals.

4.3 Lipolysis

To my knowledge, there is no information available concerning mobilization of stored fat in birds during the premigratory period. However, it is probable that fat would be mobilized to provide energy during the overnight fast, as shown in the American goldfinch, *Carduelis tristis* (Carey et al. 1978), and yellow bunting, *Emberiza citrinella* (Evans 1969).

4.4 Flight Muscle

During the premigratory period, various biochemical and morphological changes occur in the flight muscles. Increased storage of fat has been documented in the flight muscles of white-throated sparrow (Odum and Perkinson 1951), white-crowned sparrow (Farner et al. 1961; King et al. 1963), rosy pastor and white wagtail (Pilo and George 1983; Vallyathan 1963). Premigratory hypertrophy of flight muscles, measured as an increase in the fat-free dry weight, has been shown in yellow wagtail, *Motacilla flava* (Fry et al. 1972), rosy pastor (Vallyathan 1963), Swainson's thrush, *Hylocichla ustulata* (Child 1969), and gray catbird, *Dumatella carolinensis* (Marsh 1981, 1984). In many migratory species, the muscle fibers predominantly are of the fast-twitch, oxidative-glycolytic type (Lundgren and Kiessling 1988b). When preparing for migration, the aerobic capacity of the flight muscles dominates glycolytic activity in such species as reed warbler, *Acrocephalus scirpaceus*, European robin, *Erithacus rubecula*, and blackbird, *Turdus merula* (Lundgren and Kiessling 1985). In this study, the catabolic activity of the flight muscles was determined by measurements of the following enzymes: B-hydroxyacyl-CoA dehydrogenase (HO; a measure of β-oxidation), citrate synthase (CS; an indicator of the activity of Krebs cycle activity), and cytochrome oxidase (CO; a measure of oxidative phosphorylation). Similarly, in the gray catbird, Marsh (1981) identified an increase in enzymatic capacity for β-oxidation and increased total aerobic capacity of the flight muscle complex during the premigratory period. Taken together, these adjustments increase the capacity of migrating birds to utilize fat as a source of energy and to power flight. Both are necessary for supporting the high energetic demands of migration (Dawson et al. 1983).

5 Migratory Phase: Flight and Fuel Mobilization (Fig. 4)

Once birds have prepared for migration by depositing large stores of fat, initiation of migration or flight phase represents a metabolic shift from fuel storage or anabolic functions to fuel mobilization or catabolic functions. We now turn our attention to the metabolic aspects of source, production, deposition, mobilization, and utilization of fat during migration.

During migration some species of birds stop and feed in order to replenish nutrient stores that have become depleted during flight. Hyperphagia is persistent throughout the period of migration, thus allowing for rapid weight gain once a stopover site has been reached (King 1961). Rate of mass gain observed in small passerine birds during migration may reach nearly 10% body mass/day (Dolnik and Blyumental 1967; Fry et al. 1972; Cherry 1982; Biebach et al. 1986; Bairlein 1985b; Moore and Kerlinger 1987).

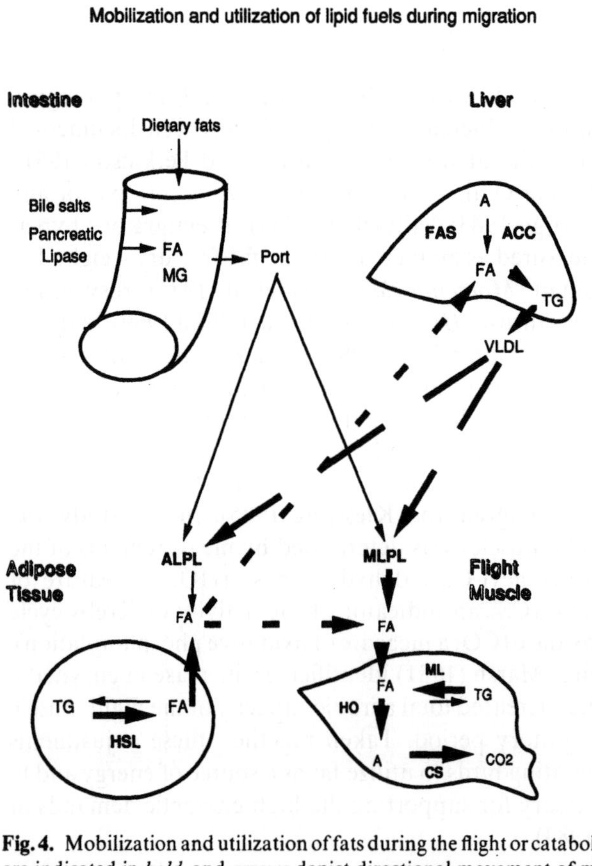

Fig. 4. Mobilization and utilization of fats during the flight or catabolic phase of migrations. Enzymes are indicated in *bold*, and *arrows* depict directional movement of metabolites. The thickness of the arrows indicates the relative flux of the metabolites through that portion of the pathway. For *abbreviations*, see Fig. 2

5.1 Liver Lipogenesis

In captive dark-eyed juncos, the increased levels of liver FAS activity observed prior to premigratory hyperphagia and fattening persisted through the migratory phase of vernal migration (M. Ramenofsky, R.S. Savard, M.R.C. Greenwood, in preparation). Thus, in addition to hyperphagia, an increased synthetic contribution by the liver may further support the rapid weight gain observed in migrant species. This presumes, of course, that sufficient food is available for migrants at a particular stopover site.

5.2 Adipose Tissue Lipoprotein Lipase

From field observations conducted during autumnal migration, Biebach (1985) recognized that availability of food at a stopover site and body condition, as determined by body mass and fat stores, affect migratory activity in the spotted flycatcher (*Muscicapa striata*). In captive spotted flycatchers (Biebach 1985) and garden warblers (Gwinner et al. 1985), rates of Zugunruhe (migratory restlessness) increased when food was restricted, an experimental protocol simulating the conditions during flight. When birds were refed, Zugunruhe initially ceased but later reappeared as body mass and stores of fat continued to increase. Serial measurements of ALPL were made in garden warblers throughout this experimental procedure (M. Ramenofsky, H. Schwabl, I. Schwabl-Benzinger, E. Gwinner, in preparation). Activity of ALPL remained stable during the periods prior to and during food restriction as well as during the later stages of refeeding, despite dramatic changes in fat deposition and body mass observed during these times. A peak of ALPL was apparent only during the early stages of refeeding, when rates of fat deposition and body mass were first elevated. These data suggest that activity of ALPL is refractory to changes in fat deposition and body mass, except when refeeding follows depletion of adipose depots. The transient nature of the increased response of ALPL to refeeding observed in garden warblers diverges from that observed in mammals (Fried et al. 1983) and suggests strict regulation of the avian ALPL. Hormonal and metabolic factors modulating levels of ALPL activity are not known for migratory birds and require further study. Nevertheless, these observations suggest an adaptation for rapidly replenishing stores of fat while refueling during migration in birds.

5.3 Lipolysis

During migration, fat stores are mobilized and, under certain conditions, depleted (Bairlein 1985a; Biebach et al. 1986; Karlsson et al. 1988; Evans, this Vol.). For example, during simulated flight in pigeons, the levels of FA in the adipose tissue decreased while those in the plasma, liver, and muscle increased (Vallyathan and George 1969). These data indicate that FA released from adipose tissue travel in the circulation to liver and flight muscles. More recently

(M. Ramenofsky, R.S. Savard, M.R.C. Greenwood, in preparation) lipolysis was measured, as determined by glycerol release from the claviculo-coracoid adipose depot in captive juncos during vernal migration. Birds were sampled before, during, and after flight activity, as indicated by Zugunruhe. Lipolysis increased with Zugunruhe but reached a peak only after the conclusion of one night of activity. These results suggest that mobilization of FA from adipose tissue increases throughout the period of Zugunruhe to supply sufficient levels of substrate for the energetic demands of flight. The rate of FA mobilization observed in captive juncos may be less than that of free-living birds, since the energy expenditure of flight activity in free-living birds exceeds that of birds held in captivity. Nevertheless, the relationship between lipolysis and Zugunruhe is reminiscent of the exercise-induced activity of aerobic enzymes found in the skeletal muscle of birds and mammals (Chaffee and Mayhew 1964; Holloszy and Booth 1976; Marsh 1981; Lundgren and Kiessling 1986). In addition, Viswanathan et al. (1987) and John et al. (1988) measured elevated levels of plasma growth hormone, catecholamines, and FA in free-living, homing pigeons that had just completed a flight of 48 km that lasted 60 to 80 min. Growth hormone, catecholamines, AVP, and glucagon have been shown to act as lipolytic agents in birds (Heald et al. 1965; O'Neill and Langslow 1978; Goodridge and Ball 1965; Grande and Prigge 1970; Langslow and Hales 1969; John et al. 1988). These results suggest increased lipolysis attending flight activity is likely affected by a variety of hormones. The mechanisms underlying this process deserve more attention.

5.4 Flight Muscle Lipoprotein Lipase

In juncos held in captivity during vernal migration, peak activity of MLPL occurred when birds exhibited Zugunruhe (M. Ramenofsky, R.S. Savard, M.R.C. Greenwood, in preparation). Once Zugunruhe is initiated, MLPL provides an increased flow of FA to working muscles. Regulation of MLPL activity is poorly understood, but administration of high physiological levels of corticosterone increased activity of MLPL in flight muscle (Yarian et al. 1985; Gray et al. in press). In addition, flight activity has been shown to increase the concentration of corticosterone within the adrenal gland of a number of migratory species (Peczely 1976). Furthermore, growth hormone, catecholamines, and FA increase in the plasma following flight activity in free-ranging homing pigeons (Viswanathan et al. 1987). Therefore, the flight-related increase of MLPL activity observed in juncos may be affected by any one of these hormonal or metabolic factors. These points need to be investigated further.

6 Conclusions

The regulation of fat storage and fat metabolism during migration has been neglected in recent years. Much of our understanding of the basic biochemistry, physiology, and bioenergetics of migratory fattening relies upon studies that were conducted at least 15 years ago. In an effort to update this important aspect of migration, a survey has been presented of the basic processes of digestion of dietary lipids, lipogenesis, deposition, mobilization, and utilization of fat, most of which has been obtained from domesticated avian species and mammals. This survey identified important enzyme systems that could provide information on the regulation of lipid metabolism during migration. Empirical studies on migratory species indicate that regulation of fat metabolism during migration involves processes that are influenced by a host of factors including environmental conditions, endogenous or genetic programs, and physiological state of the organism. In an effort to identify the types of regulatory factors involved in fat storage and metabolism during migration, I refer to the classification system of environmental factors influencing migration of Wingfield et al. (this Vol.). The physiological and behavioral properties of migration are best understood when the migratory period is divided into the two distinct phases of premigratory or fuel absorption and migratory or fuel mobilization and utilization. Taken together, it appears that initial predictive factors (e.g., photoperiod and endogenous genetic programs) influence preparation for migration by increasing activities that enhance intake, production, and storage of fuels and, in particular, fat. These activities include hepatic lipogenesis via FAS, hyperphagia, and deposition of fat in liver, muscle, and adipose tissue. Morphological and biochemical changes occur in the flight muscles during the premigratory phase. These changes include hypertrophy of flight muscles, increased capacity for both oxidation of fatty acids and total aerobic capacity for energy production. Initial predictive factors influencing these changes are not known and certainly deserve full attention. Once birds have completed the final stage of preparation for migration, supplementary factors (e.g., food intake and body condition, weather, lunar phase) influence the initiation of migratory flight. Flight itself may act as a stimulus, and possibly a supplementary factor, for secretion of such hormones as catecholamines, corticosterone, and growth hormone. These hormones promote lipolysis and mobilization of FA from adipose tissue, as well as activity of MLPL to enhance deposition of FA into muscle, and thus provide energy for flight. Finally, the mechanisms that terminate migration and accompanying metabolic activities remain unknown.

Acknowledgments. Financial support for the preparation of this manuscript was provided in part by grants from the National Science Foundation, BNS # 8403798 to M. Ramenofsky and J. Gray and from the Obesity Core Center of St. Luke's Hospital and Rockefeller University to MR. I am grateful to the following individuals for their helpful comments and advice on earlier versions of the manuscript: Lee Astheimer, William Buttemer, Greg Florant, Janet Gray, Sara Hiebert, and John Wingfield.

References

Allen WV (1976) Biochemical aspects of lipid storage and utilization in animals. Am Zool 16:631–647
Annison EF (1983) Lipid metabolism. In: Freeman BM (ed) Physiology and biochemistry of the domestic fowl, vol 4. Academic Press, London, pp 165–174

Bairlein F (1985a) Body weights and fat deposition of Palaearctic passerine migrants in the central Sahara. Oecologia (Berl) 66:141–146

Bairlein F (1985b) Efficiency of food utilization during fat deposition in the long-distance migratory garden warbler, *Sylvia borin*. Oecologia (Berl) 68:118–125

Ballantyne JS, John TM, George JC (1988) The effects of glucagon on hepatic mitochondrial metabolism in the pigeon, *Columba livia*. Gen Comp Endocrinol 72:130–135

Bensadoun A, Kompiang IP (1979) Role of lipoprotein lipase in plasma triglyceride removal. Fed Proc 38:2622–2626

Bensadoun A, Rothfield A (1972) The form of absorption of lipids in the chicken, *Gallus domesticus*. Proc Soc Exp Biol Med 141:814–817

Benson JD, Bensadoun A, Cohen D (1975) Lipoprotein lipase of ovarian follicles in the domestic chicken (*Gallus domesticus*). Proc Soc Exp Biol Med 148:347–350

Berthold P (1976) Migration: control and metabolic physiology. In: Farner DS, King JR (eds) Avian biology, vol 5. Academic Press, London, pp 77–128

Berthold P (1985) Physiology and genetics of avian migration. In: Rankin MA (ed) Migration: mechanisms and adaptive significance. Contributions in Marine Science. University of Texas, Austin, Port Aransas, Texas, pp 526–543

Biebach H (1985) Sahara stopover in migratory flycatchers: fat and food affect the time program. Experientia (Basel) 41:695–697

Biebach H, Friedrich W, Heine G (1986) Interaction of body mass, fat, foraging and stopover period in trans-Sahara migrating passerine birds. Oecologia (Berl) 69:370–379

Blem CR (1976) Patterns of lipid storage and utilization in birds. Am Zool 16:671–684

Blem CR (1980) The energetics of migration. In: Gauthreaux SA (ed) Animal migration, orientation, and navigation. Academic Press, London, pp 175–224

Borensztajn J (1987) Lipoprotein lipase. Evener, Chicago, p 336

Carey C, Dawson WR, Maxwell LC, Faulkner JA (1978) Seasonal acclimatization to temperature in cardueline finches. II. Changes in body composition and mass in relation to season and acute cold stress. J Comp Physiol 125:101–113

Chaffee RRJ, Mayhew WW (1964) Studies on chemical thermoregulation in the house sparrow (*Passer domesticus*). Can J Physiol Pharmacol 42:863–866

Chan BL, Lisanti MP, Rodriguez-Boulan E, Saltiel AR (1988) Insulin-stimulated release of lipoprotein lipase by metabolism of its phosphatidylinositol anchor. Science 241:1670–1672

Chandrabose KA, Bensadoun A (1971) Effect of hypophysectomy on some enzymes involved in lipid metabolism of the domestic chicken (*Gallus domesticus*). Comp Biochem Physiol 39:45–54

Cherry JD (1982) Fat deposition and length of stopover of migrant white-crowned sparrows. Auk 99:725–732

Child GI (1969) A study of nonfat weights in migrating Swainson's thrushes (*Hylocichla ustulata*). Auk 86:327–338

Dawson WR, Marsh RL, Yacoe ME (1983) Metabolic adjustments of small passerine birds for migration and cold. Am J Physiol 245:R755–767

De Graw WA (1975) Seasonal changes in liver malic enzyme activity in white-crowned sparrows, *Zonotrichia leucophrys gambelli*. Proc Nebr Acad Sci Affil Soc 85th Annu Meet, p 12

Dolnik V, Blyumental TI (1967) Autumnal premigratory and migratory periods in the chaffinch (*Fringilla coelebs coelebs*) and some other temperate-zone birds. Condor 69:435–368

Donaldson WE (1979) Regulation of fatty acid synthesis. Fed Proc 38:2617–2621

Eckel RH (1987) Adipose tissue lipoprotein lipase. In: Borensztajn J (ed) Lipoprotein lipase. Evener, Chicago, pp 79–132

Evans PR (1969) Winter fat deposition and overnight survival of yellow buntings (*Emberiza citrinella* L.). J Anim Ecol 38:415–423

Farner DS, Oksche A, Kamemoto FI, King JR, Cheyney HE (1961) A comparison of the effect of long daily photoperiods on the pattern of energy storage in migratory and nonmigratory finches. Comp Biochem Physiol 2:125–142

Filippo G de, Milone M, Esposito R, Caliendo MF (1988) Winter and premigratory lipid metabolism in *Erithacus r. rubecula*. Abstract from 14th Conference of European Comparative Endocrinologists, University of Salzberg, Austria, September 4–9

Florant G, Greenwood MRC (1986) Seasonal variations in pancreatic function in marmots: the role of pancreatic hormones and lipoprotein lipase in fat deposition. In: Heller HC, Musacchia XJ, Wang LCH (eds) Living in the cold: physiological and biochemical adaptations. Elsevier Science, New York, pp 273–280
Fried SK, Hill JO, Nickel M, DiGirolamo M (1983) Prolonged effects of fasting-refeeding on rat adipose tissue lipoprotein lipase activity: influence of caloric restriction during refeeding. J Nutr 113:1861–1869
Fry CH, Ash JKS, Ferguson-Lees IJ (1970) Spring weights of some Palaearctic migrants at Lake Chad. Ibis 112:58–82
Fry CH, Ferguson-Lees IF, Dowsett RJ (1972) Flight muscle hypertrophy and ecophysiological variation of yellow wagtail (*Motacilla flava*) races at Lake Chad. J Zool (Lond) 167: 293–306
Garfinkel AS, Baker N, Schotz MC (1967) Relationship of lipoprotein lipase activity to triglyceride uptake in adipose tissue. J Lipid Res 8:274–280
Garrett RL, Young RJ (1975) Effect of micelle formation on the absorption of neutral fat and fatty acids by the chicken. J Nutr 105:827–838
George JC, Vallyathan NV (1964) Lipase and succinic dehydrogenase activity of the particulate fractions of the breast muscle homogenate of the migratory starling *Sturnus roseus* in the pre-migratory and post-migratory periods. J Cell Comp Physiol 63:381–391
Gibson WR, Nalbandov AV (1966) Lipolysis and lipogenesis in liver and adipose tissue of hypophysectomized cockerels. Am J Physiol 211(6):1352–1356
Gifford CE, Odum EP (1965) Bioenergetics of lipid deposition in the bobolink, a transequatorial migrant. Condor 67:383–403
Goodridge AG (1972) Regulation of the activity of acetyl coenzyme A carboxylase by palmitoyl coenzyme A and citrate. J Biol Chem 247:6946–6952
Goodridge AG (1975) Hormonal regulation of the activity of the fatty acid synthesizing system and of the malic enzyme concentration in liver cells. Fed Proc 34:117–123
Goodridge AG (1985) Fatty acid synthesis in eucaryotes. In: Vance DE, Vance JE (eds) Biochemistry of lipids and membranes. Benjamin/Cummings, Menlo Park, CA, pp 143–178
Goodridge AG, Ball EG (1965) Studies on the metabolism of adipose tissue XVIII. In vitro effects of insulin, epinephrine, and glucagon on lipolysis and glycolysis in pigeon adipose tissue. Comp Biochem Physiol 16:367–381
Goodridge AG, Ball EG (1966) Lipogenesis in the pigeon: in vitro studies. Am J Physiol 211:803–808
Goodridge AG, Ball EG (1967) The effect of prolactin on lipogenesis in the pigeon: in vitro studies. Biochemistry 6:2335–2343
Goodridge AG, Crish JF, Hillgartner FB, Wilson SB (1989) Nutritional and hormonal regulation of the gene for avian malic enzyme. J Nutr 119:229–308
Gornall DA, Kuksis A, Morley N (1972) Lipid-metabolizing enzymes in the ovary of the laying hen. Biochim Biophys Acta 280:225–230
Grande F (1968) Effect of glucagon on plasma free fatty acids and blood sugar in birds. Proc Soc Exp Biol Med 128:532–536
Grande F (1969) Effect of catecholamines on plasma free fatty acids and blood sugar in birds. Proc Soc Exp Biol Med 13:740–744
Grande F, Prigge WF (1970) Glucagon infusion, plasma FFA and triglycerides, blood sugar and liver lipids in birds. Am J Physiol 218:1406–1411
Gray JM, Yarian D, Ramenofsky M Corticosterone, foraging behavior, and metabolism in dark-eyed juncos (*Junco hyemalis*). Gen Comp Endocrinol, in press
Gruen R, Hietanen E, Greenwood MRC (1978) Increased adipose tissue lipoprotein lipase activity during development of the genetically obese rat (fa/fa). Metabolism 27:1955–1965
Gwinner E, Biebach H, Kries I (1985) Food availability affects migratory restlessness in caged garden warblers (*Sylvia borin*). Naturwissenschaften 72:51
Hamosh M, Hamosh P (1983) Lipoprotein lipase: its physiological and clinical significance. Mol Aspects Med 6:199–289
Hazelwood RL (1971) Endocrine control of avian carbohydrate metabolism. Poult Sci 50:9–18
Heald PJ, McLaughlan PM, Rookledge KA (1965) The effects of insulin, glucagon, and adreno-

corticotrophic hormone on the plasma glucose and free fatty acids of the domestic fowl. J Endocrinol 33:83–95
Hochachka PW, Neely JR, Driedzic WR (1977) Integration of lipid utilization with Krebs cycle activity in muscle. Fed Proc 36:2009–2014
Holloszy JO, Booth FW (1976) Biochemical adaptations to endurance exercise in muscle. Annu Rev Physiol 38:273–291
Husbands DR (1972) The distribution of lipoprotein lipase in tissues of the domestic fowl and the effects of feeding and starving. Br J Poult Sci 13:85–90
John TM, George JC (1973) Influence of glucagon and neurohypophysial hormones on plasma free fatty acid levels in the pigeon. Comp Biochem Physiol A Comp Physiol 45:541–547
John TM, Viswanathan M, George JC, Scanes CG (1988) Flight effects on plasma levels of free fatty acids, growth hormone, and thyroid hormones in homing pigeons. Horm Metab Res 20:271–273
Karlsson L, Persson K, Pettersson J, Walinder G (1988) Fat-weight relationships and migratory strategies in the robin *Erithacus rubecula* at two stop-over sites in south Sweden. Ringing & Migr 9:160–168
King JR (1961) The bioenergetics of vernal premigratory fat deposition in the white-crowned sparrow. Condor 63:128–142
King JR, Farner DS (1956) Bioenergetic basis of light-induced fat deposition in the white-crowned sparrow. Proc Soc Exp Biol Med 93:354–359
King JR, Murphy ME (1985) Periods of nutritional stess in the annual cycles of endotherms: fact or fiction? Am Zool 25:955–964
King JR, Barker S, Farner DS (1963) A comparison of energy reserves during autumnal and vernal migratory periods in the white-crowned sparrow, *Zonotrichia leucophrys gambelii*. Ecology 44:513–521
Langslow DR, Hales CN (1969) Lipolysis in chicken adipose tissue in vitro. J Endocrinol 43:285–294
Leveille GA, O'Hea EK, Chakrabarty K (1968) In vivo lipogenesis in the domestic chicken. Soc Exp Biol Med 128:398–401
Leveille GA, Romsos DR, Yeh YY, O'Hea E (1975) Lipid biosynthesis in the chick: a consideration of site of synthesis, influence of diet and possible regulating mechanisms. Poult Sci 54:1075–1093
Lundgren BO, Kiessling K-H (1985) Seasonal variation in catabolic enzyme activities in breast muscle of some migratory birds. Oecologia (Berl) 66:468–471
Lundgren BO, Kiessling K-H (1986) Catabolic enzyme activities in the pectoralis muscle of premigratory and migratory juvenile reed warbler, *Acrocephalus scirpaceus* (Herm.). Oecologia (Berl) 68:529–532
Lundgren BO, Kiessling K-H (1988a) Catabolic enzyme activities in the pectoralis muscle of migratory and non-migratory goldcrests, great tits, and yellowhammers. Ornis Scand 19:190–194
Lundgren BO, Kiessling K-H (1988b) Comparative aspects of fibre types, areas, and capillary supply in the pectoralis muscle of some passerine birds with differing migratory behaviour. J Comp Physiol B Biochem Syst Environ Physiol 158:165–173
Marsh RJ (1965) Feeding behavior of the golden-crowned sparrow, *Zonotrichia atricapilla*, under natural and altered photoperiods. Master's Thesis. San Jose State College, San Jose, CA
Marsh RL (1981) Catabolic enzyme activities in relation to premigratory fattening and muscle hypertrophy in the gray catbird (*Dumetella carolinensis*). J Comp Physiol B Biochem Syst Environ Physiol 141:417–423
Marsh RL (1984) Adaptations of the gray catbird (*Dumetella carolinensis*) to long-distance migration: flight muscle hypertrophy associated with elevated body mass. Physiol Zool 57:105–117
Mattocks PW Jr (1976) The role of gonadal hormones in the regulation of the premigratory fat deposition in the white-crowned sparrow, *Zonotrichia leucophrys gambelii*. Master's Thesis, University of Washington, Seattle
McGreal RD, Farner DS (1956) Premigratory fat deposition in the Gambel white-crowned sparrow: some morphological and chemical observations. Northwest Sci 30:12–23
Moldenhauer RR (1965) An investigation of the feeding pattern of caged white-crowned sparrows, *Zonotrichia leucophrys gambelii* (Nuttall), in relationship to vernal fat deposition. Master's Thesis, Oregon State University, Corvallis
Moore FR, Kerlinger P (1987) Stopover and fat deposition by North American wood-warblers (Parulinae) following spring migration over the Gulf of Mexico. Oecologia (Berl) 74:47–54

Moore MC, Donham RS, Farner DS (1982) Physiological preparation for autumnal migration in white-crowned sparrows. Condor 84:410–419

Newsholme EA, Leech AR (1983) Biochemistry for the medical sciences. Wiley, New York, p 952

Odum EP, Perkinson JD Jr (1951) Relation of lipid metabolism to migration in birds: seasonal variation in body lipids of the migratory white-throated sparrow. Physiol Zool 24:216–230

O'Neill IE, Langslow DR (1978) The action of hydrocortisone, insulin, and glucagon on chicken liver hexokinase and glucose-6-phosphatase and on the plasma glucose and free fatty acid concentrations. Gen Comp Endocrinol 34:428–437

Patel ST, Shah PV, Pilo B (1977) Composition of hepatic lipids during post- and premigratory periods of the migratory starling (*Sturnus roseus*) and wagtail (*Motacilla alba*). Pavo 15:115–129

Pearce J (1980) Comparative aspects of lipid metabolism in avian species. Biochem Soc Trans 8:295–296

Peczely P (1976) Etude circanuelle de la fonction corticosurrenalienne chez les especes de passereaux migrants et non-migrants. Gen Comp Endocrinol 30:1–11

Pettersson J, Hasselquist D (1985) Fat deposition and migratory capacity of robins *Erithacus rubecula* and goldcrests *Regulus regulus* at Ottenby, Sweden. Ringing & Migr 6:66–76

Pilo B, George JC (1983) Diurnal and seasonal variation in liver glycogen and fat in relation to metabolic status of liver and m. pectoralis in the migratory starling, *Sturnus roseus*, wintering in India. Comp Biochem Physiol 74:601–604

Pond CM (1981) Storage. In: Townsend CR, Calow P (eds) Physiological ecology. Sinauer, Massachusetts, pp 190–219

Pond CM, Mattacks CA (1985) Cellular structure of adipose tissue in birds. J Morphol 185:195–202

Prigge WF, Grande F (1971) Effects of glucagon, epinephrine and insulin on in vitro lipolysis of adipose tissue from mammals and birds. Comp Biochem Physiol 39B:69–82

Qureshi AA, Jenik RA, Kim M, Lornitzo FA, Porter JW (1975) Separation of two active forms (holo-*a* and holo-*b*) of pigeon liver fatty acid synthetase and their interconversion by phosphorylation and dephosphorylation. Biochem Biophys Res Commun 66:344–351

Schmidt-Nielsen K (1983) Animal physiology: adaptation and environment. Cambridge University Press, Cambridge, New York, p 619

Schwabl H, Farner DS (1989a) Dependency on testosterone of photoperiodically-induced vernal fat deposition in female white-crowned sparrows. Condor 91:108–112

Schwabl H, Farner DS (1989b) Endocrine and environmental control of vernal migration in male white-crowned sparrows, *Zonotrichia leucophrys gambelii*. Physiol Zool 62:1–10

Scow RO, Blanchette Mackie EJ, Smith LC (1976) Role of capillary endothelium in the clearance of chylomicrons. A model for lipid transport from blood by lateral diffusion in cell membrane. Circ Res 39:139–162

Shah RV, Patel ST, Pilo B (1978) Glucose-6-phosphate dehydrogenase and 'malic' enzyme activities during adaptive hyperlipogenesis in migratory starling (*Sturnus roseus*) and white-wagtail (*Motacilla alba*). Can J Zool 56:2083–2087

Stetson MH, Erickson JE (1972) Hormonal control of photoperiodically-induced fat deposition in white-crowned sparrows. Gen Comp Endocrinol 19:355–362

Vallyathan NV (1963) On the lipid content and lipase activity in the breast muscle of *Sturnus roseus* (*Linnaeus*). PAVO 1:106–109

Vallyathan NV, George JC (1969) Effect of exercise on lipid levels in the pigeon. Arch Int Physiol Biochem 77:863–868

Viswanathan M, John TM, George JC, Etches RJ (1987) Flight effects on plasma glucose, lactate, catecholamines and corticosterone in homing pigeons. Horm Metab Res 19:400–402

Weis-Fogh T (1952) Fat combustion and metabolic rate of flying locusts (Schistocerca gregaria forskål). Phil Trans R Soc Ser B 237:1–36

Weise GM (1967) Castration and spring migration in the white-throated sparrow. Condor 69:49–68

Wheeland RA, Martin RJ, Meier AH (1976) The effect of prolactin and CB 154 on in vivo lipogenesis and enzyme patterns in the Japanese quail, *Coturnix coturnix japonica*, and of photostimulation on enzyme patterns in the white-throated sparrow, *Zonotrichia albicollis*. Comp Biochem Physiol B Comp Biochem 53:379–385

Yarian D, Ramenofsky M, Gray JM (1985) Corticosterone, lipid stores, and foraging behavior in dark-eyed juncos, *Junco hyemalis*. Am Zool 25:433

Endocrine Mechanisms of Migration

J.C. WINGFIELD,[1] H. SCHWABL[2], and P.W. MATTOCKS JR.[3]

1 Introduction

Migration involves complex interrelationships of anabolic and catabolic metabolism, behavior, and reproductive development. Vernal migration (the movement of individuals from the wintering area to the breeding area) occurs during a period of gonadal development and increasing levels of reproductive hormones. Autumnal migration (the movement of individuals from the breeding area to wintering grounds) occurs after the breeding season when the gonads have regressed to a completely inactive state, and circulating levels of reproductive hormones are basal (e.g., Farner 1955; Wingfield and Farner 1980). Note that juveniles also undergo autumnal migration. Since the two migratory periods occur in radically different physiologic states, it is reasonable to suspect that the factors timing and controlling at least the preparations for migration will be different. On the other hand, both migratory states have common requirements, e.g., for hyperphagia, fat deposition (as fuel for long flights), and migratory flight itself. Thus, it is likely that some of the regulatory mechanisms of migration will be similar in spring and autumn and others different.

Those factors (both external and internal) that influence migration are perceived by the central nervous system and this information is transduced into neuroendocrine and endocrine secretions that then regulate migratory processes. However, endocrine systems regulate diverse functions within an organism, and it is important first to define precisely which functions are involved in migration:

1. Deposition of fat as a fuel for prolonged migratory flight.
2. Integration of enzyme systems for storage *and* rapid utilization of energy stores.
3. Increased hematocrit, to enhance oxygen transport during prolonged flight.
4. Hypertrophy of flight muscles, presumably for increased work during prolonged flights.
5. Development, integration, and synchronization of migratory behavior, including changes in the diel pattern of activity.

[1] Department of Zoology, NJ-15, University of Washington, Seattle, Washington 98195, USA
[2] Max-Planck-Institut für Verhaltensphysiologie, Vogelwarte, D-8138 Andechs, FRG
[3] 915 E Third Ave., Ellensburg, Washington 98926, USA

E. Gwinner (Ed.) Bird Migration

Our knowledge of the hormonal bases of some of these processes in relation to migration is limited (e.g., Farner 1955, 1970; Mattocks 1976), and the data are often equivocal and suggest very complex regulatory processes. A summary of the effects of hormones on hypertrophy of flight muscles and hematocrit is presented, followed by a more extensive discussion of the endocrine control of premigratory hyperphagia, fattening and migratory behavior itself (*Zugunruhe*).

2 Muscle Hypertrophy

Hypertrophy of flight muscles (assessed by an increase in fat-free dry weight) has been shown for vernal migration in yellow wagtails (*Motacilla flava*), rose-colored starling (*Sturnus roseus*), and Swainson's thrush (*Hylocichla* = *Catharus ustulata*), and for autumnal migration in the catbird (*Dumatella carolinensis*) and Swainson's thrush (Vallyathan 1963; Child 1969; Fry et al. 1970, 1972; Marsh 1981). In contrast, no flight muscle hypertrophy was found prior to autumnal migration in the wood thrush (*H. mustelina*), veery (*H. fuscescens*), and juvenile willow warbler (*Phylloscopus trochilus*) (Hicks 1967; Baggott 1975). It seems likely that hypertrophy of flight muscles occurs in at least some species prior to migration, but the hormonal basis of this hypertrophy (and atrophy after migration) remains unknown, although John et al. (1983) suggested a role for growth hormone.

3 Hematocrit

In Gambel's white-crowned sparrow (*Zonotrichia leucophrys gambelii*), there is an increase in hematocrit of both males and females prior to vernal, but not autumnal, migration (Wingfield and Farner 1980). A similar increase in hematocrit has been demonstrated in red-headed buntings, *Emberiza bruniceps* (Thapliyal et al. 1982). Testosterone treatment increases erythropoesis and hematocrit in the domestic fowl (e.g., Domm and Taber 1946), red-winged blackbird *Aegelaius phoeniceus* (Robinzon and Rogers 1979), and the red-headed bunting (Thapliyal et al. 1983). However, Kern et al. (1972) found that injections of a synthetic androgen, depotestosterone cypionate, into female white-crowned sparrows had no effect on hematocrit. In the latter study, huge doses were used (150–1200 μg/3 days) and it is possible that the decrease of hematocrit on this treatment was a pharmacological effect. Thapliyal et al. (1983) also could find no effect of castration or treatment with testosterone propionate on hematocrit or hemoglobin concentration in red-headed buntings. The increase in hematocrit prior to vernal migration in this species appears to be regulated by thyroid hormones (Thapliyal et al. 1983). Conversely, ovariectomy prevents the rise in hematocrit of female white-crowned sparrows when photostimulated (Fig. 1), suggesting a role for gonadal hormones. Clearly the control of hematocrit prior to migration requires more research.

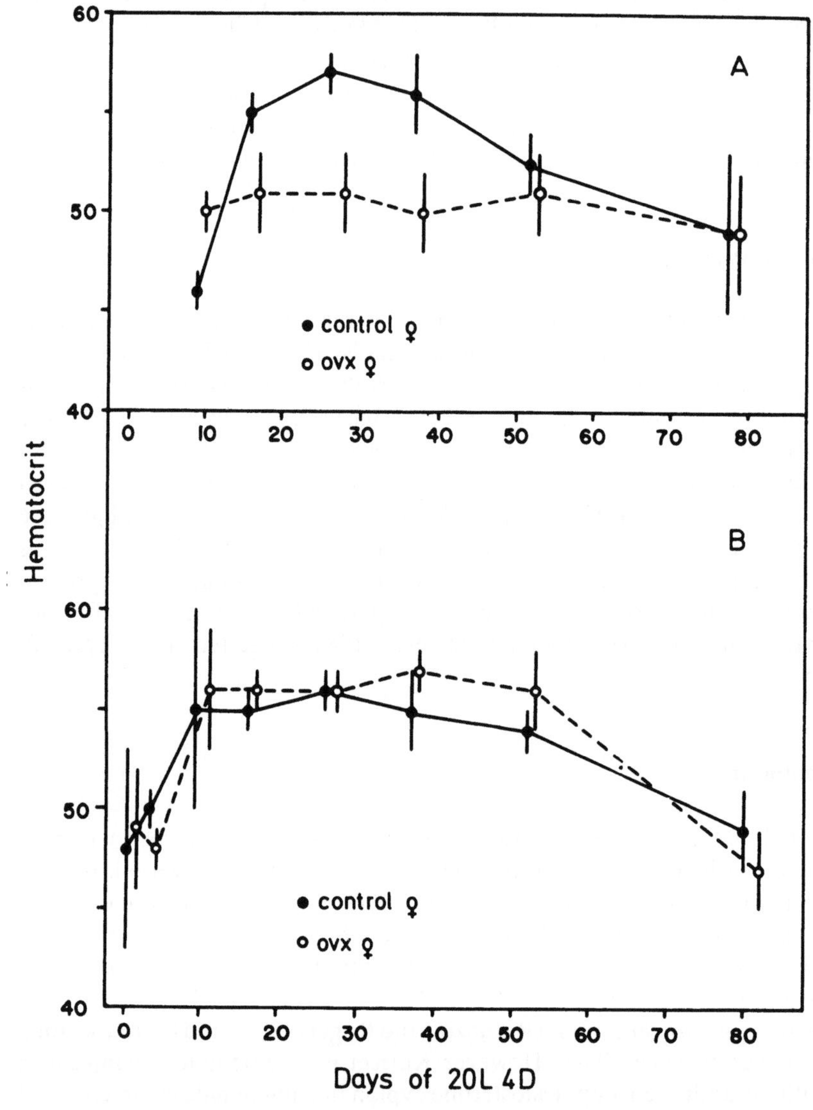

Fig. 1A,B. Effects of ovariectomy (*OVX*) on long day (*20L 4D*) induced changes in hematocrit in female white-crowned sparrows, *Zonotrichia leucophrys gambelii*. **A** Effect of ovariectomy before winter solstice; **B** effect of ovariectomy after winter solstice. *Vertical bars* are standard errors of the means (7≤1 1≤12). (H. Schwabl and D.S. Farner unpublished)

4 Endocrine Mechanisms of Premigratory Hyperphagia and Zugunruhe

4.1 The Role of Reproductive Hormones in the Control of Vernal Fattening and Zugunruhe

The possible role of gonadal hormones on vernal migration was first postulated formally by Rowan (1925) in the dark-eyed junco, *Junco hyemalis*, and later in the N. American crow, *Corvus brachyrhynchos* (Rowan 1932). Intact juncos and crows exposed to long days in midwinter showed a movement northward when released. However, the castrates moved southeast after release, as did nonphotostimulated controls. Thus the possibility arose that vernal migration is dependent upon the gonad, whereas autumnal migration is not.

Lofts and Marshall (1961) castrated 15 male bramblings (*Fringilla montifringilla*) during March and followed zugunruhe throughout the subsequent period of spring migration. They found equally high levels of zugunruhe in both normal and castrated individuals. Intact birds initiated zugunruhe about 1 week earlier than castrates, but both groups attained maximum activity simultaneously, and both declined to a minimum together. There was a tendency for normal birds to show the nightly peak of zugunruhe several hours before castrates. The authors concluded from these data that since nocturnal activity can be fully developed in the absence of the gonads, actual migration by castrates was possible. Morton and Mewaldt (1962) and King and Farner (1963) obtained similar results for golden-crowned sparrows, *Zonotrichia atricapilla*, and white-crowned sparrows, respectively. Castrates deposited less fat than intacts, and began zugunruhe about 1 week later. Otherwise castrates showed similar responses to controls.

Weise (1967) pointed out that previous studies in bramblings and sparrows had used birds castrated in late winter or early spring. Although this was well before the vernal migratory period, it was after days had begun to lengthen. Thus all subjects had been exposed to some photoperiodic stimulation before the experiments began. Weise captured white-throated sparrows, *Zonotrichia albicollis*, in mid-December and held them on 9 h of light for several weeks. Nine males were then castrated, held on short days for 3–5 more weeks, and were then transferred to long days of 15 h light. In this experiment castrates demonstrated only a slight trace of the extensive zugunruhe shown by the intact birds. Furthermore, castrates underwent no fat deposition in the manner of photostimulated controls or of spring migrants under outdoor conditions.

Stetson and Erickson (1972) confirmed and extended these results for white-crowned sparrows castrated in October and photostimulated in November (20L 4D). Intact birds responded to photostimulation with an immediate increase in body mass reaching a maximum on day 20. In contrast, castrates slowly lost mass until about day 40, later they gradually regained initial mass. Gwinner (1971) refers to similar unpublished data on golden-crowned sparrows. Thus it appears that if gonad removal is performed while birds are experiencing diminishing day length, then vernal premigratory fattening is abolished and zugunruhe greatly diminished.

Mattocks (1976) in a detailed study of white-crowned sparrows held on natural day length, again showed that castration before the winter solstice completely eliminated premigratory hyperphagia, fattening, and to some extent reduced zugunruhe the following spring (Fig. 2A). Plasma levels of testosterone were undetectable in castrates. Implants of small quantities of testosterone for a limited time in February (2 weeks) reinstated zugunruhe and fattening to varying

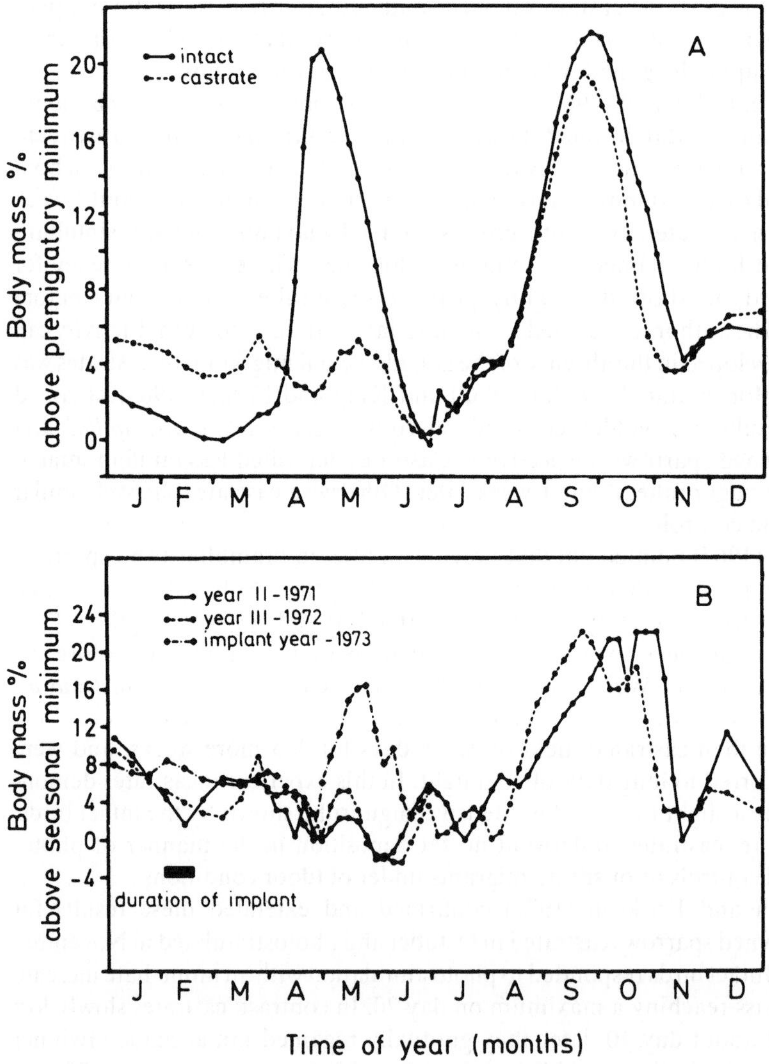

Fig. 2. A Seasonal changes in body mass in castrated and intact male white-crowned sparrows, *Zonotrichia leucophrys gambelii*; **B** in castrated males given an implant of testosterone for 2 weeks in February (Mattocks 1976)

degrees in the spring at the same time as intact controls (Fig. 2B). Schwabl et al. (1988) have obtained identical data for ovariectomized female white-crowned sparrows. Moreover, implants of testosterone just prior to photostimulation restored premigratory hyperphagia and fattening in ovariectomized females. Thus it appears that testosterone is important for regulation of vernal migration in females as well as in males. Further, Schwabl and Farner (1989a) showed that implants of 5α-dihydrotestosterone or estradiol were ineffective in reinstating vernal migratory phenomena in ovariectomized females. An androgenic inhibitor (ATD) that blocks aromatization of testosterone to estradiol, and also blocks 5 α-reduction, inhibited the effect of testosterone suggesting that there may be a synergy of testosterone metabolites acting at the target organ level.

Gwinner (1975) and Turek and Gwinner (1982) have suggested that testosterone may induce nocturnal migratory restlessness by changing the phase relationship between two circadian oscillators, one controlling morning activity and the other evening activity. As a result, one of these oscillators shifts into the night and represents zugunruhe. This hypothesis is based on observations of European starlings, *Sturnus vulgaris*, that "splitting" of locomotor activity into two components can be induced by testosterone injections both in constant light (Gwinner 1974) or in a light-dark cycle (Turek and Gwinner 1982). In the latter situation a diurnal pattern emerges reminiscent of that typical for the migratory season and thus may represent a mechanism for androgenic regulation of zugunruhe.

The next question is: do circulating levels of androgenic steroids increase prior to, or during, the vernal migratory period? Measurements of sex steroid hormones in free-living white-crowned sparrows indicate elevations of circulating levels of testosterone in late winter and early spring in males and DHT in females, while estradiol levels remained below the detection limit of the assay (Wingfield and Farner 1978).

The role of androgens (testosterone and perhaps its metabolites) in the development of premigratory hyperphagia and fattening for vernal migration are well established. However, it is equally clear that gonadal hormones have no influence over autumnal migration (e.g., Mattocks 1976; Schwabl et al. 1988; Fig. 2). Gonadectomized birds show autumnal premigratory hyperphagia, fattening, and zugunruhe identical to intact controls. Clearly autumnal migration has a very different hormonal basis but the exact nature of its endocrine control is still obscure.

Yokoyama (1976) showed that lesions of the posterior or entire median eminence decreased photoperiodically induced testicular growth, vernal premigratory fattening, and zugunruhe in *Z. l. gambelli* (Fig. 3A). Destruction of the anterior median eminence alone had no appreciable effects on testicular growth, fattening, or zugunruhe. Lesions in the basal infundibular nucleus decreased testicular growth and abolished zugunruhe, but allowed varying degrees of fattening.

Systemic administration of prolactin, testosterone propionate, or a combination in the posterior median eminence of lesioned birds induced fattening, but not zugunruhe (Fig. 3B). Thus the data of Yokoyama (1976) provide evidence

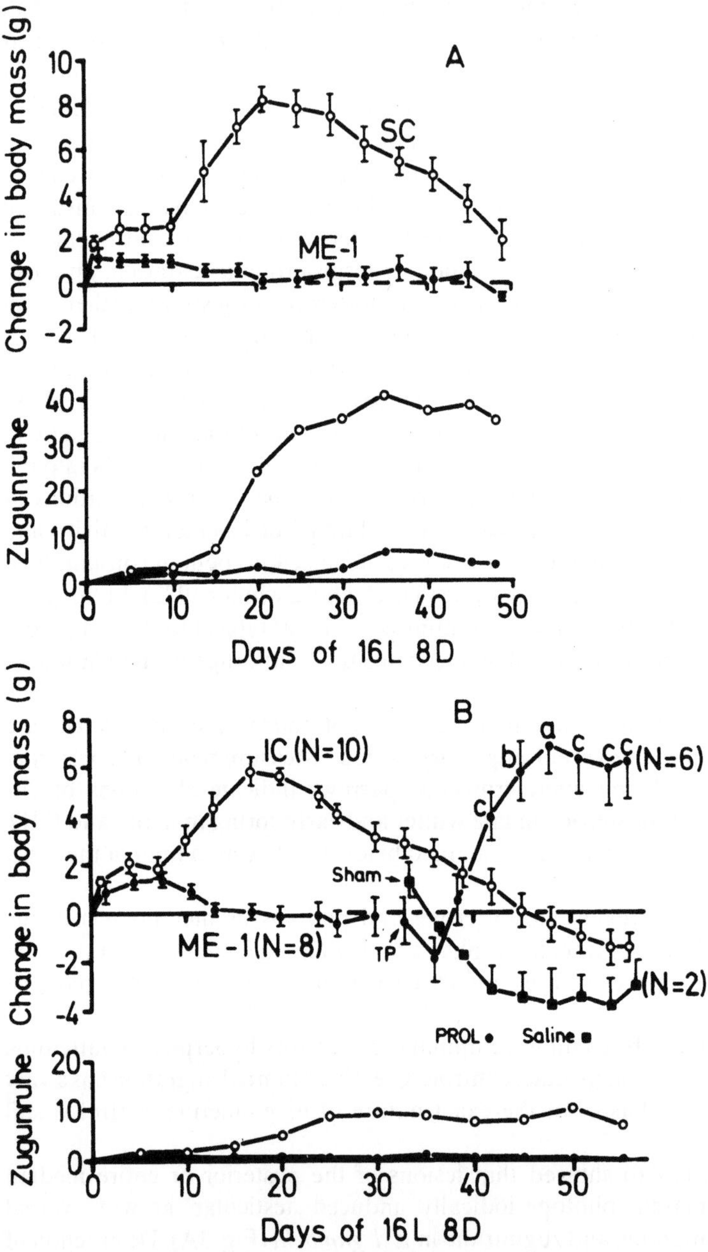

Fig. 3. A Changes in body mass and zugungruhe in male white-crowned sparrows, *Zonotrichia leucophrys gambelii* with lesions in the median eminence (*ME-1*) and sham-operated controls (*SC*) after photostimulation (*16L 8D*). **B** The effects of prolactin (*PROL*) and testosterone propionate (*TP*) injections on change in body weight and zugunruhe are presented. *IC* Control injected. *Vertical bars* are standard errors of the means; *a, b, c* indicate statistically significant differences from saline injected group. (Yokoyama 1976)

for central control by the infundibular nucleus and median eminence. Implants of testosterone in the infundibular nucleus, median eminence, and preoptic area decreased photoperiodically induced increases in LH and testicular growth. However, the birds fattened normally. Zugunruhe developed in some individuals but not all (Yokoyama 1977).

Testosterone may act by increasing the release of prolactin (see also Yokoyama 1976, 1977). Subcutaneous implants of testosterone propionate and daily injections of prolactin in median eminence lesioned birds increased food intake and body weight but did not induce zugunruhe. Furthermore, photostimulation through fiber optics inserted in the basal hypothalamus to stimulate encephalic photoreceptors, increased zugunruhe even though the ambient photophase was only 8 h (Yokoyama and Farner 1978). In white-throated sparrows, lesions of the ventral medial hypothalamus (thought to be a satiety center) increased fattening (Kuenzel and Helms 1967) and thus may be a focus for hormone action.

4.2 Thyroid Hormones

Putzig (1938b) showed an increase in thyroid activity by histological measures during vernal migration in lapwings, *Vanellus vanellus*. Elevation of thyroid activity was also noted during fall migration. Another histological study of the thyroid gland in rose-colored starlings indicated a decrease in colloidal material and increased secretory activity just prior to vernal migration (George and Naik 1964). Wilson and Farner (1960) found increased thyroid activity in white-crowned sparrows to occur when temperature declined. Histological signs of enhanced secretory activity were not correlated with migratory periods. In contrast, Oakeson and Lilley (1960) working with a California population of white-crowned sparrow, showed that thyroid activity increased during the vernal migratory period. In yet another species, the Canada goose, *Branta canadensis*, there were no histological changes in thyroid tissue during vernal or autumnal periods (Mori and George 1978).

Wagner (1930) was able to induce zugunruhe in whitethroats, *Sylvia communis*, with thyroid extracts. Merkel (1937, 1938, 1940) obtained the same effects in whitethroats and European robins, *Erithacus rubecula*, with small doses of thyroxine. However, high doses inhibited zugunruhe. He suggested that a very slight increase in thyroid activity could be responsible for initiation of both vernal and autumnal migratory behavior. Furthermore, he suggested that a slight increase in thyroid activity was necessary for stimulation of migration, and a greater amount was necessary to cause molting. This makes the two processes mutually exclusive. However, Putzig (1938a) injected large doses of thyrotropic hormone into European robins and enhanced zugunruhe, thus contradicting Merkel.

More recent investigations on the migratory red-headed bunting show that thyroidectomy inhibits nocturnal restlessness, and injections of T3 and T4 increased locomotor activity (Fig. 4B; Pathak and Chandola 1982b). Similarly, in the European robin, treatment with methyl thiouracil (a thyroid blocker) resulted

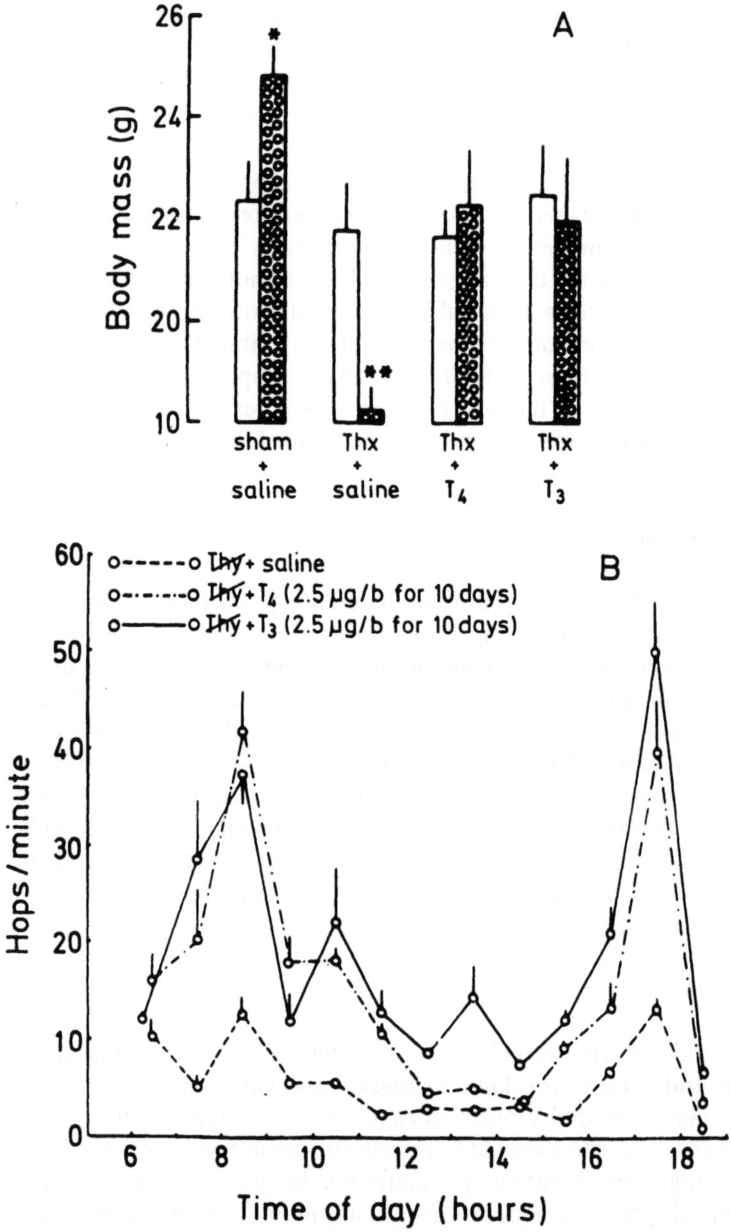

Fig. 4. **A** Changes in body mass and zugunruhe (**B** perch hopping) in thyroidectomized, and thyroxine and triiodothyronine injected male red-headed buntings, *Emberiza bruniceps*. *Vertical bars* are standard errors of the means. (Pathak and Chandola 1982a)

in decreased zugunruhe. Small injections of thyroid stimulating hormone (TSH) enhanced night activity associated with migration (Ieromnimon 1977, 1978). Treatment with a thyroid blocker (methyl thiouracil) decreased zugunruhe, and small doses of TSH increased night activity (Ieromnimon 1977). It would thus appear that TSH and thyroid hormones are intimately involved in the expression of zugunruhe. However, in the latter study, small numbers of birds were used and thyroid hormones may just fine-tune zugunruhe rather than initiate it. Schildmacher (1952) suggests that the beginning of spring migration occurs as thyroid activity increases, whereas autumnal migration becomes possible with the decline of sex hormones and possibly a decline in thyroid activity after molt. However, caution should be exercised here since Thompson (1936) and Farner (1955) emphasize that zugunruhe may not always be truly representative of migratory restlessness.

Thyroid hormones also appear to be involved in the regulation of premigratory fattening. Schildmacher and Rautenberg (1952) injected 0.2 mg of thyroxine into bramlings and chaffinches, *Fringilla coelebs*, and obtained an increase in food intake resulting in an elevation of body mass despite higher metabolic rate. However, Srivastava and Saxena (1979) showed that chemical thyroidectomy with neo-mercazole had no effect on the mass of a migratory bush quail, *Perdicula asiatica*. More detailed studies of the red-headed bunting show that thyroidectomy inhibits long-day induced increases in body mass and injections of T4 reversed this (Fig. 4A; see also Lal 1988). Note that castration in this species blocked the vernal increase in body mass, and testosterone propionate and T4 injections together or alone were effective in restoring mass in thyroidectomized birds. Testosterone in the absence of thyroid hormones was ineffective (Thapliyal et al. 1983).

A role for thyroid hormones in the regulation of migratory behavior is suggested by measurement of circulating plasma levels in white-crowned sparrows. Tri-iodothyronine (T3) levels increase in April at the height of vernal migration, whereas there is no change during spring in the sedentary house sparrow, *Passer domesticus* (Smith 1982). Changes in plasma levels of thyroxine (T4) were less clear-cut. The converse was found for T4 levels in the blood of snow geese (*Anser caerulescens*) and T3 levels showed no change (Campbell and Leatherland 1980). In rose-colored starlings, both T3 and T4 levels increased in spring just prior to vernal migration (Pathak and Chandola 1984), and similar results were obtained in Canada geese (*Branta canadensis*), in which T4 levels were highest in the vernal postmigratory period. T4 levels were lowest during the postmigratory period in autumn. Plasma levels of tri-iodothyronine (T3), on the other hand, were maximal during spring migration (John and George 1978).

In the red-headed bunting, injections of T3 and T4 increased locomotor activity, and restored loss in body mass to pretreatment levels (Fig. 4; Pathak and Chandola 1982b). Furthermore, the circulating T3/T4 ratio increased prior to vernal migration possibly through an increase in extrathyroidal conversion of T4 to T3 (Fig. 5; Chandola and Pathak 1980). In fact, there is a higher conversion of T4 to T3 in the premigratory period in March versus the postmigratory period in June. Thus, the increase in T3 levels in spring could be due to increased

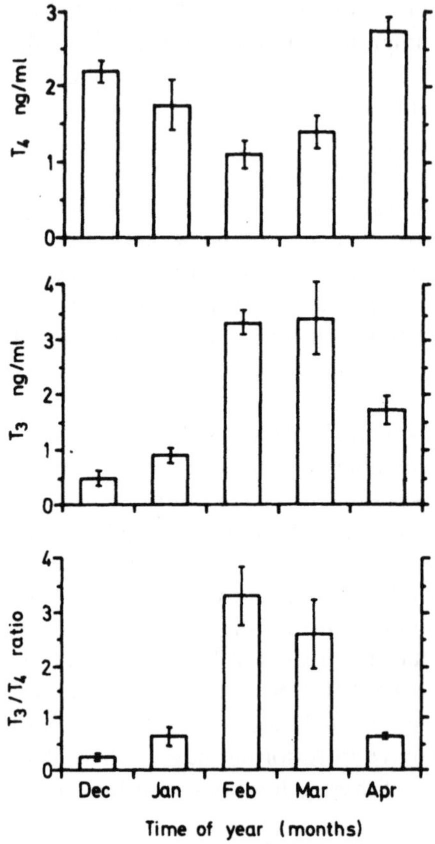

Fig. 5. Changes in plasma levels of thyroxine (*T4*) and triiodothyronine (*T3*) and the *T3/T4* ratio in male red-headed buntings, *Emberiza bruniceps. Vertical bars* are standard errors of the means. (After Chandola and Pathak 1980)

conversion rather than to an increase in secretion (Pathak and Chandola 1982a). Moreover, inhibition of the T4-T3 conversion enzyme by propylthiouracil in thyroidectomized birds prevented T4-induced fattening during the nonmigratory period (January) but failed to do this in the vernal premigratory period (Pathak and Chandola 1982a). The authors suggest that the deiodinating enzyme is more active in spring and doses similar to that which suppressed the action of T4 in January were not effective later in spring. Additionally, it is entirely possible, indeed probable, that although thyroid hormones may have a role in migratory processes, other factors are involved. It is also of considerable interest that there was no increase in the T3/T4 ratio prior to autumnal migration (Pathak and Chandola 1982b).

4.3 Growth Hormone

In adult pigeons, *Columba livia*, growth hormone increases lipolysis and elevates plasma levels of free fatty acids (FFA) (John et al. 1973). Goodridge and Ball (1967a) claimed that growth hormone injections into pigeons resulted in enlarged liver and enhanced lipogenic activity. Reasons for such apparently contradictory results are not clear. However, circulating levels of growth hormone are not correlated with vernal migratory periods in green-winged teal, *Anas crecca*, and Canada geese (Scanes et al. 1980; John et al. 1983).

The role of growth hormone on migration may be associated with thyroid hormones. Thyrotropin releasing factor (TRF) increases plasma levels of T4 within 20 min of injection in the snow goose, and injections of TSH are effective within 15 min. Increases in circulating T3 lag behind those of T4 (Campbell and Leatherland 1979). However, TRF also results in an elevation of growth hormone in the duck, *Anas platyrhynchos*, and domestic fowl (Pethes et al. 1979; Harvey et al. 1978) but does not effect prolactin, at least in snow geese (Campbell et al. 1981). In the green-winged teal there is a significant correlation of T4 and growth hormone with highest levels of both hormones occurring in August just before onset of autumnal migration (Scanes et al. 1980).

4.4 Neurohypophysial Peptides

Reports on the effects of neurohypophysial hormones on migration are few, but John and George (1986) present evidence that arginine vasotocin can release FFAs from adipose tissue in a nonmigratory bird, the pigeon. What implications this may have for hormonal regulation of migration remains to be determined.

4.5 Glucocorticosteroids

Both white wagtails, *Motacilla alba*, and yellow wagtails, *M. flava*, show an increase in histochemical activity of interrenal tissue toward the vernal migratory phase suggesting increased corticosteroid production (John 1965). Similar results were obtained for the rose-colored starling (Naik and George 1963). Lorenzen and Farner (1964) found histological signs of enhanced interrenal activity during spring in male and female white-crowned sparrows, a marked reduction in apparent activity in summer and during molt, followed by a resurgence of activity in the autumnal migratory period. Conversely, there were no histological changes in the interrenal tissue of Canada geese either pre- or post migration (Mori and George 1978). In European blackbirds, *Turdus merula*, interrenal activity appeared highest in summer and lower during migration (Fromme-Bouman 1962). Thus morphological data appear to provide no clear-cut trend for involvement of adrenocortical secretions on migration.

Experimental evidence suggests a role for adrenocorticosteroids in regulation of fattening and migratory activity. Injections of cortisol into chaffinches induced abrupt increases in migratory activity and a slight increase in fat in autumn (Dolnik and Blyumental 1967). In a migratory population of song sparrow, *Melospiza melodia*, implants of corticosterone in summer induced marked fattening in males. However, body mass did not change, suggesting that fat was deposited at the expense of protein (Fig. 6; Wingfield and Silverin 1986).

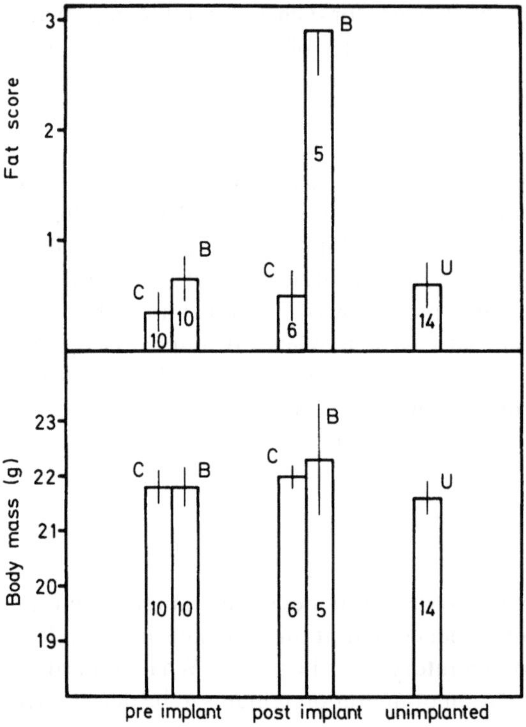

Fig. 6. Effects of subcutaneous implants of corticosterone on fat score and body mass of free-living, territorial, male song sparrows, *Melespiza melodia*. *C* control; *B* corticosterone-treated; *U* unmanipulated birds. *Vertical bars* are standard errors of the means. Sample size = *numbers* within columns (Wingfield and Silverin 1986)

In white-crowned sparrows, there is evidence that corticosterone blockers such as metyrapone will reduce premigratory hyperphagia and corticosterone replacement therapy will restore foraging levels to normal (Fig. 7; see also Wingfield 1988). Thus, it is possible that corticosteroids could regulate fattening, although this is likely to be a fine-tuning effect.

Péczely (1976) has presented extensive correlative evidence for a relationship between adrenocortical activity and migration. In these experiments he measured corticosteroid production in vitro using adrenal glands collected from several species in the field. In nonmigrants such as the house sparrow, yellowhammer, *Emberiza citrinella*, linnet, *Acanthis cannabina*, and great tit, *Parus major*, corticosterone production is lowest in spring and high during summer, including the postbreeding period. In migratory species such as the stonechat, *Saxicola torquata*, whitethroat, red-backed shrike, *Lanius collurio*, and brambling, corticosterone production is also high during the breeding period, but even higher levels occur during both migratory periods.

Curiously, changes in circulating corticosterone levels do not always support a role in fattening and migration. In the white-crowned sparrow, males show an increase in corticosterone during vernal migration, while females (undergoing the same migration) do not. Neither sex showed a marked increase in corticosterone levels during autumn migration (Wingfield and Farner 1978). Schwabl et al.

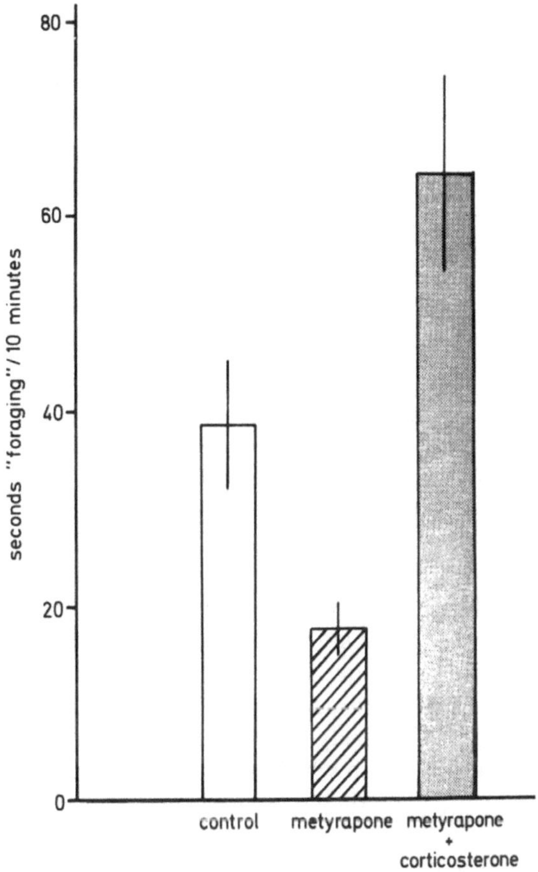

Fig. 7. Effects of metyrapone (a corticosterone inhibitor) and corticosterone replacement on foraging behavior in photostimulated male white-crowned sparrows, *Zonotrichia leucophrys gambelii*. *Vertical bars* are standard errors of the means. (Gearhart et al. in Wingfield 1988)

(1984a,b) obtained similar results in migratory European blackbirds. However, as pointed out by Péczely (1976), it is possible that any increased activity of the adrenal during migration may be a result of intense activity rather than a cause. In the nonmigratory pigeon, plasma levels of corticosterone increase markedly during long distance flights (up to 557 km) compared with controls held within the loft (Haase et al. 1986). A possible reason for the apparent contradictory results obtained from different species may be due to the ways in which samples were collected. Wingfield and Farner (1978) collected blood samples from white-crowned sparrows during the migratory period, but did not know whether individuals had just completed a flight, were in the midst of a flight, or were about to begin a migratory bout. Investigations in the future would benefit considerably by collecting samples from birds of known migratory state.

4.6 Prolactin

Meier and his associates (e.g., Meier and Dusseau 1978; Meier and Farner 1964; Meier and Ferrell 1978; Meier et al. 1980) suggest that the phase relationship of two circadian rhythms (i.e., the rhythm of plasma levels of corticosterone and prolactin varies as a function of day length and/or an endogenous circannual rhythm) regulates premigratory hyperphagia, fattening, and even migratory direction in the white-throated sparrow. The phase angle between these two rhythms, one usually associated with dawn, the other with dusk, then would determine the migratory direction and fattening versus no fattening response. However, it must be emphasized that this temporal synergism has not, as yet, been replicated in other migratory species (see also Farner and Gwinner 1980; Follett 1984). However, the effects of prolactin on fattening are well established and appear to depend upon time of day of administration (e.g., Meier et al. 1980; Vleck et al. 1980). In other cases, prolactin may actually inhibit migratory activity (e.g., European robin; Ieromnimon 1977, 1978).

Administration of 1 mg of ovine prolactin per day into pigeons caused the liver to double in size and enhanced lipogenic activity (Goodridge and Ball 1967a,b); and increased lipoprotein lipase activity in the crop sac and adipose tissue, especially when the injections were given in the afternoon (Garrison and Scow 1975). However, prolactin injections did not accelerate lipid synthesis in house sparrows and white-crowned sparrows (Goodridge 1964), and failed to increase lipogenesis in fasting pigeons (Goodridge and Ball 1965). Further, prolactin did not increase lipogenesis in vitro in white-crowned sparrows, but could nonetheless influence fattening by actions directly on the liver (Farner et al. 1968; Goodridge 1965).

Yokoyama (1976, 1977) suggests that in white-crowned sparrows, testosterone acts on vernal premigratory fattening by increasing the release of prolactin. Subcutaneous implants of testosterone propionate and daily injections of prolactin in median eminence lesioned birds (this treatment abolishes normal premigratory events) increased food intake and body weight but did not induce zugunruhe. Furthermore, central administration of prolactin into the nonmigratory ring dove, *Streptopelia risoria*, increased feeding behavior (Buntin and Tesch 1985), and specific binding of prolactin has been demonstrated in hypothalamic, diencephalic, and telencephalic sites in the dove brain. These loci are thought to regulate food intake (Buntin and Ruzycki 1987). Application of these techniques to truly migratory species would be of great interest.

Measurements of circulating prolactin in relation to migration of female white-crowned sparrows revealed that although ovariectomy abolished vernal fattening if the surgery was performed before photostimulation, there was no effect on circulating prolactin (Schwabl et al. 1988). In snow geese, no changes in circulating prolactin were found during both vernal and autumnal migratory periods (Campbell et al. 1981). Again, it is possible that changes in daily rhythms of prolactin could occur that were not detected by Schwabl et al. (1988). Joseph and Meier (1973) provide evidence that pituitary content of prolactin, as measured by bioassay, changes with season and may provide a basis for pro- and anti-fattening effects of prolactin at different times of day. However, these data should be interpreted with caution since changes in pituitary content do not parallel plasma levels.

As is the case with growth hormone, the actions of prolactin and corticosteroids on fattening are complex and many investigations are contradictory. Future studies would benefit by bearing in mind the effects of time of day as demonstrated by Meier and his colleagues.

4.7 Catecholamines

Goodridge and Ball (1965) showed that epinephrine stimulated lipolysis in adipose tissue of the pigeon and that tissue from females was less sensitive than that from males. The situation in migratory species is less clear. In snipe, *Capella gallinago*, sampled in India, epinephrine levels in interrenal tissue are low in winter, increase in early spring and then decline sharply as migration begins (possibly due to increased release). Norepinephrine levels decreased in winter and increased before migratory departure (Ghosh and Bhakta 1980). In the rose-colored starling, there were no histological changes in the chromaffin cells in winter, but just prior to vernal migration chromaffin tissue hypertrophied and total content of epinephrine and norepinephrine increased (Naik and George 1965). In the chaffinch, a general amine blocker, reserpine, reduced migratory activity in the fall but had no effect on feeding behavior (Dolnik and Blyumental 1967). Treatment of male white-crowned sparrows with α-methyl-para-tyrosin (α MPT), which blocks synthesis of catecholamines, inhibited the long day-induced increase in body mass (Fig. 8). The effect becomes apparent after about 10 days of photostimulation and when long day-induced fat deposition begins (see Fig. 3). Whether these effects occur centrally, e.g. on feeding behavior, or peripherally, remains to be determined. Similarly, although these data implicate catecholamines in the regulation of migratory processes (i.e., zugunruhe and fattening), cause and effect have yet to be established.

4.8 Pancreatic Hormones

Insulin had no effect on the metabolism of fat pads in house sparrows and white-crowned sparrows, and did not accelerate lipogenesis (Goodridge 1964). In the chaffinch, injections of insulin depressed locomotor activity during autumnal migration (Dolnik and Blyumental 1967). Glucagon abolished synthesis of fatty acids from acetate and Goodridge (1964) suggests that a high output of glucagon inhibits lipid synthesis during nonmigratory periods. A decline in glucagon during the migratory period may allow lipid synthesis to proceed at a more rapid rate but the role of glucagon in lipid mobilization remains to be determined.

In nonmigratory pigeons, glucagon stimulates lipolysis, whereas insulin had no effect (Goodridge and Ball 1965). In domestic geese, *Anser anser*, glucagon increased plasma levels of glucose and triglycerides (Bulla et al. 1980). Avian

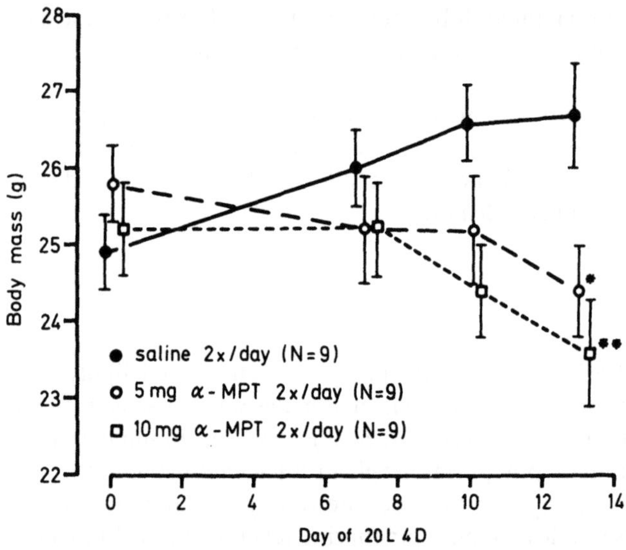

Fig. 8. Effect of treatment with α-methyl-para-tyrosin (*αMPT*, an inhibitor of catecholamine synthesis) on long day-induced (*20L 4D*) increase of body mass of white-crowned sparrow, *Zonotrichia leucophrys gambelii*. * $p < 0.02$ vs controls, ** $p < 0.01$ vs controls. *Vertical bars* are standard errors of the means. (K.S. Matt, H. Schwabl, D.S. Farner unpublished work)

pancreatic polypeptide (APP) reduces plasma levels of glycerol, stimulates hepatic glycogenolysis, and decreases plasma FFA but leaves plasma levels of glucose unchanged (Hazelwood et al. 1973; Hazelwood and Langslow 1978). Clearly, the roles of pancreatic hormones in migratory processes deserve more study.

5 Termination of Migratory Periods

Very little is known about the endocrine basis (if any) of the termination of migratory periods, often accompanied by a return to a lean state. Wagner (1961) suggested that high levels of testosterone associated with territorial behavior upon arrival on the breeding grounds may terminate vernal migration. Schwabl and Farner (1989b) tested this hypothesis in white-crowned sparrows and found that implantation of testosterone during vernal migration failed to disrupt fattening and zugunruhe. On the other hand, seasonal prolongation of high circulating levels of testosterone maintained zugunruhe probably because this steroid is known to delay onset of photorefractoriness and thus termination of all vernal reproductive stages (Schwabl and Farner 1989b). The mechanisms that terminate spring migration remain to be elucidated.

Even less is known about termination of autumnal migration. Schüz (1952) has suggested that in some species autumnal zugdisposition is maintained throughout the winter, thus allowing facultative movements of individuals in relation to local weather conditions (see also Terrill, this Vol.). Possible endocrine bases for these phenomena have not been explored.

6 Conclusions and Roles of Environmental Cues that Influence Migration

Although physiologists have been studying migration in birds for almost 70 years, much of the data is contradictory and confusing. Part of this confusion may be due to differences in experimental design and also whether investigations were performed in spring or autumn, during preparations for migration, or during the migratory period itself. All of these variables could have profound influences on the regulatory endocrine systems. Additionally, it is not always clear whether endocrine changes are a cause or an effect of migratory processes. It may be possible to gain some insight into this complexity by considering the ways in which the environment regulates migratory phenomena (and thus regulates changes in endocrine function that acts as a link between environmental and physiological mechanisms). A scheme is presented below for classifying ways in which environmental signals influence migration through endocrine mechanisms. This scheme follows that presented for the regulation of reproductive cycles in birds (Wingfield 1980, 1983), and may also be useful for other nonreproductive processes in the avian annual cycle.

6.1 Initial Predictive Information

These cues initiate preparations for migration (such as fat deposition, muscle hypertrophy, and associated migratory behavior) in advance of the actual migratory period. This state of readiness to migrate is termed zugdisposition (Farner 1955; Schüz 1952). Initial predictive information then maintains zugdisposition and terminates the migratory period. Examples are the annual cycle of day length and endogenous circannual rhythms (e.g., Farner 1985; Gwinner 1986; Berthold 1989; Schwabl and Farner 1989b, see also Gwinner and Berthold, both this Vol.). It has been proposed that in some species (e.g., waterfowl and shorebirds) zugdisposition may be maintained throughout winter (Vluegel 1948; Schüz 1952). This would constitute a basis for winter movements in response to unpredictable severe weather (see also Terrill, this Vol.).

6.2 Supplementary Information

This type of information regulates day to day adjustments in migratory state and thus supplements initial predictive information (see also Marshall 1961; Wagner 1930; Evans 1970). The degree of premigratory fattening is largely a function of migration speed, magnitude of the barriers to be crossed, and aerodynamic considerations (Farner et al. 1961; King et al. 1963; Odum et al. 1961; Blem 1976).

However, endocrine adjustments to these variables remain unknown. Weather has profound effects on day to day variation in migratory state (see Gauthreaux 1971, 1982; Richardson this Vol. for reviews). In general, increasing temperature in spring tends to elevate zugunruhe in European robins, and white-crowned sparrows (Palmgren 1937; Putzig 1937, 1938a; Schildmacher 1938; Farner and Mewaldt 1953). Conversely, low ambient temperatures in spring tended to decrease fattening in yellow wagtails (Fry et al. 1970). Inclement weather may decrease available food and thus retard migration (Dolnik and Blyumental 1967; Evans 1970). In autumn, decreasing temperature fine tunes onset of zugunruhe (once zugdisposition has developed) in goldcrests, *Regulus regulus* (Palmgren 1936), song thrush, *Turdus philomelos* (Siivonen and Palmgren 1936), linnet and yellowhammer (Stolt 1969), and European robin (Putzig 1938a, 1939). Additionally, temperature and wind velocity may affect short-term migratory behavior while en route (Evans 1970). Changes in air pressure may also be important. Stolt (1969) found that activity decreased when barometric pressure increased (possibly indicative of fair weather) in linnets, yellowhammers, and redpolls, *Acanthis flammea*, whereas activity increased as barometric pressure declined (possibly indicative of an approaching storm). Alternatively, Gauthreaux (1971) found that no weather conditions (with the possible exception of heavy rain) affected autumnal departure of migrants from southern Louisiana.

Other factors may also fine tune migratory processes. In the Canada goose, there were two periods of feeding (morning and evening) on sunny days, whereas on cloudy days (similar temperature range) there was only one long feeding bout. Low temperatures decreased food intake and activity, and these birds did not migrate. Snow melt opens up foraging areas and feeding rates increase accordingly. Thus, hyperphagia in the Canada goose may be fine tuned by temperature and food availability in the spring (McLandress and Raveling 1981a). Finally, Gwinner (1967) found that the intensity of zugunruhe in European robins and redstarts, *Phoenicurus phoenicurus*, was greater on moonlit nights than on totally dark nights.

Clearly, local weather conditions and food availability can have pronounced influences on whether an individual migrates that day or rests and replenishes fat stores, etc. The hormonal mechanisms underlying these processes have only been studied superficially.

6.3 Synchronizing Information

Many species of waterfowl (e.g., Anseriformes, Ciconiiformes, Gruiformes) migrate in flocks or even family parties. In these cases there must be some degree of behavioral synchronization of premigratory fattening, especially during migration itself. In the Canada goose, paired females spent more time feeding than unpaired females. Paired geese precede unpaired geese to feeding areas in spring, and unpaired geese initially weighed less, and gained less weight during the period of fattening (McLandress and Raveling 1981b). There is also evidence for synchrony in fattening and onset of autumnal migration in several species of north European birds, although this may be an effect of decreasing photoperiod rather than a result of behavioral interactions (Dolnik and Blyumental 1967). However, these authors do present evidence for social facilitation of feeding and zugunruhe during wave migration in the chaffinch.

It seems likely that behavioral interactions that synchronize the migration of groups of individuals would act through the endocrine system. As far as we are aware, this aspect of hormonal control has not been explored.

6.4 Modifying Information

This type of information tends to disrupt the migratory process and may result in considerable delay or even high mortality. Very severe storms tend to disrupt migration to an extent that may preclude attempts at breeding. In some cases reverse migrations may occur (e.g., Canada geese; Wege and Raveling 1983). Low availability of food may also deplete fat and protein stores to an extent that prevents further migration and breeding, or results in death. Although there are no hormonal data addressing this phenomenon at present, it seems likely that such severe conditions would induce a classical stress response resulting in high circulating levels of glucocorticosteroids. These in turn tend to redirect the individual's behavior toward survival (see Wingfield 1988). Once unfavorable conditions pass, then the premigratory events need to be recycled and a state of zugdisposition regained. The hormonal mechanisms here are likely to be different from those of supplementary cues that accelerate or retard day to day migratory events without the individual being stressed. Only further research will clarify this interesting problem.

The scheme presented above may be of heuristic value in separating possible conflicting variables in the design of experiments to elucidate endocrine mechanisms. Clearly, the distinction between preparations for migration (regulated by initial predictive information) and adjustments made before, during, and after a migratory flight (regulated by supplementary factors) must be made. There is no reason to expect that any one combination of hormones is going to regulate all migratory processes. Indeed, examining the scheme above it is clear that the endocrine mechanisms underlying the effects of initial predictive information and supplementary information are entirely different. Superimpose upon this the mechanisms by which synchronizing and modifying factors may influence migration, it is easy to see why there is so much conflict and confusion in the literature presented here. Hopefully future investigations will bear in mind the role of different environmental factors and the migratory state of individual birds when designing experiments to elucidate hormonal regulation.

Acknowledgements. The senior author was supported by National Science Foundation Grant number DCB 8616189 while preparing this communication. The authors are grateful to Lynn Erckmann, Heidrun Bamberg and Ingrid Schwabl who drafted the figures.

References

Baggott GK (1975) Moult, flight muscle "hypertrophy" and premigratory lipid deposition of the juvenile willow warbler *Phylloscopus trochilus*. J Zool (Lond) 175:299–314

Berthold P (1989) The control of migration in European warblers. In: Quellet H (ed) Proc XIX Int Ornithol Congr, Ottawa, Univ Ottawa Press, Ottawa, pp 215–249

Blem CR (1976) Patterns of lipid storage in birds. Am Zool 16:671–686

Bulla J, Dobalova M, Zelnik J, Granat J (1980) The influence of glucagon on plasma levels of glucose, triglycerides, total lipids and cholesterol in geese. Adv Physiol Sci 33:409–419

Buntin JD, Ruzycki E (1987) Characteristics of prolactin binding sites in the brain of the ring dove (*Streptopelia risoria*). Gen Comp Endocrinol 65:243–253

Buntin JD, Tesch D (1985) Effects of intracranial prolactin administration on the maintenance of incubation readiness, ingestive behavior, and gonadal condition in ring doves. Horm Behav 19:188–203

Campbell RR, Leatherland JF (1979) Effect of TRH, TSH and LH-RH on plasma thyroxine and triiodothyronine in the lesser snow goose (*Anser caerulescens caerulescens*) and plasma thyroxine in the Rouen duck (*Anas platyrhynchos*). Can J Zool 57:271–274

Campbell RR, Leatherland JF (1980) Seasonal changes in thyroid activity in the lesser snow goose (*Anser caerulescens caerulescens*). Can J Zool 58:1144–1150

Campbell RR, Etches RJ, Leatherland JF (1981) Seasonal changes in plasma prolactin concentration and carcass lipid levels in the lesser snow goose (*Anser caerulescens caerulescens*). Comp Biochem Physiol 68:653–658

Chandola A, Pathak VK (1980) Premigratory increase in circulating triidothyronine/thyroxine ratio in the red-headed bunting (*Emberiza bruniceps*). Gen Comp Endocrinol 41:270–273

Child GI (1969) A study of non-fat weights in migrating Swainson's thrushes (*Hylocichla ustulata*). Auk 86:327–338

Dolnik VR, Blyumental TI (1967) Autumnal premigratory and migratory periods in the chaffinch (*Fringilla coelebs coelebs*) and some other temperate zone passerine birds. Condor 69:435–468

Domm LV, Taber E (1946) Endocrine factors controlling erythrocyte concentrations in the blood of the domestic fowl. Physiol Zool 19:258–281

Evans PR (1970) Timing mechanisms and the physiology of bird migration. Sci Prog Oxford 58:263–275

Farner DS (1955) The annual stimulus for migration: experimental and physiologic aspects. In: Wolfson A (ed) Recent studies in avian biology. University Illinois, Urbana, pp 198–237

Farner DS (1970) Predictive function in the control of annual cycles. Environ Res 3:119

Farner DS (1985) Annual rhythms. Annu Rev Physiol 47:65–82

Farner DS, Gwinner E (1980) Photoperiodicity, circannual and reproductive cycles. In: Epple A, Stetson MH (eds) Avian endocrinology. Academic Press, London, pp 331–366

Farner DS, Mewaldt LR (1953) The recording of diurnal activity patterns in caged birds. Bird-Banding 24:55–65

Farner DS, Oksche A, Kamemoto FI, King JR, Cheyney HE (1961) A comparison of the effect of long daily photoperiods on the pattern of energy storage in migratory and non-migratory finches. Comp Biochem Physiol 2:125–142

Farner DS, King JR, Stetson MH (1968) The control of fat metabolism in migratory birds. Prog Endocrinol Ecerp Med Int Cong Ser 184:152–157

Follett BK (1984) Birds. In: Laming GE (ed) Marshall's physiology of reproduction, vol 1. Reproduction of vertebrates. Churchill Livingstone, London, pp 283–350

Fromme-Bouman H (1962) Jahresperiodische Untersuchungen an der Nebennierenrinde der Amsel (*Turdus merula* L.). Vogelwarte 21:188–198

Fry CH, Ash JS, Ferguson-Lees IJ (1970) Spring weights of Palaearctic migrants at Lake Chad. Ibis 112:58–82

Fry CH, Ferguson-Lees IJ, Dowsett RJ (1972) Flight muscle hypertrophy and ecophysiological variation of yellow wagtails, *Motacilla flava* races at Lake Chad. J Zool (Lond) 167:293–306

Garrison MM, Scow RO (1975) Effect of prolactin on lipoprotein lipase in crop sac and adipose tissue of pigeons. Am J Physiol 228:1542–1544

Gauthreaux SA Jr (1971) A radar and direct visual study of passerine spring migration in southern Louisiana. Auk 88:343–365

Gauthreaux SA Jr (1982) The ecology and evolution of avian migration systems. In: Farner DS, King JR, Parkes KS (eds) Avian biology, vol 6. Academic Press, London, pp 93–168

George JC, Naik DV (1964) Cyclic changes in the thyroid of the migratory starling (*Sturnus roseus*) Linnaeus. Pavo 2:37–47

Ghosh A, Bhakta D (1980) Behavior of catecholamines in the migratory process of *Capella gallinago*. Adv Physiol Sci Hung 33:393–399

Goodridge AG (1964) The effect of insulin, glucagon and prolactin on lipid synthesis and related metabolic activity in migratory and non-migratory finches. Comp Biochem Physiol 13:1–26

Goodridge AG (1965) Hormonal regulation of the activity of the fatty acid synthesizing system and of malic enzyme concentrations in liver cells. Fed Proc 34:117–123

Goodridge AG, Ball EG (1965) Studies on the metabolism of adipose tissue XVIII. In vitro effects of insulin, epinephrine, and glucagon on lipolysis and glycolysis in pigeon adipose tissue. Comp Biochem Physiol 16:367–381

Goodridge AG, Ball EG (1967a) Lipogenesis in the pigeon: in vivo studies. Am J Physiol 213:245–249

Goodridge AG, Ball EG (1967b) The effect of prolactin on lipogenesis in the pigeon. Biochemistry 6:2335–2343

Gwinner E (1967) Wirkung des Mondlichtes auf die Nachtaktivität von Zugvögeln. Lotsenversuch an Rotkehlchen (*Erithacus rubecula*) und Gartenrotschwänzen (*Phoenicurus phoenicurus*). Experentia Basel 23:1–5

Gwinner E (1971) Endogenous timing factors in bird migration. In: Galler SR (ed) Animal orientation and navigation. NASA Washington DC, pp 321–338

Gwinner E (1974) Testosterone induces "splitting" of circadian locomotor activity rhythms in birds. Science 185:72–74

Gwinner E (1975) Effects of season and external testosterone on the free-running circadian activity rhythm of European starlings (*Sturnus vulgaris*). J Comp Physiol 103:315–328

Gwinner E (1986) Circannual rhythms. Springer, Berlin Heidelberg New York, p 154

Haase E, Rees A, Harvey S (1986) Flight stimulates adrenocortical activity in pigeons (*Columba livia*). Gen Comp Endocrinol 61:424–427

Harvey S, Scanes CG, Chadwick A, Bolton NJ (1978) The effect of thyrotropin-releasing hormone (TRH) and somatostatin (GHRIH) on growth hormone and prolactin secretion in vitro and in vivo in the domestic fowl (*Gallus domesticus*). Neuroendocrinology 26:249–260

Hazelwood RL, Langslow DR (1978) Intrapancreatic regulation of hormone secretion in the domestic fowl, *Gallus domesticus*. J Endocrinol 76:449–459

Hazelwood RL, Turner SD, Kimmel JR, Pollock HG (1973) Spectrum effects of a new polypeptide (third hormone?) isolated from chicken pancreas. Gen Comp Endocrinol 21:485–490

Hicks DL (1967) Adipose tissue composition and cell size in fall migratory thrushes. Condor 69:387–399

Ieromnimon V (1977) Beobachtungen über die Wirkung von Hormonen auf das Zugverhalten bei Rotkehlchen (*Erithacus rubecula*). I. Die Wirkung von Prolaktin (LtH = Hpr) im Jahreszyklus. Vogelwarte 29:126–134

Ieromnimon V (1978) Beobachtungen über die Wirkung von Hormonen auf das Zugverhalten bei Rotkehlchen (*Erithacus rubecula*). II. Die Wirkung von Pharmaka. Vogelwarte 29:221–230

John TM (1965) A histochemical study of adrenal corticoids in the pre- and post-migratory wagtails, *Motacilla alba* and *M. flava*. Pavo 4:9–14

John TM, George JC (1978) Circulating levels of thyroxine (T4) and triiodothyronine (T3) in the migratory Canada goose. Physiol Zool 51:361–370

John TM, George JC (1986) Arginine vasotocin induces free fatty acid release from avian adipose tissue in vitro. Arch Int Physiol Biochem 94:85–89

John TM, McKeown BA, George JC (1973) Influence of exogenous growth hormone and its antiserum on plasma free fatty acid level in the pigeon. Comp Biochem Physiol 46:497–504

John TM, Georce JC, Scanes CG (1983) Seasonal changes in circulating levels of luteinizing hormone and growth hormone in the migratory Canada goose. Gen Comp Endocrinol 51:44–49

Joseph MM, Meier AH (1973) Daily rhythms of plasma corticosterone in the common pigeon (*Columba livia*). Gen Comp Endocrinol 20:326–330

Kern MD, Graw WA De, King JR (1972) Effects of gonadal hormones on the blood composition of white-crowned sparrows. Gen Comp Endocrinol 18:43–52

King JR, Farner DS (1963) The relationship of fat deposition to zugunruhe and migration. Condor 65:200–223

King JR, Barker S, Farner DS (1963) A comparison of energy reserves during autumnal and vernal migratory periods in the white-crowned sparrow, *Zonotrichia leucophrys gambelii*. Ecology 44:513–521

Kuenzel WJ, Helms CW (1967) Obesity produced in a migratory bird by hypothalamic lesions. Bioscience 17:395–396

Lal P (1988) Role of thyroid in sexual and body weight cycles of the migratory red-headed bunting (*Emberiza bruniceps*). Gen Comp Endocrinol 70:291–300

Lofts B, Marshall AJ (1961) Zugunruhe activity in castrated bramblings, *Fringilla montifringilla*. Ibis 103:189–194

Lorenzen LC, Farner DS (1964) An annual cycle in the adrenal gland of the white-crowned sparrow, *Zonotrichia leucophrys gambelii*, Gen Comp Endocrinol 4:253–263

Marsh RL (1981) Catabolic enzyme activities in relation to premigratory fattening and muscle hypertrophy in the gray catbird (*Dumatella carolinensis*). J Comp Physiol B Biochem Syst Environ Physiol 141:417–424

Marshall AJ (1961) Breeding seasons and migration. In: Marshall AJ (ed) Biology and comparative physiology of birds, vol 2. Academic Press, London, pp 307–339

Mattocks PW Jr (1976) The role of gonadal hormones in the regulation of the premigratory fat deposition in the white-crowned sparrow, *Zonotrichia leucophrys gambelii*. MS Thesis, University of Washington, Seattle

McLandress MR, Raveling DG (1981a) Hyperphagia and social behavior of Canada geese prior to spring migration. Wilson Bull 93:310–324

McLandress MR, Raveling DG (1981b) Food habits and body composition changes of giant Canada geese prior to spring migration. Auk 98:65–79

Meier AH, Dusseau JW (1968) Prolactin and the photoperiodic gonadal response in several avian species. Physiol Zool 41:95–103

Meier AH, Farner DS (1964) A possible endocrine basis for premigratory fattening in the white-crowned sparrow, *Zonotrichia leucophrys gambelii*. Gen Comp Endocrinol 4:584–595

Meier AH, Ferrell BR (1978) Avian endocrinology. In: Florkin H, Scheer H, Brush A (eds) Chemical Zoology, vol 10. Academic Press, London, pp 213–271

Meier AH, Ferrell BR, Miller LJ (1980) Circadian components of the circannual mechanism in the white-throated sparrow. In: Nöhring R (ed) Acta XVII Cong Int Ornithol. Deutschen Omith Gesellschaft Berlin, pp 458–462

Merkel FW (1937) Zur Physiologie des Vogelzugtriebes. Zool Anz 117:297–308

Merkel FW (1938) Zur Physiologie der Zugunruhe bei Vögeln. Ber Ver Schles Ornithol 23:1–71

Merkel FW (1940) Neuere Untersuchungen über die Ursachen des Vogelzug-Triebes. Natur Volk 70:167–178

Mori JG, George JC (1978) Seasonal histological changes in the gonads, thyroid and adrenal of the Canada goose (*Branta canadensis interior*). Acta Anat 101:304–324

Morton ML, Mewaldt LR (1962) Some effects of castration on a migratory sparrow (*Zonotrichia atricapilla*). Physiol Zool 35:237–247

Naik DV, George JC (1963) Histochemical demonstration of increased corticoid level in the adrenal of *Sturnus roseus* (Linneus) towards the migratory phase. Pavo 1:103–105

Naik DV, George JC (1965) Certain cyclic changes in the histology and histochemistry in the adrenal of the migratory starling, *Sturnus roseus* Linnaeus. Pavo 3:121–131

Oakeson BB, Lilley BR (1960) Annual cycle of thyroid histology in two races of white-crowned sparrow. Anat Rec Suppl 136:41–57

Odum EP, Connell CE, Stoddard HL (1961) Flight energy and estimated flight ranges of some migratory birds. Auk 78:515–527

Palmgren P (1936) Über den Massenwechsel bei *Regulus r. regulus* (L). Ornis Fenn 13:159–164

Palmgren P (1937) Auslösung der Frühlingszugunruhe durch Wärme bei gekäfigten Rotkehlchen, *Erithacus rubecula* (L). Ornis Fenn 14:71–73

Pathak VK, Chandola A (1982a) Involvement of thyroid gland in the development of migratory disposition in the red-headed bunting, *Emberiza bruniceps*. Horm Behav 16:46–59

Pathak VK, Chandola A (1982b) Seasonal variations in extrathyroidal conversion of thyroxine to triiodothyronine and migratory disposition in red-headed bunting. Gen Comp Endocrinol 47:433–440

Pathak VK, Chandola A (1984) Variations in circulating thyroxine and tri-iodothyronine concentration in relation to spring migration in rosy pastor, *Sturnus roseus*. Horm Behav 18:111–116

Péczely P (1976) Étude circanuelle de la fonction corticosurrènalienne chez les éspeces de passereaux migrants et non-migrants. Gen Comp Endocrinol 30:1–11

Pethes G, Scanes CG, Rudas P (1979) Effect of synthetic thyrotropin-releasing hormone on the circulating growth hormone concentration in cold and heat-stressed ducks. Acta Vet Acad Sci Hung 27:175–177

Putzig P (1937) Weitere Versuche über die Beziehungen der Keimdrüsen zum Zugverhalten. Vogelzug 9:189–200

Putzig P (1938a) Beobachtungen über Zugunruhe beim Rotkehlchen (*Erithacus rubecula*). Vogelzug 9:10–14

Putzig P (1938b) Der Frühwegzug des Kiebitzes. J Ornithol 86:123–165

Putzig P (1939) Beiträge zur Stoffwechselphysiologie des Zugvogels. Vogelzug 10:139–154

Raveling DG (1976) Migration reversal: a regular phenomenon of Canada geese. Science 193:153–154

Robinzon B, Rogers JG Jr (1979) The effect of gonadal and thyroidal hormones on the regulation of food intake, adiposity, and on various endocrine glands in the red-winged blackbird (*Aegelaius phoeniceus*). Gen Comp Endocrinol 38:135–147

Rowan W (1925) Relation of light to bird migration and development changes. Nature 115:494–495

Rowan W (1932) Experiments in bird migration. III. The effects of artificial light, castration and certain extracts on the autumn movements of the American crow (*Corvus brachyrhynchos*). Proc Natl Acad Sci 18:639–654

Scanes CG, Jallageas M, Assenmacher I (1980) Seasonal variations in the circulating concentrations of growth hormone in male Pekin ducks (*Anas platyrhynchos*) and teal (*Anas crecca*): correlations with thyroidal function. Gen Comp Endocrinol 41:76–89

Schildmacher H (1938) Hoden und Schildrüse des Gartenrotscharzes, *Phoenicurus phoenicurus phoenicurus* (L.) unter dem Einfluss zusätzlicher Belichtung im Herbst und Winter. Biol Zentralbl 58:464–472

Schildmacher H (1952) Untersuchungen zur Zugdisposition der Zugvögel. Die Wissenschaft vom Vogel und die Volkswirtschaft. Herausgegeben vom Kulturbund zur demokratischen Erneuerung Deutschlands, Zentralkommission Ornithologie und Vogelschutz, Berlin, Reprint, 9 pp

Schildmacher H, Rautenberg W (1952) Über die Wirkung kleiner Mengen von Thyroxin auf das Körpergewicht bei Finkenvögeln. Biol Zentralbl 71:397–405

Schüz E (1952) Vom Vogelzug. Grundriss der Vogelzugskunde. Schops, Frankfurt am Main

Schwabl H, Farner DS (1989a) Dependency on testosterone of photoperiodically-induced vernal fat deposition in female white-crowned sparrows. Condor 91:108–112

Schwabl H, Farner DS (1989b) Endocrine and environmental control of vernal migration in male white-crowned sparrows, *Zonotrichia leucophrys gambelii*. Physiol Zool

Schwabl H, Wingfield JC, Farner DS (1984a) Endocrine correlates of autumnal behavior in sedentary and migratory individuals of a partially migratory population of the European blackbird (*Turdus merula*). Auk 101:499–507

Schwabl H, Schwabl-Benzinger I, Farner DS (1984b) Relationship between migratory disposition and plasma levels of gonadotropins and steroid hormones in the European blackbird. Naturwissenschaften 71:329–330

Schwabl H, Schwabl-Benzinger I, Goldsmith AR, Farner DS (1988) Effects of ovariectomy on long day-induced premigratory fat deposition, plasma levels of luteinizing hormone and prolactin, and molt in white-crowned sparrows, *Zonotrichia leucophrys gambelii*. Gen Comp Endocrinol 71:398–405

Siivonen L, Palmgren P (1936) Über die Einwirkung der Temperatursenkung auf die Zugstimmung bei einer gekäfigten Singdrossel (*Turdus philomenos philomenos* Brehm). Ornis Fenn 13:64–67

Smith JP (1982) Annual cycle of thyroid hormones in the plasma of white-crowned sparrows and house sparrows. Condor 84:160–167

Srivastava AK, Saxena AK (1979) Effect of long and short days on body weight and gonads of chemically thyroidectomized jungle bush quail (*Perdicula asiatica*). Pavo 17:85–90

Stetson MH, Erickson JE (1972) Hormonal control of photoperiodically-induced fat deposition in white-crowned sparrows. Gen Comp Endocrinol 19:355–362

Stolt BO (1969) Temperature and air pressure experiments on activity in passeriforme birds with notes on seasonal and circadian rhythms. Zool Bidr Upps 38:175–231

Thapliyal JP, Pati AK, Singh VK, Lal P (1982) Thyroid, gonad and photoperiod in the hemopoesis of the migratory red-headed bunting, *Emberiza bruniceps*. Gen Comp Endocrinol 46:327–332

Thapliyal JP, Lal P, Pati AK, Gupta BBP (1983) Thyroid and gonad in the oxidative metabolism, erythropoesis, and light response of the migratory red-headed bunting, *Emberiza bruniceps*. Gen Comp Endocrinol 51:444–453

Thompson AL (1936) Recent progress in the study of bird migration: a review of the literature 1926–1935. Ibis 78:472–530

Turek FW, Gwinner E (1982) Role of hormones in the circadian organization of vertebrates. In: Aschoff J, Daan S, Groos GA (eds) Vertebrate circadian systems. Springer, Berlin Heidelberg New York, pp 173–182

Vallyathan NV (1963) On the lipid content and lipase activity in the breast muscle of *Sturnus roseus*. Pavo 1:106–110

Vleck CM, Wingfield JC, Farner DS (1980) No temporal synergism of prolactin and corticosterone influencing reproduction in white-crowned sparrows. Am Zool 20:899

Vluegel DA (1948) Enkele waarnemingen over "vorstvlucht" en "randtrek" in het Sloeschengengebied tijdens de winters van 1935/1936 en 1936/1937. Ardea 36:143–162

Wagner HO (1930) Über Jahres- und Tagesrhythmus bei Zugvögeln. (1 Mitteilung). Z Vgl Physiol 12:703–724

Wagner HO (1961) Beziehungen zwischen dem Keimdrüsenhormon Testosteron und dem Verhalten von Vögeln in Zugstimmung. Z Tierpsychol 18:302–319

Wege ML, Raveling DG (1983) Factors influencing the timing, distance, and path of migrations of Canada geese. Wilson Bull 95:209–221

Weise CM (1967) Castration and spring migration in the white-throated sparrow. Condor 69:49–68

Wilson AC, Farner DS (1960) The annual cycle of thyroid activity in white-crowned sparrows of eastern Washington. Condor 62:414–425

Wingfield JC (1980) Fine temporal adjustment of reproductive functions. In: Epple A, Stetson MH (eds) Avian endocrinology, Academic Press, London, pp 367–389

Wingfield JC (1983) Environmental and endocrine control of reproduction: an ecological approach. In: Mikami S-I, Ishii S, Wada M (eds) Avian endocrinology: environmental and ecological aspects. Tokyo, Jap Sci Soc. Springer, Berlin Heidelberg New York Tokyo, pp 265–288

Wingfield JC (1988) Changes in reproductive function of free-living birds in direct response to environmental perturbations. In: Stetson MH (ed) Processing of environmental information in vertebrates. Springer, Berlin Heidelberg New York, pp 121–148

Wingfield JC, Farner DS (1978) The annual cycle of plasma irLH and steroid hormones in feral populations of the white-crowned sparrow, *Zonotrichia leucophrys gambelii*. Biol Reprod 19:1046–1056

Wingfield JC, Farner DS (1980) Control of seasonal reproduction in temperate zone birds. Prog Reprod Biol 5:62–101

Wingfield JC, Silverin B (1986) Effects of corticosterone on territorial behavior of free-living male song sparrows, *Melospiza melodia*. Horm Behav 20:405–417

Yokoyama K (1976) Hypothalamic and hormonal control of photoperiodically induced vernal functions in the white-crowned sparrow, *Zonotrichia leucophrys gambelii*. I. The effects of hypothalamic lesions and exogenous hormones. Cell Tissue Res 174:391–416

Yokoyama K (1977) Hypothalamic and hormonal control of photoperiodically induced vernal functions in the white-crowned sparrow, *Zonotrichia leucophrys gambelii*. II. The effects of hypothalamic implantation of testosterone propionate. Cell Tissue Res 176:91–108

Yokoyama K, Farner DS (1978) Induction of zugunruhe by photostimulation of encephalic receptors in white-crowned sparrows. Science 201:76–79

Circannual Rhythms in Bird Migration: Control of Temporal Patterns and Interactions with Photoperiod

E. GWINNER[1]

1 Introduction

The idea that endogenous timing mechanisms may play an important role in the control of avian migrations is about as old as the insight that birds do indeed migrate. As early as 1702 Baron von Pernau (1702) suggested that migratory birds were "driven at the proper time by a hidden drive" ("*durch einen verborgenen Zug zur rechten Zeit getrieben*"), and similar propositions were made by other early investigators of migrations like Naumann (1822); Brehm (1828), and von Homeyer (1881). The participation of endogenous factors in the timing of migrations seemed especially likely in long-distance migrants that spend the winter close to the equator. These birds molt and start homeward migration at rather well-defined times in winter and early spring, in spite of the apparent absence of regular seasonal environmental changes. In view of this situation even Rowan (1926), who was so successful in explaining many aspects of avian annual cycles on the basis of photoperiodic effects, could not escape the conclusion that "those species that breed in the northern hemisphere and winter on the equator or cross it and winter in the southern hemisphere, make necessary the assumption that there is another and internal factor, a physiological rhythm".

In spite of these early intuitions it was not before the late 1960s that, following similar discoveries in hibernating mammals (Pengelley and Fisher 1963), an endogenous annual rhythmicity was first rigorously demonstrated for a migratory bird (Gwinner 1967). Previous attempts at testing the endogenous rhythm hypothesis had been unsuccessful because they suffered from a lack of guidance by a model that could be used to describe the phenomena, to define the relevant questions, and to design the appropriate experiments. Such a model became available in the early 1960s, when Aschoff (1960) and Pittendrigh (1960) proposed to treat endogenous diurnal rhythms like physical self-sustaining oscillators that free-run with their own natural ('circa-') period in constant conditions and are synchronized with appropriate environmental zeitgebers in the natural situation. This model could be easily adjusted to biological annual rhythms and has encouraged a series of experiments leading to the demonstration of endogenous annual rhythms in many organisms including several species of migratory birds (Gwinner 1986a,b, for reviews).

[1]Max-Planck-Institut für Verhaltensphysiologie, Vogelwarte, D-8138 Andechs, FRG

E. Gwinner (Ed.) Bird Migration

In addition to prompting experimental support for the existence of circannual rhythmicities, the introduction of the general oscillator model has also facilitated the development of two specific concepts in the field:

1. In proposing continuously ongoing repetitive processes within the organism as the major basis of annual biological rhythmicities, the general oscillator model suggests a concrete mechanism for the functioning of endogenous *programs*. Such circannual programs, in addition to determining the onset of certain seasonal events, might also control their temporal and even spatial patterns. Programs of this kind have been suggested on and off in various contexts (e.g., Streseman 1934), but only the rigorous application of the oscillator model has allowed unambiguous and testable formulations of these concepts.
2. The general oscillator model has led to the description and analysis of the mechanism through which endogenous circannual rhythms are *synchronized* by environmental zeitgebers, particularly photoperiod. Recent investigations in this area have provided information about the subtle and biologically meaningful ways in which photoperiod affects the circannual oscillator, in addition to grossly adjusting its period (e.g., Gwinner 1988, 1989b).

The present chapter will focus on these two aspects of circannual rhythm control.

2 Circannual Rhythms in Migratory Birds and Their Synchronization with Photoperiod

In a previous review (Gwinner 1986a), 13 species of migratory birds have been listed in which circannual rhythms persisting for at least two cycles with a period different from 1 year, have been clearly demonstrated. Moreover, there are at least five additional species of migrants, in which circannual rhythms are likely to be involved, although experimental evidence is not quite as convincing as in the other cases. Among the species investigated, circannual rhythms have been shown to control migratory activity, migratory fattening, migratory orientation, food preference and utilization, molt, and reproductive condition (Gwinner 1986a for a review; see also Berthold and Bairlein, this Vol.). An example of the circannual rhythm in testicular width, molt, and zugunruhe in a garden warbler (*Sylvia borin*) is shown in Fig. 1 (upper diagram).

Although circannual rhythms usually have periods slightly different from 1 year under constant conditions (Fig. 1, upper diagram), their period is exactly 1 year under natural conditions. The necessary corrections of the period to 1 year are accomplished by the annual photoperiodic cycle which provides the most powerful zeitgeber for circannual rhythms in most species yet investigated. The effectiveness of photoperiod as a zeitgeber is illustrated by the results in Fig. 1 (middle and lower diagram): male garden warblers exposed to one photoperiodic cycle per year, went through one testicular cycle, molted twice a year, and showed migratory restlessness twice a year (middle diagram). The period of these rhythms however, could be halved by exposing birds to a photoperiodic cycle with a period

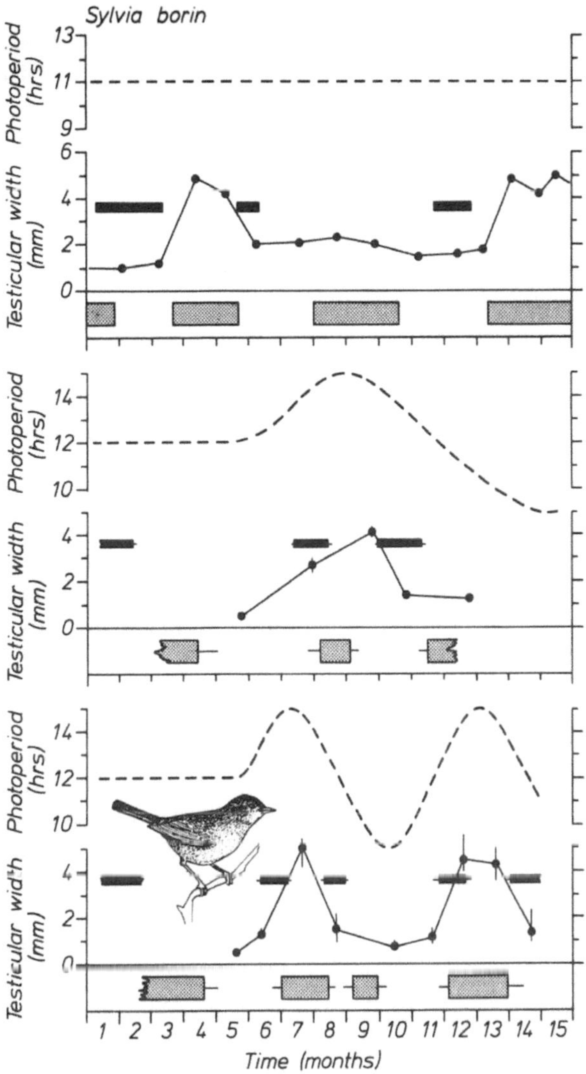

Fig. 1. Changes in testicular width (-●-●-) and occurrence of molt (-▬-) and migratory restlessness (- ▭-) in garden warblers held either in a constant photoperiod (LD 11:11, *upper diagram*) or in sinusoidal photoperiodic cycles with a period of either 12 months (*middle diagram*) or 6 months (*lower diagram*). The upper diagram shows the performance of an individual representative bird; the middle and lower diagrams that of groups of birds. *Horizontal lines* at the bars: standard errors of the means. *Vertical lines* at the curve points: standard errors of the means (middle diagram) or ranges (lower diagram) (After Gwinner 1987b and 1989b)

of 6 instead of 12 months (lower diagram). In some investigations up to six biological cycles could be compressed into one natural year, indicating a wide range of entrainment in this class of biological rhythms (Gwinner 1981, 1986a).

3 Endogenous Programs for Migration

Circannual rhythms not only play a role in determining the onset and cessation of seasonal activities, but may also be involved in producing seasonal patterns. Thus, variations in body weight, mainly reflecting changes in the amount of fat deposited, are partly endogenously preprogrammed in a species-specific circannual pattern (Gwinner 1986b). There is also evidence that seasonal changes in the migratory direction are determined by the circannual system (Gwinner and Wiltschko 1978, 1980). The following account will focus on the role of circannual programs in controlling the time course of migratory activity.

Early investigations of caged, first-year willow warblers (*Phylloscopus trochilus*) revealed a seasonal pattern of fall nocturnal migratory restlessness at least grossly similar to the temporal pattern of actual migration in free-living conspecifics (Gwinner 1968). This finding suggested that a circannual mechanism might be the basis of the previously unexplained ability of first-year birds to terminate migratory activity in their specific wintering areas. It was proposed (1) that young birds were equipped with a temporal program controlled by a circannual rhythmicity that determines duration and temporal variations of migratory activity; and (2) that this temporal program may be organized in a species- or population-specific fashion in such a way as to allot just enough time for migration to reach the specific wintering areas. According to this hypothesis then, the wintering area would be determined by a vector whose direction would be given by a compass direction for flight and whose length would be determined by the time spent flying (Gwinner 1968).

This hypothesis has been extensively investigated since its initial proposition in 1968. The results have been repeatedly summarized and critically evaluated (e.g., Gwinner 1977, 1986a,b; Berthold 1984, 1988, and this Vol.), so that the following account can be rather general except where new results are considered.

Consistent with the hypothesis of an endogenous time program for migratory distance are several series of results indicating a positive correlation between the migratory distance normally covered by a species or population on fall migration, and the overall duration of nocturnal activity developed by caged individuals of the same species or population. Although there are a few exceptions, the data obtained from more than 12 taxa, mainly *sylviinine* warblers, are generally quite impressive in that they show that the further a species or population migrates, the longer and the more intense the fall migratory restlessness developed by caged individuals of that species or population (e.g., Gwinner 1986b; Berthold 1988 for reviews). Some results also suggest similarities in the temporal pattern of fall migration in the field and fall migratory activity in the cage, but generally these data are not quite as striking as those on overall distance and duration (Berthold 1984; Gwinner 1977, 1986a,b, for reviews).

Although the above mentioned correlative data are consistent with the hypothesis that a time program may determine migratory distance, they do by no means prove such a mechanism. It is possible that the circannual migratory time programs do not actually determine migratory distance but rather provide just a "temporal window" within which other mechanisms can act. In fact, fall migratory restlessness of caged birds often tends to last longer in late fall and winter than actual migration and the overall time spent by caged birds for their autumnal nocturnal activity often seems to exceed the time required for these birds to reach their wintering area (Gwinner 1986a).

Some of these discrepancies may reflect deficiencies in the recording technique for migratory restlessness. In most studies, perch-hopping activity was measured but, at least in warblers, whirring (wing fluttering in a sitting position) may be a more relevant measure (Czeschlik 1974; Biebach et al. 1985; Berthold and Querner 1988). By recording wing whirring activity with the help of an infrared videorecorder, Berthold and Querner (1988) have recently found that garden warblers held in a 12-h photoperiod produced an average of about 165 h of whirring per fall migratory season. Assuming a flight speed of 25–30 km/h this overall flight time would be equivalent to a flight distance of about 4500 to 5000 km, approximately corresponding to the actual flight distance of this population. Berthold and Querner interpreted these results as direct evidence for the control of migratory distance by a time program, but it is more likely that this relatively good fit is only the result of the particular data reduction technique used by these authors. They counted whirring activity in successive 1-min bins and correspondingly interpreted all rest phases between whirring bouts, even those as short as 1 min, as migratory pauses. Thus they obtained a much shorter overall time with migratory restlessness per season than they would have, had they sampled migratory restlessness with the more conventional 30-min bins. This standard procedure does not recognize the short interruptions of nocturnal activity in the minute range. This is reasonable, because the actual nocturnal migration does not proceed in very short bouts interrupted by frequent rest phases, but rather consists of continuous flight for hours.

The preceding discussion reflects a major dilemma of present-day research in this field. A rigorous testing of the hypothesis using caged birds must be based on the comparison of the performance of caged birds with that of free-living conspecifics. Unfortunately, however, very little is known about the temporal and spatial patterns of actual migration in most long-distance migrants, especially those traveling at night. For instance, although reasonable estimates of flight speeds in calm air can be made for all the species studied (see Butler and Woakes, Nachtigall, this Vol.), the actual speed of migratory flights is very hard to estimate as it depends heavily on the wind conditions which are usually not known in sufficient detail. For the same reasons the actual distance covered by a bird on its fall migration is very hard to estimate. Depending on wind and topographic conditions (and the bird's strategy to cope with them) this distance may be much longer than the beeline distance between the breeding and the wintering area (see Gwinner 1977, 1986a for further details).

It appears as though further progress in this field depended on the gathering of more information about the detailed way in which migration proceeds in free-living birds. Such information would also facilitate interpretations of the increasing amount of data indicating that endogenous components can be heavily affected by acute climatic and nutritional conditions (see e.g., Richardson and Terrill, this Vol.). Definite models on the mechanisms determining migratory distance will clearly depend on the final evaluation of the relative roles of endogenous timing mechanisms and the exogenous cues modifying them.

4 Fine Tuning of Seasonal Activities by Photoperiod

A factor influencing migratory activity in a very dramatic way is photoperiod, the circannual zeitgeber which, as can be seen in Fig. 1 (middle and lower diagrams), can synchronize circannual rhythms. The effect of photoperiod, however, often goes beyond that of gross synchronization. This has been shown in studies in which birds were exposed to more realistic photoperiodic simulations than the sinusoidal cycles used in the experiment of Fig. 1. An example is provided by results obtained from the garden warbler, a species that can spend the winter either in regions close to the equator, or up to 20° south of it. Figure 2 shows that birds held in winter under photoperiodic conditions characteristic for these two latitudinal regions, showed conspicuous differences in the progress of their seasonal activities. Particularly, there was a clear tendency for molt, testicular growth, and spring zugunruhe to occur earlier in the group held in the longer 20°S photoperiod than in the group exposed to the equatorial photoperiod. This acceleration of the system by long photoperiods is probably biologically significant for those individuals of a population which, in autumn, happen to migrate to more southerly wintering areas. In order to return to their breeding grounds at the same time as conspecifics wintering further north, it is necessary for these birds to speed up molt and to begin northward migration and the associated gonadal growth earlier than the conspecifics wintering further north (Gwinner 1987a, 1989b).

In addition to this conclusion, the data presented in Fig. 2 also suggest another functionally significant aspect of photoperiodic action. It shows that although the longer 20°S photoperiod accelerates the system relative to the situation in the 0° photoperiod, this acceleration is small enough to prevent the birds from becoming reproductively active in the wintering area. A similar kind of delayed responsiveness to long photoperiods has also been found in two other transequatorial migrants, the bobolink (*Dolichonyx oryzivorus*; Engels 1969) and the pied flycatcher (*Ficedula hypoleuca*; Gwinner 1989a; see Fig. 3). A detailed analysis of the situation in garden warblers has revealed that in this species the circannual system responding to photoperiod has qualitatively and quantitatively just the properties required to (1) prevent gonadal recrudescence in long photoperiods, and (2) still enable birds wintering further south to start earlier with spring migration (Gwinner et al. 1988).

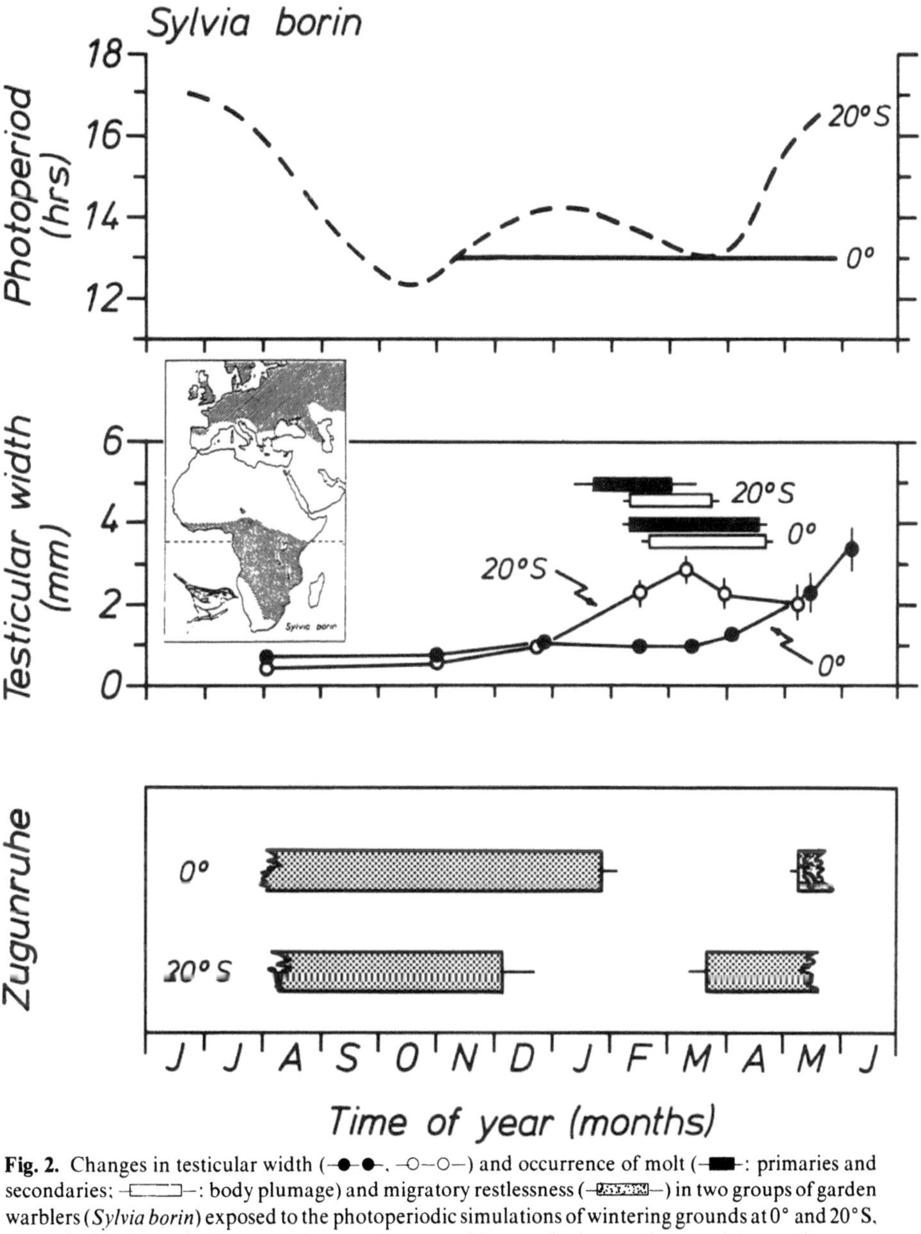

Fig. 2. Changes in testicular width (–●–●–, –○–○–) and occurrence of molt (–■–: primaries and secondaries; –□–: body plumage) and migratory restlessness (–▒–) in two groups of garden warblers (*Sylvia borin*) exposed to the photoperiodic simulations of wintering grounds at 0° and 20° S, respectively, shown in the *upper diagram*. *Horizontal lines* at the bars and *vertical lines* at the curve points: standards errors of the means (After Gwinner 1987a, 1989b)

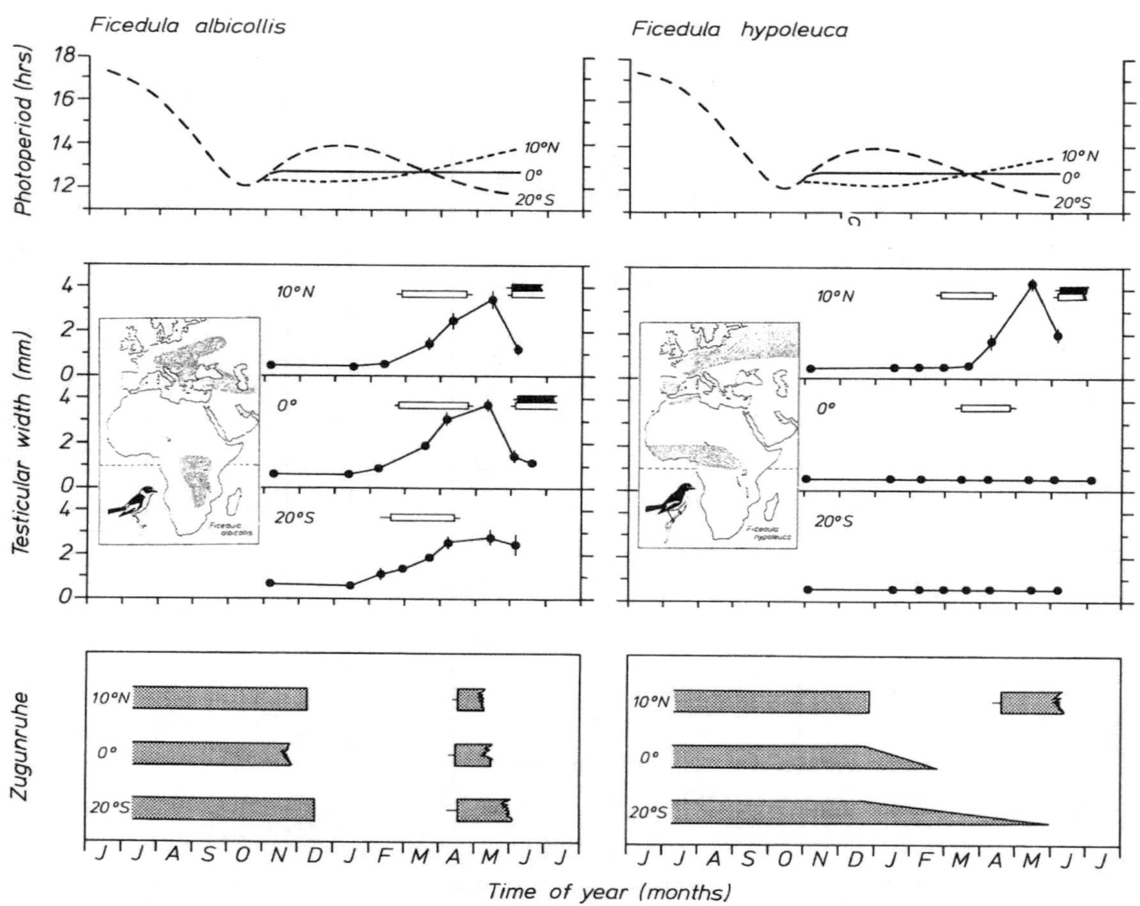

Fig. 3. Changes in testicular width (–●–●–) and occurrence of molt (–■–: primaries and secondaries; –□–: body plumage) and migratory restlessness (–▨–) in 3 groups of collared flycatchers (*Ficedula albicollis*), *left*, and pied flycatchers (*F. hypoleuca*), *right*, exposed to photoperiodic simulations shown in the *upper two diagrams*. *Horizontal lines* at the bars and *vertical lines* at the curve points: standards errors of the means (After Gwinner 1989a,b)

Whereas long photoperiods are effective in advancing various circannual processes in winter, the opposite is true in summer and early autumn. Garden warblers as well as several other species of birds have been shown to start fall migratory restlessness earlier and to finish the preceding processes of plumage development earlier if held in shorter photoperiods than in longer photoperiods (Berthold et al. 1970, 1972; Gwinner et al. 1971). The acceleration of the system by short photoperiods during this part of the year is biologically significant for such individuals that hatch late in the year and hence grow up in shorter photoperiods. To be ready for fall migration at the appropriate time, these birds must initiate fall zugunruhe following completion of postjuvenile development at an earlier age than birds hatched earlier in the year. Taken together then it appears as though, depending on circannual phase, the relative effects of long and short days may be qualitatively and quantitatively different, but the circannual system seems to always respond to day lengths such that an optimal adjustment to season and latitude is achieved.

5 Photoperiodic Constraints to the Expression of Circannual Rhythms Under Constant Conditions

The results presented in the previous section clearly indicate that a proper understanding of the functional significance of circannual systems in migratory birds depends on the detailed knowledge of their interactions with photoperiod. This is most strikingly demonstrated by the fact that in some species circannual rhythms continue only if the (constant) photoperiod has values close to those normally experienced by the species or population in the field. If photoperiod is too long or too short, the rhythm becomes arrested.

Pied and collared flycatchers (*Ficedula hypoleuca* and *F. albicollis*) breeding in western Europe are long-distance migrants whose wintering areas in Africa differ markedly (Fig. 3). Whereas the pied flycatcher normally stays north of the equator, on the average about 10° N, the collared flycatcher spends the winter in central and southern Africa, to about 20° S. A comparative investigation of these two species revealed that the collared flycatchers showed relatively normal patterns of molt, gonadal growth, and zugunruhe irrespective of whether they were exposed to photoperiodic simulations of wintering areas at 10° N, 0°, or 20° S. This was different in pied flycatchers which showed normal patterns only if held in their "own" 10° N photoperiodic simulation. Under the longer 0° and 20° S conditions, in contrast, the rhythm stopped: there was no gonadal growth and spring zugunruhe and winter molt were performed only by a few of the birds held in the 0° simulation. A supplementary study revealed that the failure of the collared flycatchers to continue cyclicity under 0° and 20° S conditions was due to their failure to terminate photorefractoriness in photoperiods that long: a 4-week exposure of birds held initially in a 0° photoperiod to a short 8-h photoperiod restored their capacity to respond to long days so that rhythmicity continued with gonadal growth and spring zugunruhe when the birds were returned to the longer equatorial photoperiod (Gwinner 1989a).

The results then suggest that these two species of long-distance migrants have adjusted their photoperiodic response system in a subtle and species-specific way to the conditions normally experienced by them. The species migrating to the more southerly wintering areas can terminate photorefractoriness under much longer photoperiods than the species wintering further north. In more general terms these results suggest that circannual rhythms may go through photoperiodic "bottlenecks" once or several times per cycle during which certain photoperiodic conditions may have to be met for the underlying processes to continue. Once again these data show that the functional meaning – and even the mechanisms – of circannual rhythms can only be properly understood and analyzed if the interactions between circannual rhythmicity and zeitgebers are taken into account.

6 Conclusions

Most of the earlier investigations on circannual mechanisms in the control of avian migrations have focused on the behavior of the system under constant environmental conditions. Much has been learned from these studies, particularly about the performance of such migrants that spend the winter in a seasonally constant or unpredictable tropical environment. However, just like circadian rhythms, circannual rhythms can exert their function only in conjunction with environmental zeitgebers, the most important zeitgeber being the seasonal cycle of photoperiod (Sect. 2). As shown in Section 4, photoperiod not only synchronizes the endogenous rhythm with the natural year, but also serves to adjust particular fractions of the endogenous cycle in a biologically meaningful manner to the birds' requirements. If photoperiod is longer or shorter than that normally experienced by a species, circannual rhythmicity may become arrested (Sect. 5). Obviously then, endogenous circannual components and exogenous synchronizing cues constitute a functional entity which, as a whole, provides for adaptive temporal organization. Future investigations aimed at analyzing the effects of various photoperiods on different phases of the endogenous mechanisms will certainly provide a deeper insight into the adaptive time structure of seasonal control mechanisms; they should supplement or replace the traditional studies on birds held for extended periods in constant photoperiods.

Whereas photoperiod affects the circannual system by influencing its basic structure, other stimuli modify its output without altering the underlying mechanisms. As known from numerous field observations, a variety of external stimuli related to food, climate, and social cues are known to modify the actual migratory performance of a bird (e.g., see Richardson, this Vol.). These results provide a challenge for those who have proposed, on the basis of laboratory investigations, that the temporal patterns, the duration, and perhaps even the distance of migration are determined by rather rigid circannual programs (Sect. 3). Presumably both groups of factors have their share in controlling these various features of migration, but the way in which they interact can be complex and different among species (e.g., Sect. 5). Detailed empirical studies are required to clarify the situation.

Acknowledgement. This work was supported by the Deutsche Forschungsgemeinschaft.

References

Aschoff J (1960) Exogenous and endogenous components in circadian rhythms. Cold Spring Harbor Symp Quant Biol 25:11-28

Berthold P (1984) The endogenous control of bird migration: a survey of experimental evidence. Bird Study 31:19-27

Berthold P (1988) The control of migration in European warblers. Proc 19th Int Ornithol Congr Ottawa, pp 215-249

Berthold P, Querner U (1988) Was Zugunruhe wirklich ist – eine quantitative Bestimmung mit Hilfe von Video-Aufnahmen bei Infrarotbeleuchtung. J Ornithol 129:372-375

Berthold P, Gwinner E, Klein H (1970) Vergleichende Untersuchung der Jugendentwicklung eines ausgeprägten Zugvogels, *Sylvia borin*, und eines weniger ausgeprägten Zugvogels, *S. atricapilla*. Vogelwarte 25:297-331

Berthold P, Gwinner E, Klein H (1972) Circannuale Periodik bei Grasmücken. I. Periodik des Körpergewichts, der Mauser und der Nachtunruhe bei *Sylvia atricapilla* und *S. borin* unter verschiedenen konstanten Bedingungen. J Ornithol 113:170-190

Biebach H, Wegner H, Habersetzer J (1985) Measuring migratory restlessness in captive birds by an ultrasonic system. Experientia (Basel) 41:411-412

Brehm CL (1828) Der Zug der Vögel. Isis 21:912-922

Czeschlik D (1974) A new method of recording migratory restlessness in caged birds. Experientia (Basel) 30:1490

Engels WL (1969) Photoperiodically induced testicular recrudescence in the transequatorial migrant *Dolichonyx* relative to natural photoperiods. Biol Bull 137:256-264

Gwinner E (1967) Circannuale Periodik der Mauser und der Zugunruhe bei einem Vogel. Naturwissenschaften 54:447

Gwinner E (1968) Circannuale Periodik als Grundlage des jahreszeitlichen Funktionswandels bei Zugvögeln. Untersuchungen am Fitis (*Phylloscopus trochilus*) und am Waldlaubsänger (*Ph. sibilatrix*). J Ornithol 109:70-95

Gwinner E (1977) Circannual rhythms in bird migration. Annu Rev Ecol Syst 8:381-405

Gwinner E (1981) Circannual systems. In: Aschoff J (ed) Handbook of behavioral neurobiology, vol 5. Plenum, New York, pp 391-410

Gwinner E (1986a) Circannual rhythms. Zoophysiology 18. Springer, Berlin Heidelberg New York

Gwinner E (1986b) Circannual rhythms in the control of avian migrations. Adv Study Behav 16:191-228

Gwinner E (1987a) Annual rhythms of gonadal size, migratory disposition and molt in garden warblers, *Sylvia borin*, exposed in winter to an equatorial or a southern hemisphere photoperiod. Ornis Scand 18:251-256

Gwinner E (1987b) Photoperiodic synchronization of circannual rhythms in gonadal activity, migratory restlessness, body weight, and molt in the garden warbler (*Sylvia borin*). In: Pévet P (ed) Comparative physiology of environmental adaptations, vol 3. Karger, Basel, pp 30-44

Gwinner E (1988) Photoprefractoriness in equatorial migrants. Proc 19th Int Ornithol Congr Ottawa, pp 626-623

Gwinner E (1989a) Einfluß der Photoperiode auf das circannuale System des Halsbandschnäppers (*Ficedula albicollis*) und des Trauerschnäppers (*F. hypoleuca*). J Ornithol 130:1-13

Gwinner E (1989b) Photoperiod as a modifying and limiting factor in the expression of avian circannual rhythms. J Biol Rhythms 4:237–250

Gwinner E, Wiltschko W (1978) Endogenously controlled changes in migratory direction of the garden warbler, *Sylvia borin*. J Comp Physiol 125:267–273

Gwinner E, Wiltschko W (1980) Circannual changes in migratory orientation of the garden warbler, *Sylvia borin*. Behav Ecol Sociobiol 7:73–78

Gwinner E, Berthold P, Klein H (1971) Untersuchungen zur Jahresperiodik von Laubsängern, II. Einfluß der Tageslichtdauer auf die Entwicklung des Gefieders, des Gewichtes und der Zugunruhe bei *Phylloscopus trochilus* und *Ph. collybita*. J Ornithol 112:253–265

Gwinner E, Dittami JP, Beldhuis JA (1988) The seasonal development of photoperiodic responsiveness in an equatorial migrant, the garden warbler *Sylvia borin*. J Comp Physiol A Sens Neural Behav Physiol 162:389–396

Homeyer EF von (1881) Die Wanderungen der Vögel. Grieben Leipzig

Naumann J (1822) Naturgeschichte der Vögel Deutschlands. Fleischer, Leipzig

Pengelley ET, Fisher KC (1963) The effect of temperature and photoperiod on the yearly hibernating behavior of captive golden-mantled ground squirrels (*Citellus lateralis tescorum*). Can J Zool 41:1103–1120

Pernau FA von (1702) Unterricht was mit dem lieblichen Geschöpff, denen Vögeln, auch ausser dem Fang, nur durch Ergründung deren Eigenschaften und Zahmmachung oder anderer Abrichtung man sich vor Lust und Zeitvertreib machen könne. Nürnberg

Pittendrigh CS (1960) Circadian rhythms and the circadian organization of living systems. Cold Spring Harbor Symp Quant Biol 25:159–184

Rowan W (1926) On photoperiodism, reproductive periodicity, and the annual migrations of birds and certain fishes. Proc Boston Soc Nat Hist 38:147–189

Stresemann E (1934) Aves. In: Kükenthal W, Krumbach T (eds) Handbuch der Zoology, Bd 7. Gruyter, Berlin Leipzig

Genetics of Migration

P. Berthold[1]

1 Introduction

Until quite recently, evidence for immediate genetic effects on migration was scarce although there were a considerable number of suggestive findings in support of a strong genetic role (e.g., Berthold 1983). In the last 10 years, however, both experimental results as well as data obtained in field studies have accumulated favouring the hypothesis that immediate genetic effects are involved in the control of all important aspects of avian migration.

2 Endogenous Annual Cycles – Circannual Rhythmicity

Circannual rhythms have now been demonstrated in about 15 migratory bird species. Biological rhythms play an essential role in the periodic, seasonal occurrence of migratory activity (migratory restlessness or Zugunruhe in caged individuals) as well as migratory disposition (which is primarily expressed as fat deposition and by specific hormonal constitutions). Moreover, circannual rhythms also appear to be involved in the control of seasonal food preferences and seasonal ecomorphological adaptations of the alimentary tract, as well as other factors (for review see Gwinner, this Vol.). Although not yet unequivocally demonstrated, there is little doubt that avian circannual rhythms are directly genetically controlled. Young birds, hand-raised in constant experimental conditions from their first days after hatching, readily develop normal circannual rhythms (Gwinner, this Vol.). A crucial experiment to demonstrate the genetic basis of circannual rhythms would be to cross-breed individuals whose circannual rhythms have different properties (e.g. periods) and to demonstrate, appropriately, their inheritance.

[1]Max-Planck-Institut für Verhaltensphysiologie, Vogelwarte, D-7760 Radolfzell-Möggingen, FRG

E. Gwinner (Ed.) Bird Migration

3 Migratory Activity

3.1 Occurrence

There is evidence from a cross-breeding experiment that the occurrence of migratory activity can be rapidly transmitted genetically. When exclusively migratory blackcaps (*Sylvia atricapilla*, from southern Germany) and resident conspecifics (from the Cape Verde Islands) were cross-bred, 30% of the F_1-hybrids exhibited migratory activity (Berthold et al. 1989). Thus, migratory activity (or the urge to migrate) is a behavioural trait that can be inherited. Selective breeding experiments with partially migratory blackcaps from southern France have shown that in this case the urge to migrate is sex-dependent, i.e., it is expressed more in females than males (Berthold 1986).

3.2 Onset

There is direct experimental, and highly suggestive statistical evidence that the onset of migratory activity in inexperienced young birds is under direct genetic control. (1) Hybrids of migratory and nonmigratory warblers partially displayed migratory activity (Sect. 3.1). The migratory-active hybrids developed Zugunruhe at the same age as their migratory parents demonstrating that the timing of migration onset is highly heritable (Berthold et al. 1989). (2) Parent-offspring relatioships from a selective breeding experiment with blackcaps (Sect. 5) also show that the date of the onset of migratory activity is inherited (Berthold et al., in prep.). (3) When the dates of the onset of migratory activity of caged groups of birds from 19 populations were compared with the initiation of actual migration in wild conspecifics of the same populations, a highly significant correlation coefficient ($r = 0.967$) was obtained. This relationship indicates that the stimulus to migrate in free-living birds is, essentially, a result of the same genetic factors as those controlling the onset of migratory restlessness in captive individuals (Berthold 1989).

3.3 Amount

Comparative investigations in various species of passerine birds have indicated that the amount of Zugunruhe developed during the first fall migratory season is closely related to the distance covered between the breeding grounds and the winter quarters (e.g. Berthold 1988a; Gwinner, this Vol.). These differences may be due to genetic differences as shown by the results of cross-breeding experiments with blackcaps. When birds from southern Germany (which display large amounts of Zugunruhe consistent with their long migratory routes) were cross-bred with African conspecifics from the Canary Islands (which display small amounts of Zugunruhe consistent with their short migratory routes), hybrids exhibited intermediate amounts of restlessness (Fig. 1). Thus, Zugunruhe is a

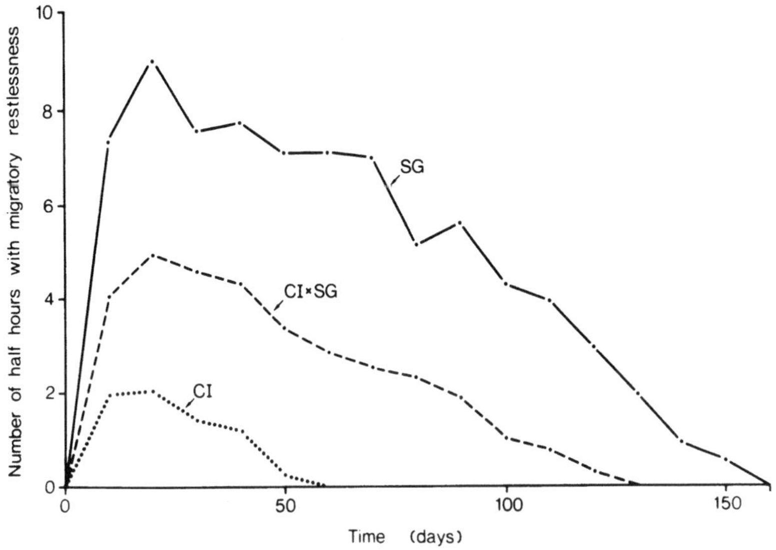

Fig. 1. Time course of nocturnal migratory activity in groups of hand-raised blackcaps, *Sylvia atricapilla*, from two populations (*SG* southern Germany; *CI* Canary Islands, Africa) and their hybrids. *Vertical lines*: SE (After Berthold and Querner 1981)

quantitatively inherited characteristic and appears most likely to depend on a fairly large number of genes. In another experiment with blackcaps in which individuals from migratory and resident populations were cross-bred (Sect. 3.1) the hybrids also showed more or less intermediate amounts of migratory restlessness.

3.4 Termination

The cross-breeding experiments referred to in the previous sections have demonstrated that hybrids show a tendency to terminate their first fall migratory restlessness more or less at an intermediate age relative to their parental stocks. However, experiments such as those that have demonstrated convincingly genetic influences on migratory onset have not been carried out with respect to termination of migratory activity.

3.5 Seasonal Patterns

In a number of passerines tested under various experimental conditions, the seasonal patterns of migratory activity appear to be adaptively preprogrammed to meet the demands brought about by physiographic characteristics of the migratory journey (for review, see Berthold 1983). There is no doubt that these patterns have a significant genetic component (for review, see Berthold 1985). The cross-breeding experiment with blackcaps from southern Germany and the Canary Islands (Fig. 1) has also demonstrated intermediacy of the pattern of Zugunruhe in the hybrids.

3.6 Diel Patterns

In the white-throated sparrow (*Zonotrichia albicollis*) it was shown that migratory restlessness is controlled by a circadian rhythmicity (McMillan et al. 1970). It is likely, then, that Zugunruhe is closely linked to a genetic control mechanism like other features of circadian rhythms (e.g. Aschoff 1981). There is also evidence from European buntings (Berthold 1983) and other species (Brensing 1987) that diurnal patterns of migration related activities are genetically determined. An experiment demonstrating genetic influences in a classical way, for example, by cross-breeding individuals of populations with different diurnal migration times, however, is still needed.

4 Resting Behaviour

4.1 Diurnal Patterns

A detailed analysis of diurnal activity patterns of resting migrant passerine communities at stopover sites in central Europe coupled with comparisons of patterns of locomotor activity by caged conspecifics (Berthold et al. 1986; Brensing 1987) yielded the following results: there are species-specific differences in activity peaks that appear to be preprogrammed in order to reduce competition among members of commonly resting guilds. In some species, specific peaks are related to evening flights to common roosts (Fig. 2). These data strongly suggest that genetic differences are responsible for a series of species-specific characteristics of diurnal patterns, at least during the migratory season. However, considerable variation in patterns of diurnal activity by migrants at stopover sites is known to occur, for example, as a function of the amount of fat deposited. Nevertheless, in these cases a relatively narrow genetically determined frame may limit variability.

4.2 Habitat Preferences

Comparative studies of habitat partitioning and habitat preferences of resting passerine communities at stopover sites in central Europe (Berthold et al. 1976; Bairlein 1981; Berthold 1988a) have shown: (1) resting individuals choose habitats species-specifically and thus show species-characteristic patterns of distribution within habitat scales (Fig. 3); (2) these patterns are surprisingly constant over many years (Fig. 3); (3) they are clearly not a result of tradition (Berthold and co-workers, in prep.); (4) they are not obtained by prowling around

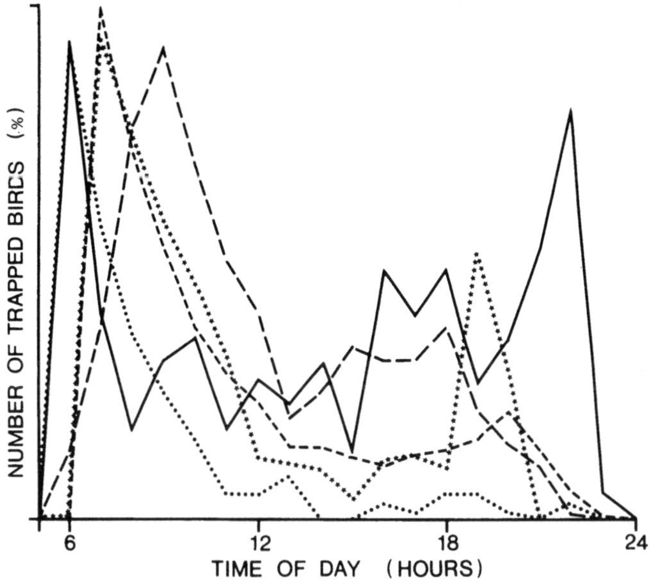

Fig. 2. Diurnal activity patterns of five species of passerines, obtained from trapping data during the autumn migratory period in a southern German staging area. *Solid line* spotted flycatcher, *Muscicapa striata*; *dotted lines* grashopper warbler, *Locustella naevia*; *broken, short lines* reed warbler, *Acrocephalus scirpaceus*; *broken*, long *lines* blue tit, *Parus caeruleus*; *asterisks* reed bunting, *Emberiza schoeniclus* (after Brensing 1987)

because clear-cut habitat segregation is quickly reached even by new arrivals in the early morning. Most likely migrants are equipped with some type of innate search image related to their morphological structure that enables them to make rapid, appropriate habitat choices. It is likely that such a process includes genetically controlled imprinting mechanisms during the early stages of behavioural ontogeny.

5 Partial Migration

5.1 Control

In addition to theoretical treatments of the possible control mechanisms of partial migration (e.g. Lundberg 1987, 1988) there now exists a solid empirical base. Selective breeding experiments with partially migratory blackcaps from Southern France for up to six generations have clearly shown that both behavioural traits – migratoriness and residency – have a strong and direct genetic basis (Berthold and Querner 1982a; Berthold 1988b; Fig. 4; for heritability estimates, see Sect. 7). Most likely, however, both behavioural traits are so-called threshold characters which are determined by mutiple loci (for details,

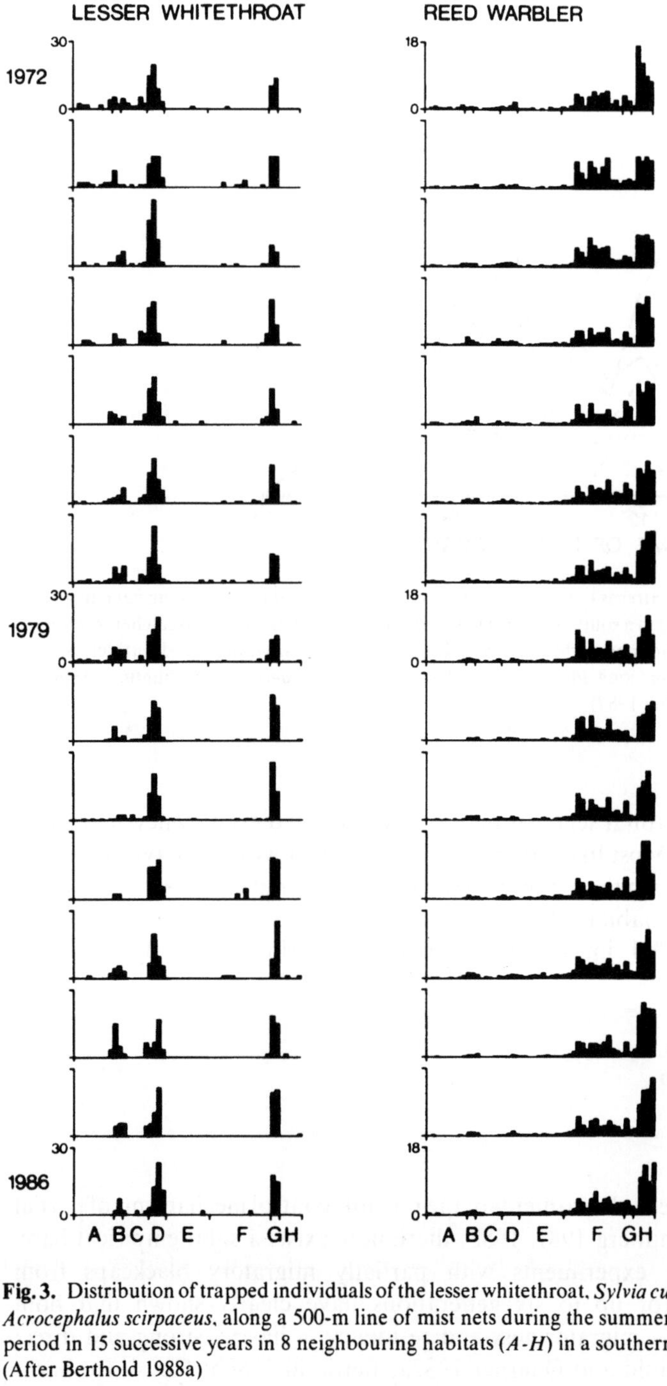

Fig. 3. Distribution of trapped individuals of the lesser whitethroat, *Sylvia curruca*, and reed warbler, *Acrocephalus scirpaceus*, along a 500-m line of mist nets during the summer and autumn migratory period in 15 successive years in 8 neighbouring habitats (*A-H*) in a southern German trapping area (After Berthold 1988a)

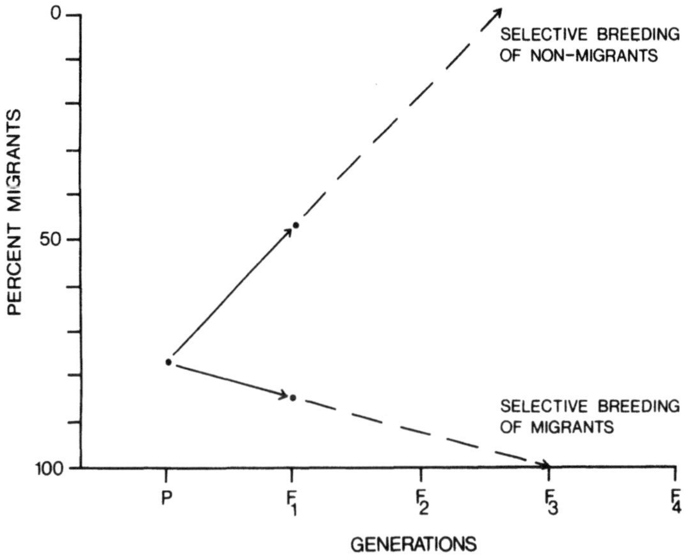

Fig. 4. Selective breeding data to the F_1-generation from hand-raised, southern France blackcaps, *Sylvia atricapilla*. *Solid lines* response to bidirectional selection for 20 migrants and 19 nonmigrants from a parental stock (of 102 partially migratory individuals). The ordinate shows the percentage of migratory behaviour, the *broken lines* indicate further theoretical changes from continued two-way selection assuming a uniform, quasi-linear rate of alteration (After Berthold 1988b)

see Berthold 1988b). The concept that the genetic influences in the investigated populations are high is derived not only from the heritability estimates (Sect. 7), but also from consistency tests. When 40 blackcaps were repeatedly checked for migratoriness or sedentariness in subsequent years none changed from being previously nonmigratory to being later migratory active, and only six switched from migratory to nonmigratory behaviour (unpublished data). The latter situation, commonly observed in many bird species in the wild, is not necessarily due to weak genetic control. Rather, it may be an effect of maturation, and thus this age-dependent change may also have a genetic base.

There is experimental and observational evidence that partial migration is essentially genetically controlled in three other species: the European robin (*Erithacus rubecula*), European blackbird (*Turdus merula*) and the stonechat (*Saxicola torquata*, for review, see Berthold 1984). A recalculation of the data obtained by Nice (1934) for the song sparrow (*Melospiza melodia*) indicates the same for that species (for review, see Berthold 1984). In agreement with this interpretation, Terrill et al. (1989) recently observed spontaneous individual variation in migratory behaviour of song sparrows that was not related to age, sex, food availability or social rank. There is presently no investigation demonstrating that obligate partial migration in any other bird species would be controlled without an essential genetic basis. In species performing irregular irruptive movements ("facultative partial migrants", see Terrill and Able 1988) as the blue tit (*Parus caeruleus*) control mechanisms are certainly different (e.g. Schüz 1971; Smith and Nilsson 1987).

Thus, genetic factors seem to play an important role in the widespread migratory strategy of partial migration.

5.2 Evolutionary Potential

The two-way selective breeding experiment with blackcaps from southern France (Sect. 5.1) has shown that a partially migratory population can become either almost completely migratory, or sedentary, in only two to five generations. The rapid selection response observed implies a strikingly high evolutionary potential with respect to strong selection pressures (Berthold 1988b; Fig. 4). Thus, one can predict that a population such as this might change rapidly with respect to migratory behaviour in the event of rather dramatic environmental changes such as those brought about by a new ice age, warm period, etc. in a short time. Similarly, high alteration rates of migratoriness and residency as in the blackcap were also found in the European robin (Biebach 1983).

6 Evolution and Possible Inheritance of Novel Migratory Habits

Beginning about 25 years ago, blackcaps from central Europe have been migrating in increasing numbers to the NW, creating a novel migratory direction and new wintering areas mainly situated in England and Ireland. Previously, they had oriented exclusively southward to the Mediterranean area and to Africa. The rapid growth of the new wintering community in the north suggests several hypothetical positive selection factors which have been listed in a recent paper (Berthold and Terrill 1988). In the meantime it has been experimentally demonstrated that blackcaps kept under photoperiodic conditions that mimicked those of the new northern winter quarters showed earlier spring migratory activity and gonadal development than conspecifics kept under conditions of traditional winter quarters (Terrill and Berthold, in prep.). Earlier arrival in the breeding grounds and advanced breeding maturity might well lead to assortative mating among members of that pioneer population and may also increase breeding success relative to later arrivals. Because a preference for migratory direction can be inherited (Sect. 8) it is not unlikely that the novel migratory direction is maintained via the transmission of genetic information from successful northern winterers to their offspring. Increased fitness and immediate inheritance of the novel migratory direction could well explain the rapid growth of that new wintering population. A special study to test these possibilities is under way.

7 Heritability Values

Thus far, calculated heritability values are only available for selection responses from the two-way selective breeding experiment with blackcaps from southern France. Heritability estimates were obtained in the range from 0.58 to about 1 and hence are fairly high. For more details, see Berthold (1988b).

8 Inheritance of Migratory Directions

Quite a number of experiments demonstrating spontaneous preferences of appropriate migratory directions by inexperienced migrants suggests innate knowledge of proper orientation (for review, see Berthold 1977; Helbig 1989). Garden warblers (*Sylvia borin*) show the normally occurring seasonal shift in migratory direction, even when kept under constant experimental conditions. This indicates that migratory directional preference is closely linked to endogenous annual rhythms (Gwinner and Wiltschko 1978; Sect. 2). In a cross-breeding experiment with blackcaps using birds from the exclusively migratory population from southern Germany and from the resident African population from the Cape Verde Islands, we were able to demonstrate inheritance of the migratory direction. Hybrids, to which migratory activity had been genetically transmitted, displayed this activity along an axis (NW-SE) that is normally used for migration by the migratory parents (Berthold et al. 1989).

9 Morphological Adaptations

Wing length, wing shape, body weight, etc. are well-known adaptive morphological correlates of migratory behaviour. A cross-breeding experiment with exclusively migratory blackcaps from southern Germany (with long wings and high body weight) and poorly migratory African conspecifics (from the Canary Islands, with short wings and low body weight) demonstrated the strict genetic control of both characteristics, i.e., wing length as well as body weights, were intermediately expressed in the hybrids (Berthold and Querner 1982b). Presumably, further studies that explore other pertinent characteristics such as fat deposition and wing shape will demonstrate genetic influences on these as well (see also Berthold 1985).

10 Adaptation of Juvenile Development

Various annual processes are known to be adaptations to migration, and juvenile development, above all juvenile moult, is one of the most prominent. As a rule, early migrants show early, rapid moult; late migrants start moult later and proceed more slowly. The accelerated moult in the first group is probably a result of selection for separation of moult and migration. Despite the fact that juvenile moult is strongly influenced by photoperiod ("calendar reaction", for review, see Berthold 1988a; Berthold and Querner 1989; Gwinner, this Vol.), immediate genetic influences on its temporal pattern have been demonstrated as well. In a cross-breeding experiment with early migrating (and consequently, early and rapidly moulting) blackcaps from southern Germany and late migrating African conspecifics from the Canary Islands (with later and slower moult) the hybrids were intermediate with respect to the onset, termination and duration of their moult (Berthold and Querner 1982b; Fig. 5). In a similar experiment Gwinner and Neußer (1985) demonstrated intermediacy in the temporal course of moult intensity in hybrids from African and European stonechats. A cross-breeding experiment with German blackcaps and African conspecifics from the Cape Verde Islands also showed that other population differences in juvenile development, such as the growth of the so-called second set of body feathers of the juvenile plumage, can have an immediate genetic basis (Berthold and Querner 1989; for the garden warbler see also Berthold et al. 1974; Berthold 1977, 1988a).

11 Prospects

This review suggests that many characteristics related to and important for migration can be inherited. Genetic influences on migration may be advantageous because: (1) The short life span of small birds precludes a strong role for learning and experience in migratory behaviour and (2) with sufficient genetic variation within populations it allows for rapid adaptations to altered environmental conditions; the latter point is illustrated by the evolutionary potential of partial migration (Sect. 5.2) and the development of novel migratory habits (Sect. 6). It is not unlikely that the relatively high variability of migratory events in short-distance and late migrants is due to high genetic variation that provides a broad base for individual genetic adaptations in that group of migrants experiencing the most variable environmental conditions.

Direct genetic control of annual processes of course does not mean that the modificatory effectiveness of environmental factors is low. The fairly extensively studied, and thus well-understood control of the temporal course of the juvenile

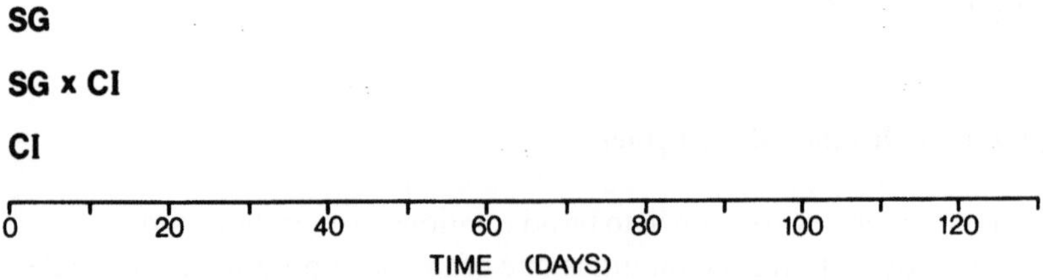

Fig. 5. Time course (and SE) of juvenile moult in groups of hand-raised blackcaps, *Sylvia atricapilla*, from two populations (*SG* southern Germany; *CI* Canary Islands, Africa) and their hybrids (After Berthold and Querner 1982b)

moult in passerines demonstrates nicely that, under given environmental conditions, different populations display population-specific patterns and hybrids behave intermediately. These results demonstrate genetic factors as the basis for the observed differences (Sect. 10; Fig. 5). If, however, environmental conditions change substantially patterns may change substantially too, e.g., juvenile moult can be accelerated by about one-third in extremely short day lengths (Berthold 1988a). This example shows that the genetic control of juvenile moult observed under one environmental condition does not mean that the same pattern of moult is also found under other conditions. On the other hand, course and degree of that variability may be so highly predictable under various conditions (as is the case in the juvenile moult) that the potential variability (reaction norm) is again to be seen under fairly strong genetic control. Hence, migratory events underlying immediate and strong genetic control may at the same time be well open to considerable environmental variation. The degree of genotype-environment interaction remains to be established in all cases so far demonstrated.

Acknowledgement. This work was supported by the Deutsche Forschungsgemeinschaft.

References

Aschoff J (1981) Handbook of behavioral neurobiology, vol 4. Plenum , New York

Bairlein F (1981) Ökosystemanalyse der Rastplätze von Zugvögeln: Beschreibung und Deutung der Verteilungsmuster von ziehenden Kleinvögeln in verschiedenen Biotopen der Stationen des "Mettnau-Reit-Illmitz-Programmes". Ökol Vögel 3:7–137

Berthold P (1977) Endogene Steuerung desVogelzuges. Vogelwarte 29: (Sonderh) 4–15

Berthold P (1983) Genetic basis of bird migration. Ornis Fenn Suppl 3:14–16

Berthold P (1984) The control of partial migration in birds: a review. Ring 10:253–265

Berthold P (1985) Endogenous components of annual cycles of migration and moult. Acta XVIII Congr Int Ornithol. Moscow 1982, pp 922–929

Berthold P (1986) Wintering in a partially migratory Mediterranean blackcap (*Sylvia atricapilla*) population: strategy, control, and unanswered questions. In: Farina A (ed) Proc First Conf Birds Wintering Mediterranean Region. Ric Biol Selvaggina Suppl 10:33–45

Berthold P (1988a) The control of migration in European warblers. Acta XIX Int Ornithol Congr, Ottawa 1986, pp 215–249

Berthold P (1988b) Evolutionary aspects of migratory behavior in European warblers. J Evol Biol 1:195–209

Berthold P (1989) Wegzugbeginn und Einsetzen der Zugunruhe bei 19 Vogelpopulationen – eine vergleichende Untersuchung. J Ornithol (Sonderh) 130: 217–222

Berthold P, Querner U (1981) Genetic basis of migratory behavior in European warblers. Science 212:77–79

Berthold P, Querner U (1982a) Partial migration in birds: experimental proof of polymorphism as a controlling system. Experientia (Basel) 38:805

Berthold P, Querner U (1982b) Genetic basis of moult, wing length, and body weight in a migratory bird species, *Sylvia atricapilla*. Experientia (Basel) 38:801–802

Berthold P, Querner U (1989) The annual cycle of the blackcap (*Sylvia atricapilla*) on the Cape Verde Islands: characteristics, environmental and genetic control. Proc Seventh Pan-Afr Ornithol Congr, Nairobi 1988 (in press)

Berthold P, Terrill SB (1988) Migratory behaviour and population growth of blackcaps wintering in Britain and Ireland: some hypotheses. Ringing & Migr 9:153–159

Berthold P, Gwinner E, Querner U (1974) Vergleichende Untersuchung der Jugendentwicklung südfinnischer und südwestdeutscher Gartengrasmücken *Sylvia borin*. Ornis Fennica 51:146–154
Berthold P, Bairlein F, Querner U (1976) Über die Verteilung von ziehenden Kleinvögeln in Rastbiotopen und den Fangerfolg von Fanganlagen. Vogelwarte 28:267–273
Berthold P, Brensing D, Heine G (1986) Tageszeitliche "Fangmuster" von Kleinvögeln und deren Bedeutung. J Ornithol 127:515–517
Berthold P, Wiltschko W, Miltenberger H, Querner U (1990) Genetic transmission of migratory behaviour in birds. Experientia (Basel) 46:107–108
Biebach H (1983) Genetic determination of partial migration in the European robin (*Erithacus rubecula*). Auk 100:602–606
Brensing D (1987) Ökophysiologische Untersuchungen der Tagesperiodik von Kleinvögeln. Universität, Tübingen
Gwinner E, Neußer V (1985) Die Jugendmauser europäischer und afrikanischer Schwarzkehlchen (*Saxicola torquata* und *axillaris*) sowie von F_1-Hybriden. J Ornithol 126:219–220
Gwinner E, Wiltschko W (1978) Endogenously controlled changes in migratory direction of the garden warbler, *Sylvia borin*. J Comp Physiol 125:267–273
Helbig AJ (1989) The orientation system of *Sylvia* warblers: a selective review. Proc seventh Pan-African Ornithol Congr, Nairobi 1988 (in press)
Lundberg P (1987) Partial bird migration and evolutionarily stable strategies. J Theor Biol 125:351–360
Lundberg P (1988) The evolution of partial migration in birds. Tree 3:172–175
MacMillan JP, Gauthreaux SA, Helms CW (1970) Spring migratory restlessness in caged birds: a circadian rhythm. Bioscience 20:1259–1260
Nice MM (1934) Zur Naturgeschichte des Singammers. J Ornithol 82:1–96
Schüz E (1971) Grundriß der Vogelzugskunde. Parey, Berlin
Smith HG, Nilsson JA (1987) Intraspecific variation in migratory pattern of a partial migrant, the blue tit (*Parus caeruleus*): an evaluation of different hypotheses. Auk 104:109–115
Terrill SB, Able KP (1988) Bird migration terminology. Auk 105:205–206
Terrill SB, Able KP, Holberton RL, Tuttle EM (1989) Partial migration in song sparrows (*Melospiza melodia*): an experimental approach (in preparation)

Chapter 4 Avian Flight

The Mechanics of Flight and Bird Migration Performance

J.M.V. Rayner[1]

1 Introduction

Flight is a demanding adaptation. It requires a high degree of morphological specialization, with a body structure capable of generating and withstanding considerable aerodynamic and mechanical stresses. The physiological systems must be developed to support the considerable energy rates associated with flight activity. Parallelling these must be behavioural and ecological specializations allowing the animal to exploit the flexibility and agility that flight confers. In particular, flying animals travel relatively fast, and therefore – provided they can obtain sufficient food and water – have the opportunity to travel long distances in relatively short times. It is against this background that avian migration has evolved.

In this chapter I describe the mechanics and energetics of flapping flight, explore some of the implications of migration for flight morphology and migration strategy, and suggest that bounding flight is an important adaptation for those small migrants which store considerable levels of fat prior to departure.

2 Mechanics of Flapping Flight

The major forces which a flying animal experiences are its own weight, reflecting the downwards acceleration of gravity, and the frictional drag of the airflow over the body (parasite drag) and wings (profile drag); mean drag acts parallel to the direction of flight and tends to retard the animal (Rayner 1987, 1988; Fig. 1a). For reasonably streamlined bodies drag at cruising flight speeds is about one-tenth of weight. Selective pressures to reduce drag are evident in the smooth body and wing profiles of most birds: only species which fly slowly, and thus do not experience high drag, are poorly streamlined.

In an aircraft the forces to overcome weight and drag are essentially generated separately: engines provide thrust to counter drag, and wings produce a vertical lift force by *aerofoil* action. The aerofoil is the name given to the cross-sectional profile of a typical wing, with a rounded leading edge, sharp trailing edge, and asymmetry relative to the external airflow (Fig. 1b). The

[1] Department of Zoology, University of Bristol, Woodland Road, Bristol BS8 1UG, UK

E. Gwinner (Ed.) Bird Migration

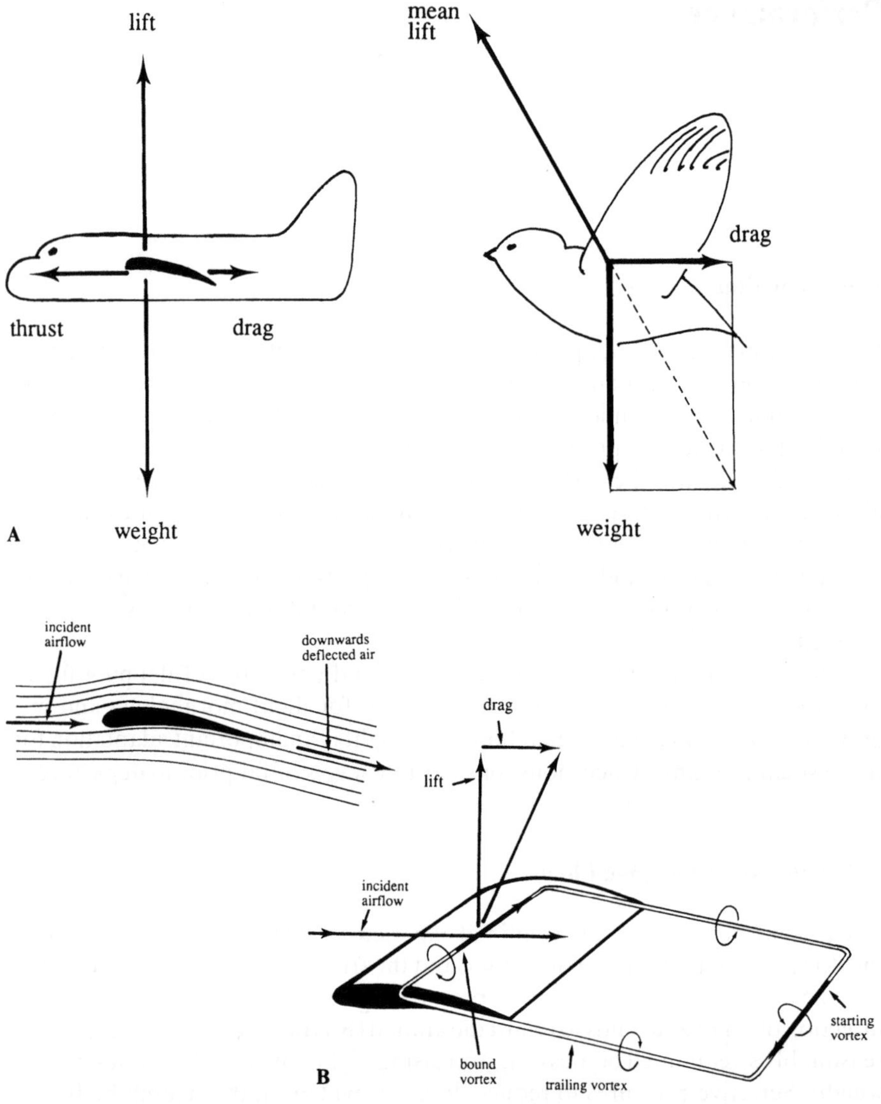

Fig. 1. A The forces on a flying aircraft and a flying animal. The aircraft uses separate mechanisms to generate vertical lift to balance weight and horizontal thrust to counter drag. The animal must configure the wing-beat so that mean aerofoil lift has both vertical and positive horizontal components, producing both thrust and weight support. It may or may not choose to use lift from the upstroke to do this. **B** The streamlined, tapered *aerofoil* section deflects incident air downwards, and generates a transverse lift force by a reaction to the downward momentum. A steadily moving aerofoil is associated with a vortex system consisting of a *bound* vortex on the wing and *trailing* vortices behind the wing tips. A transverse *starting* vortex is shed whenever the magnitude of the circulation (the strength of the bound vortex) changes. The lift force is proportional to the product of circulation and velocity. There is also a drag force, parallel to the incident airflow, representing the energy required to generate the wake vortices and to overcome friction on the aerofoil surface. (Rayner 1988a)

rounded leading edge and tapered profile ensure low drag, while the asymmetry ensures a difference in path length, and therefore in speed, for air passing above or below the profile. The higher speed induced just above the aerofoil is associated with a fall in pressure, and the pressure difference between the sides of the profile is experienced as a vertical lift force. It is this force which balances the weight.

The lift-generating aerofoil creates a distribution of vortices (localized rotational movements of the air); these vortices are responsible for the change in fluid speed around the wing, and hence for the pressure difference between the wing surfaces. Vorticity is essential for flight because transport of momentum within a fluid is only possible with closed vortex loops, and momentum must flow in the air away from and below a wing matching the lift force the wing generates. (I have described the generation and distribution of aerofoil vortices in animals and aircraft elsewhere: Rayner 1980, 1985a, 1986, 1988.) The vortices behind an aircraft wing may be seen as the "vapour trail" under appropriate atmospheric conditions.

All vertebrate wings act as aerofoils, but the analogy to aircraft wings breaks down in powered flight because animal fliers have no "engine" with which to generate thrust. The only force available is the aerofoil lift, and to obtain a horizontal force they must flap the wings. Lift varies periodically in both strength and direction, but with appropriate wing-beat geometry the mean force can act forwards and upwards, balancing weight and drag (including the additional *induced* drag required to generate the wake vortices; Fig. 1b). *Bird wings are flapped not to support the weight, but to provide a mean horizontal thrust to balance drag.*

Lift is impossible without wake vorticity, and the distribution of vorticity in the wake is diagnostic of the action of the wings: the strength of the wake vortex is related to the magnitude of the lift force at the corresponding phase of the wing-beat, and when there is no wake vortex, the wing did not generate lift. My group at Bristol has developed a method of visualizing the vortex wakes of flying birds and bats (Spedding et al. 1984; Spedding 1986, 1987; Rayner 1985a, 1988). Example wake patterns for a slow-flying pigeon *Columba livia* and a kestrel *Falco tinnunculus* are shown in Fig. 2. In the pigeon, at least at the low flight speeds of this observation, the upstroke is inactive; all force generation is confined to the downstroke, when the wing moves forwards and downwards through the air. The wake is a sequence of toroidal ring vortices, each generated by a single down-stroke; each vortex ring convects air below and behind the bird. In the kestrel the situation is rather different: the vortex wake is now continuous, with air being forced downwards behind the bird throughout the wing-beat; the upstroke generates lift, but it induces air to move *forwards* and downwards (Fig. 2b). The upstroke contributes to weight support, but at the cost of a reduction in mean thrust.

By analogy to terrestrial quadrupedal movement these patterns of wing movement and the corresponding wake vortices are referred to as *gaits* (Rayner 1986, 1988); the various gaits are distinguished by the degree of use of the upstroke for lift generation. Flight speed and wing morphology are both implicated in gait selection. The vortex ring gait appears to be used in slow flight in

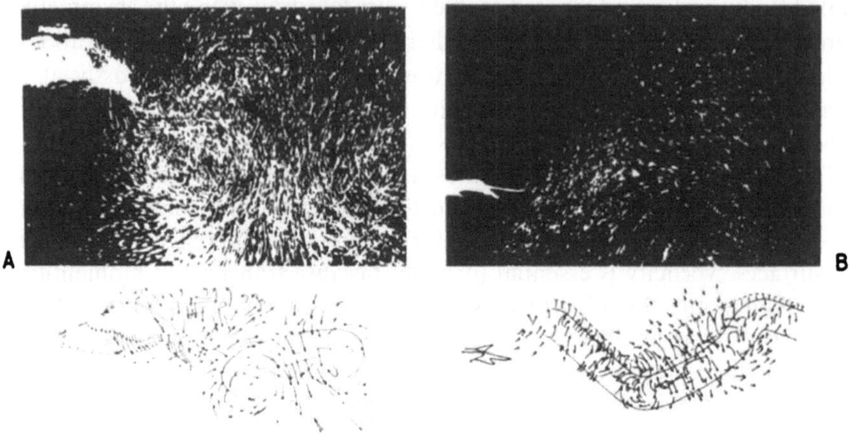

Fig. 2. Wake vortex structures in **A** pigeon *Columba livia* and **B** kestrel *Falco tinnunculus*. The pigeon demonstrates the vortex ring gait typical of slow flight and small-winged birds, while the kestrel wake is continuous with lift generated during the upstroke; note that the wake is narrower during the upstroke than the downstroke. To generate these photographs, the bird is flown through a cloud of neutrally-bouyant helium-filled soap bubbles, and is photographed with a series of flash guns at brief time intervals. Where the bubbles appear as streaks the fluid is moving fastest. Stereo images of the wake enable the three-dimensional flow patterns to be determined, and from these the positions of the vortex cores can be identified. (Rayner 1985, 1988a, after Spedding et al. 1984 and Spedding 1987)

all birds and bats, and in faster flight in animals with small or rounded wings. The continuous wake gait is confined to cruising flight in animals with longer, more pointed wings. Wing-beat kinematics are simpler in the continuous wake than the vortex ring gait, but there are various reasons why the continuous wake is not appropriate for slow flight or for small-winged animals. Most important, in these circumstances retardation from a lifting upstroke would outweigh extra weight support, and with small wings the downstroke cannot produce sufficient extra thrust to balance the retardation. Moreover, flight speeds rise with the continuous wake gait, the wing-beat frequency and amplitude can decrease, and the moment required to elevate the wing during the upstroke can be reduced or eliminated (Rayner 1986, 1987, 1988); with a non-lifting upstroke a moment (a force rotating the wing about the humeral joint) must always be applied to raise the wing.

3 Estimation of Flight Energetics

3.1 Mechanical Modelling

Knowledge of the configuration of the wake vortex and its relation to wing-beat kinematics permits computation of the mechanical energy requirements for flying birds. The pectoralis muscles and the wings are responsible for all work to increase the kinetic energy of the air in the form of both ordered vortex structures and turbulence associated with drag. It is this component of energy output in flight which reflects aerodynamic performance, and which is sensitive to mechanical design and wing morphology, and to the flight strategy (speed etc.) adopted by the bird. Additional metabolic energy is radiated from the bird as heat.

I have detailed the calculations involved elsewhere: see Rayner (1979) for the vortex ring gait, and Rayner (1986) for the continuous vortex gait. Study of the wing-beat kinematics of a range of birds confirms that the continuous vortex gait is more normal in cruising flight, and I use that model here. The vortex ring gait is common (and possibly universal) in smaller passerines using bounding flight (Sect. 5).

A number of predictions from mechanical energetics, originally pointed out by Pennycuick (1969, 1975), are of great significance for the migration problem. First, mechanical power output varies with speed according to a U-shaped curve (Fig. 3). At minimum power speed V_{mp} the power per unit time is least, and the bird should fly at this speed to maximize time in the air. A migrant is likely to aim at maximizing range, or minimizing energy for given distance, and should select maximum range speed V_{mr}, at which the cost of transport $C = P(V)/MgV$ is minimum. The bird should increase its speed to compensate for head winds (Pennycuick 1978), although it may choose not to reduce speed in a tail wind but to have a higher ground speed and thus reduce flight time. To maximize flight speed (minimize time) the bird should select the highest speed possible within

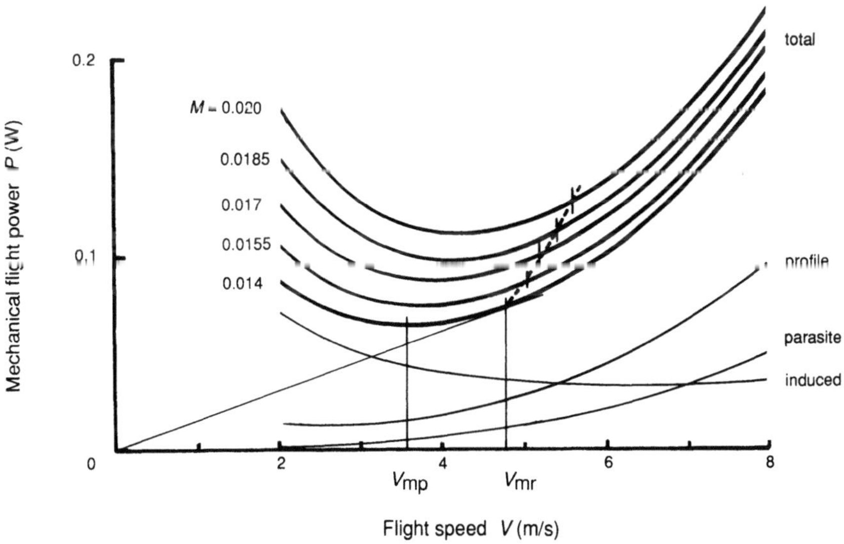

Fig. 3. Variation of mechanical power with flight speed for a swallow *Hirundo rustica* of mass $M = 0.014$ kg, wing span $B = 0.325$ m and wing area $S = 0.0128$ m^2, showing parasite, profile and induced components of drag. Maximum range (minimum cost of transport) speed V_{mr} is defined by the tangent from the origin to the curve; cost of transport is proportional to the gradient of this tangent. 0.014 kg is the fat-free or minimum survival mass of this bird: wild populations will usually be somewhat heavier, migrants reaching up to 0.020 kg before departure. Other curves show total mechanical power for birds of higher mass but the same wing morphology. There is a modest rise in maximum range speed with mass, but a more marked rise in power and cost of transport at that speed

physiological limits. (The predictions discussed below assume that birds can select speed or gait freely according to mechanical criteria: this has not been tested.)

It is also possible to relate performance to morphology. The three relevant quantities are body mass M (kg), wing span B (m) and wing area S (m^2). By calculating V_{mr}, $P_{mr} = P(V_{mr})$ and C_{mr} for a range of birds, and relating the results to morphology, the following equations were derived for birds (Rayner 1988; cf. Norberg and Rayner 1987):

$$P_{mr} = 14.95M^{1.161} \quad = 27.21M^{1.590}B^{-1.818}S^{0.275}\ \mathrm{W}, \tag{1}$$
$$V_{mr} = 12.49M^{0.199} \quad = 10.00M^{0.413}B^{-0.553}S^{-0.095}\ \mathrm{m/s}, \tag{2}$$
$$C_{mr} = 0.109M^{-0122} \quad = 0.278M^{0.178}B^{-1.267}S^{0.371}. \tag{3}$$

The bivariate relations, which apply to interspecific size variation, reproduce the familiar conclusion (e.g. von Helmholtz 1874; see Rayner 1988 for details) that mechanical flight power increases with mass as $M^{7/6}$, and that speed increases as $M^{1/6}$. There are small discrepancies since wing span increases with positive allometry against size ($M^{0.400}$ rather than $M^{1/3}$ for isometric design; Rayner 1988); as a result the cost of transport falls slightly against size. Mechanical power increases more rapidly with mass than does resting and feeding metabolism (which scale as $M^{0.75}$), and large birds experience tight constraints on the amount of energy they can devote to flight. They have responded to this by evolving relatively longer wings, by devoting less time to flight (and ultimately becoming flightless, perhaps), or by gliding and soaring as often as possible. For these reasons the majority of migrants are relatively small birds.

The equations also reveal how performance changes as the mass of an individual bird varies. It is invalid to model this by the interspecific bivariate relation $P_{mr} = 14.95M^{1.161}$ (or an equivalent with index – usually around 0.75 – derived from regression analysis of metabolism, e.g. Summers and Waltner 1979; Davidson 1984), since bivariate scaling relations implicitly incorporate interspecific size-dependent variation of wing span and wing area. In an individual bird wing morphology is constant, even though mass may fluctuate widely. Equation (1) predicts that mechanical power varies as $M^{1.590}$ *in an individual*; the index is close to the theoretical value of 1.5 for isometric wing morphology (Pennycuick 1975; 1978).

From these equations it is also possible to predict the optimal flight morphology for a bird according to various criteria. Design evolves within a complex mosaic of interacting selective forces (Rayner 1988), but if migration performance is paramount selection should favour reduced cost of transport (maximum range). This is achieved primarily by evolving long, thin wings of high aspect ratio. Fast flight may also be important to migrants, and correlates with small wings. Morphological design is hard to interpret in Eqs. (1)–(3) since the three quantities M, B, S are highly correlated. I have used principal components analysis to derive uncorrelated measures of design (Rayner 1988) which are easier to interpret. The components Q measure respectively size, wing size (or wing loading) and wing shape (similar to aspect ratio). Power may be expressed in terms of these components as $\log(P_{mr}) = 0.56 + 1.91Q_1 + 0.58Q_2 - 0.21Q_3$; cost of transport depends more closely on Q_3 than on Q_2 (Fig. 15A in Rayner 1988).

3.2 Direct Measurement

A number of methods are available for the direct determination of total energy consumption in a flying bird (Butler and Woakes, this Vol.), and during the past 20 years there have been successive refinements in the accuracy of the techniques. Three reliable approaches have been used: (1) recording of gas exchange, usually oxygen and carbon dioxide, and usually in birds flying in wind tunnels, (2) determination of mass loss, either in populations of long-distance migrants or in individuals flying in captivity, and (3) a combination of isotopically-labelled water and time budgeting. All have advantages and disadvantages, and all pose considerable technical problems.

To date metabolic flight power has been measured in 111 birds (and 8 bats) (Fig. 4). Many of the bird data were listed by Masman and Klaasen (1987). I have

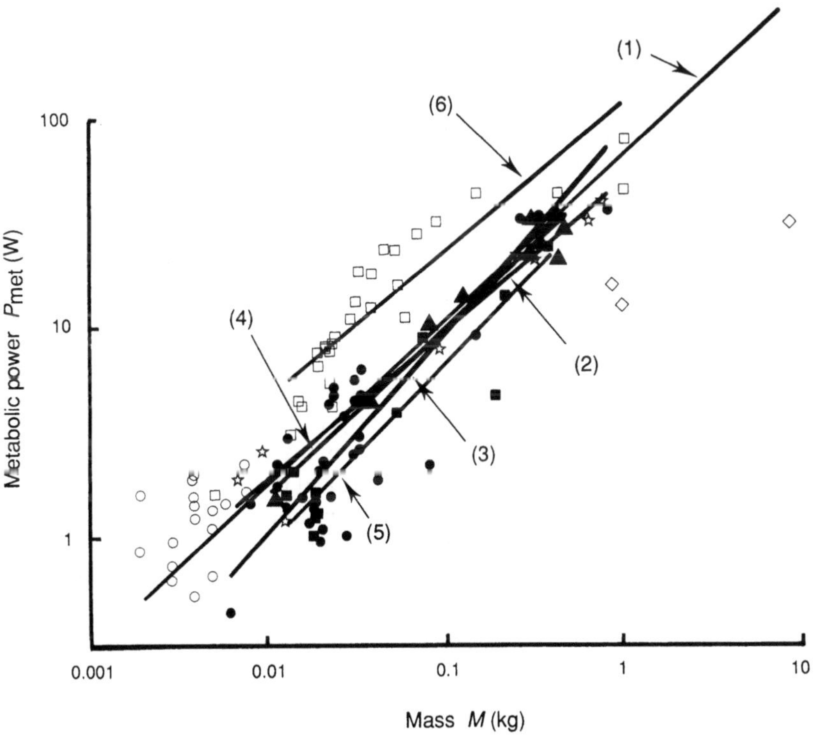

Fig. 4. Measured metabolic power against mass for birds and bats (*stars*) on logarithmic scales. *Solid symbols* are measurements included in the cruising flight sample, as follows: *triangles*, gas exchange including wind tunnel; *squares*, labelled isotope; *circles*, mass loss. *Open symbols: squares*, slow flight and take-off; *circles*, hovering hummingbirds; *diamonds*, large birds gliding or soaring. The *fitted lines* (reduced major axis; Rayner 1985b) show various subsets of the data (Table 1); *1* birds; *2* bats; *3* cruising flight; *4* cruising flight gas exchange (including wind-tunnel measurements); *5* cruising flight labelled isotopes; *6* slow flight. The cruising flight mass loss line virtually coincides with the line for the pooled cruising flight sample

reanalyzed these data to obtain bivariate correlation relations with body mass and multivariate relations with body mass, wing span and wing area for various classes of birds, for comparison with the prediction of Eq. (1). The results are shown in Table 1. Interpretation of these data suffers from a number of problems, and although the equations shown are valid for statistical comparison of the various groups and measurement techniques, their value as predictors of metabolism is untested. There are further complications. In the multiple allometric equations the three quantities involved (mass, wing span and wing area) are highly correlated: the scaling index for wing span is always negative (implying as expected that longer, thinner wings permit a reduction in flight cost), but the index for area is generally, though not always, positive. Comparison with principal components of wing morphology (Rayner 1988 and above) shows a more consistent pattern, power falls with increasing aspect ratio (Q_3), but increases with wing size (Q_2) owing to a rise in profile drag; this conforms closely to the mechanical prediction. (The small positive coefficient for Q_3 for labelled isotope sustained flight measurements arises because the range of aspect ratio of birds tested by this technique is small, most birds considered to date being efficient fliers with relatively thin wings.)

A number of conclusions can be drawn from comparison of these results. There is no significant difference between birds and bats (analysis of covariance, $F_{1,115} = 0.106$), but both mean power and gradient are significantly lower for non-passerine (excluding hummingbirds) than for passerine birds (means: $F_{1,91} = 40.7$; gradients: $F_{1,89} = 9.72$). This may be an artefact of the relatively small number of non-passerines tested, and of the facts that many of those measured used gliding flight, and that two relatively heavy ducks in the sample were measured immediately after take-off. It might, however, indicate a real physiological difference, as found for BMR by Aschoff and Pohl (1970).

As might be expected from the U-shaped power-velocity curve, the speed of measurement is also important, and speed is hard to control in measurements on free-flying birds. In some species measured in wind tunnels power varies parabolically with speed, as expected from mechanical performance, but in others there is little or no variation. This problem remains unexplained, and may be an artefact of the wind-tunnel technique. Free flight observations do show evidence of variation with speed: birds recorded in slow flight or at take-off have significantly higher metabolic power than birds in steady or cruising flight ($F_{1,86} = 23.15$). This may also arise because there is a gradual decline in RQ at the start of a flight (Butler et al. 1977; Rothe et al. 1987; Nachtigall, this Vol.), and metabolic power requirements may take some time to stabilize. Hovering flight (confined in this sample to hummingbirds) also shows significantly higher energy demands than cruising flight ($F_{1,80} = 18.49$).

It is a matter of some concern that the measurement technique has a major influence on the implied morphology-power relation (Fig. 4). Not surprisingly the mass loss technique is rather inaccurate (multiple correlation coefficient r_{35} = 0.879 compared with r_8 = 0.991 and r_9 = 0.961 for gas exchange and labelled isotope techniques). There is no significant difference between the labelled isotope and mass loss techniques, but gas exchange (mainly wind tunnel) meas-

Table 1. Correlations of measured flight power and morphology. The following allometric expressions relate measured metabolic flight power (P_{met}, watts) and morphology [M mass (kg), B span (m), S area (m^2)] in birds, for various classes of data. Metabolic flight power for 71 birds was listed by Masman and Klaasen (1987); additional records analyzed here include more recent publications. Principal components are transformed wing morphology (Rayner 1988); the three components measure size (Q_1), loading or relative wing size (Q_2) and aspect or wing shape (Q_3); formulae given are numerically identical to those for mass, span, area. Bivariate relations with mass M are reduced major axes (Rayner 1985b) to allow for error in each variate. Correlation coefficients r for bivariate regressions and product-moment correlations for multiple regressions are shown. Where necessary power equivalents in the original sources were converted to watts by using the following relations: 1 l O_2/h = 5.58 W, 1 l CO_2/h = 6.88 W, 1 kcal/h = 1.161 W; metabolism of 1 g fat releases 39.7 kJ. The last line of the Table reproduces the mechanical predictions of Section 3.1

	n	$r_{n\text{-}2}$	$P_{met}(M)$	n	$r_{pmc,n\text{-}2}$	$F_{3,n\text{-}4}$	$P_{met}(M,B,S)$	$\log[P_{met}(Q_1,Q_2,Q_3)]$
All data	119	0.853	$64.54M^{0.782}$	90	0.879	98.86	$162.87M^{1.312}B^{-2.572}S^{0.505}$	$2.56+1.17Q_1+0.50Q_2-0.29Q_3$
Bats	8	0.978	$50.40M^{0.724}$	6	0.968	9.98	$884.90M^{0.281}B^{-1.691}S^{1.594}$	$2.16+1.35Q_1-0.04Q_2-0.33Q_3$
Birds	111	0.840	$66.37M^{0.790}$	84	0.881	92.46	$98.39M^{1.505}B^{-2.539}S^{0.236}$	$2.54+1.19Q_1+0.59Q_2-0.26Q_3$
Non-passerines	26	0.756	$44.54M^{0.682}$	24	0.884	23.90	$1875.76M^{0.834}B^{-4.067}S^{1.732}$	$2.83+0.82Q_1+0.24Q_2-0.54Q_3$
Hummingbirds[a]	18	0.347	$733.79M^{1.165}$					
Passerines	67	0.721	$542.73M^{1.355}$	56	0.788	28.36	$39.84M^{1.930}B^{-1.690}S^{-0.553}$	$2.89+1.54Q_1+0.80Q_2-0.10Q_3$
Slow flight	24	0.876	$116.31M^{0.708}$	18	0.925	27.54	$4565.90M^{0.860}B^{-4.134}S^{1.919}$	$3.11+1.06Q_1+0.24Q_2-0.56Q_3$
Cruising flight	64	0.916	$72.05M^{0.912}$	58	0.930	114.53	$114.61M^{1.145}B^{-1.225}S^{0.523}$	$2.55+1.50Q_1+0.41Q_2-0.15Q_3$
Gas exchange	11	0.985	$62.43M^{0.800}$	10	0.991	111.57	$124.31M^{0.913}B^{-0.869}S^{0.299}$	$2.70+1.40Q_1+0.30Q_2-0.12Q_3$
Wind tunnel[b,c]	9	0.992	$66.17M^{0.814}$	8	0.993	98.69	$75.87M^{0.866}B^{-0.175}S^{0.037}$	$2.73+1.48Q_1+0.29Q_2-0.04Q_3$
Labelled isotopes	11	0.926	$52.31M^{0.886}$	11	0.961	28.06	$5.04M^{1.716}B^{-0.437}S^{-1.145}$	$2.38+1.29Q_1+0.75Q_2+0.08Q_3$
Mass loss	39	0.870	$88.55M^{0.970}$	37	0.879	37.36	$278.31M^{0.983}B^{-1.644}S^{0.708}$	$2.52+1.50Q_1+0.32Q_2-0.24Q_3$
Excluding gas exchange	53	0.864	$73.78M^{0.922}$	48	0.893	57.93	$107.47M^{1.153}B^{-1.378}S^{0.279}$	$2.47+1.44Q_1+0.41Q_2-0.16Q_3$
Prediction[d]			$14.35M^{1.161}$				$27.21M^{1.590}B^{-1.818}S^{0.275}$	$0.56+1.91Q_1+0.58Q_2-0.21Q_3$

[a] Hummingbirds: all hovering flight, measured by gas exchange.
[b] Wind tunnel: all measurements by gas exchange.
[c] Masman and Klaasen (1987) erroneously included many cage or captive slow-flight measurements in their 'wind-tunnel' sample.
[d] Mechanical components of flight energy only: see Sect. 3.1.

urements are on average 32.5% higher ($F_{2,56} = 11.70$) than the other techniques; the discrepancy is particularly great in smaller birds. There are presently only two cruising flight gas exchange measurements made outside a wind tunnel, and it is not possible to test whether this discrepancy is symptomatic of a flaw in the metabolic measurements, or of the stress placed on birds flying in wind tunnels. Given that gas exchange measurement techniques are well established, it appears likely that the wind tunnels are responsible. This is not surprising since wing-beat kinematics of birds flying in wind tunnels often differ substantially from those of the same species in free flight (JMV Rayner, pers. obs.).

Notwithstanding these possible limitations of the wind-tunnel method, the pooled records of the three techniques seem to give the most reliable estimate available. I use the "cruising flight" relations below. Some *caveats* must, however, be noted in the general application of these formulae:

1. The heaviest birds included in the sample are pigeons and crows, with mass below 0.5 kg. Extrapolation to birds outside the size range measured is not justified.
2. Many large birds adopt gliding flight to reduce power demands (Pennycuick 1975). As far as possible I have excluded birds which did not use flapping flight for the bulk of the time. Labelled isotope measurements of marine birds using extensive gliding flight (Flint and Nagy 1984; Adams et al. 1986) imply significantly lower energy expenditure in activity than would be predicted by these equations.
3. Many groups of birds, including Procellariiformes, Pelecaniformes, waterfowl and most waders, are not represented in the cruising flight sample. The bulk of the sample are Passeriformes. Flight physiology in other groups may differ widely.
4. Other flight strategies such as bounding may distort total power consumption.
5. All of these equations are statistical abstractions; they take no account of special physiological or mechanical adaptations in an individual bird.

3.3 Reconciliation of Mechanical and Metabolic Estimates

Both approaches to flight energetics suffer from problems. The mechanical approach is well founded in experimental aerodynamics, but only estimates a fraction of the energy consumption in flight. Muscle efficiencies of the order of 0.25 are commonly quoted, and if these values are realistic (but see below) three-quarters of total power demands are ignored by the aerodynamic models. On the other hand, direct measurements are difficult to carry out, hard to extrapolate, and the existing data base is too small to permit reliable estimation of migration energetics for a wide range of bird species.

Some authors (e.g. Pennycuick 1975; Tucker 1973) have attempted to modify their aerodynamic estimates to account for muscular heat loss and for "postural" flight-independent energy demands. This approach is probably the best method currently available for obtaining an indication of migration energetics for an individual. However, the method requires a value for muscular efficiency; hitherto this has been assumed constant. There is no justification for this assumption, and violation of it may be the source of the discrepancy in shape between measured and predicted power-speed curves which has already been noted (Walsberg, this Vol.). Efficiency may be expected to vary with flight speed, since it is determined by muscle contraction strain rate and stress; these quantities are related to wing-beat frequency and amplitude, which for mechanical reasons must vary with speed (Rayner 1986, 1988). Moreover, attempts to measure efficiency in tilting wind tunnels have not correctly accounted for the change in mechanical flight power with flight angle (Rayner 1986).

At the simplest, the bivariate relations for mechanical [Eq. (1)] and metabolic powers (Table 1, cruising flight) can be reconciled by taking efficiency as the ratio of mechanical input to metabolic output powers as $0.207M^{0.249}$; for small passerines this would imply efficiency between 0.07 and 0.15, and these values do not seem unreasonable. There are indications from muscle ultrastructure and physiology in flying birds (Rezende 1985), insects (Ellington 1985) and terrestrial vertebrates (Heglund and Cavagna 1985) that efficiency values of this magnitude are typical, and that efficiency can be expected to increase with size. However, considerably more research is required on the physiology of avian flight muscle and on metabolic energy flows in flight before the magnitude and size dependence of flight muscle efficiency can be fully established.

4 Migration Performance and Wing Morphology

4.1 Performance Changes During a Long Flight

Equations (1)–(3) predict flight performance (power, flight speed and cost of transport) for maximum range in flying birds. These may be used to estimate the dependence of flight range on pre-migration mass hypertrophy, and to predict how performance varies during a flight as fuel is consumed and mass decreases. Models of this kind have been presented before (e.g. Pennycuick 1975), but these revised estimates of mechanical performance allow direct modelling of the influence of wing morphology. Some attempts to perform this calculation have erroneously applied interspecific scaling relations to an individual bird (Summers and Waltner 1979; Davidson 1984), while others have simply ignored performance changes with mass, thereby underestimating range. The change in mass as fuel is consumed can be large, and can have a pronounced effect on range and migration time, particularly for species with high pre-migration mass hypertrophy.

To simplify the algebra, write maximum range speed as $V = vM^a$ and total metabolic power in flight as $\Pi = P/e = pM^b/e$, where p and v depend on wing span B and wing area S; both vary between individuals or between species, but in an individual may be taken to be constant. Metabolic power is assumed to be related to mechanical power by the efficiency e, which is taken to be constant in an individual. According to Eqs. (1) and (2), $a = 0.413$ and $b = 1.590$; Pennycuick's model (1969, 1975), which neglected wing allometry, gave $a = 0.5$ and $b = 1.5$; from measured flight metabolism (Sect. 3.2) I found $b = 1.145$; range predictions using any of these figures do not differ significantly from one another. The efficiency e is the weakest part of this analysis; Pennycuick's (1975) estimate of 0.23 appears generous; comparison between aerodynamic theory and measurement (above) suggested that it could be significantly lower than this, and may vary with size. Whether it varies with hypertrophy or with pre-migration changes in muscle physiology is not known.

To simplify further, neglect basal metabolism (flight power is between 10 and 20 times BMR), and assume that all mass loss in flight is associated with fat metabolism (ignore evaporative water loss and defecation), and that the bird flies in still air at sea level without the advantages of wind or of high altitude flight. Metabolism of fat releases energy $q = 39.7$ kJ/g, which is in part consumed as mechanical energy and in part radiated as heat. The rate of change of mass with time can be related to power by

$$\mathrm{d}M/\mathrm{d}t = -\Pi/q = -(p/qe)M^b; \tag{4}$$

if mass at the start of the flight is taken to be M_0, then the mass after time t (seconds) is given by

$$M(t) = [M_0^{\,1-b} - (1-b)pt/qe]^{1/(1-b)}. \tag{5}$$

Flight speed at any instant is obtained by substitution of $M(t)$ in Eq. (2) for flight speed (or equivalently in $V = vM^a$).

Flight range can be related to mass change by integrating the implied speed-time relation, and mass after distance d is given by

$$M(d) = [M_0^{\,1+a-b} - (1+a-b)pd/qev]^{1/(1+a-b)}. \tag{6}$$

To simulate migration, suppose that the fat-free mass (or the minimum mass for survival after the flight) is m, and that before departure the initial mass is enhanced by fat and protein deposition to the level fm, where $f > 1$. Then the maximum range r accessible on combustion of all of this additional tissue can be estimated as

$$r = (qev/p)\,(f^{1+a-b} - 1)\,m^{1+a-b}/(1+a-b), \tag{7}$$

[the result is slightly different if $b = 1+a$; cf. Pennycuick (1969), Eq. (48)]. By substituting for p and v from Eqs. (1) and (2) range can be determined from wing dimensions B and S as

$$r = 82430\, e\, m^{-0.177}\, B^{1.265}\, S^{-0.370}\,(1 - f^{-0.177})\ \mathrm{km}. \tag{8}$$

Figure 5 shows the relation of range to pre-migration mass for a swallow. Efficiency of 0.09 was determined as the ratio between predicted mechanical power [Eq. (1)] and measured power (labelled isotope method; Hails 1979).

Bairlein (1985) records mean mass of swallows as 17.5 g before, and 15 g after, a trans-Sahara flight of approximately 500 km. This is in close agreement with the predicted range for this mass change (Fig. 5), and as the minimum survival mass

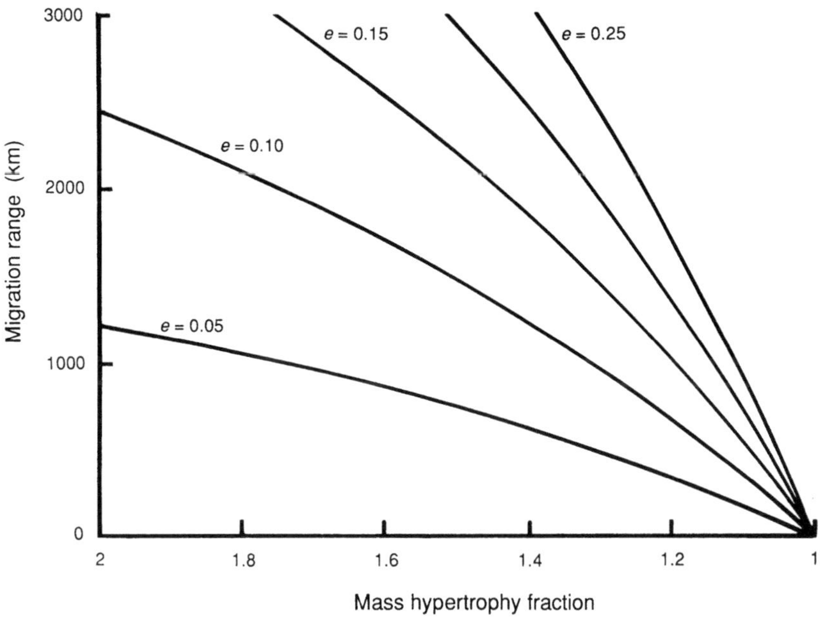

Fig. 5. Relation of migration range to pre-migration mass hypertrophy for a swallow with fat-free (or post-migration) mass 0.014 kg, and various values of efficiency. The effect of various initial fat and protein loads, and the decline in mass with range, may be read from this graph. Swallows generally enhance mass by less than 40%; doubling the load to 80% would increase range by only 70% because of the enhanced flight costs incurred when total mass is high; since the bird has the opportunity of foraging at intervals during the flight, additional mass hypertrophy may not be cost-effective. Efficiency values of the order of 0.09 are indicated for the swallow from comparison of mechanical and metabolic power estimates; a 40% fat load represents a range of just over 1000 km; a fall in mass from 0.0175 to 0.015 kg (Bairlein 1985) coincides with a range of approximately 500 km, which was the distance between the two measurement sites

of a swallow is approx. 14 g (G. Jones, pers. comm.) gives the bird a comfortable margin. South-bound migrating swallows may increase fat by up to 35% prior to crossing the Sahara (Moreau and Dolp 1970): consumption of all of this fat gives an estimated range of 1050 km; this implies that feeding stopovers during the complete flight are obligatory for this species. Other migratory passerines begin migration with a considerably larger proportion of the body as fat, and thus have a much greater range [they must use bounding flight (Sect. 5) to cope with performance changes as fat is consumed]; most passerines have significantly smaller wings, and lower aspect ratio, than swallows, which are relatively efficient (low cost of transport) in flight.

Range estimates are founded on the prediction that flight performance changes as mass is consumed over a long flight. Clearly the mechanical component of power will vary (predicted as $M^{1.490}$), but whether total metabolic flight power (as opposed to BMR) varies proportionately remains untested. Although the mechanical model does not draw a distinction between them, it is probably necessary to distinguish the effect of external load (e.g. Rayner et al. 1990) and of mass hypertrophy. For instance, pre-migratory physiological changes in the flight muscles (Marsh 1981, 1984) presumably match optimal migration performance. For this reason experiments on weighted birds may not reproduce migration conditions.

Performance changes induced by fuel consumption should be most obvious in flight speed. Mean speed over the whole flight is usually only slightly greater than speed at minimum (post-migration) mass, and changes in speed in an individual are unlikely to be resolved since wind drift and course changes would distort arrival/departure records. A more realistic test could be made by recording body mass and flight speeds at the start and finish of migration, and exploring whether the ratio of these speeds behaves as predicted as $f^{0.5}$.

4.2 Morphological Design for Migration

Equation (8) predicts the relation between range r and pre-migratory total mass f. This relation suggests only weak dependence of migration range on size, since with isometric design $B \propto M^{1/3}$ and $S \propto M^{2/3}$, and hence $r \propto M^0$. Range would be higher in larger birds if efficiency increases with size, as suggested above. But mechanical limits prevent large birds from carrying large amounts of extra fuel, and therefore f is not size-independent; moreover, the topography of fat deposition around the body may influence parasite drag.

Among birds of the same mass and wing size, range is higher with larger aspect ratio, owing to the greater mechanical efficiency of flight and lower cost of transport (Sect. 3.1) in long-winged birds; increased wing loading has the effect of reducing range, despite the associated higher flight speed. There is a clear trend for migratory bats to have relatively high aspect ratios (Norberg and Rayner 1987), but birds do not always respond in the same way (Niethammer 1937; Rayner 1988). It may be that a requirement for flexibility, or for optimized mechanical and ecological performance on winter or summer grounds or during migratory stopovers, have comparable importance to migration in morphological selection (Leisler, this Vol.).

However, one group of birds, the waterfowl, shows clear specialization for migration (Rayner 1988). Ducks and swans, and to a lesser extent geese, have a relatively high aspect ratio, as expected, but also have notably small wings. This trend is most marked in species breeding in high arctic latitudes such as eider ducks (Somaterini). I suggest that this reflects a compromise adaptation between migration economy and flight speed. Fast flight may be advantageous in the climatic conditions of the arctic: it permits the bird to move rapidly in response to adverse weather, and migration flights may be sustained even in high wind. There is a significant positive interspecific correlation between relative wing loading and breeding latitude in Palaearctic ducks (Rayner 1988). Some temperate or tropical coastal waders also have relatively small, pointed wings: this may also be a response to migration performance, or may reflect pressures for fast flight owing to tidal limitations on feeding times.

5 Bounding Flight

As the mass of a migrant varies during a long flight the balance between the mechanical power output by the flapping wings and the power input by the flight muscles will shift. It has been suggested that birds may compensate for this by consuming protein from the flight muscles, rather than having to transport excess muscle at the end of the flight (Pennycuick 1978). However, this response cannot entirely resolve the imbalance, since wings of a given size are optimized to output a given level of mechanical power, and this level cannot be appropriate throughout a sustained flight. This problem is particularly acute in small passerines, where pre-migration mass hypertrophy can result in an initial total mass double the lean mass.

These birds have adopted a unique solution to this problem. Rather than evolving muscles able to provide a range of power to the wings, they have developed muscles which are histologically homogeneous and which give a constant quantity of energy with each contraction. The bird simply introduces pauses between bouts of flapping, so that the mean power output is at the required level. The result is the characteristic "bounding" flight, in which the wings are folded against the body for a portion of the flight time. I have discussed this adaptation in detail elsewhere (Rayner 1985c), showing in particular how bounding is determined primarily by muscle physiology and not by aerodynamic factors, and that for scaling reasons it is restricted to smaller birds. Bounding is typical of many small passerines, and of other relatively small species including hummingbirds, kingfishers, small owls and some woodpeckers. It is confined to species with low aspect ratio wings, and is not observed for instance in hirundines where aspect ratio is high and wing loading low. Interestingly, premigration mass increase, in hirundines, and in other similar species such as swifts and wood-swallows, is relatively small (see above), and never reaches the 100% mass increase found in some passerines and waders.

Bounding flight appears to be all but essential for a bird to cope with large volumes of fat in the early phases of a migratory flight. It is not clear whether bounding evolved as a response to migration or as a means of allowing a bird some flexibility in the way it exploits its habitat, but it is of great benefit in both circumstances as it allows gait and muscle efficiency to remain unaltered, and therefore most efficient, whether in cruising flight or in take-off or hovering. With compensatory changes in flight muscle mass and protein content, gait can remain optimal throughout a sustained migration flight. The adaptation does not appear to reduce the energy costs of migration, but it does permit greater fuel load, and hence greater range, than would be possible with steady flapping flight.

6 Summary

Flow-visualization experiments on flying birds permit the distribution and intensity of wake vortices to be determined and related to wing-beat kinematics and other determinants of gait. From this information it is possible to estimate mechanical energy costs in flight in relation to size and wing morphology. Similar variations of power with morphology are found by analysis of metabolic measurements of flight power, although the data base is small and there is evidence that some techniques (gas exchange in wind tunnels) are less reliable. Comparisons of metabolic and mechanical estimates indicate flight muscle efficiencies around 0.10 in small birds, increasing with size. Predictions of migration range must take into account change in mass and hence in flight speed and power as fat is consumed; calculations for trans-Sahel migration in a swallow *Hirundo rustica* are in agreement with observed mass dynamics. Bounding flight is a method of avoiding imbalance between mechanical and metabolic energy flows while mass changes during migration.

Acknowledgements. My work on vertebrate flight performance has been funded by grants from SERC and by a research fellowship from the Royal Society of London. I am grateful to Gareth Jones for comments and discussion on the manuscript.

References

Adams NJ, Brown CR, Nagy KA (1986) Energy expenditure of free-ranging wandering albatrosses *Diomedea exulans*. Physiol Zool 59:583–591

Aschoff J, Pohl H (1970) Der Ruheumsatz von Vögeln als Funktion der Tagezeit und der Körpergrösse. J Ornithol 111:38–47

Bairlein F (1985) Body weights and fat deposition of Palearctic passerine migrants in the central Sahara. Oecologia (Berl) 66:141–146

Butler PJ, West NH, Jones DR (1977) Respiratory and cardiovascular responses of the pigeon to sustained level flight in a wind-tunnel. J Exp Biol 71:7–26

Davidson NC (1984) How valid are flight range estimates for waders? Ringing & Migr 5:49–64

Ellington CP (1985) Power and efficiency of insect flight muscle. J Exp Biol 115:293–304

Flint EN, Nagy KP (1984) Flight energetics of free-living sooty terns. Auk 101:288–294

Hails CJ (1979) A comparison of flight energetics in hirundines and other birds. Comp Biochem Physiol A Comp Physiol 63:581–586

Heglund NC, Cavagna GA (1985) Efficiency of vertebrate locomotory muscles. J Exp Biol 115:283–292

Helmholtz H von (1874) Über ein Theorem, geometrisch ähnliche Bewegungen flüssiger Körper betreffend, nebst Anwendung auf das Problem, Luftballons zu lenken. Mber k Akad Wiss Berl (1873), pp 501–514

Marsh RL (1981) Catabolic enzyme activities in relation to premigratory fattening and muscle hypertrophy in the gray catbird (*Dumetella carolinensis*). J Comp Physiol B Biochem Syst Environ Physiol 141:417–423

Marsh RL (1984) Adaptations of the gray catbird *Dumetella carolinensis* to long distance migration: flight muscle hypertrophy associated with elevated body mass. Physiol Zool 57:105–117

Masman D, Klaasen M (1987) Energy expenditure during free flight in trained and free-living Eurasian kestrels (*Falco tinnunculus*). Auk 104:603–616

Moreau RE, Dolp RM (1970) Fat, water, weights and wing-lengths of autumn migrants in transit on the northwest coast of Egypt. Ibis 112:209–228

Niethammer G (1937) Über die Beziehung zwischen Flügellänge und Wanderstrecke bei einigen europäischen Singvögeln. Arch Naturgesch 6:519–525

Norberg UM, Rayner JMV (1987) Ecological morphology and flight in bats (Mammalia; Chiroptera): wing adaptations, flight performance, foraging strategy and echolocation. Philos Trans R Soc Lond B Biol Sci 316:335–427

Pennycuick CJ (1969) The mechanics of bird migration. Ibis 111:525–556

Pennycuick CJ (1975) Mechanics of flight. In: Farner DS, King JR (eds) Avian biology, vol 5. Academic Press, London, pp 1–75

Pennycuick CJ (1978) Fifteen testable predictions about bird flight. Oikos 30:165–176

Rayner JMV (1979) A new approach to animal flight mechanics. J Exp Biol 80:17–54

Rayner JMV (1980) Vorticity and animal flight. In: Elder HY, Trueman ER (eds) Aspects of animal movement. Cambridge University Press, pp 177–189 (Semin Ser Soc Exp Biol, vol 5)

Rayner JMV (1985a) Vorticity and propulsion mechanics in swimming and flying vertebrates. In: Riess J, Frey E (eds) Konstruktionsprinzipien lebender und ausgestorbener Reptilien. Universität Tübingen, pp 89–118 (Konzepte SFB230, vol 4)

Rayner JMV (1985b) Linear relations in biomechanics: the statistics of scaling functions. J Zool (Lond) 206:415–439

Rayner JMV (1985c) Bounding and undulating flight in birds. J Theor Biol 117:47–77

Rayner JMV (1986) Vertebrate flapping flight mechanics and aerodynamics, and the evolution of flight in bats. In: Nachtigall W (ed) Bat flight-Fledermausflug. Fischer, Stuttgart, pp 27–74 (Biona Report 5)

Rayner JMV (1987) The mechanics of flapping flight in bats. In: Fenton MB, Racey PA, Rayner JMV (eds) Recent advances in the study of bats. Cambridge University Press, pp 23–42

Rayner JMV (1988) Form and function in avian flight. Curr Ornithol 5:1–77

Rayner JMV, Jones G, Hughes P (1990) Load carrying in flying bats. Proc 4th European Bat Research Symposium, Carolinum, Prague (in press)

Rezende MA (1985) The aerobic power output of the chief flight muscle of birds (*M. pectoralis*): a theoretical and experimental investigation. PhD Thesis, University of Bristol

Rothe H-J, Biesel W, Nachtigall W (1987) Pigeon flight in a wind tunnel. II. Gas exchange and power requirements. J Comp Physiol B Biochem Syst Environ Physiol 157:99–109

Spedding GR (1986) The wake of a jackdaw (*Corvus monedula*) in slow flight. J Exp Biol 125:287–307

Spedding GR (1987) The wake of a kestrel (*Falco tinnunculus*) in flapping flight. J Exp Biol 127:59–78

Spedding GR, Rayner JMV, Pennycuick CJ (1984) Momentum and energy in the wake of a pigeon (*Columba livia*) in slow flight. J Exp Biol 111:81–102

Summers RW, Waltner M (1979) Seasonal variations in the mass of waders in southern Africa, with special reference to migration. Ostrich 50:21–37

Tucker VA (1973) Bird metabolism during flight: evaluation of a theory. J Exp Biol 58:689–709

The Physiology of Bird Flight

P.J. BUTLER and A.J. WOAKES[1]

1 Introduction

Flight is the major means of transport in birds and the study of locomotion in these animals has an extra dimension compared with that in some other vertebrates, for many species of birds migrate twice a year over long distances, sometimes without stopping and sometimes at high altitude. Such efforts clearly impose large demands on the respiratory, cardiovascular and muscular systems.

It is not easy to obtain physiological data from a flying animal and it is doubtful if any one experimental method would adequately give all of the necessary data for a full analysis of the energetics and metabolic requirements of flight. This would entail not only measurements of oxygen consumption (power input) together with respiratory and cardiovascular variables but also details of the behaviour of the bird and of the prevailing meterological conditions, especially strength and direction of the wind (Pennycuick 1969; Alerstam 1979). Techniques which allow determinations of metabolism during migration are indirect and details of the birds' flight behaviour have to be assumed. Direct measurements of power input and of respiratory and cardiovascular variables can be obtained under controlled conditions in a wind tunnel, but long flights in excess of a day, closely resembling the more extreme conditions of natural migration, have not yet been achieved.

2 Techniques

Some of the earliest values of metabolic rate during flight were estimated from the mass loss recorded during a long flight with the assumption that fat constitutes by far the major part of this loss (Nisbet et al. 1963). Using the calorific value of fat (9.5 kcal g^{-1}, 39.7 kJ g^{-1}), these authors calculated that a 19 g blackpoll warbler, *Dendroica striata*, flying at 6–12°C has a power input of approximately 1 kcal h^{-1} (1.16 W or 3.5 ml O_2 min^{-1}). Berger and Hart (1974) have criticized this method as they believe that total water loss, and also mass loss, may exceed the production of metabolic water.

The production of CO_2 during flight can be determined by loading the bird with D_2O and $H_2{}^{18}O$ and measuring the levels of the isotopes in the blood before

[1] School of Biological Sciences, University of Birmingham, P.O. Box 363, Birmingham B15 2TT, UK

E. Gwinner (Ed.) Bird Migration

and after the flight. The method is based on the observation that the oxygen in respiratory CO_2 is in isotopic equilibrium with the oxygen in body water (Lifson et al. 1949). Thus, the hydrogen of body water is lost primarily as water, whereas the oxygen is lost both as water and as CO_2; carbon dioxide production can be estimated from the difference in turnover rates of the ^{18}O and D. Although there are certain assumptions associated with this method (Lifson and McClintock 1966), errors in field measurements of CO_2 production can be reduced to < 10% (Nagy 1980; Tatner and Bryant 1989). The use of this method by LeFebvre (1964) with pigeons, of mean mass 38.4 g, flying a distance of at least 480 km, demonstrated that the value of power input obtained (22 kcal h^{-1}, 25.6 W, 76.4 ml O_2 min^{-1}) is similar to that calculated from estimated fat loss (i.e. the difference between estimated fat content at the beginning of flight, based on measurements from other birds, and actual fat content at the end of the flight). It also happens to be similar to the value measured for pigeons flying at 10 m s^{-1} in a wind tunnel (Butler et al. 1977). Thus, measurement of fat loss and CO_2 production using the $D_2\,^{18}O$ method seem to give reasonable values of power input during long flights and the latter technique has been used on a range of smaller birds (Utter and LeFebvre 1970; Hails 1978). It has not yet been used with migrating birds.

It is the use of wind tunnels that has provided the most comprehensive data on the physiology of flight in birds. This technique, however, does have its drawbacks. The bird has to fly in a restricted space at a set velocity which may well affect its natural flight pattern, it must tolerate the noise of the fan motor and is in an optically motionless and monotonous environment (Butler et al. 1977; Rothe and Nachtigall 1980). In addition, it may also have to carry a face mask and trailing wires and cannulae (Fig. 1), all of which impose an extra demand as a

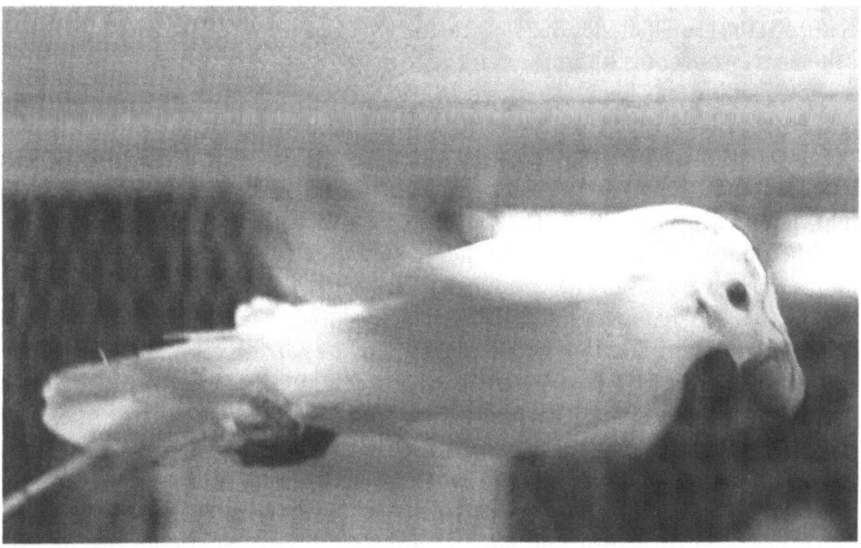

Fig. 1. A pigeon flying in a wind tunnel while wearing a mask for measuring gas exchange and respiratory frequency. Note the sample tube and thermocouple lead from the mask positioned above the head and along the back of the animal (Butler et al. 1977)

result of their mass and drag (Tucker 1968, 1972). These latter factors can be taken into account, although it has been argued that Tucker's estimate of the increase in the cost of flight resulting from a mask and tube was inadequate (Greenewalt 1975). Attempts to use radiotelemetry to obtain physiological data from free-flying birds have either produced a reasonable number of variables during very short (> 10 s) flights (Hart and Roy 1966; Berger et al. 1970a) or a limited number of variables (heart rate and respiratory frequency) during flights of several minutes' duration (Butler and Woakes 1980). Thus, despite its shortcomings, the wind tunnel is currently the only practicable means of studying the physiological aspects of flight in birds and most of the data presented in this review were obtained from birds flying in wind tunnels. The capabilities of the locomotory muscles that are most active during flight, viz. the pectorales, can be assessed by histochemical and morphometric analyses. Details of recent advances in these areas will be presented.

3 Metabolic Rate

A flying animal has to provide lift, to overcome gravity, as well as to provide forward propulsion and, as the mass of birds increases prior to migration (the result of a large deposition of fat which is the main source of energy during long-distance flight), so the power output of the pectoralis muscle has to increase in order to overcome the extra drag and to keep the animal airborne. During long flights, power is produced almost entirely from aerobic metabolism, which means that the respiratory and cardiovascular systems have to supply the necessary fuel and oxygen and to remove metabolic end-products. Four to five times more oxygen than is required for mechanical work has to be supplied to the muscles because of the low efficiency of chemical to physical energy transformation. Much of the energy output is, therefore, in the form of heat.

Power input has been measured (as oxygen consumption) during flapping flight from at least seven species of birds ranging in mass from the budgerigar, *Melopsittacus undulatus* (35 g), to the white necked raven, *Corvus cryptoleucus* (480 g). Figure 2 shows power input (in watts) for six of these species during level flight at different velocities. The budgerigar and pigeon display the classical U-shaped curve (Tucker 1968; Rothe et al. 1987) as predicted by theoretical models for power output (Pennycuick 1969; Rayner 1979), whereas the starling has a particularly flat metabolic rate vs velocity curve, which may result from the change in body attitude that occurs during flight (Torre-Bueno and Larochelle 1978). The relationship between *minimum* power input (P_i), in W, of seven species of birds during level flight in a wind tunnel (after taking into acount the mass and drag of mask, tube, etc.) and body mass (m), in kg, is $P_i = 50.4\ m^{0.73}$ (Butler and Woakes 1985). If the data obtained by Rothe et al. (1987) from the pigeon are added, the equation becomes $P_i = 53.1\ m^{0.74}$. The corresponding equations for *non-passerine* birds at rest and small mammals during *maximum* sustainable exercise are: $P_i = 5.4\ m^{0.72}$ and $P_i = 22.6\ m^{0.73}$ respectively (Prinzinger and Hanssler 1980; Pasquis et al. 1970, respectively). [It should be noted that Prin-

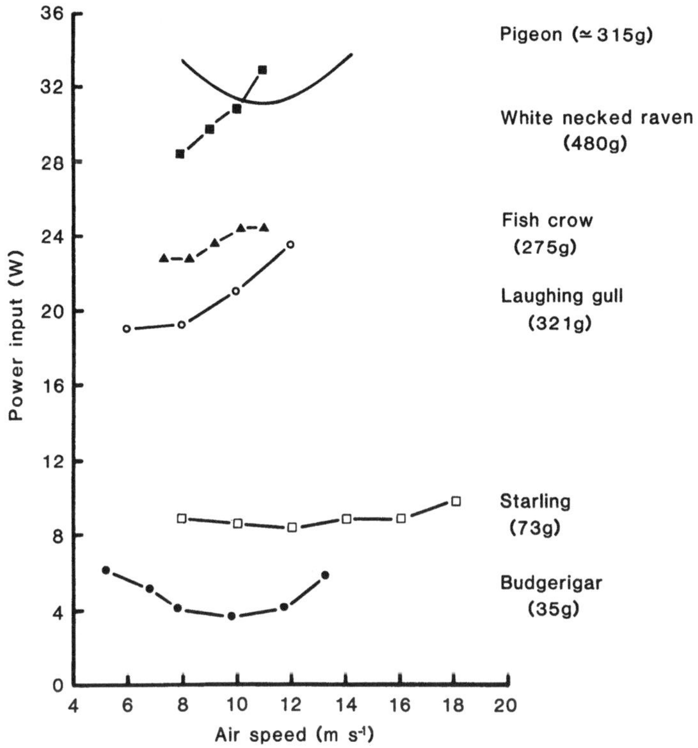

Fig. 2. Power input (W) at different air speeds for birds of different mass during horizontal flapping flight in a wind tunnel (modified, with the inclusion of data from Rothe et al. 1987; from Butler and Woakes 1985)

zinger and Hanssler's equation for non-passerines is almost identical to that of Aschoff and Pohl (1970) for passerines – $P_i = 5.6\, m^{0.73}$.] Thus, *minimum* power input during flapping flight in birds is 9.8 times their resting level and 2.3 times the *maximum* rate of exercising small mammals.

Despite the high rate at which oxygen is used during flight, this form of locomotion is still relatively attractive as far as long migrations are concerned. The energy cost of flight (i.e. the energy required to transport a given mass a given distance) is considerably less than the energy cost of walking or running (Tucker 1970; Schmidt-Nielsen 1972) because higher metabolic rate during flight is coupled to a higher air speed. Clearly, flying into a head wind will increase the energy cost, whereas a tail wind will reduce it and it would be expected that birds would fly at altitudes with the most favourable winds (Alerstam 1981). Another way of reducing the energy cost of migration is for a bird to glide or soar whenever possible. In herring gulls, *Larus argentatus*, power input during gliding is only twice the resting value (Baudinette and Schmidt-Nielsen 1974). However, a number of migrating birds do not have the necessary wing conformation (e.g. high aspect ratio) to permit effective gliding, and/or do not find themselves with the necessary rising air currents for soaring. They have to rely on flapping flight with or without wind assistance, which means that optimum performance of the pectoral muscles is of the utmost importance.

4 Locomotory Muscles

The major flight muscles in birds are the pectoralis and supracoracoideus and they have been described, from anatomical considerations, as the muscles chiefly responsible for depression and elevation, respectively, of the wing (George and Berger 1966). Studies involving electromyographic recordings have indicated a rather more complex situation. The pectoralis consists principally of the pars thoracicus (Raikow 1985) which, in many birds such as the pigeon, is divisible into two heads, the sternobrachialis (SB) and the thoracobrachialis (TB). These have different origins, insertions and fibre orientations and are innervated by distinct branches of the brachial plexus (Fig. 3). They also have different functions (Dial et al. 1987, 1988). Both are involved in depression of the humerus, but the SB assists in protraction, whereas the TB assists in retraction of the humerus. The TB is particularly active during landing and, perhaps, hovering, whereas the SB appears to be important during all modes of flight.

Both heads of the pectoralis and the supracoracoideus begin to generate tension before the downstroke and upstroke phases, respectively, of the wing beat cycle (Fig. 4). This not only serves to "brake" the wing, it may also provide a mechanism for energy conservation during flight, partly by elastic storage (Dial

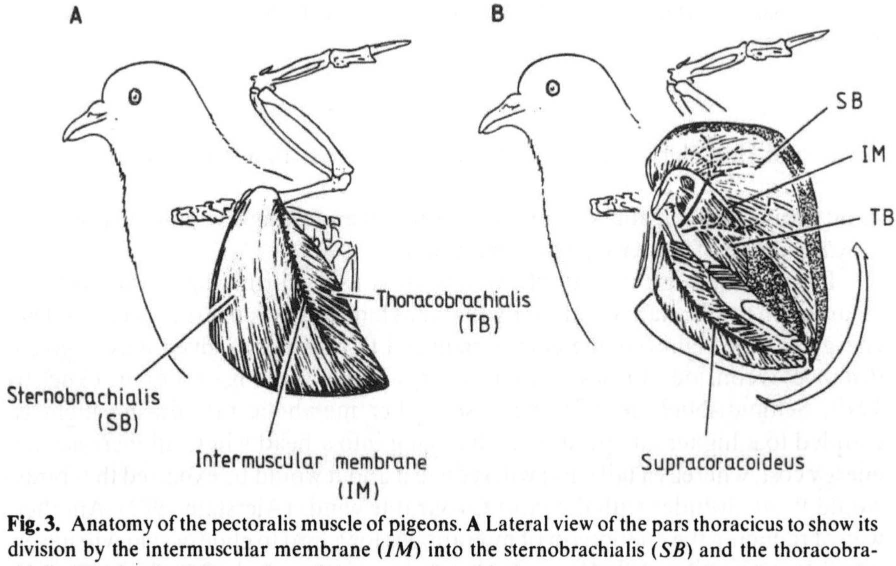

Fig. 3. Anatomy of the pectoralis muscle of pigeons. **A** Lateral view of the pars thoracicus to show its division by the intermuscular membrane (*IM*) into the sternobrachialis (*SB*) and the thoracobrachialis (*TB*) heads. **B** Deep view of the pectoralis, with the pars thoracicus deflected, to show the innervation of the *SB* and *TB* regions by distinct branches of the brachial plexus. The anterior nerve innervates the *SB*, the posterior nerve the *TB*. (Redrawn from Dial et al. 1987)

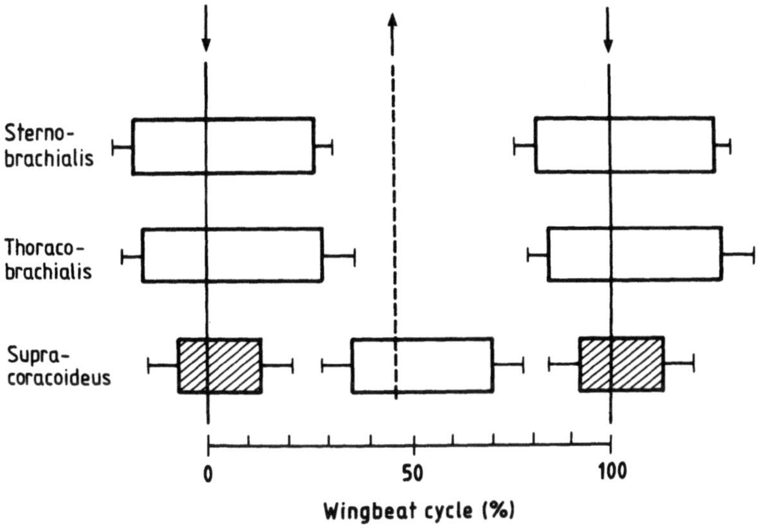

Fig. 4. Summary of the activity of the two major heads of the pectoralis, sternobrachialis (SB) and thoracobrachialis (TB) and of the supracoracordeus (SC) during flapping flight for 10 pigeons flying at 3-12 m s^{-1}. Wing-beat cycles are normalized to 100% with the beginning of the downstroke represented as zero and the end of the upstroke as 100%. *Arrows* indicate the end of the downstroke (*downward arrow*) and upstroke (*upward arrow*). The *bars* represent the average period of activity for each muscle + 1 SD. Sample sizes are 54 for SB, 28 for TB and 33 for SC. *Cross-hatching* represents muscle activity that occurred in approximately 50% of observations (After Dial et al. 1988)

et al. 1987). In approximately 50% of the observations by Dial et al. (1988), the supracoracoideus was active at the same time as the pectoralis (Fig. 4). Although homologous muscles in other vertebrates show a similar phenomenon, its functional significance is unclear. Other muscles are also associated with the wing-beat cycle, e.g. the deltoideus major is active during the elevation of the wing in the Canada goose, *Branta canadensis* (Weinstein et al. 1984).

The pectoralis may constitute up to 35% of total body mass (Greenewalt 1962). In volant birds it consists predominantly of fast oxidative glycolytic (FOG) fibres, together with a small percentage of fast glycolytic (FG) fibres. In a few species of birds there are also some slow oxidative (SO) fibres which are normally involved in postural support. The presence of these was thought to be associated with the gliding or soaring behaviour of some birds, e.g. herring gull (Talesara and Goldspink 1978). This appears to be the case for the turkey vulture, *Cathartes aura* (Rosser and George 1986a), although Rosser and George (1986b) found SO fibres in the pectoralis muscle of only 2 of the 42 volant species which they studied and these did not include the herring or ring-billed gulls. SO fibres are present in the pectoralis of the flightless emu, *Dromaius novaehollandiae* (Rosser and George 1984), and in the tiny deep red strip of the chicken pectoralis (Suzuki 1978), which comprises < 1% of the adult muscle and which may be a rudiment of an earlier developmental stage of the pectoralis (Matsuda et al. 1983). The reported presence of a large percentage (12%) of SO fibres in the dorsal areas of the pectoralis of the grey catbird, *Dumetella carolinensis* (Marsh 1984), which is a migrating species, is perplexing in view of the fact that Rosser and George (1986b) found no indication of such fibres in the pectoralis muscles of 17 other species of passerines. So, the functional significance of SO fibres, when they are present, is not clear in all cases.

FG fibres are relatively large, have few mitochondria and make up anything from 0 to approximately 50% of the fibres in the pectoralis of volant species (Rosser and George 1986b). They metabolize glycogen anaerobically and are used predominantly during take-off, landing and when the birds undergo sudden changes in direction (Parker and George 1975; Dial et al. 1987).

The FOG fibres have smaller diameters than FG fibres and, despite the fact that they possess glycolytic enzymes and can presumably produce ATP anaerobically, they are highly oxidative. In free-living tufted ducks, *Aythya fuligula*, the mass specific activity of citrate synthase (CS), an indicator of the capacity for oxidation of acetyl CoA in the citric acid cycle, is higher in the pectoralis muscle than in the heart (Butler and Turner 1988). Marsh (1981) found an even higher level of CS in the pectoralis of the grey catbird, just prior to migration and claimed that it was "the highest reported for any vertebrate skeletal muscle". It must be remembered, however, that CS activity most likely scales with body mass in a similar fashion to oxygen uptake (Somero and Childress 1980). The proportion of FOG fibres varies throughout the pectoralis of any one species, being greater in the deeper regions, and also between (volant) species from 100% in starlings, owls, hawks and kingfishers to approximately 50% in the ruffed grouse (Rosser and George 1986b). Thus, the pectoralis muscle normally consists of a preponderance of fast contracting but highly oxidative fibres, which are required to beat the wings continuously at high frequencies, sometimes for many hours.

Related to the greater potential oxygen demand in the pectoralis of flying birds is a greater capillary/fibre ratio than that in the leg muscles of birds and mammals (Pietschmann et al. 1982; Turner and Butler 1988). Along with the smaller cross-sectional area of the FOG fibres in the pectoralis, there are, therefore, shorter diffusion distances from capillaries to fibres. In fact, the smallest fibres and the greatest capillary densities are found in birds which migrate the longest distances (Lundgren and Kiessling 1988).

Associated with premigratory fattening is an increase in the mass of the pectoralis muscle (Marsh 1984). In the grey catbird, there is no associated change in the specific activity of CS although there is an increase in activity of 3-hydroxyacyl-CoA-dehydrogenase (HAD), which is a key enzyme in the β-oxidation of fatty acids (Marsh 1981). There is, however, an increase in CS and cytochrome oxidase (COX) activities in a number of other species prior to migration (Lundgren and Kiessling 1985, 1986).

It is not clear to what extent endocrine changes themselves and increased locomotor activity (training) just prior to migration effect the adaptations seen in the pectoralis muscle. It would seem that hypertrophy of the pectoralis is a direct result of the increase in body mass associated with premigratory fattening in the gray catbird (Marsh 1984). Endurance training can increase muscle capillarity and mass specific activity of CS, and cause a transformation of muscle fibres from

FG to FOG in the leg muscles of tufted ducks (Butler and Turner 1988). Also, it has been proposed that the increased oxidative capacity in migratory reed warblers, *Acrocephalus scirpaceus*, is a training effect (Lundgren and Kiessling 1986). These authors do point out, however, that both increased locomotor activity and hyperphagia are initiated by the hormones corticosterone and prolactin (Meier and Martin 1971). Nonetheless, studies on the levels of thyroid hormones (T3 and T4) and growth hormone in migratory Canada geese have suggested that they are directly involved in the hypertrophy of the pectoral muscles prior to migration (John and George 1978; John et al. 1983). Also, high levels of T4 and T3 can cause increases in CS and cytochrome c in the soleus muscle of rats (Winder 1979).

"Conditioning" of the pectoral muscles before migration takes on even greater significance when it is realized that periods of relative inactivity, as occur during moulting and incubation, cause significant reductions in muscle mass, myoglobin concentration and in CS and HAD activities (Pages and Planas 1983; George et al. 1987; Butler and Turner 1988; Piersma 1988).

5 Cardiovascular and Respiratory Adjustments

It is the combined functions of the cardiovascular and respiratory systems to supply the required oxygen and metabolic substrates to the flight muscles and to remove the waste products of metabolism (e.g. CO_2 and heat).

5.1 Ventilation

Respiratory data are given for four species of birds in Table 1. It can be seen that at relatively low ambient temperatures ($< 23\,°C$) ventilation volume ($\dot{V}_I$) increases by a similar proportion as oxygen uptake ($\dot{V}_{O_2}$) and that the proportion of oxygen extracted from the inspired gas (O_2ext) during flight is similar to that in resting birds. The relative contributions of respiratory frequency (f_{resp}) and tidal volume (V_T) to the increase in $\dot{V}_I$ during flight vary between species. In white-necked ravens, V_T does not change at all, in fish crows it almost doubles and in starlings there is a four-fold increase. Thus, in the first two species, frequency makes the greater contribution, whereas in the third, volume predominates. Respiratory frequency and tidal volume in the fish crow (Bernstein 1976) and respiratory frequency in the barnacle goose, *Branta leucopsis* (Butler and Woakes 1980), are independent of flight velocity. This could mean that, like the starling and the fish crow, oxygen uptake of the barnacle goose is largely independent of flight velocity. Certainly in the budgerigar, both oxygen uptake and respiratory frequency change with velocity in a U-shaped fashion (Tucker 1968).

Despite the apparent matching between ventilation volume and oxygen uptake in the four species listed in Table 1, there does appear to be an increase in effective lung ventilation above that required by metabolic rate (hyperventilation) during flight in starlings (Torre-Bueno 1978a).

Table 1. Values of respiratory frequency, f_{resp} (min^{-1}); tidal volume, V_T (ml); ventilation volume, $\dot{V}_I$ ($1\ min^{-1}$ BTPS, except for fish crow where it is $1\ min^{-1}$ STPD)[a]; oxygen extraction, O_2ext, and oxygen uptake, $\dot{V}_{O_2}$ (ml O_2 min^{-1} STPD) during rest and while flying in a wind tunnel for 4 species of birds. The values for $\dot{V}_{O_2}$ during flight are the minima that have been recorded and have been corrected for the drag and mass of mask etc., where necessary (see Butler et al. 1977 for further details)

Data[b]		Rest					Flight					
	Mass (kg)	f_{resp}	V_T	$\dot{V}_I$	O_2ext	$\dot{V}_{O_2}$	f_{resp}	V_T	$\dot{V}_I$	O_2ext	$\dot{V}_{O_2 min}$
(a) Budgerigar (18–20°C) *Melopsittacus undulatus*	0.035	–	–	0.047	0.27[c]	2.62	199	1.15	0.232	0.26[c]	10.9
(b) Starling (10–14°C) *Sturnus vulgaris*	0.073	92	0.67	0.061	0.28	3.16	180	2.8	0.504	0.31	28.1
(c) Fish crow (12–22°C) *Corvus ossifragus*	0.275	27.3	8.2	0.223	0.19	8.5	120	14.9	1.79	0.19	68.0
(d) White-necked raven (14–22°C) *Corvus cryptoleucus*	0.48	32.5	10.5	0.34	0.24[c]	17.0	140	10.7	1.40	0.29[c]	84.9

[a] BTBS = Body temperature, pressure saturated; STPD = standard temperature, pressure dry.

[b] Data from (a) Tucker 1968, 1972; (b) Torre-Bueno and Larochelle 1978; Torre-Bueno 1978a; (c) Bernstein et al. 1973; Bernstein 1976; (d) Hudson and Bernstein 1981, 1983.

[c] Calculated using the formula: $O_2ext = \dfrac{\dot{V}_{O_2}}{\dot{V}_I \times F_I O_2}$ (Bernstein and Schmidt-Nielsen 1974).

5.2 Respiratory Heat Loss and Temperature Control

Partial pressure of oxygen (P_{O_2}) in the air sacs increases and P_{CO_2} decreases during flight in starlings (Torre-Bueno 1978a). An explanation of this could be that hyperventilation results from the increase in body temperature, T_b (by at least 2°C), that occurs in all birds so far studied, even at low (0°C) environmental temperatures (see, for example, Torre-Bueno 1976; Hudson and Bernstein 1981; Hirth et al. 1987). However, when the usual elevation in T_b during running is prevented in ducks and domestic fowl there are still signs of hyperventilation (Kiley et al. 1982; Brackenbury and Gleeson 1983). In ducks there was an increase in lactic acid and hence a slight acidosis which could have stimulated ventilation and in the fowl hyperventilation seemed to occur only at higher work loads. Thus, it appears as if factors, other than hyperthermia, contribute to the overall hyperventilation. At relatively low ambient temperatures, T_a (< 23°C), T_b does not change with varying T_a at constant velocity nor with varying velocity at constant T_a in flying starlings, white-necked ravens and pigeons respectively (Torre-Bueno 1976; Hudson and Bernstein 1981; Hirth et al. 1987).

At relatively high T_a (> 23°C), T_b increases with increasing T_a at constant velocity and with increasing velocity at constant T_a in the same three species and in the raven ventilation increases with increasing T_b. Thus, in this species and in the fish crow, at constant velocity, ventilation progressively rises above that required for adequate oxygen uptake as T_a increases (Fig. 5) and O_2ext falls. At 25°C, O_2ext is 0.13 in the fish crow (Bernstein 1976).

The physiological significance of the hyperthermia during flapping flight, even at low ambient temperatures, is uncertain, although Torre-Bueno (1976) concluded that birds adjust their insulation to allow a certain increase in body temperature during flight in order to improve muscle efficiency and increase maximal work output. At higher ambient temperatures, heat stored in the body can be a sizeable proportion of the total production during short flights. Whatever the significance of the increase in core temperature, brain temperature is maintained at a lower level. In fact, even in birds under resting thermoneutral conditions, there is approximately a 1°C difference between brain and body temperatures (Phillips et al. 1985). This difference is at least maintained as the birds become hyperthermic during flight and may even increase (Bernstein et al. 1979). The structure responsible for this phenomenon is the *rete mirabile opthalmicum*, RMO. It is closely associated with the circulatory system of the eye and warm arterial blood from the body is thought to be cooled by the countercurrent exchange with venous blood returning from the relatively cool beak, evaporative surfaces of the upper respiratory tract and the eye (Mitgård 1983).

Hyperventilation during flight no doubt serves a thermoregulatory function and as T_a increases, a greater proportion of total heat loss during flight is by respiratory evaporation. However, even at a T_a of 30°C this is only approximately 20% for the budgerigar (Tucker 1968) and 30% for the fish crow, white-necked raven and pigeon (Hudson and Bernstein 1981; Biesel and Nachtigall 1987). So, most metabolic heat must be dissipated by means other than respiratory

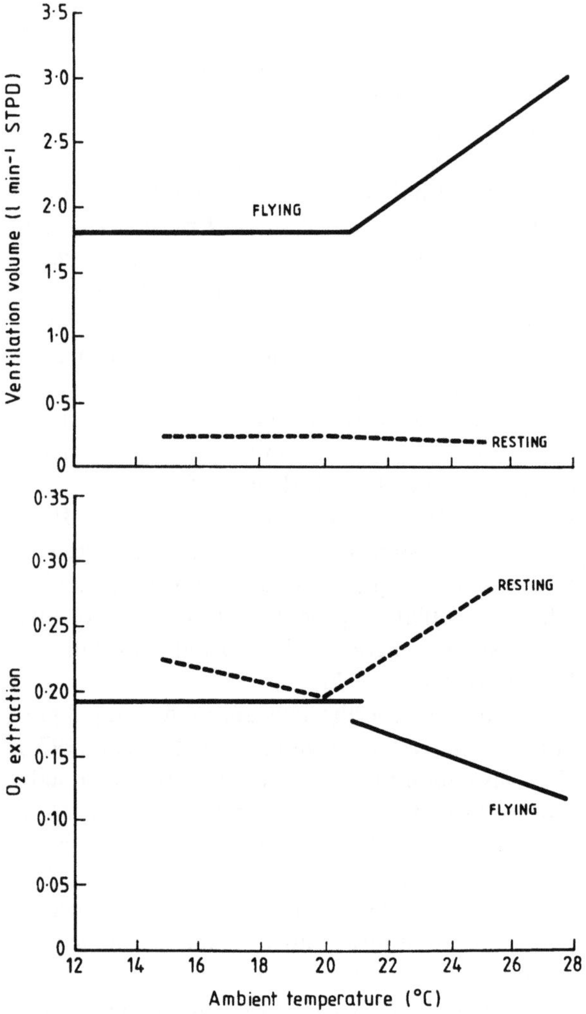

Fig. 5. *Above* Ventilation volume; *below* oxygen extraction plotted against ambient temperature in fish crows at rest and while flying in a wind tunnel at constant speed (between 7.4–11.0 m s^{-1}). Both of these variables are independent of flight speed (After Bernstein 1976)

evaporation. Indeed, herring gulls can lose up to 80% of their total heat production during flight through their feet (Baudinette et al. 1976).

5.3 Respiratory Water Loss

At moderate to high T_a water is lost at a greater rate than its metabolic production, i.e. the birds are dehydrating (Tucker 1968; Biesel and Nachtigall 1987) and the starling is in water balance only at temperatures below 7° (Torre-Bueno 1978b). This author suggested that during migrations, birds ascend to altitudes where the air is cool enough to enable a greater proportion of heat to be dissipated by non-evaporative means, thus keeping them in water balance.

As already seen for the starling, hyperventilation at relatively low T_a can cause low levels of P_{CO_2} (hypocapnia) in the respiratory system which is also reflected in P_{CO_2} in blood (Butler et al. 1977). At high ambient temperatures the extra increase in ventilation would tend to cause further hypocapnia and alkalosis. It has been demonstrated that the white-necked raven resorts to so-called compound ventilation when flying at high T_a. A high frequency, shallow ventilatory component is superimposed upon a deeper, lower frequency component and may serve to reduce hypocapnia during thermal panting (Hudson and Bernstein 1978).

5.4 Ventilation and Wing Beating

Ever since Marey's (1890) pioneering studies on bird flight it has been known that lung ventilation may be co-ordinated with the beating of the wings. In crows and pigeons there is a 1:1 correspondence between the two, whereas ratios as high as 5:1 (wing-beat frequency:respiratory frequency) have been reported for the black duck, quail and pheasant (Hart and Roy 1966; Berger et al. 1970b). However, as mentioned earlier, the flights reported by these authors were of only a few seconds duration and it was concluded that co-ordination was not obligatory. During flights of up to 10-min duration, pigeons in a wind tunnel showed a very close relationship between these two activities. Wing beating occurred in bursts and, although respiratory frequency was often either slower or faster than the wing-beat frequency between bursts of wing beating, there was always close co-ordination between the two when the wings were flapping (Butler et al. 1977). Almost immediately upon landing pigeons panted at a frequency identical to the mean resonant frequency of the respiratory system, i.e. 10 Hz (Kampe and Crawford 1973). During flight, however, respiratory frequency was approximately 7 Hz, i.e. the same as wing-beat frequency.

With a 1:1 correspondence, it is possible to imagine how contractions of the flight muscles could assist respiratory air flow, with higher ratios this is not so obvious. Nevertheless, a correspondence of 3:1 has been found in the free-flying barnacle goose and it is clear that the wing beat is tightly locked to fixed phases of the respiratory cycle (Fig. 6) during flights of relatively long duration. These phase relationships can be maintained even during transient changes in one of the activities (Butler and Woakes 1980). As Tucker (1968) comments "It is hard to believe that the contractions of the flight muscles have no influence on ventilation . . .". The precise significance of such close co-ordination between the two has yet to be discovered, but represents a fascinating area for further study.

6 Circulation

The role of the various components of the circulatory system in presenting oxygen to (and removing CO_2 from) the exercising muscles can best be described by Fick's formula:

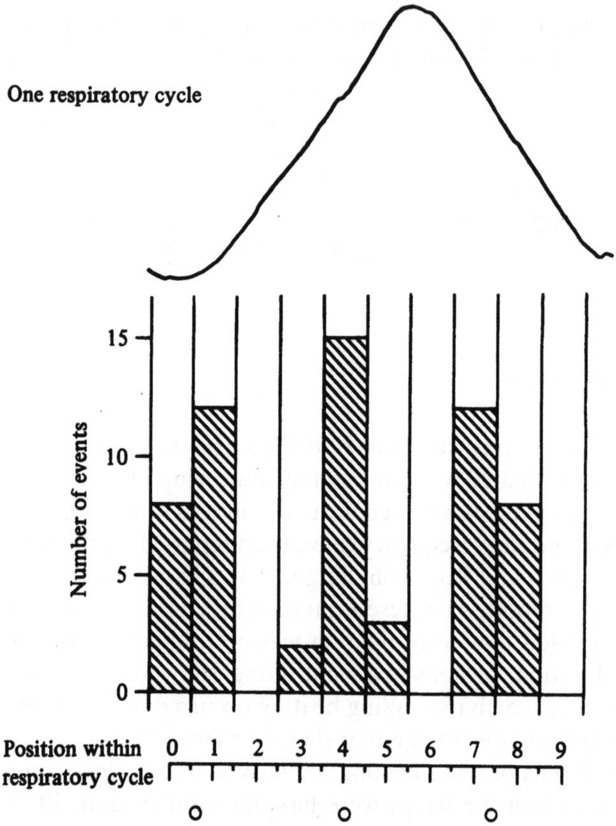

Fig. 6. Histogram showing the positions during the respiratory cycle at which the wings were fully elevated (events) during flight of a barnacle goose. Twenty respiratory cycles of equal duration were chosen and each was divided into 10 equal intervals. Below the histogram is plotted the mean position of each group of events (O) and above the histogram is a trace of one of the chosen respiratory cycles (inspiration, up) (Butler and Woakes 1985)

$$\dot{V}_{O_2} = HR \times SV \cdot (C_aO_2 - C_{\bar{v}}O_2)$$

where:

$\dot{V}_{O_2}$ = rate of oxygen consumption
HR = heart rate
SV = cardiac stroke volume
C_aO_2 = oxygen content of arterial blood
$C_{\bar{v}}O_2$ = oxygen content of mixed venous blood

The only study in which data have been collected on all but one of these variables in flying birds (thus allowing the absent one to be calculated) is that by Butler et al. (1977) working with pigeons. When flying in a wind tunnel at 10 m s^{-1}, oxygen uptake in these birds was ten times the resting value (Table 2). C_aO_2 was maintained at slightly below the resting value but $C_{\bar{v}}O_2$ was halved, giving a 1.8 times increase in ($C_aO_2 - C_{\bar{v}}O_2$). There was no significant change in SV, so the

Table 2. Mean values of respiratory and cardiovascular variables measured in pigeons at rest and after 6 min of steady level flight in a wind tunnel at a speed of 10 m s^{-1}. Resting values were obtained with the wind tunnel turned off (Butler et al. 1977)

	Rest	Flying at 10 m s^{-1}
Heart rate (beats min^{-1})	115	670
Mean arterial blood pressure (kPa)	18.9	19.6
Mean venous blood pressure (kPa)	0.2	0.3
Respiratory frequency (min^{-1})	19.7	411
Oxygen uptake (ml min^{-1} STPD)	9.0	88.4
Carbon dioxide production (ml min^{-1} STPD)	7.6	81.3
Respiratory exchange ratio	0.85	0.92
Partial pressure of oxygen in arterial blood (kPa)	11.6	12.7
Oxygen content of arterial blood (vol %)	15.1	13.7
pH of arterial blood	7.43	7.36
Partial pressure of oxygen in mixed venous blood (kPa)	7.6	5.6
Oxygen content of mixed venous blood (vol %)	10.5	5.4
pH of mixed venous blood	7.36	7.24
Partial pressure of CO_2 in arterial blood (kPa)	3.6	2.1
Partial pressure of CO_2 in mixed venous blood (kPa)	4.7	4.3
Haematocrit (%)	42	41
Blood lactate concentration (mg %)	9.0	59.8
Body temperature (°C)	41.3	43.3
Mass (kg)	0.442	

major factor in transporting the extra oxygen to the muscles was the six-fold increase in heart rate. As mean arterial blood pressure did not change, total peripheral vascular resistance must have declined by the same proportion as cardiac output (HR × SR) increased.

Birds have larger hearts and lower resting heart rates than mammals of similar body mass (Lasiewski and Calder 1971; Grubb 1983). They also have a greater cardiac output for a given oxygen consumption than similarly sized mammals (Grubb 1983). In other words, $(C_aO_2 - C_{\bar{v}}O_2)$ is less in birds than in similarly sized mammals, because $C_{\bar{v}}O_2$ is not normally reduced to as low a level in birds as it is in mammals. Thus, the higher cardiac output in birds may be an important factor in their attaining a higher maximum oxygen uptake ($\dot{V}_{O_2max}$) during flight than similarly sized mammals do when running. It is certainly interesting to note that bats have larger hearts and a higher blood oxygen-carrying capacity than other mammals of similar size (Jürgens et al. 1981), and that their $\dot{V}_{O_2}$ during flight is similar to that of birds of similar mass (Butler 1981; Butler and Woakes 1985).

7 High Altitude

Most passerines migrating at night fly below 2000 m, however, some birds have been observed at extremely high altitudes. A flock of 30 (probably whooper) swans, *Cygnus cygnus*, was located by radar off the west coast of Scotland at an altitude of 8000–8500 m (Stewart 1978). Bar-headed geese, *Anser indicus*, have been observed flying at altitudes of approximately 9000 m (where the P_{O_2} is, at 6.9 kPa, approximately one-third of that at sea level) during their migration across the Himalayas (Swan 1961). Unfortunately, there have been no physiological studies on birds flying at high altitudes (real or simulated), but work on inactive animals has indicated that the adaptations may be of great importance in the altitude performance of birds.

Although the lungs of birds do not appear to confer any particular advantages with respect to maintaining an elevated $\dot{V}_{O_2}$ during flapping flight their greater effectiveness may contribute significantly to the performance of birds at high altitude (Scheid 1985). It is, however, the cardiovascular system which shows most adaptations to the conditions of high altitude. Unlike Pekin ducks, bar-headed geese do not increase their haematocrit (packed cell fraction of the blood) and haemoglobin (Hb) concentration when exposed to simulated high altitude (Black and Tenney 1980). This means that there is no increase in blood viscosity, which could impair circulation of the blood, but also that there is no increase in the oxygen-carrying capacity of the blood. This is more than counter balanced by the higher affinity for oxygen (low P_{50}) of the Hb of the goose (approx. 5 kPa at pH 7.5 compared with 7.5 kPa in the duck) which allows the maintenance of a higher C_aO_2 (and hence $C_aO_2 - C_vO_2$) at high altitudes than in the duck. A comparison between greylag and bar-headed geese indicated that the difference in oxygen affinity of the Hb is the result of a very small intrinsic difference which is magnified by the presence of inositol pentaphosphate (Rollema and Bauer 1979).

It is also clear from the data of Black and Tenney (1980) that the respiratory system of the bar-headed goose is able to maintain a very small difference between the P_{O_2} in inspired air (P_IO_2) and arterial blood (P_aO_2) when at high altitude. At sea level ($P_IO_2 - P_aO_2$) is 7 kPa, whereas at a simulated altitude of 10 668 m it is a mere 0.5 kPa, giving proof of the effectiveness of the avian lung under these conditions. The large increase in ventilation that is required to maintain such a small difference between P_IO_2 and P_aO_2 also causes a decline in P_aCO_2 and an increase in arterial pH (pHa); the bird becomes hypocapnic and alkalotic.

In a number of mammals (dog, monkey, rat, man) hypocapnia causes a reduction in cerebral blood flow. This is not so in ducks and bar-headed geese (Grubb et al. 1977; Faraci and Fedde 1986). Also, hypoxia causes a greater increase in cerebral blood flow in ducks than in dogs, rats and man (Grubb et al. 1978). At high altitude, of course, both these factors occur together (hypocapnic hypoxia) and hypocapnia attenuates the increase in cerebral blood flow induced by hypoxia (Grubb et al. 1979). However, these authors found that blood flow is similar at a given O_2 *content*, in both normocapic and hypocapnic ducks. This is because during hypocapnia (and alkalosis) there is a leftward shift of the oxygen equilibrium curve so that a given oxygen content is achieved at a lower P_{O_2}. It has also been suggested that the RMO, as well as keeping the brain cool, may serve to enhance the supply of oxygen to the brain, particularly at high altitudes (Bernstein et al. 1984; Pinshow et al. 1985).

It has not been easy to obtain physiological data from birds flying under sea level conditions, and it will clearly be even more difficult to do so under conditions which mimic those at high altitude. The design of an appropriate wind tunnel has been presented (Linderoth 1975) but to the authors' knowledge no data have been reported on birds flying at simulated high altitudes. Nonetheless, this is clearly an area of great significance. Man is only just capable of reaching the top of Mt. Everest without supplementary oxygen (West 1983); whereas the bar-headed goose can fly over it.

References

Alerstam T (1979) Wind as selective agent in bird migration. Ornis Scand 10:76–93

Alerstam T (1981) The course and timing of bird migration. In: Aidley DJ (ed) Animal migration. Cambridge University Press, Cambridge, pp 10–54

Aschoff J, Pohl H (1970) Der Ruheumsatz von Vögeln als Funktion der Tageszeit und der Körpergrösse. J Ornithol 111(1):38–47

Baudinette RV, Schmidt-Nielsen K (1974) Energy cost of gliding flight in herring gulls. Nature 248:83–84

Baudinette RV, Loveridge JP, Wilson KJ (1976) Heat loss from feet of herring gulls at rest and during flight. Am J Physiol 230:920–924

Berger M, Hart JS (1974) Physiology and energetics of flight. In: Farner DS, King JR, Parkes KC (eds) Avian biology, vol 4. Academic Press, London, pp 415–477

Berger M, Hart JS, Roy OZ (1970a) Respiration, oxygen consumption and heart rate in some birds during rest and flight. Z Vgl Physiol 66:201–214

Berger M, Roy OZ, Hart JS (1970b) The co-ordination between respiration and wing beats in birds. Z Vgl Physiol 66:190–200

Bernstein MH (1976) Ventilation and respiratory evaporation in the flying crow, *Corvus ossifragus*. Respir Physiol 26:371–382

Bernstein MH, Schmidt-Nielsen K (1974) Ventilation and oxygen extraction in the crow. Respir Physiol 21:393–401

Bernstein MH, Thomas SP, Schmidt-Nielsen K (1973) Power input during flight of the fish crow, *Corvus ossifragus*. J Exp Biol 58:401–410

Bernstein MH, Curtis MB, Hudson DM (1979) Independence of brain and body temperatures in flying American kestrels, *Falco sparverius*. Am J Physiol 237:R58–R62

Bernstein MH, Duran HL, Pinshow B (1984) Extrapulmonary gas exchange enhances brain oxygen in pigeons. Science 226:564–566

Biesel W, Nachtigall W (1987) Pigeon flight in a wind tunnel. IV. Thermoregulation and water homeostasis. J Comp Physiol B Biochem Syst Environ Physiol 157:117–128

Black CP, Tenney SM (1980) Oxygen transport during progressive hypoxia in high-altitude and sea-level water fowl. Respir Physiol 39:217–239

Brackenbury JH, Gleeson M (1983) Effects of P_{CO_2} on respiratory pattern during thermal and exercise hyperventilation in domestic fowl. Respir Physiol 54:109–119

Butler PJ (1981) Respiration during flight. In: Hutas I, Debreczeni LA (eds) Advances in physiological sciences, vol 10. Pergamon, Oxford pp 155–164

Butler PJ, Turner DJ (1988) Effect of training on maximal oxygen uptake and aerobic capacity of locomotory muscles in tufted ducks *Aythya fuligula*. J Physiol 401:347–359

Butler PJ, Woakes AJ (1980) Heart rate, respiratory frequency and wing beat frequency of free flying barnacle geese *Branta leucopsis*. J Exp Biol 85:213–226
Butler PJ, Woakes AJ (1985) Exercise in normally ventilating and apnoeic birds. In: Gilles R (ed) Circulation, respiration and metabolism. Springer, Berlin Heidelberg New York, pp 40–45
Butler PJ, West NH, Jones DR (1977) Respiratory and cardiovascular responses of the pigeon to sustained, level flight in a wind-tunnel. J Exp Biol 71:7–26
Dial KP, Kaplan SR, Goslow GE Jr, Jenkins FA Jr (1987) Structure and neural control of the pectoralis in pigeons: Implications for flight mechanics. Anat Rec 218:284–287
Dial KP, Kaplan SR, Goslow GE Jr (1988) A functional analysis of the primary upstroke and downstroke muscles in the domestic pigeon (*Columba livia*) during flight. J Exp Biol 134:1–16
Faraci FM, Fedde MR (1986) Regional circulatory responses to hypocapnia and hypercapnia in bar-headed geese. Am Physiol 250:R499–R504
George JC, Berger AJ (1966) Avian myology. Academic Press, London
George JC, John TM, Minhas KJ (1987) Seasonal degradative, reparative and regenerative ultrastructural changes in the breast muscle of the migratory Canada goose. Cytobios 52:109–126
Greenewalt CH (1962) Dimensional relationships for flying animals. Smithson Misc Collect 144:1–46
Greenewalt CH (1975) The significant dimensions, their departure from the requirements for dimensional similarity and the effect on flight aerodynamics of that departure. Trans Am Philos Soc 65:5–67
Grubb B (1983) Allometric relations of cardiovascular functions in birds. Am J Physiol 245:H567–H572
Grubb B, Mills CD, Colacino JM, Schmidt-Nielsen K (1977) Effect of arterial carbon dioxide on cerebral blood flow in ducks. Am J Physiol 232:H596–601
Grubb B, Colacino JM, Schmidt-Nielsen K (1978) Cerebral blood flow in birds: effect of hypoxia. Am J Physiol 234:H230–H234
Grubb B, Jones JH, Schmidt-Nielsen K (1979) Avian cerebral blood flow: influence of the Bohr effect on oxygen supply. Am J Physiol 236:H744–H749
Hails CJ (1978) A comparison of flight energetics in hirundines and other birds. Comp Biochem Physiol A Comp Physiol 63:581–585
Hart JS, Roy OZ (1966) Respiratory and cardiac responses to flight in pigeons. Physiol Zool 39:291–306
Hirth K-D, Biesel W, Nachtigall W (1987) Pigeon flight in a wind tunnel III Regulation of body temperature. J Comp Physiol B Biochem Syst Environ Physiol 157:111–116
Hudson DM, Bernstein MH (1978) Respiratory ventilation during steady-state flight in white-necked raven, *Corvus cryptoleucus*. Fed Proc 37:472
Hudson DM, Bernstein MH (1981) Temperature regulation and heat balance in flying white-necked ravens, *Corvus cryptoleucus*. J Exp Biol 90:267–281
Hudson DM, Bernstein MH (1983) Gas exchange and energy cost of flight in the white-necked raven, *Corvus cryptoleucus*. J Exp Biol 103:121–130
John TM, George JC (1978) Circulating levels of thyroxine (T_4) and triiodothyronine (T_3) in the migratory Canada goose. Physiol Zool 51:361–370
John TM, George JC, Scanes CG (1983) Seasonal changes in circulating levels of luteinizing hormone and growth hormone in the migratory Canada goose. Gen Comp Endocrinol 51:44–49
Jürgens KD, Bartels H, Bartels R (1981) Blood oxygen transport and organ weights of small bats and small non-flying mammals. Respir Physiol 45:243–260
Kampe G, Crawford EC (1973) Oscillatory mechanics of the respiratory system of pigeons. Respir Physiol 18:188–193
Kiley JP, Kuhlmann WD, Fedde MR (1982) Ventilatory and blood gas adjustments in exercising isothermic ducks. J Comp Physiol 147:107–112
Lasiewski RC, Calder WA Jr (1971) A preliminary allometric analysis of respiratory variables in resting birds. Respir Physiol 11:152–166
LeFebvre EA (1964) The use of D_2O^{18} for measuring energy metabolism in *Columba livia* at rest and in flight. Auk 81:403–416
Lifson N, McClintock R (1966) Theory of use of the turnover rates of body water for measuring energy and material balance. J Theor Biol 12:46–74
Lifson N, Gordon GB, Visscher MB, Nier AO (1949) The fate of utilised molecular oxygen and the

source of the oxygen of respiratory carbon dioxide, studied with the aid of heavy oxygen. J Biol Chem 180:803–811
Linderoth LS Jr (1975) A wind tunnel for bird flight studies at high altitudes. J Appl Physiol 39:501–502
Lundgren BO, Kiessling K-H (1985) Seasonal variation in catabolic enzyme activities in breast muscle of some migratory birds. Oecologia (Berl) 66:468–471
Lundgren BO, Kiessling K-H (1986) Catabolic enzyme activities in the pectoralis muscle of premigratory and migratory juvenile reed warblers *Acrocephalus scirpaceus* (Herm). Oecologia (Berl) 68:529–532
Lundgren BO, Kiessling K-H (1988) Comparative aspects of fibre types, areas, and capillary supply in the pectoralis muscle of some passerine birds with differing migratory behaviour. J Comp Physiol B Biochem Syst Environ Physiol 158:165–173
Marey EJ (1980) Le vol des oiseaux. Masson, Paris
Marsh RL (1981) Catabolic enzyme activities in relation to premigratory fattening and muscle hypertrophy in the gray catbird (*Dumetella carolinensis*). J Comp Physiol 141:417–423
Marsh RL (1984) Adaptations of the gray catbird *Dumetella carolinensis* to long-distance migration: flight muscle hypertrophy associated with elevated body mass. Physiol Zool 57(1):105–117
Matsuda R, Bandman E, Strohman RC (1983) Regional differences in the expression of myosin light chains and tropomyosin subunits during development of chicken breast muscle. Dev Biol 95:484–491
Midtgard U (1983) Scaling of the brain and the eye cooling system in birds: a morphometric analysis of the *Rete Ophthalmicum*. J Exp Zool 225:197–207
Meier AH, Martin DD (1971) Temporal synergism of corticosterone and prolactin controlling fat storage in the white-throated sparrow, *Zonotrichia albicollis*. Gen Comp Endocrinol 17:311–318
Nagy KA (1980) CO_2 production in animals: analysis of potential errors in the doubly labeled water method. Am J Physiol 238:R465–473
Nisbet ICT, Drury WH Jr, Baird J (1963) Weight-loss during migration Part I: Deposition and consumption of fat by the blackpoll warbler *Dendroica striata*. Bird Banding 34:107–159
Pages T, Planas J (1983) Muscle myoglobin and flying habits in birds. Comp Biochem Physiol A Comp Physiol 74:289–294
Parker GH, George JC (1975) Effects of short and long term exercise on intracellular glycogen and fat in pigeon pectoralis. Jpn J Physiol 25:175–184
Pasquis P, Lacaisse A, Dejours P (1970) Maximal oxygen uptake in four species of small mammals. Respir Physiol 9:298–309
Pennycuick CJ (1969) The mechanics of bird migration. Ibis 111:525–556
Phillips JG, Butler PJ, Sharp PJ (1985) Physiological strategies in avian biology. Blackie, Glasgow
Piersma T (1988) Breast muscle atrophy and constraints on foraging during the flightless period of wing moulting Great Crested Grebes. Ardea 76:96–106
Pietschmann M, Bartels H, Fons R (1982) Capillary supply of heart and skeletal muscle of small bats and non-flying mammals. Respir Physiol 50:267–282
Pinshow B, Bernstein MH, Arad Z (1985) Effects of temperature and P_{CO_2} on O_2 affinity of pigeon blood: implications for brain O_2 supply. Am J Physiol R758–R764
Prinzinger R, Hanssler I (1980) Metabolism – weight relationship in some small nonpasserine birds. Experientia (Basel) 36:1299–1300
Raikow RJ (1985) Locomotor system. In: King AS, McLelland J (eds) Form and function in birds, vol 3. Academic Press, London, pp 57–147
Rayner JMV (1979) A new approach to animal flight mechanics. J Exp Biol 80:17–54
Rollema HS, Bauer C (1979) The interaction of inositol pentaphosphate with the hemoglobins of highland and lowland geese. J Biol Chem 254:12038–12043
Rosser BWC, George JC (1984) Some histochemical properties of the fiber types in the pectoralis muscle of an emu (*Dromaius novaehollandiae*). Anat Rec 209:301–305
Rosser BWC, George JC (1986a) Slow muscle fibres in the pectoralis of the turkey vulture (*Carthartes aura*): an adaptation for soaring flight. Zool Anz 217:252–258
Rosser BWC, George JC (1986b) The avian pectoralis: histochemical characterization and distribution of muscle fiber types. Can J Zool 64:1174–1185
Rothe H-J, Nachtigall W (1980) Physiological and energetic adaptations of flying birds, measured by

the wind tunnel technique. A survey. In: Proceedings XVII International Ornithological Congress. Deutsche Ornithologen-Gessellschaft, Berlin, pp 400–405

Rothe H-J, Biesel W, Nachtigall W (1987) Pigeon flight in a wind tunnel. II. Gas exchange and power requirements. J Comp Physiol B Biochem Syst Environ Physiol 157:99–109

Scheid P (1985) Significance of lung structure for performance at high altitude. In: Ilyichev VD, Gavrilov VM (eds) Acta XVIII Congressus Internationalis Ornithologici, vol 11. Nauka, Moscow, pp 976–977

Schmidt-Nielsen K (1972) Locomotion: Energy cost of swimming, flying and running. Science 177:222–229

Somero GN, Childress JJ (1980) A violation of the metabolism – size paradigm: activities of glycolytic enzymes in muscle increase in larger-size fish. Physiol Zool 53:322–337

Stewart AG (1978) Swans flying at 8,000 metres. Br Birds 71:459–460

Zusuki A (1978) Histochemistry of the chicken skeletal muscles. II. Distribution and diameter of three fiber types. Tohoku J Agric Res 29:38–43

Swan LW (1961) The ecology of the High Himalayas. Sci Am 205:68–78

Talesara GL, Goldspink G (1978) A combined histochemical and biochemical study of myofibrillar ATPase in pectoral, leg and cardiac muscle of several species of bird. Histochem J 10:695–710

Tatner P, Bryant DM (1989) The doubly-labelled water technique for measuring energy expenditure. In: Bridges CR, Butler PJ (eds) Techniques in comparative respiratory physiology. An experimental approach. Cambridge University Press, Cambridge, pp 77–112

Torre-Bueno JR (1976) Temperature regulation and heat dissipation during flight in birds. J Exp Biol 65:471–482

Torre-Bueno JR (1978a) Respiration during flight in birds. In: Piiper J (ed) Respiratory Function in Birds, Adult and Embryonic. Springer, Berlin Heidelberg New York, pp 89–94

Torre-Bueno JR (1978b) Evaporative cooling and water balance during flight in birds. J Exp Biol 75:231–236

Torre-Bueno JR, Larochelle J (1978) The metabolic cost of flight in unrestrained birds. J Exp Biol 75:223–229

Tucker VA (1968) Respiratory exchange and evaporative water loss in the flying budgerigar. J Exp Biol 48:67–87

Tucker VA (1970) Energetic cost of locomotion in animals. Comp Biochem Physiol 34:841–846

Tucker VA (1972) Metabolism during flight in the laughing gull, *Larus atricilla*. Am J Physiol 222:237–245

Turner DL, Butler PJ (1988) The aerobic capacity of locomotory muscles in the tufted duck, *Aythya fuligula*. J Exp Biol 135:445–460

Utter JM, LeFebvre EA (1970) Energy expenditure for free flight by the purple martin (*Progne subis*). Comp Biochem Physiol 35:713–719

Weinstein GN, Anderson C, Steeves JD (1984) Functional characterization of limb muscles involved in locomotion in the Canada goose *Branta canadensis*. Can J Zool 62:1596–1604

West JB (1983) Maximal exercise at extreme altitudes on Mount Everest. J Appl Physiol 55:688–698

Winder WW (1979) Time course of the T_3- and T_4-induced increase in rat soleus muscle mitochondria. Am J Physiol 236:C132–C138

Wind Tunnel Measurements of Long-Time Flights in Relation to the Energetics and Water Economy of Migrating Birds

W. Nachtigall[1]

1 Introduction

Statements on the energetics and water economy of flying birds must be based on reliable measurements from long-distance flights (long-time flights) and gas exchange. These measurements can be made in free-flying birds by radio-tracking (Butler 1985), by catching exhaled air in bags which are then dropped off (Polus 1985), with double-labelled water (LeFebvre 1964) and finally using wind-tunnel methods (Tucker 1968; Torre-Bueno and Larochelle 1978 and others). The apparent advantages of free flight in natural surroundings are limited by two factors: the technical problems of radio-tracking have not yet been completely solved and one cannot be absolutely sure what the bird really does during its flight (exact flight speed, headwinds, gliding phases, turning flight, pauses etc.). With Polus's method (1985), one obtains only a single data point. Thus wind-tunnel experiments are indispensible for many problems. We improved the wind-tunnel technology to attain long-time flights ranging from 1–3 h (Nachtigall and Rothe 1978; Rothe and Nachtigall 1987; Nachtigall 1987).

2 Methods and Material

2.1 Wind-Tunnel Method

The Eiffel-type wind tunnel (Fig. 1) was 9 m long with a flight chamber measuring 1 × 1 × 1.4 m, and reached maximum wind speeds of 24 m s^{-1} (86 km h^{-1}).

The wind speed within the flight chamber varied no more than 3% while turbulence factors were equal to or less than 1.4 so that very good flow conditions were obtained.

2.2 Training and Rearing of Experimental Animals

The following pigeon breeds were tested for their suitability for wind-tunnel flights: English Tippler, Russian Griwuni and Nikolajew, Hannoveranians,

[1] Fachrichtung Zoologie, Universität des Saarlandes, D-6600 Saarbrücken, FRG

E. Gwinner (Ed.) Bird Migration

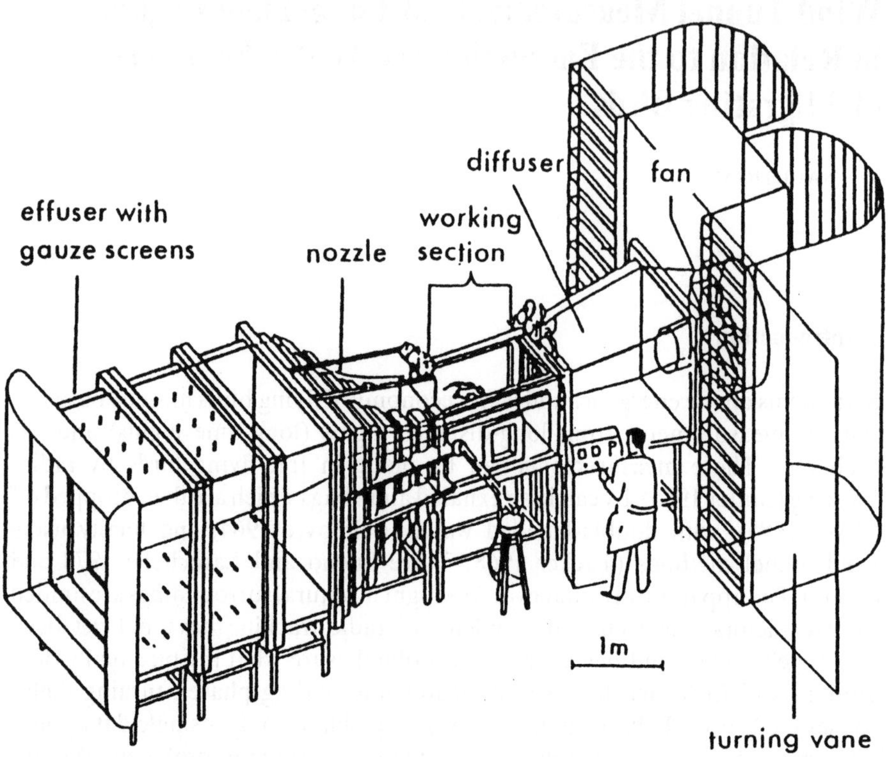

Fig. 1. Wind tunnel used by the work group Nachtigall at the Zoological Institute, University of Saarland, Saarbrücken, to study bird flight (Rothe and Nachtigall 1987)

Viennese and Serbian. Specimens from a new breed derived from Tippler and Griwuni and named Grippler turned out to be the most suitable fliers. These birds were trained, following a five-step program, to fly in the wind tunnel for up to 3 h without a break (equivalent to more than 200 km) while wearing a respiratory mask.

2.3 Flight Energetics

In order to measure the oxygen consumption and carbon dioxide production and calculate the respiratory quotient, part of the exhaled air was sucked (similar to Tucker 1968) from a light mask and dried after measuring the humidity. Then it was pumped into a paramagnetic oxygen analyzer (Servomex-OA 1845) and an infra-red CO_2 analyzer (H + B Uras MT).

3 Results and Discussion

3.1 Type of Fuel in Relation to Flight Duration

Early in flight, pigeons used carbohydrates as fuel (RQ near 1.00), and during approximately 1 h they gradually switched to fat fuel (RQ near 0.7; see Fig. 2A) which they continued to consume for the remainder of the flight. Since the energy content of fat is around 2.5 times that of carbohydrates, it would appear that fat fuel is especially suitable for long-distance flights.

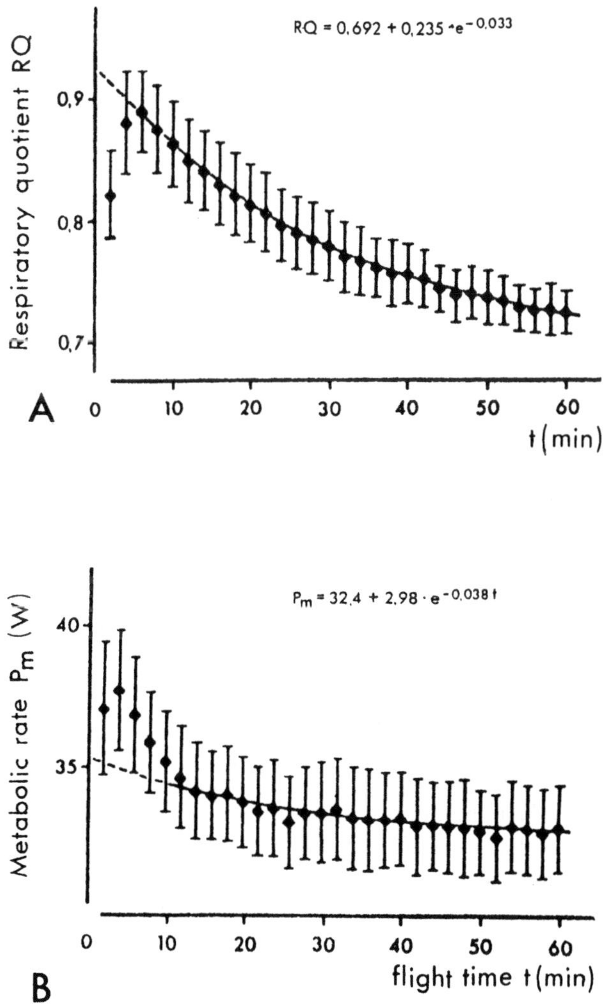

Fig. 2. Respiratory measurements and data of pigeon flight in a wind tunnel plotted as a function of time. *Bars* Standard deviations. **A** Respiratory quotient; **B** metabolic rate (Rothe et al. 1987)

3.2 Metabolic Power in Relation to Flight Duration

Following an initial decrease just after flight began, metabolic rate reached a value of about 32 W within the first hour of flight (Fig. 2B). This gives a specific metabolic power of 100 W kg^{-1} for a body mass of 0.33 kg. With a correction factor of 1.3 (taking into account additional drag of mask etc.; Rothe et al. 1987) this corresponds to a specific metabolic power of 77 W kg^{-1} for a free, long-distance flight out of doors. This means that a pigeon consumes a minimum of 24 W during a long-distance flight at the optimum flight speed of 11 to 12 m s^{-1}. Assuming a muscle efficiency of 0.25 power output equals 6 W. Therefore 18 W are liberated as waste heat.

3.3 Metabolic Power in Relation to Flight Speed

A U-shaped dependency of the metabolic power upon flight speed (Rothe et al. 1987; Fig. 3A) with a minimum at 11 to 12 m s^{-1} was found in agreement with Tucker's (1968 etc.) measurements and Pennycuick's (1968) theoretical approach. Mass loss behaved in a similar manner (Fig. 3B). This represents the fat fuel consumption only under certain marginal conditions (e.g. RQ = 0.72, maintained water homeostasis etc.; Nachtigall 1987). Mass loss as well as waste heat production is minimal at a flight speed of 11 to 12 m s^{-1}. (For studies examining heat dissipation mechanisms during flight see Hirth et al. 1987, and the following sections).

3.4 Mechanisms of Heat Dissipation in Relation to Ambient Temperature

At higher ambient temperatures that proportion of the heat produced that can be dissipated by non-evaporative mechanisms (conduction, convection) declines according to simple physical laws so that evaporative mechanisms (water content of the air exhaled) must be employed to a greater extent (Fig. 4A). The pigeons try to use the safer non-evaporative mechanisms as long as possible. Thus, at higher ambient temperatures, the heat conduction number increases. Flying under air temperatures in the range of $7.5°C < T_a < 25°C$ pigeons let their legs, which are usually held under their feathers at low temperatures, hang down further and further at successively higher temperatures thus using them as coolers (Fig. 4B). Since convection mechanisms of heat dissipation are not physically possible at $T_a > T_b$ (T_b body temperature) evaporative mechanisms must be used to a greater extent. These mechanisms are very effective (1 g water dissipates 2.4 kJ), but are also dangerous because excessive evaporation may lead to dehydration.

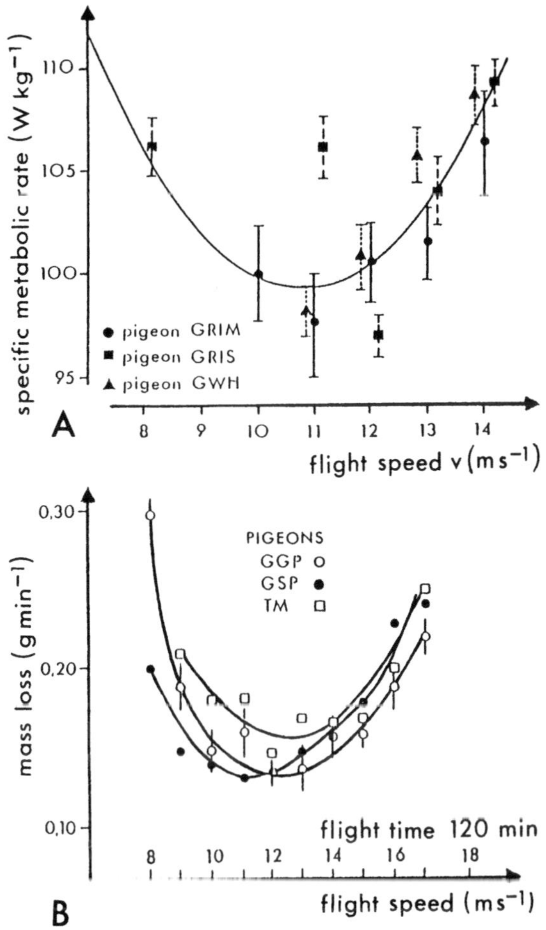

Fig. 3A,B. Metabolic and physiological data as a function of flight speed. **A** Metabolic rate; **B** mass loss (**A** from Rothe et al. 1987; **B** from Nachtigall 1987)

3.5 Limits of Heat Dissipation

The use of fat during flight produces a metabolic energy of 38.9 kJ and 1.06 g metabolic water per gram. Under the conditions discussed the metabolic water is able to dissipate 8.7% of the heat produced; the remaining 91.3% is dissipated by non-evaporative mechanisms (Biesel and Nachtigall 1987). At all ambient temperatures T_a and flight speeds v analyzed ($7.5 < T_a < 25\,°C$; $8\ m\ s^{-1} < v < 16\ m\ s^{-1}$) the water loss necessary for evaporative cooling is higher than the metabolic water production. Only at $T_a \leq 5\,°C$ does metabolic water production equal or exceed water loss due to evaporative cooling (cf. Fig. 4A). In all other cases flight time and thus flight distance is limited by dehydration processes rather than by lack of fuel (cf. Yapp 1956, 1962).

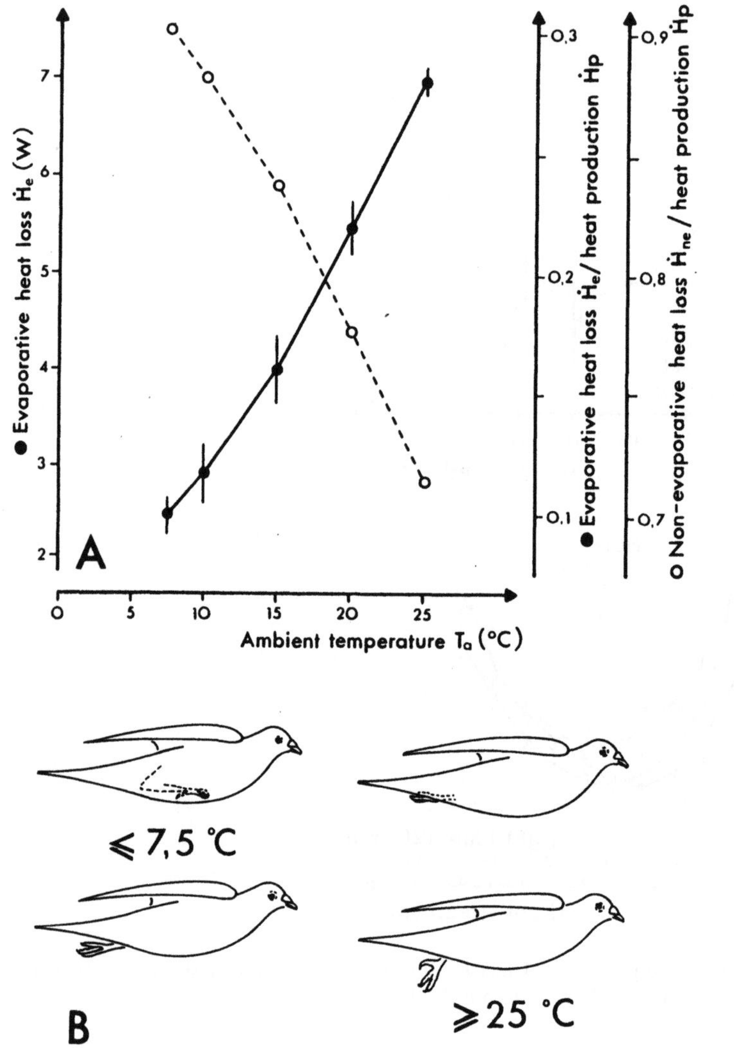

Fig. 4. **A** Evaporative ($\dot{H}_e$) heat loss (*solid line*) and the ratio of evaporative and non-evaporative ($\dot{H}_{ne}$) heat of the total heat production ($\dot{H}_p$) (*solid* and *broken lines*) as a function of ambient temperature. Detailed explanation in text. **B** Changes in leg positions according to ambient temperatures of a pigeon flying in a wind tunnel (Biesel and Nachtigall 1987)

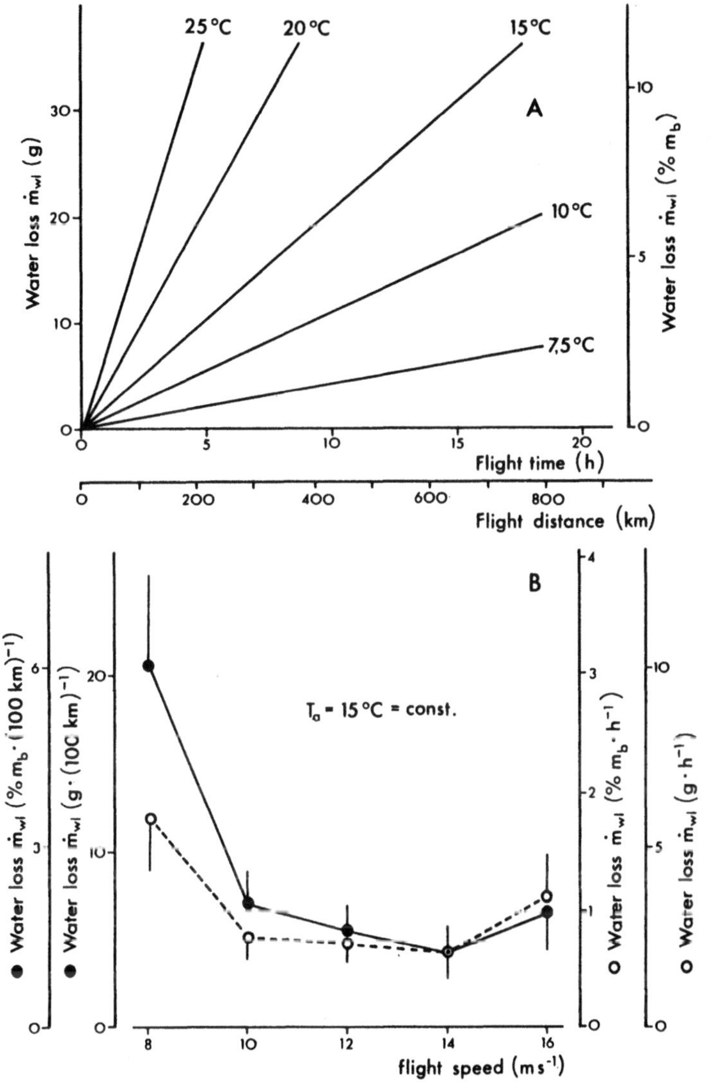

Fig. 5. A Limitation of flight time and distance due to water loss at different ambient temperatures. Detailed explanation in text. **B** Absolute and relative water loss per 100 km flight distance ("Water cost of transport") as a function of flight speed (Biesel and Nachtigall 1987)

3.6 Maximum Flight Times and Distances in Relation to Ambient Temperature

If one assumes that a water loss of 5% of body mass is tolerable (our pigeons only drank after wind-tunnel flights if water loss was more than approximately 3%), then the following marginal values can be expected from well-trained pigeons (Fig. 5A) which are able to fly 20 h non-stop (ca. 900 km at v = 12 m s^{-1}). At T_a = 25 °C, the pigeon would be able to remain for 2 h in the air and fly 90 km; at 10 °C it could fly about 15 h (650 km). Figure 5B shows measurements and calculations for an ambient temperature of 15 °C.

4 Final Remarks

Measurements of this kind can be carried out on pigeons, which are relatively easy to handle and supply us with data on the energetics of bird flight that are important for aspects of bird migration. If ambient temperatures during migration exceed those at which a high proportion of the generated heat can be dissipated non-evaporatively, the birds will be obliged to interrupt their flight despite sufficient fuel supplies. The evolution of mechanisms to reduce the danger of dehydration (e.g. flying at high altitudes and at night) is therefore an inevitable necessity for long-distance fliers migrating over oceans and deserts. Alternatively, a chain of drinking stations at regular intervals would be required. Ecophysiologists have found that hospitable places exist, even in apparently inhospitable regions of the Sahara desert (Bairlein 1985 etc.; Biebach 1985 etc.). If such places were to be destroyed by environmental disturbances, birds might face serious problems on their migrations.

Data obtained from wind-tunnel measurements may compliment other experimental approaches, or may offer the only practicable approach to the study of flight energetics. Data obtained in pigeons which may be one of the few avian species suitable for such studies, can be converted tentatively for birds with different body masses. Wind-tunnel experiments with migrating birds are principally possible as has been found in preliminary flight tests with garden warblers.

References

Bairlein F (1985) Body weights and fat deposition of Palaearctic passerine migrants in the central Sahara. Oecologia (Berl) 66:141–146

Biebach H (1985) Sahara stopover in migratory flycatchers: fat and food affect the time program. Experientia (Basel) 41:695–697

Biesel W, Nachtigall W (1987) Thermoregulation and water homeostasis in flying pigeons. J Comp Physiol B Biochem Syst Environ Physiol 157:117–128

Butler PJ (1985) Telemetry from free flying birds. In: Nachtigall W (ed) BIONA-report 3:391–196. Akad Wiss Mainz; Fischer, Stuttgart New York

Hirth D, Biesel W, Nachtigall W (1987) Pigeon flight in a wind tunnel. III. Regulation of the body temperature. J Comp Physiol B Biochem Syst Environ Physiol 157:111–116

LeFebvre EA (1964) The use of D_2 O^{18} for measuring energy metabolism in *Columbia livia* at rest and in flight. Auk 81:403–416

Nachtigall W (1987) Vogelflug und Vogelzug. Rasch & Röhring, Hamburg

Nachtigall W, Rothe HJ (1978) Eine Methode, Tauben für den freien Flug im Windkanal zu trainieren. Naturwissenschaften 65:266

Pennycuick CJ (1968) Power requirements for horizontal flight in the pigeon *Columba livia*. J Exp Biol 49:527–555

Polus M (1985) Quantitative and qualitative respiratory measurements on unrestrained free-flying pigeons by AMACS (airborne measuring and control systems). In: Nachtigall W (ed) BIONA-report 3:293–301. Akad Wiss Lit Mainz; Fischer, Stuttgart New York

Rothe HJ, Nachtigall W (1987) Pigeon flight in a wind tunnel. I. Aspects of wind tunnel design, training methods and flight behaviour of different pigeon races. J Comp Physiol B Biochem Syst Environ Physiol 157:91–98

Rothe HJ, Biesel W, Nachtigall W (1987) Pigeon flight in a wind tunnel. II. Gas exchange and power requirements. J Comp Physiol B Biochem Syst Environ Physiol 157:99–109

Torre-Bueno JR, Larochelle J (1978) The metabolic cost of flight in unrestrained birds. J Exp Biol 75:223–229

Tucker VA (1968) Respiratory exchange and evaporative water loss in the flying budgerigar. J Exp Biol 48:67–87

Yapp WB (1956) Two physiological considerations in bird migration. Wilson Bull 68:312–319

Yapp WB (1962) Some physical limitations on migration. Ibis 104:86–89

Chapter 5 Strategies and Tactics of Migration

Optimal Bird Migration: The Relative Importance of Time, Energy, and Safety

T. ALERSTAM and Å. LINDSTRÖM[1]

1 Introduction

"Optimization is the process of minimizing costs or maximizing benefits, or obtaining the best possible compromise between the two. Evolution by natural selection is a process of optimization" (R. McNeill Alexander 1982).

The use of optimization methods is not new in bird migration research. The theory of bird flight, as first developed by Pennycuick (1969, 1975) and Tucker (1974), has been and still is of great importance as a basis for interpreting flight adaptations in migratory birds. The use of wind by migrating birds, and the problem of drift and compensation, have been elucidated by simple optimization arguments (Alerstam 1979a,b). An optimality approach also underlies evaluations of the evolution of geographical migration patterns and partial migration, involving assumptions about frequency-dependent selection processes, but in this contribution we will restrict ourselves to consider adaptations that help birds to accomplish their migratory journey as successfully as possible.

What are the critical costs and benefits for migratory birds? Perhaps it is of foremost importance to travel as fast as possible, being rewarded by arriving earlier at the destination than many competitors. Alternatively, if food resources are sparse, the highest premium may be on energy economy during migration. A third possibility is that predation constitutes such a significant hazard to migrants that they are primarily adapted to minimize the associated mortality risk during migration.

We will show that predicted migratory adaptations, in terms of fat deposition strategies, flight behaviour as well as habitat selection, differ depending on which of these optimization criteria is valid. Appropriate tests will make it possible to distinguish which selective forces – time, energy or predation minimization – are primarily responsible for the migratory habits of different species or categories of individuals under different seasons and environmental conditions. Much of our reasoning is influenced by developments in foraging theory (cf. Stephens and Krebs 1986).

[1] Department of Ecology, University of Lund, Ecology Building, S-223 62 Lund, Sweden

E. Gwinner (Ed.) Bird Migration

2 Optimal Migration

2.1 Fat Load and Flight Distance

The potential flight distance a bird can attain will be a negatively accelerated function of its fat load. This function may be estimated on the basis of flight mechanical theory (Pennycuick 1975). During migratory flight a bird's body mass (M) becomes progressively reduced as fat reserves are used up. The increment in flight distance (Y) with the decrement in body mass (M) due to fat consumption will follow the condition:

$$dY/dM \propto -M^{-1} (L/D), \quad (1)$$

where (L/D) is the effective lift:drag ratio (cf. Pennycuick 1975: Eq. 34). Assuming that the lift:drag ratio decreases with increasing fat reserves in proportion to $M^{-1/2}$, as would be expected for an ideal bird (Pennycuick 1975), the following proportionality emerges:

$$dY/dM \propto -M^{-3/2}. \quad (2)$$

Integrating this equation with respect to M gives the total flight distance a bird can attain on a given fat load:

$$Y \propto [1 - (1+f)^{-1/2}], \quad (3)$$

where f is the relative amount of fat at departure, expressed as the ratio of fat mass to lean body mass of the bird.

Although we use the term fat load for the amount of fuel stored relative to lean body mass, it is of no critical importance that all of the fuel mass consists of fat. In fact, our reasoning and predictions apply with f corresponding to any type of excess mass stored and used as fuel for the migratory flights.

The relationship between potential flight distance and fat load according to Eq. (3) is shown in Fig. 1. The utility of additional fat decreases with an increasing fat load, because heavy fat reserves cause high flight costs. By way of example, a

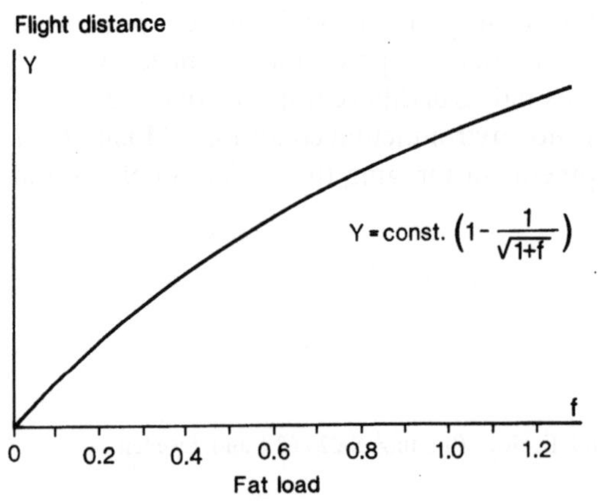

Fig. 1. Flight range in still air as a function of fat load (relative to lean body mass) at departure. The proportionality constant in this function differs between species and individuals depending on size, morphology and flight efficiency

given amount of additional fat will allow a bird that already has deposited fat reserves equal to its lean body mass (f = 1) to fly an extra distance of only 35% [$= 2^{-3/2}$, Eq. (2)] of the corresponding distance attained without any prior fat reserves.

The relationship in Fig. 1 is fundamental for the derivations and predictions of optimal fat accumulation strategies in this work. It should be pointed out that the above-mentioned assumption about a reduction in effective lift:drag ratio with increasing fat load is somewhat uncertain and provisional (Pennycuick 1975, p. 18–20). However, even if the reduction in the effective lift:drag ratio with increasing fat load is less pronounced than assumed, our qualitative conclusions will not be affected, and the quantitative predictions will be modified only to a small degree.

2.2 Fat Deposition Rules for Maximum Migration Speed

2.2.1 Optimal Strategies Under Different Average Conditions

A bird accumulating fat at a constant rate k, corresponding to a daily gain in fat mass relative to lean body mass, will increase its fat load with fat deposition time t (days) as f = kt. Inserting this into Eq. (3) gives

$$Y \propto [1 - (1+kt)^{-1/2}]. \tag{4}$$

This equation allows us to calculate the gain in potential flight distance over t as illustrated in Fig. 2 for different rates of fat accumulation.

The rate of fat deposition achieved by a migrant at a stopover site may become reduced with time because of resource depression (Charnov et al. 1976), and because of increased energy costs during foraging as the bird has to carry progressively heavier burdens of fat reserves. These effects will add to the negative acceleration of the Y(t) functions. Since there are no empirical data to indicate the possible importance of these effects, we restrict ourselves to consider cases where birds maintain a constant rate of fat accumulation.

The optimal strategy, in terms of stopover time and departure fat load, for maximizing the overall speed of migration, depends on the average expected rate of fat deposition (k) at stopover sites along the migration route as well as on the average expected time cost for search/settling (t_o), before efficient fat deposition can start, at each new stopover site. The optimal time of fat deposition at each stopover site, maximizing total speed of migration, can be found by solving:

$$d\,[Y(t+t_o)^{-1}]/dt = 0. \tag{5}$$

Differentiating and rearranging, this equation is satisfied when:

$$dY/dt = Y/(t+t_o). \tag{6}$$

This condition can be found graphically by drawing a tangent to the Y(t) curve from the relevant point on the search/settling time axis, as illustrated in Fig. 2 for

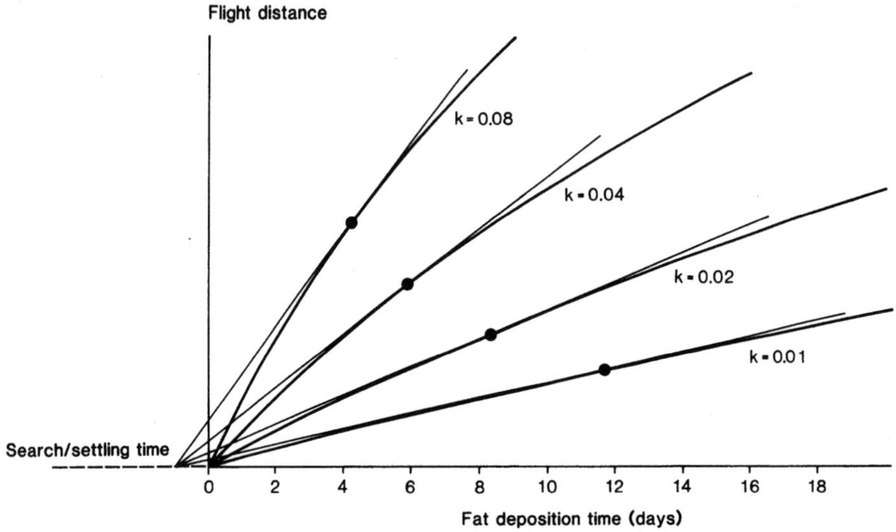

Fig. 2. Gain in potential flight distance with time for different rates of fat deposition (k = daily gain in fat mass relative to lean body mass). Given a certain search/settling time, optimal stopover time and flight range can be found by drawing a tangent to the gain curves, identifying the points (*solid dots*) at which speed of migration (ratio distance: time) is maximized. The search/settling time is indicated along the abscissa, increasing from the origin to the left

different average rates of fat deposition. The slope of this tangent shows the maximum speed of migration.

Combining Eqs. (4) and (6) the optimal fat load at departure, $\hat{f}(= k \cdot \hat{t})$, and the optimal staging time, $\hat{t} + t_0$, for maximizing the total speed of migration may be calculated for migrants travelling under different regimes of typical k and t_0 at stopover sites along their migration route, as presented in Fig. 3. It will be noticed that the proportionality constant in Eq. (4) disappears in these calculations, and the quantitative predictions in Fig. 3 apply to widely different types of birds, irrespective of size, morphology or flight efficiency.

2.2.2 *Effect of Wind, Flight Duration and Fat Loss During Search/Settling*

The above predictions are valid under calm conditions or when an equal balance between tail- and headwinds is to be expected on the different flights along the migration route. Wind conditions will modify optimal strategies to a small degree, because of differences in expected flight speed for birds with different fat loads (cf. Pennycuick 1978). Migrants which can expect wind assistance during their migration flights will have somewhat smaller optimal fat loads and shorter staging times than predicted in Fig. 3, and, conversely, slightly larger fat loads and longer staging times will be optimal for migrants travelling in predominating headwinds.

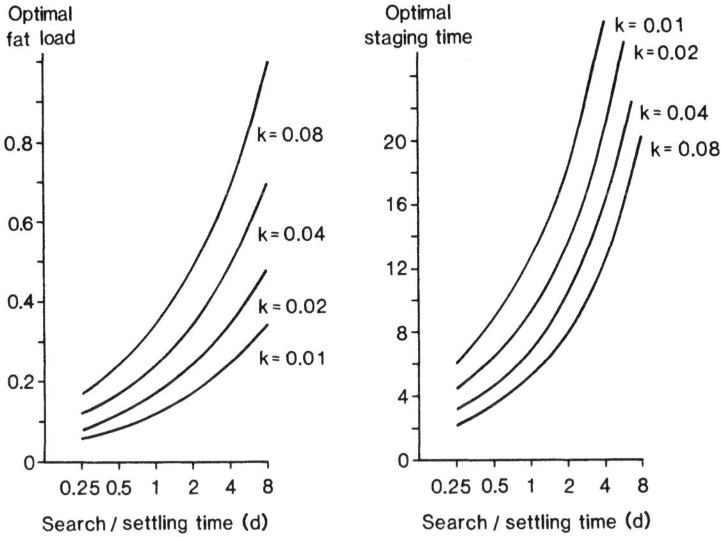

Fig. 3. Optimal departure fat load (relative to lean body mass) and staging time (days), maximizing total speed of migration, for birds travelling under different average conditions of expected fat deposition rate (k = daily gain in fat mass relative to lean body mass) and search/settling time cost at stopover sites along the migration route

In the above calculations of total speed of migration, flying time, which is short in comparison with staging time, has been neglected. This is permissible if the migratory flights are rapidly completed during "free" time that cannot be used for efficient fat deposition or search/settling, as is the case for many nocturnal migrants. However, if flight time entails lost opportunities for foraging and fattening, as in diurnal migrants, the time costs for flying should be taken into account. Doing so, the gain in flight distance should be related to the total time for fat deposition plus flight rather than to fat deposition time alone. The effect will be to reduce optimal fat load and staging time in comparison with the predictions in Fig. 3. However, the differences will be small, except with a very high rate of fat deposition. By way of example, with $t_0 = 2$ and $k = 0.02$ $\hat{f}$ becomes reduced from 0.24 to 0.22 and $\hat{t}$ from 12 to 11 days by incorporating the influence of flight time, while the corresponding effect with $t_0 = 2$ and $k = 0.08$ will be to reduce $\hat{f}$ from 0.48 to 0.37 and $\hat{t}$ from 6 to 4.6 days.

In many cases it is probably adaptive for migrants not to use up all fat during the flights, but to land with a safety margin of fat reserves, sufficient to cover energy losses before efficient refuelling can start at the new stopover site. Such an expected fat cost for search/settling, f_0, in addition to the above-mentioned time cost, t_0, will lead to larger optimal departure fat loads and longer staging times than indicated in Fig. 3. Graphically the optimal solution for maximizing migration speed is found by shifting the abscissa $Y(f_0)$ distance units upwards and drawing a tangent to the distance gain curve from t_0 on the new abscissa (cf. Fig. 2).

2.2.3 Stopover Decisions

The above considerations show how average conditions along the migration route determine the overall optimal fat accumulation strategy. Under which circumstances should a bird stop for fat deposition and at which fat load should it continue its journey, when encountering stopover sites of different qualities, allowing faster or slower fattening rates than the expected average rate? The answer accords with the marginal value theorem (Charnov 1976; Stephens and Krebs 1986): In order to maximize migration speed, a migrant should depart from a stopover site when the marginal rate of gain in flight distance drops to the expected average speed of movement along the migration route.

By differentiating Eq. (4) it could be shown that a migrant with fat status f_i, encountering a site allowing a fat deposition rate k_i, should stop for fat accumulation if:

$$k_i(1+f_i)^{-3/2} > k(1+\hat{f})^{-3/2}, \qquad (7)$$

where k is the expected fattening rate at alternative staging sites and $\hat{f}$ is the optimal departure fat load under average conditions, as defined earlier. If this condition is not met the bird should move on to alternative staging sites. Similarly, a migrant should leave a refuelling site when f_i has increased or k_i decreased to the extent that the above condition is no longer fulfilled. Different conditions, in terms of k_i and f_i, when a migrant should decide to stop for fat deposition or continue migration are exemplified in Fig. 4.

These results show that in time-selected migration, stopover decisions should be governed by a combination of two major factors, the migrant's actual fat status and the prospects for rapid fat deposition at the potential stopover site.

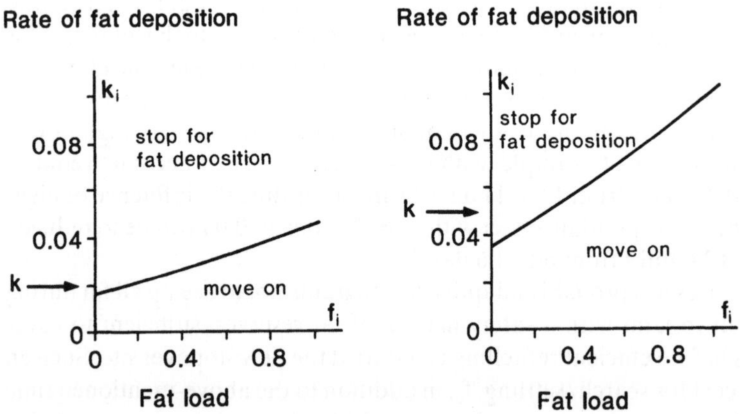

Fig. 4. Stopover decision for maximizing migration speed is determined by the bird's actual fat status, f_i, and by the rate of fat deposition, k_i, at the potential stopover site. Two different examples are illustrated for migrants with an expected average rate of fat deposition along the migration route, k = 0.02, and with expected t_o = 1 day (giving $\hat{f}$ = 0.17), and for migrants with k = 0.05 and t_o = 1 day ($\hat{f}$ = 0.26), respectively

2.3 *Fat Deposition Rules for Minimum Energy Expenditure*

The cost for migration that appears in the energy budget of a migratory bird over and above the permanent existence metabolism costs, to a large extent is reflected by the cumulative amount of fat deposited and consumed during migration. Using minimum total fat consumption during the migratory journey as an optimization criterion for energy-selected migration, the optimal departure fat load, associated with the maximum ratio of distance:fat, may be determined from the relationship:

$$dY/df = [Y - Y(f_0)]/f, \qquad (8)$$

where f_0 = expected fat loss during search/settling at a new stopover site, and the distance Y is a function of fat load f according to Eq. (3). The graphical solution is illustrated in Fig. 5, as are the resulting quantitative predictions of optimal departure fat load, $\hat{f}$, for different fat costs during search/settling.

Predicted fat loads are independent of fat deposition rate and various time costs. However, wind will affect the predictions, so that migrants expecting wind assistance during their migratory flights should depart with smaller fat reserves than those which often have to travel into opposing winds.

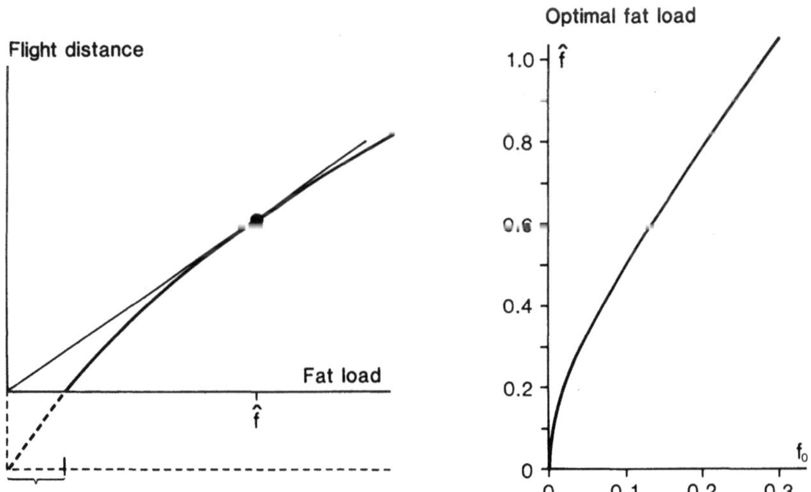

Fig. 5. Optimal fat load for minimizing energy/fat expenditure during migration. Given an expected fat loss during search/settling at each new stopover site (f_0), the point (*solid dot*) of departure fat load/flight range at which the ratio distance:fat is maximized, can be found by drawing a tangent to the distance gain curve (cf. Fig. 1) as shown. Resulting predictions of optimal fat load at departure are presented for different f_0 to the *right*. Fat load (f) and fat loss (f_0) are expressed relative to lean body mass

2.4 Flight Speed

Not only fat deposition strategies but also flight speeds are expected to differ between birds minimizing energy expenditure or duration of their migratory journey. Energy cost for flight transport over distance Y equals:

$$E = YP/V, \tag{9}$$

where P and V refer to flight power and speed, respectively. P is a function of V according to the power curve for flapping flight (Pennycuick 1975). Solving for minimum energy cost by setting the derivative with respect to flight speed equal to zero, gives the relationship

$$dP/dV = P/V, \tag{10}$$

which is the usual condition for the maximum range speed V_{mr} (Pennycuick 1975).

Adding up flying time and staging time gives the total duration of migration:

$$T = Y/V + YP/VK, \tag{11}$$

where K is the rate of energy/fat gain (converted into the same units as P) at the stopover sites. Solving for T_{min} by differentiating with respect to V gives

$$dP/dV = (P+K)/V. \tag{12}$$

This condition is fulfilled at a higher flight speed than V_{mr} as can be seen from Fig. 6. A similar derivation applies also to the optimal flight speed in birds when feeding young (Norberg 1981). Flight speeds associated with E_{min} (i.e. V_{mr}) and

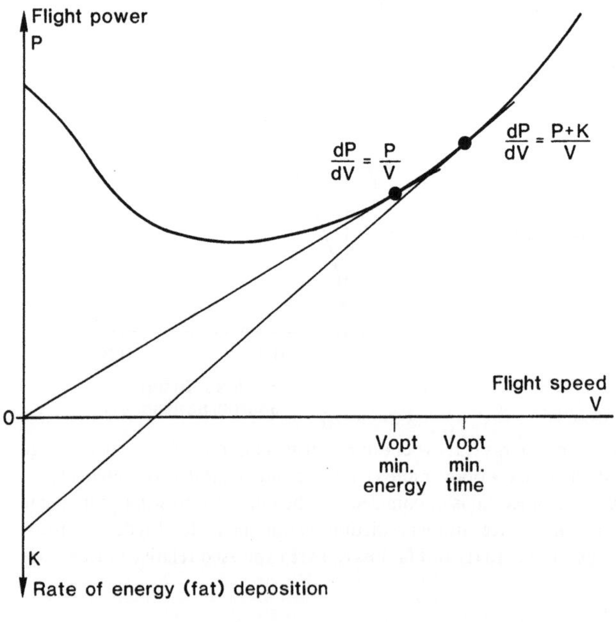

Fig. 6. Optimal flight speed for minimizing total cost of energy or time, respectively, for the migratory journey, can be found from the power curve for flapping flight by drawing tangents according to the different conditions in Eqs. (10) and (12), respectively

T_{min}, respectively, are affected by winds in a similar way; optimal speed increases with headwinds and decreases with tailwinds (Pennycuick 1978; Alerstam 1985). Hence, detailed radar measurements of the flight speed of birds (cf. Alerstam 1981, 1985; Schnell and Hellack 1979) may help to reveal the importance of time minimization during migration.

Cross-country soaring normally is slower than flapping flight, but may still be advantageous from an energetic point of view if thermal conditions are sufficiently good to yield a lower cost of transport (ratio P:V) than in flapping flight. Referring to the graphical presentation in Fig. 6, cross-country soaring is beneficial if the associated point P,V falls to the right of the tangent from the origin to the power curve for flapping flight. In comparison, to be advantageous for maximizing migration speed, the point P,V for cross-country soaring flight must fall to the right of the tangent from the energy deposition axis to the power curve (Fig. 6). Hence, in time-selected migration the probability that cross-country soaring migration will be advantageous over flapping flight becomes a bit more restricted, to an increasing extent with an increasing average rate of fat deposition.

2.5 *Predictions for Time- and Energy-Selected Migration*

Besides the quantitative predictions about optimal fat loads and stopover times, presented in Figs. 3, 4 and 5, the most important divergent predictions for time- and energy-selected migration are summarized in Table 1.

The differences in predictions generally fall back on the basic fact that birds maximizing migration speed are expected to be sensitive to variation in the rate of fat deposition (k), while this is not so for migrants minimizing energy expenditure. Of course, it is unrealistic to claim that energy-minimizers behave without paying any regard whatsoever to k. One must assume that they avoid the most unsuitable stopover habitats where fattening rate is very slow and they incur

Table 1. Some divergent predictions for time- and energy-selected migration

	Time-selected migration	Energy-selected migration
1. Departure fat load	$\hat{f}$ increasing with average k	$\hat{f}$ independent of k
	$\hat{f}_i$ increasing with stopover site quality (k_i)	
	$\hat{f}$ dominants (high k) $>$ $\hat{f}$ subdom. (low k)	$\hat{f}$ dominants (small f_0) $<$ $\hat{f}$ subdom. (large f_0)
	$\hat{f}$ dominants (short t_0) $<$ $\hat{f}$ subdom. (long t_0)	
	$\hat{f}$ diurnal migrants $\leq$ $\hat{f}$ nocturnal migrants	$\hat{f}$ independent of diurnal/ nocturnal flights
2. Stopover decision	Dependent on f_i and k_i	Independent of k_i
3. Flight speed	$\hat{V} > V_{mr}$, increasing with k	$\hat{V} = V_{mr}$, independent of k
4. Flight behaviour	Soaring migration less advantageous with increasing k	Advantage of soaring migration independent of k

the risk of not being able to complete their journey in time during the migration season. In many cases it is not obvious how a migrant benefits from minimal energy expenditure per se. Associated gains in fitness are indirect and mainly related to a reduction in necessary foraging time under hazardous conditions. Below we will consider the importance of predation more closely.

Of particular interest is the contrasting predictions about optimal fat loads in dominants and subdominants. If dominance is associated with an enhanced rate of fat accumulation one should expect that, among time-minimizers, dominants migrate with larger fat loads than subdominants. In contrast, for energy-minimizers, where the primary benefit for dominants is a reduced energy cost for search/settling, the expected difference in fat load between dominants and subdominants is exactly the opposite. One should note, however, that the latter relationship may occur also in time-selected migration, if the major benefit for dominants is not a faster rate of fattening, but a reduced time cost for search/settling.

Under identical conditions (t_o, f_o, k), time-selected migration is associated with a larger optimal fat load than energy-selected migration.

2.6 *Response to Variability*

Are migrating birds expected to be sensitive not only to mean availability but also to variability in resources? If so, under which conditions should migrants minimize uncertainty in fat deposition rates and time delays and when should they prefer uncertain stopover sites and variable food resources?

Caraco et al. (1980) demonstrated that juncos *Junco phaeonotus* prefer constant reward (risk aversion) when their expected daily energy budget is positive, while they change to preference for variable rewards (risk proneness) with a negative energy budget, according to the "expected daily energy budget rule" (cf. Real and Caraco 1986; Stephens and Krebs 1986).

In time-selected migration there may be a disproportionately high premium on early arrival at the destination, before the majority of conspecific competitors arrive. In such a case, the utility function of migration speed (which in turn is closely correlated with fat deposition rate) is convex, and risk-prone foraging during fat deposition will be predicted.

In energy-selected migration, risk proneness could be advantageous if the expected energy budget for the migratory journey is insufficient (during the allotted migration season). If individuals, for which a risk-prone decision is optimal, show flexibility and change to risk aversion if they succeed in obtaining sufficient resources, the probability for these birds of avoiding a shortfall may rise substantially above 50% (Stephens and Krebs 1986).

To have surplus energy reserves after completion of the migratory flight is of utmost importance for some birds, not least migrants arriving in spring at arctic breeding sites. During fat accumulation before such a flight, risk-prone foraging may be optimal under certain premises, according to the "achieved mean effect" (Stephens and Krebs 1986).

Moore and Simm (1986) demonstrated by feeding experiments that yellow-rumped warblers *Dendroica coronata* preferred variable food rewards while accumulating migratory fat reserves, whereas they chose constant rewards when not in migratory disposition. Furthermore, individuals in migratory disposition switched from being risk-prone to risk aversion when they had deposited maximum fat reserves. These interesting findings indicate that bird migration indeed is associated with specific risk-sensitive foraging preferences.

Birds migrating and foraging solitarily, and with the aim of establishing a territory at the destination, may be especially inclined towards attaining an exceptionally high migration speed through a risk-prone foraging behaviour. In contrast, preference for low variability may be associated with migration and foraging in flocks (cf. Ekman and Hake 1988).

2.7 Fat Deposition and Predation Hazard

For flight mechanical reasons one can expect that fat birds are more vulnerable to predation than lean ones. Estimating the effect of an added weight of fat reserves on flight performance, in a similar way as Andersson and Norberg (1981) investigated relations between overall body size and flight, one finds that, according to aerodynamic theory, flight capacity will become impaired in several important ways. The capacity for acceleration as well as the maximum possible flight speed will be reduced with increasing fat reserves. Furthermore, rate of climb becomes restricted to an increasing extent and manoeuvrability, with respect to maximum roll acceleration or minimum turning radius, will be impeded. Empirical results supporting these conclusions have been reported by Blem (1975) and Videler et al. (1988).

Hence, there are good grounds to assume that increasing fat loads in birds bring as a consequence a clearly diminishing chance of escaping a pursuing predator in flight. To militate against mortality risks birds should be inclined to depart with smaller fat reserves than is optimal in time-selected migration (cf. Lima 1986).

In predation-selected migration, birds should behave as to keep the total mortality risk during the entire migratory journey as small as possible. This is to say that mortality per distance covered should be minimized.

A bird accumulating fat at a certain rate will increase its potential speed of migration, $s(=Y/(t+t_0)$; cf. Fig. 2), with increasing fat load up to a maximum speed. With a further increase in fat load, s will fall off again. Of course, the fat level associated with maximum speed of migration corresponds to the optimal departure fat load, $\hat{f}$, in time-selected migration, as derived earlier. If predation risk is positively correlated with fat load, one will expect that mortality rate, m, is related to potential speed of migration, s, as illustrated in Fig. 7. Mortality per distance covered is given by m/s. Minimum for this quotient is associated with the condition:

$$dm/ds = m/s, \tag{13}$$

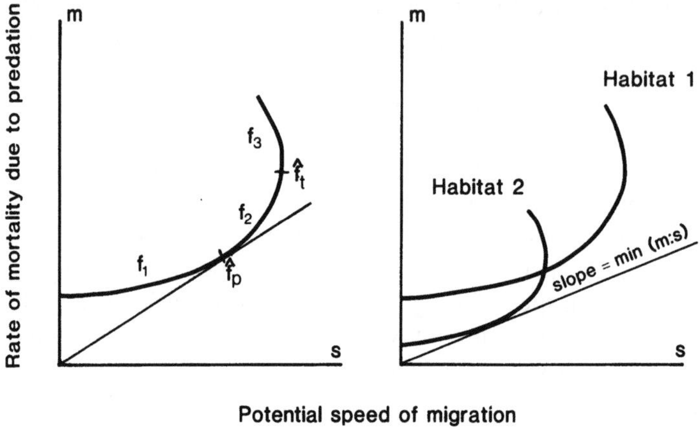

Fig. 7. Hypothetical relationship between rate of mortality due to predation (*m*) and potential speed of migration (*s*). The associated departure fat loads, f_i as indicated on the curve, increase with increasing i. Hence, optimal fat load for minimizing predation during migration, $\hat{f}_p$, is smaller than optimal $\hat{f}_t$ for minimizing migration time. Comparing habitats with different levels of predation and fattening rates, birds in predation-selected migration are expected to choose the habitat associated with the minimum possible ratio of m to s, as illustrated to the *right*

and can be found graphically by drawing a tangent from the origin to the m(s)-curve, as shown in Fig. 7.

This solution demonstrates that optimal departure fat load, to maximize safety from predation during the migratory journey, is clearly smaller than the corresponding fat level for maximizing migration speed. Furthermore, as also shown in Fig. 7, birds in predation-selected migration are expected to switch from staging sites allowing a fast rate of fat accumulation to more protected sites with a slower rate of fat gain according to which sites or habitats provide the minimum possible ratio of m to s (cf. Gilliam and Fraser 1987).

3 Patterns of Fat Accumulation in Migrating Birds

3.1 Fat Loads and Deposition Rates

Compiling data from many different studies gives overall distributions of estimated departure fat loads for waders and different categories of passerine migrants, as presented in Fig. 8 (recalculated from Lindström 1986). In the vast majority of cases, fat loads have been estimated from weight data, assuming that the mass gain, which is added to the typical lean body mass prior to departure, corresponds to fat reserves. For migrants which put on fat and leave a staging site almost synchronously, as happens in many waders, we have used the population mean of final body mass before departure to reflect the fat load. For migrants studied at a site where they arrive, put on fat and depart asynchronously, we have estimated departure fat load primarily on the basis of the mean body mass or mass

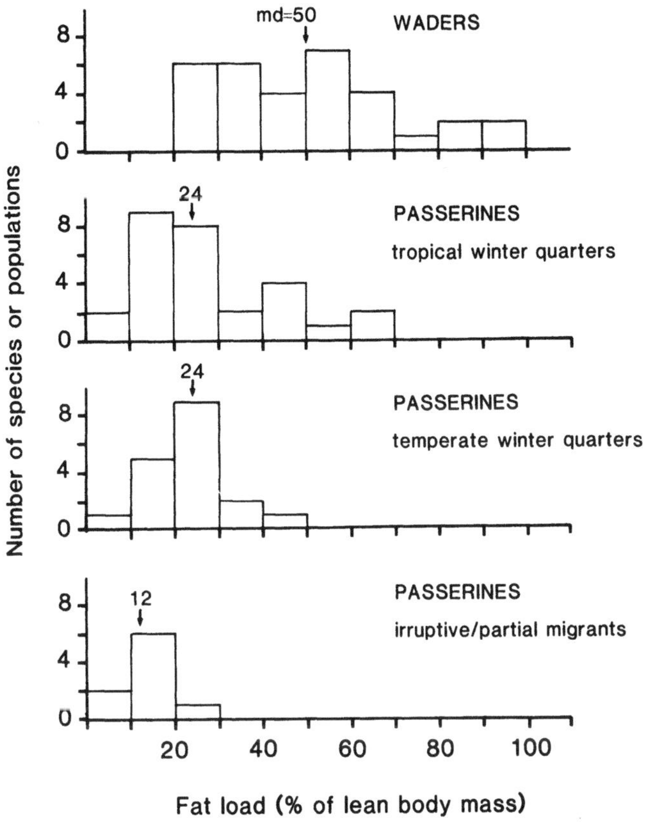

Fig. 8. Distributions of estimated fat loads in migrating waders and passerines. Fat loads from studies of many different species or populations, at different places and seasons, are included, with the exception of data on fat reserves associated with long flights across obvious ecological barriers

increase for the 25% fraction of heaviest/fattest individuals, or on the basis of the total mean body mass plus one standard deviation for the whole sample of individuals.

Our predictions about fat loads do not apply to migrants confronted with wide ecological barriers devoid of suitable refuelling sites, and such cases have been omitted from the data in Fig. 8. At ecological barriers, migrants are forced to store extra large fat reserves to accomplish the impending long flights. Hence, most passerines departing across the Sahara desert and the Mediterranean Sea, the Gulf of Mexico or the West Atlantic Ocean have fat loads (mass of fat reserves relative to lean body mass) between 40 and 70% (median = 50%, n = 63), approaching 90% in a few species. Similarly, waders attain very large fat loads, in the range 50 to 90% (median = 66%, n = 17) with maximum about 100%, before long, non-stop flights across the West Atlantic Ocean, the North Atlantic including the Greenland ice cap, or the Pacific Ocean.

Figure 8 shows that partial migrants, including different species of tits *Parus* sp., goldcrest *Regulus regulus* and redpoll *Carduelis flammea*, travel with the smallest fat reserves. The majority of regular passerine migrants depart with fat loads amounting to 20 or 30% of their lean body weight, although there is a larger scatter with more cases of relatively heavy fat loads in long-distance migrants as compared to species wintering in the temperate zone. Waders, of which the great majority of species or populations included in Fig. 8 refer to long-distance migrants, show the largest fat loads.

Reported average rates of fat deposition by migrating birds, expressed as the daily (24 h) gain in mass relative to lean body mass, range from less than 1 up to 7% (n = 58 different species or populations). The range is much the same for passerines and waders, but the median rate is somewhat higher for passerines (2.4%, n = 31) than for the waders (1.3%, n = 27).

Many investigators have noted that there is often a time delay before a migrant, newly arrived at a stopover site, starts to put on fat and gain in mass. During this time delay birds may also lose significantly in weight. For passerines there are enough data to indicate that mean time delay most often is 1 or 2 days (overall average 1.5 days, n = 11), and mean loss of mass (fat) during this period ranges from 0 to 13% of lean body mass (mean 4.4%, n = 11).

With the above estimates in hand, we can make a provisional comparison between departure fat loads predicted by our models for time- and energy-selected migration and observed fat loads. With a fat deposition rate at 2% ($k = 0.02$) and a time cost for search/settling at 2 days ($t_o = 2$) the predicted optimal fat load in time-selected migration is 0.24 (cf. Fig. 3). The corresponding prediction for energy-selected migration, with 5% expected fat loss during search/settling at a new refuelling site ($f_o = 0.05$), is 0.32 (cf. Fig. 5). These predictions accord well with observed fat loads among passerines (Fig. 8), and we conclude that our models yield realistic predictions. Critical tests as to which of the two models provides the best quantitative fit to observations must be based on detailed empirical data for single populations.

The large fat reserves among migrating waders are surprising in the light of the modest rates of fat accumulation found in many studies. Provided that available data correctly reflect characteristic fattening rates and fat levels, the observed large fat loads can only be explained, on the basis of our models, if waders are subjected to extensive time and/or energy costs during search/settling. Sites offering favourable staging conditions may be far apart, making long, energy-demanding flights between traditional resting sites advantageous.

3.2 Correlation Between Rate of Gain and Departure Load

There is a significant positive correlation between fat load at departure and fat deposition rate in both waders (Spearman rank corr. coeff. $r_s = 0.57$, $p < 0.05$, n = 15) and passerines ($r_s = 0.79$, $p = 0.01$, n = 10), as seen in Fig. 9. This indicates that there is an important element of time reduction involved in bird migration (cf. Table 1).

Comparing individuals from the same species and population, and of similar status (and thus with the same expected speed of migration), the model for

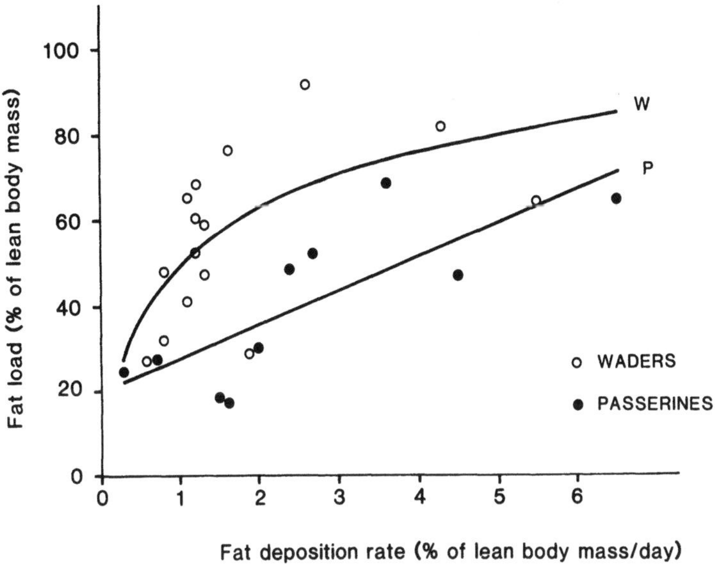

Fig. 9. Relationship between estimated fat load at departure and fat deposition rate for different species and populations of migrating waders and passerines, respectively. Lines of logarithmic and linear regression, as illustrated, provide the best fit for the two data sets

time-selected migration can be used to calculate the predicted relationship between fat load at departure (f) and rate of fat gain (k) as

$$dY/dt \propto k(1+f)^{-3/2} = \text{constant} \tag{14}$$

(cf. Sect. 2.2.3). The detailed study by Carpenter et al. (1983) of the weight gain in different individuals of migrant rufous hummingbirds *Selasphorus rufus* comes closest to fulfilling these requirements. Unfortunately, the number of individuals are too few ($n = 7$) to allow a critical numerical evaluation, but a positive correlation between f and k for the different individuals is indicated.

Biebach (1985) demonstrated by laboratory experiments that migratory activity in spotted flycatchers *Muscicapa striata* was governed by a combination of fat status and feeding possibilities.

3.3 Adult and Juvenile Migrants

There are a few studies indicating significant differences in fat load between juveniles and adults of the same species. Yearling chaffinches *Fringilla coelebs* seem to travel with somewhat larger fat reserves, 21–28% of lean body mass, than adult conspecifics (with average fat loads of 15–17%) during spring and autumn migration in southern Sweden (Lindström, unpublished work). A similar situation, with a mean fat load of about 20% in adults and 30% in juveniles, holds for yellow-rumped warblers on autumn migration in New Jersey (Murray 1979) and western sandpipers *Calidris mauri* in southwestern Canada (Butler et al. 1987). This kind of difference between adults and juveniles may occur both in time- and energy-selected migration (cf. Table 1).

There are also examples of the reverse relationship. Adults seem to put on larger fat reserves than juveniles among pied flycatchers *Ficedula hypoleuca* during autumn in Spain (mean fat loads of about 43 and 33%, respectively, Veiga 1986), among semipalmated sandpipers *Calidris pusilla* in Canada (67 vs 44%, Page and Middleton 1972) and curlew sandpipers *C. ferruginea* (96 vs 64%, Stanley and Minton 1972) in England. Such differences are expected in time-selected migration, if adults achieve (because of a relatively high capacity for efficient foraging or because of superior feeding conditions associated with age differences in seasonal timing of migration) a faster rate of fat accumulation than juveniles.

4 Speed of Migration

How fast do birds migrate? A correct measure of overall migration speed must take into account not only the time required for flying a certain distance but also, and even more important, for accumulating the necessary amount of fat at stopover or "premigratory" staging sites.

Data about the start of fat deposition by arctic waders wintering in southern Africa (Summers and Waltner 1978) and arrival times at their breeding destinations in Siberia (Glutz von Blotzheim et al. 1975) allow us to estimate the overall mean speed of spring migration in the range 150–200 km/day for knots *Calidris canutus*, sanderlings *C. alba*, curlew sandpipers and little stints *C. minutus* over a total distance of 13 000–15 000 km. Similar migration speeds hold also for long-distance wader migration in the New World, e.g. for autumn migration from N. Canada to Argentina by buff-breasted *Tryngites subruficollis* and white-rumped sandpipers *Calidris fuscicollis* (great circle distance about 13 000 km, cf. Glutz von Blotzheim et al. 1975; Myers and Myers 1979). For arctic terns *Sterna paradisaea* migrating 19 000 km from Greenland to Antarctica (Salomonsen 1967; Glutz von Blotzheim and Bauer 1982) we calculate a mean migration speed of about 150 km/day. Interestingly, wader populations performing less far-reaching migrations between arctic and temperate regions do not attain such high migration speeds. For example, knots, sanderlings and turnstones *Arenaria interpres* migrating from Great Britain to Canada or Greenland (2500–4500 km), or dunlins *Calidris alpina* migrating to the Soviet tundra (3500 km), show mean migration speeds in the range of 60–85 km/day (Glutz von Blotzheim et al. 1975; Cramp and Simmons 1983).

For long-distance passerine migration, ringing data as well as information about seasonal timing of passage, indicate that swallows *Hirundo rustica* complete the 10 000 km journey between northern Europe and southern Africa with an average speed of at least 150 km/day (a few ringing recoveries indicate exceptional speeds of up to 260–310 km/day; Glutz von Blotzheim and Bauer

1985). Furthermore, the mean migration speed of the spotted flycatcher between southern Africa and Scandinavia is almost equally fast, 140 km/day (Fransson 1986). Ringing recoveries in Africa of Fennoscandian willow warblers *Phylloscopus trochilus* show a mean speed for autumn migration of 85 km/day, with the five fastest records averaging 196 km/day, over distances up to 10 000 km (Hedenström and Pettersson 1987).

Hildén and Saurola (1982) calculated mean autumn migration speeds through Europe on the basis of recoveries of birds ringed in Finland. In Fig. 10 we have plotted these data for waders and different categories of passerines in relation to the seasonal timing of migration at Ottenby in Sweden (Edelstam 1972; Enquist and Pettersson 1986; Pettersson, pers. com.). Two aspects are of particular interest.

First, waders seem to attain higher migration speeds (median 79 km/day, n = 13 species in Fig. 10) than most passerines. The above estimates for long-distance migration also indicate that waders often migrate fast. This is surprising in view of the relatively slow rates of fat deposition recorded in several wader studies. We expect migration speed to be almost directly proportional to fat deposition rate and inversely proportional to the cost of transport (P/V, cf. Sect. 2.4). With cost of transport proportional to body mass in widely different species with isometric scaling (Pennycuick 1975), the speed of migration should be directly reflected by the rate of fat gain, expressed in relation to lean body mass.

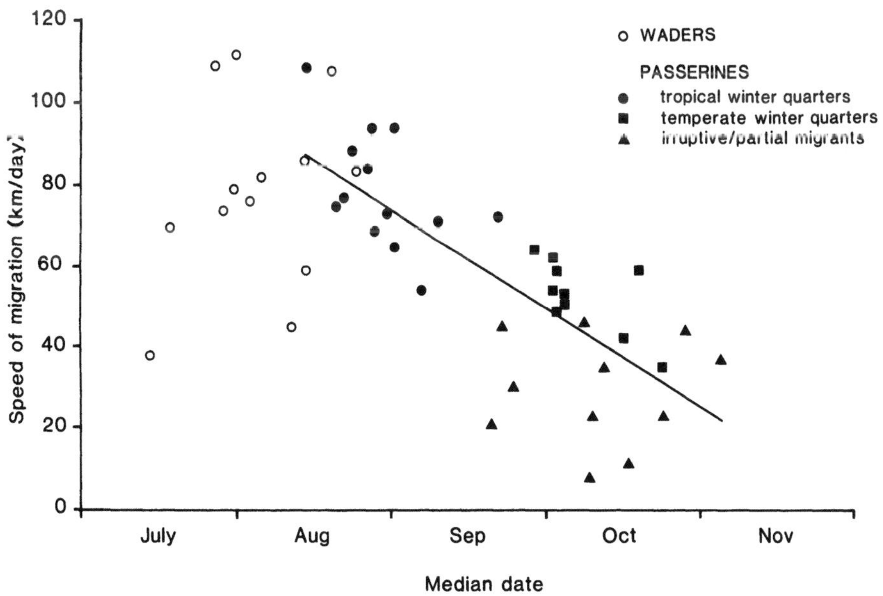

Fig. 10. Total speed of autumn migration for waders and passerines ringed in Finland (Hildén and Saurola 1982) in relation to seasonal timing of migration according to data from Ottenby in Sweden. The regression line emphasizes the distinct reduction of migration speed in the course of autumn among passerines

However, waders and passerines are not scaled isometrically, and the fact that waders are the more efficient flyers with relatively longer and more pointed wings (reducing the cost of transport) may be an important reason why they achieve high migration speeds in spite of modest fat accumulation rates. Furthermore, waders may attain high migration speeds by being particularly selective of tailwinds for their flights and by climbing to altitudes with the most favourable winds.

Second, migration speeds among passerines decrease in the course of autumn, being considerably higher for species with tropical winter quarters migrating early (median speed 75 km/day, n = 13), than for regular (median 53 km/day, n = 19) or partial (median 27 km/day, n = 11) temperate-zone migrants, completing their journey later in autumn. The explanation for this pattern remains an open question: Could it be that late autumn migrants with moderate or short migration distances can afford to travel slowly to minimize energy expenditure and predation risks, while long-distance tropical migrants primarily are selected to maximize migration speed? Or perhaps a high speed of travel is of widespread importance, and the pattern in Fig. 10 primarily reflects deteriorating food and fattening conditions in the course of autumn?

Estimating the total duration of autumn migration we find that tropical migrants (with mean migration distances in the range 6000–10 000 km according to ringing data and other sources), in spite of their high migration speeds, are travelling during a much longer time (median duration 88 days, n = 13) than regular temperate-zone migrants (median 42 days, n = 10, distances 1700–3000 km) or partial migrants (median 32 days, n = 11, distances 200–3000 km). Obviously, long-distance migration involves a considerable cost in the annual time budget. Timing and speed of migration may indeed be of greatest importance for flight and stopover strategies, as well as for the evolution of large-scale migration systems with a more or less distinct segregation between competing populations.

5 Stopover Habitat and Predation

Although predation pressure on migrating birds sometimes is demonstrably high (e.g. Rudebeck 1950–51; Walter 1979; Lindström 1989), data allowing a critical evaluation of the importance of predation for the evolution of migration strategies are lacking. To indicate one kind of observations that may be useful in this respect, we have plotted in Fig. 11 preliminary estimates of predation and energy intake rates in two different stopover habitats for migrating bramblings *Fringilla montifringilla* in southern Sweden. Typically, the finches concentrate during autumn migration to feed at fields with spill seed from summer rape. Here they are exposed to frequent attacks from raptors, mainly sparrow hawks *Accipiter nisus* (Lindström 1989). In certain autumns, with a more or less regular interval of 2 to 4 years, there is a rich crop of beech mast. In such autumns many (but far from all) bramblings switch from the rape fields to feed in the beech forests. This is surprising because, according to our field observations, their energy intake rate

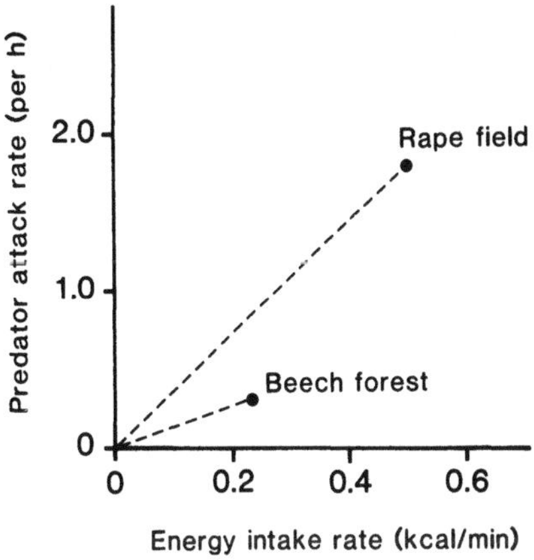

Fig. 11. Predator attack rate versus energy intake rate for bramblings in two different stopover habitats during autumn migration in southern Sweden. Data are based on field observations of bramblings feeding on rape seeds in open fields during the autumns 1985–87, and exploiting beech mast during autumn of the mast year 1987. Predator attack data are compared for the same finch flocks size, about 1000 individuals, in the two habitats. The slopes of the *broken lines* indicate the ratio of predation rate to energy intake rate for the two habitats, respectively (Lindström in prep.)

from beech mast is less than half of that which they attain at the rape fields. However, in compensation they are much better protected from predators in the beech forests.

The habitat selection by the finches may be understood if they behave in order to minimize mortality from predation during their migratory journey. In such a case they should prefer the habitat with the minimum ratio of predator attack rate (assumed to be proportional to mortality rate) to rate of energy/fat gain (assumed to be proportional to migration speed), which indeed is the beech forest (Fig. 11). In contrast, if the premium is on time rather than predation minimization, the rape fields constitute the most favourable stopover habitat.

Acknowledgements. We are grateful to Steffi Douwes and Kerstin Persson for drawing the illustrations, and to Inga Rudebeck for typing the manuscript. This study was supported by grants from the Swedish Natural Science Research Council.

References

Alerstam T (1979a) Wind as selective agent in bird migration. Ornis Scand 10:76–93
Alerstam T (1979b) Optimal use of wind by migrating birds: combined drift and overcompensation. J Theor Biol 79:341–353
Alerstam T (1981) The course and timing of bird migration. In: Aidley DJ (ed) Animal migration. Cambridge University Press, pp 9–54
Alerstam T (1985) Strategies of migratory flight, illustrated by arctic and common terns, *Sterna paradisaea* and *Sterna hirundo*. In: Rankin MA (ed) Migration: mechanisms and adaptive significance. Contrib Mar Sci Suppl 27:580–603
Alexander RMcN (1982) Optima in animals. Arnold, London
Andersson M, Norberg RÅ (1981) Evolution of reversed sexual size dimorphism and role partitioning among predatory birds, with a size scaling of flight performance. Biol J Linn Soc 15:105–130
Biebach H (1985) Sahara stopover in migratory flycatchers: fat and food affect the time program. Experientia (Basel) 41:695–97
Blem CR (1975) Geographic variation in wing-loading of the house sparrow. Wilson Bull 87:543–549

Butler RW, Kaiser GW, Smith GEJ (1987) Migration chronology, length of stay, sex ratio, and weight of western sandpipers *Calidris mauri* on the south coast of British Columbia. J Field Ornithol 58:103–111
Caraco T, Martindale S, Whittam TS (1980) An empirical demonstration of risk-sensitive foraging preferences. Anim Behav 28:820–830
Carpenter FL, Paton DC, Hixon MA (1983) Weight gain and adjustment of feeding territory size in migrant hummingbirds. Proc Natl Acad Sci USA 80:7259–7263
Charnov EL (1976) Optimal foraging, the marginal value theorem. Theor Popul Biol 9:129–136
Charnov EL, Orians GH, Hyatt K (1976) Ecological implications of resource depression. Am Nat 110:247–259
Cramp S, Simmons KEL (1983) Handbook of the birds of Europe, the Middle East and North Africa, vol 3. Oxford University Press, Oxford
Edelstam C (1972) The visible migration of birds at Ottenby, Sweden. Vår Fågelvärld, Stockholm
Ekman J, Hake M (1988) Avian flocking reduces starvation risk: an experimental demonstration. Behav Ecol Sociobiol 22:91–94
Enquist M, Pettersson J (1986) The timing of migration in 104 bird species at Ottenby – an analysis based on 39 years trapping data (English summary). Special Report from Ottenby Bird Observatory no 8, Ottenby, Sweden
Fransson T (1986) The migration and wintering area of Nordic spotted flycatcher, *Muscicapa striata* (English summary). Vår Fågelvärld 45:5–18
Gilliam JF, Fraser DF (1987) Habitat selection under predation hazard: test of a model with foraging minnows. Ecology 68:1856–1862
Glutz von Blotzheim UN, Bauer KM, Bezzel E (1975) Handbuch der Vögel Mitteleuropas, Band 6. Akademische Verlagsgesellschaft, Wiesbaden
Glutz von Blotzheim UN, Bauer KM (1982) Handbuch der Vögel Mitteleuropas Band 8/II. Akademische Verlagsgesellschaft, Wiesbaden
Glutz von Blotzheim UN, Bauer KM (1985) Handbuch der Vögel Mitteleuropas, Band 10/I. AULA-Verlag, Wiesbaden
Hedenström A, Pettersson J (1987) Migration routes and wintering areas of willow warblers *Phylloscopus trochilus* ringed in Fennoscandia. Ornis Fenn 64:137–143
Hildén O, Saurola P (1982) Speed of autumn migration of birds ringed in Finland. Ornis Fennica 59:140–143
Lima SL (1986) Predation risk and unpredictable feeding conditions: determinants of body mass in birds. Ecology 67:377–385
Lindström Å (1986) Fat deposition in migrating birds. Introductory paper no 44, Department of Ecology, University of Lund
Lindström Å (1989) Finch flock size and risk of hawk predation at a migratory stopover site. Auk 106:225–232
Moore FR, Simm PA (1986) Risk-sensitive foraging by a migratory bird *Dendroica coronata*. Experientia (Basel) 42:1054–1056
Murray GB Jr (1979) Fall migration of blackpoll and yellow-rumped warblers at Island Beach, New Jersey. Bird-Banding 50:1–11
Myers JP, Myers LP (1979) Shorebirds of coastal Buenos Aires province, Argentina. Ibis 121:186–200
Norberg RÅ (1981) Optimal flight speeds in birds when feeding young. J Anim Ecol 50:473–477
Page G, Middleton ALA (1972) Fat deposition during autumn migration in semipalmated sandpipers. Bird-Banding 43:85–96
Pennycuick CJ (1969) The mechanics of bird migration. Ibis 111:525–556
Pennycuick CJ (1975) Mechanics of flight. In: Farner DS, King JR (eds) Avian biology, vol 5. Academic Press, London, pp 1–75
Pennycuick CJ (1978) Fifteen testable predictions about bird flight. Oikos 30:165–176
Real L, Cararo T (1986) Risk and foraging in stochastic environments. Annu Rev Ecol Syst 17:371–390
Rudebeck G (1950–51) The choice of prey and modes of hunting of predatory birds with special reference to their selective effect. Oikos 2:67–88, 3:200–231
Salomonsen F (1967) Migratory movements of the arctic tern *Sterna paradisaea* in the Southern Ocean. Kgl Dan Vid Biol Medd 24 no 1

Schnell GD, Hellack JJ (1979) Bird flight speeds in nature: optimized or a compromise? Am Nat 113:53–66
Stanley PI, Minton CDT (1972) The unprecedented westward migration of curlew sandpipers in autumn 1969. Br Birds 65:365–380
Stephens DW, Krebs JR (1986) Foraging theory. Princeton University Press, Princeton
Summers RW, Waltner M (1978) Seasonal variations in the mass of waders in southern Africa, with special reference to migration. Ostrich 50:21–37
Tucker VA (1974) Energetics of natural avian flight. In: Paynter RA (ed) Avian energetics. Publ Nuttall Ornithol Club no 15, Cambridge Mass, pp 298–328
Veiga JP (1986) Settlement and fat accumulation by migrant pied flycatchers in Spain. Ringing & Migr 7:85–98
Videler JJ, Vossebelt G, Gnodde M, Groenewegen A (1988) Indoor flight experiments with trained kestrels. J Exp Biol 134:173–199
Walter H (1979) Eleonora's falcon. University of Chicago Press, Chicago London

Strategies of Trans-Sahara Migrants

H. BIEBACH[1]

1 Introduction

Major deserts of the world, like the Sahara, the Middle Asian desert, and the central Australian desert, are potential ecological barriers for migrating birds. In terms of size, annual rainfall, and temperature the most severe of these is the Sahara, which stretches from the Atlantic coast in the west to the Saudi Arabian peninsula in the east and covers an area of 12 million km². Mean daily maximum temperature during the migratory period in spring and autumn is between 25° and 38°C. Surface temperature may reach 70°C. During the day, relative humidity drops to 10%. Birds breeding in the west- and mid-Palearctic region and wintering in tropical Africa have to fly across the Sahara, a distance of at least 1500 km. The ecophysiological problems encountered by these birds during the desert crossing and their strategies overcoming these problems are discussed in this chapter.

2 Strategies of Desert Crossing

Four possible strategies of desert crossing can be suggested (Fig. 1):

1. Gradual migratory movement along vegetated areas in a north-south direction, with flight at night and stopover during the day. This is possible only by following either the Nile, with its narrow border of agricultural land and swamp vegetation, or by following the Atlantic coast with its broad belt of sparse desert vegetation.
2. Flight from oasis to oasis at night and stopover during the day. Such a strategy is possible only along a row of oases and mountains that stretches southeast from Marocco.
3. Flight over the desert at night and with daytime stopover in the desert.
4. Non-stop flight over the entire Sahara desert.

These four possibilities are not mutually exclusive and combinations are also possible.

[1]Max-Planck-Institut für Verhaltensphysiologie, Vogelwarte, D-8138 Andechs, FRG

E. Gwinner (Ed.) Bird Migration

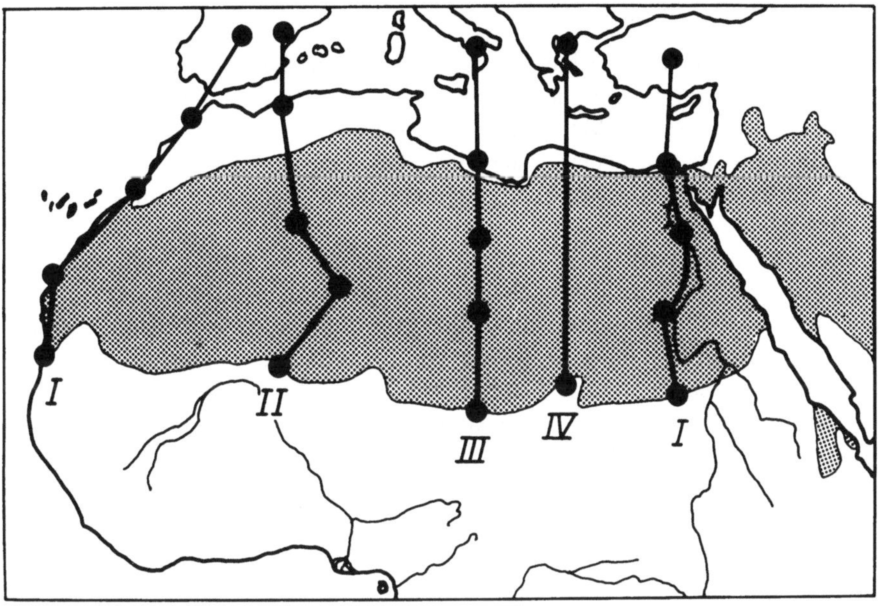

Fig. 1. Hypothetical strategies for crossing the Sahara. *I* and *II* Intermittent migratory strategies with stopover in vegetated areas where birds can feed (Nile, Atlantic coast, oasis); *III* intermittent migratory strategy with stopover in the desert without feeding possibilities; *IV* non-stop flight

No single strategy is used by all species and a single species may use different strategies in autumn and spring. For theoretical reasons, one would not expect, for example that a stork, a typical gliding bird that depends on thermal updrafts during migration would have the same needs, e.g. as an 8-g willow warbler, which depends exclusively on flapping flight. Similarly, because a wader has very different feeding requirements from a swift the two should select stopover places based on the criterion of different food supply. There is, however, a group of small passerines similar enough with respect to size, feeding ecology and flight mechanics that they can be expected to behave similarly. This group includes four families: the Sylviidae (warblers), Muscicapidae (flycatchers), Motacillidae (wagtails and pipits) and Turdidae (thrushes). The ten most numerous species of these families comprise about 50% of all passerine birds crossing the Sahara desert (Moreau 1972). Among the passerines the willow warbler (*Phyloscopus trochilus*) contributes about 20%. The following discussion is based largely on data from these species.

Birds are confronted with very different conditions during the spring and autumn migrations. For example in spring the probability of encountering a tailwind is lower, the desert is about 300 km narrower from north to south and the air temperature is generally lower than in autumn. It is therefore important to examine each season separately, but because there is so little information about birds in the desert during spring, I will concentrate mainly on autumn migration.

2.1 *Flight Along Vegetated Areas*

No mass migration of passerines has been recorded in the Nile valley in Egypt or in the Sudan, but systematic observations are scarce (Valverde 1958). In Karthoum during spring Mathiasson (1963) saw only small numbers of passerines following the Nile, whereas the Nile valley seems to form a flyway for birds that are ecologically attached to aquatic habitats such as waders, ducks, storks, and herons.

Because the belt of vegetation along the Atlantic coast is much wider than the small strip along the Nile, it is much more difficult to judge the importance of this area as a flyway. Observations of passerines in autumn indicate regular migration through the Spanish Sahara (Valverde 1957) and the western part of Morocco (Vaughan 1960; pers. comm. D. Schmidl and G. Rheinwald). Despite this regular migration it seems doubtful that the Atlantic coast could serve as a major flyway for most passerine birds because such mass migration has not been observed.

2.2 *Flight from Oasis to Oasis*

The distribution of oases in the eastern part of the Sahara makes short flights from oasis to oasis impossible. Over a distance of 600 km there are only five oases, all in the northern third of the Sahara. A potentially better situation is found in the western Sahara, where a row of oases and mountains stretches in a southeastern direction from Morocco to the southern border of the Sahara. Bairlein et al. (1984) favours the hypothesis that some species preferentially fly along this route flying during the night and stopping over during the day, but the available data are too sparse to be conclusive. Until better evidence is obtained, one has to accept Moreau's (1961) view that the number of birds observed in oases is minor compared with the great number of migrants expected.

2.3 *Non-Stop Versus Intermittent Migratory Strategy*

There are two other strategies, the non-stop flight (Moreau 1961, 1972) and the intermittent migratory flight (Bairlein 1985; Biebach et al. 1986). Several approaches can be used to verify which is occurring. The most direct way, continuous observation of individuals on their way over the desert, is for technical reasons not yet possible. Therefore only circumstantial evidence is available. I discuss here two lines of evidence, the energy budget and the water budget.

Energy Budget, Reserves. Fat is the main energy source during migration (Berthold 1975; Blem 1980). How could amounts of fat reserves be used to distinguish non-stop migrants from intermittent migrants? Except for the few oases and wadis that are assumed to play no important role, the desert is basically devoid of habitat suitable for foraging. Both strategies therefore require the accumulation of energy reserves sufficient to fuel the entire crossing and the amount of premigratory fattening cannot be used to distinguish between these two strategies. Similarly, birds found in the desert with little fuel could be failures either from a non-stop migration or an intermittent migration. However, fat birds in the desert should represent an intermittent flight strategy, as long as it can be shown that they were not found to make a landfall for some other reason.

The willow warbler on its autumn migration across the eastern Sahara will be taken as a representative example of the energetic problems during the desert crossing (Fig. 2; Table 1). As for birds in general, migratory fat is accumulated in addition to normal body mass and is therefore a good predictor of fat reserves. Body mass before the crossing, measured in the Mediterranean area on two isles of Greece (Dragonada, Karpathos) and Cyprus averages 10–11 g, which means the birds have accumulated considerable fat reserves, because the fat-free body mass of willow warblers is about 7 g. Body mass of birds caught in the desert is generally lower than before the crossing and body mass at the end of the crossing is close to the fat-free value. A more detailed look at the data shows two phenomena: First, the body mass taken from the same place in different years shows little variation, at the most 0.4 g. Second, body mass of willow warblers from nearby banding stations at about the same latitude or about the same

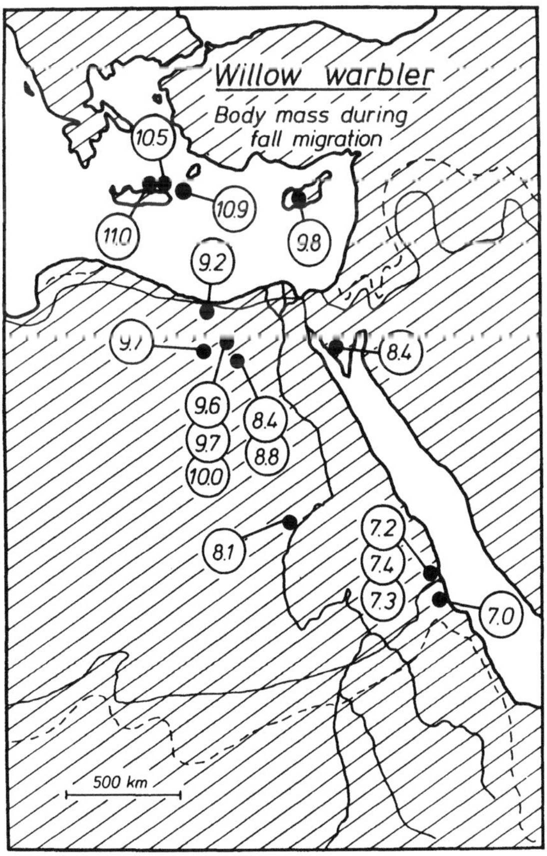

Fig. 2. Mean body mass of willow warblers during fall migration in the eastern Mediterranean and eastern Sahara. Different means for a single study site are from different years. For references, see Table 1

Table 1. Body mass of willow warblers on autumn migration in the eastern Mediterranean and Sahara

Trapping site	Date	N	Body mass (g) mean ± SD	Body mass range	Reference
Cyprus	Autumn	271	9.8	7.0–13.5	Moreau (1969)
Greece					
Karpathos	7.8.–26.9.80	82	10.9 ± 1.28	6.7–13.9	Biebach unpublished data
Dragonada	27.8.–21.10.74	1489	10.5	6.5–14.0	Phillips and Round (1975)
Egypt					
Coast	8.9.–13.10.	43	9.2 ± 1.31	7.1–11.1	Moreau and Dolp (1970)
Sadat Farm	13.9.–1.10.82	253	8.4 ± 1.06	5.7–11.7	Biebach et al. (1986)
Sadat Farm	3.8.–29.10.83	426	8.8 ± 1.30	6.2–13.7	Biebach et al. (1986)
Desert 122	23.9.–3.10.82	143	9.6 ± 1.00	6.2–12.7	Biebach et al. (1986)
Desert 122	6.9.–27.9.83	427	9.7 ± 1.22	7.7–12.2	Biebach et al. (1986)
Desert 122	30.8.–16.9.85	144	10.0 ± 1.03	6.6–13.3	Biebach unpublished data
Djebl Sprosser	9.9.–21.9.85	50	9.7 ± 0.85	6.5–11.5	Biebach unpublished data
Sinai,					
Chaterines M.	7.8.–27.10.	87	8.4 ± 1.2		Safriel and Lavée (1988)
Sudan					
Khor Arba'at	Aug.–Sept. 82	45	7.2	6.0–8.5	Nikolaus, pers. comm.
Khor Arba'at	27.8.–13.9.83	18	7.3	6.5–8.5	Nikolaus, pers. comm.
Khor Arba'at	29.8.–23.9.84	205	7.4	5.5–10.5	Nikolaus, pers. comm.
Port Sudan	2.9.–18.9.81	95	7.0	5.0–8.5	Nikolaus, pers. comm.
Wadi Halfa	August	55	8.1	6.5–10.5	Nikolaus, pers. comm.

distance from the desert border can be very different. The two stations "desert 122" and "Sadat Farm" are only 70 km apart, but are found in desert rock and oasis habitat respectively. Willow warblers found at these two sites differ in mass by about 0.8 g, or approximately 10%. Biebach et al. (1986) have suggested that leaner birds prefer stopover habitats with foraging possibilities and fat birds rest wherever at least shade or wind-exposed places are found. The same phenomenon was found by Bairlein (1987), who observed that birds resting in a very small oasis in the western Sahara tended to be fatter than those in a normally sized oasis. A similar tendency was also reported by Dolnik (this Vol.) for the Middle Asian desert who found a high fat load in birds at bare stopover sites in the desert and a low fat load in birds at stopover sites along banks of rivers or lakes in the desert. The question is now, what kind of strategy do the birds from the different places in the Eastern Sahara represent or, in other words, what distances can be covered with the different fat reserves and are these distances long enough to reach the southern border of the Sahara without refueling.

The *maximum distance* a bird can fly without refueling can be predicted if three parameters are known:

1. The available energy reserves;
2. The energetic costs during flight and at rest; and
3. The ground speed of flight, which is a composite of the speed of the bird relative to the air and the speed and direction of the wind relative to the ground.

Such calculations have been made for birds crossing the Sahara (Biebach et al. 1986; Biebach 1989; Dolnik 1989, this volume; Moreau 1961, 1972; Bairlein 1985; Fry et al. 1970). However, since new empirical information on the energy reserves and flight speed are now available (see Table 2 and Jenni and Bruderer, unpublished data), a recalculation is possible. I present this calculation here for birds of three stations: Dragonada in the Mediterranean Sea and two sites in the Libyan desert about 200 km south of the Mediterranean Sea, "desert 122" and "Sadat Farm".

The frequency distribution of body mass for the three stations is shown in Fig. 3. Birds were heaviest at Dragonada before the crossing of the ecological barrier, lightest at the oasis and medium at "desert 122". The correlation between body mass and extractable fat from birds caught on autumn migration is used to convert body mass into energy reserves (Fig. 4; Biebach in prep.). A change in body mass of 1 g thus corresponds to a change in fat content of 0.70 and 0.30 g of protein and water. This means an energy equivalent of 29.8 kJ per change of 1 g body mass. These values are close to the findings in four species of waders, where the added body mass consists for 60–70% fat (Brederode and Piersma 1989). It is assumed that all the extractable fat is available as substrate for metabolism except for 0.2 g, which might be structural and does not form part of the fat depot (Davidson and Evans 1982).

Calculations of maximum flight range for birds at each station were based on empirical data on air speed (Jenni and Bruderer, unpublished data) and theoretical assumptions about the power output during flight (Tucker 1974 in Calder 1984). The maximum range of willow warblers flying in still air with different initial body mass is shown in Table 2. Birds from Greece and the desert have to cross a distance of 1800 and 1300 km respectively. None of the Greece birds, and only a minor proportion of the desert birds have enough fat reserves to cross in still air without refueling (Fig. 3). There are several ways of interpreting these results.

First, the calculation might be based on invalid assumptions. Of all the different formulas proposed to calculate maximum range, the method used here makes predictions that are smaller than the largest predicted ranges by a factor of 0.5 (Summers and Waltner 1979; for summary see Davidson 1984). There are not

Table 2. Maximum range at still air

Body mass (g)	6.2	6.7	7.2	7.7	8.2	8.7	9.2	9.7	10.2	10.7	11.2	11.7	12.2	12.7	13.2
Fat[a] (g)	0.0	0.35	0.70	1.06	1.41	1.76	2.11	2.46	2.81	3.16	3.51	3.86	4.21	4.57	4.92
Maximum range[b] (km)	0	60	240	420	580	720	860	1000	1120	1260	1360	1480	1580	1680	1780
Flight time[c] (h)	0	1.7	6.9	12.0	16.6	20.6	24.6	28.6	32.1	35.5	38.9	42.4	45.2	48.1	51.0

[a] Fat = –4.35 + 0.702 body mass, see Fig. 4.

[b] Based on a simulation, taking into account decreasing body mass during flight (consumed fat and protein) and an equivalent of 29.8 kJ/g body mass. Minimum body mass = 6.5 g $\hat{=}$ 0.2 g fat. Power consumption = $(6.43 \times 10^{-3}h + 94.15)m^{0.974}$ (W) in Calder (1984). Table 7–2 from Tucker (1974). h = height = 1000 m; m = body mass (kg).

[c] Air speed = 9.7 m/s (L. Jenni and B. Bruderer, unpubl. data).

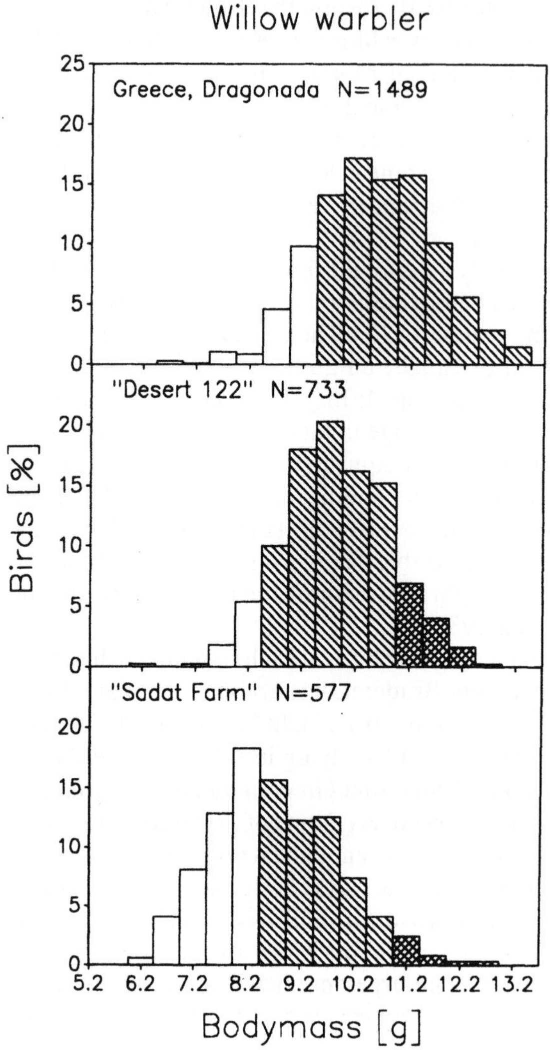

Fig. 3. Frequency distribution of body mass of willow warblers on autumn migration, before the desert crossing on the isle of Dragonada in Greece (Phillips and Round 1975) and from the two study sites in the Libyan desert (Biebach et al. 1986). In still air only birds represented by the *cross-hatched* mass classes, with a tailwind of 8 m/s those of the *dashed* and *cross-hatched* mass classes are expected to manage the desert crossing without refueling (see also Table 2 and Fig. 5)

enough empirical data on the energy expenditure during flight of bird species in the size range of willow warblers to decide which one of the formulas produces the more realistic range (for a review of the empirical data see Masman and Klaassen 1987).

A second way of interpreting the data is to take into account that no effect of wind was included in the calculation. In autumn, birds migrating over the Sahara

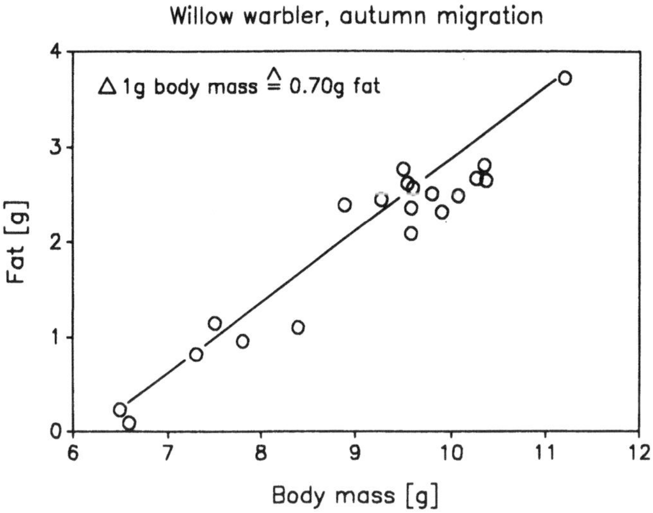

Fig. 4. Relationship between extractable fat and fresh body mass of willow warblers caught in a stopover place on autumn migration in the Libyan desert (Biebach unpubl. data). Fat was extracted from carcasses dried at 60°C, using petroleum ether in a Soxhlet apparatus. The line of linear regression, using the method of least squares is: $y = -4.35 \pm 0.702\ x$, correlation coefficient = 0.961 ($n = 20, p < 0.001$)

have a 95% chance of encountering tailwinds of about 8 m/s at a height of about 1000 m. This tailwind has to be added to the ground speed of the birds (Jenni and Bruderer, unpublished data) and results in much higher maximum ranges. Taking the most likely wind speed of 8 m/s into account, most of the Greece and the desert-122 birds are able to cross the desert without refueling, but only about 60% of the oasis birds can do so (Fig. 5). At least some of the remaining 40% of the oasis birds have the chance to refuel in the oasis (Biebach et al. 1986; Bairlein 1985), but, as pointed out earlier, regular refueling in oases is not possible for most passerines.

I conclude that the energy reserves of the Greece birds allow either a non-stop or an intermittent flight. Because they have sufficient energy reserves to complete the crossing without refueling, the desert-122 birds do not represent stragglers from a non-stop flight; more likely, these birds represent an intermittent migratory strategy. Some of the oasis birds represent potential stragglers which, in the special case of the oasis, have the chance to forage and accumulate fat to complete migration.

From a simple energetic point of view, stopover in the desert without the possibility of refueling is a waste of time and energy. But energy is not the only important factor during desert crossing. The water budget may be a critical factor (Biesel and Nachtigall 1987; Yapp 1956, 1962).

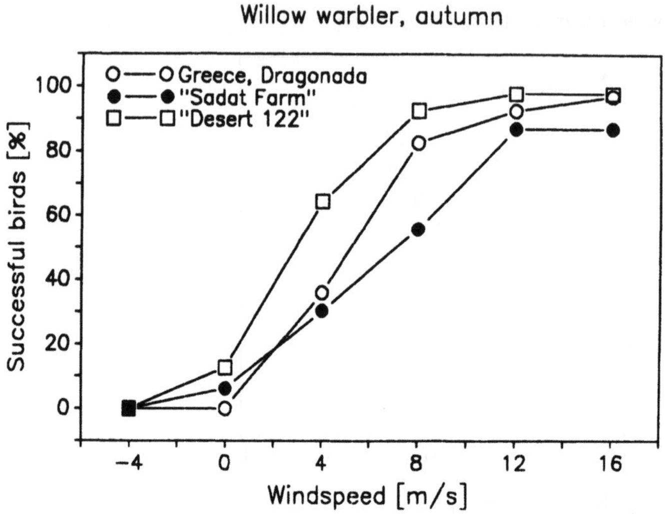

Fig. 5. Effect of wind speed during the desert crossing on the percentage of successful birds (birds that can reach the southern border of the Sahara without refueling) at three study sites: one before the crossing (Dragonada) and two in the desert (see Table 2 and Fig. 3). Wind speed was assumed to be additive to air speed (Jenni and Bruderer, unpublished data). Positive and negative values of wind speed are tail- and headwinds respectively

3 Water Budget

Water Content. Water content of small migratory birds is about 67% of the fat-free wet weight (Fig. 6). Differences between species cannot be attributed to sex, season, family or general body size (Child and Marshall 1970) but are probably to some extent due to different techniques of analysis. Within one species the fattest individuals have a higher relative water content than the leaner ones (Fry et al. 1970).

The tolerance to dehydration is more difficult to generalize. There is only one study on reed warblers (*Acrocephalus scirpaceus*) fattening in the Sahel before the northward spring crossing (Fogden 1972). Because of an extreme draught at that time of year, the birds accumulated fat but had a very low water content of only 57%. Davidson (1984) has proposed that premigratory dehydration is adaptive because it lowers the wing load and cost of flight in waders. Empirical data on an adaptive premigratory dehydration in passerines are not available. Some information about tolerance of dehydration comes from laboratory studies on resident desert birds. Under normal hydration conditions, the water content is about 66% as in the migrants, but under dehydration conditions the desert residents can tolerate values down to 56% (Skadhauge 1974). What can the water budget of birds captured in the desert tell us about the migratory flight patterns of these birds? Many anecdotes are reported for birds found in the desert close to death and attempting to get water, but measurements of water content are scarce.

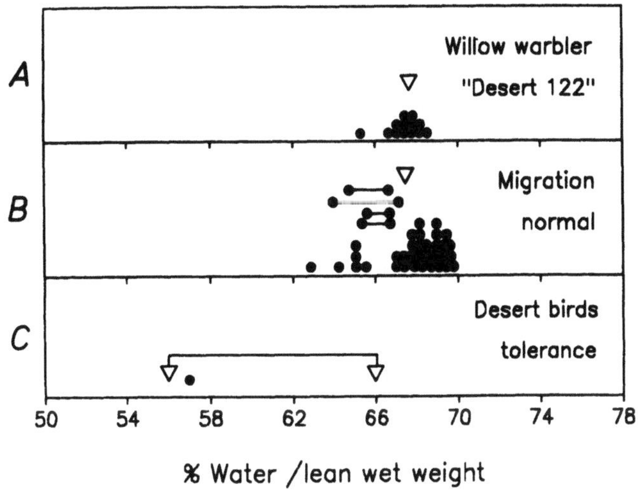

Fig. 6A-C. Water content and tolerance to dehydration. *Triangles* are means. **A** *Circles* represent individual values of willow warblers resting in the Libyan desert (Biebach unpubl. data). **B** *Circles* represent mean values of small migratory bird species. *Connected circles* are water contents of lean (*left*) and fat (*right*) birds of the same species. Data are from: Child and Marshall (1970); Fry et al. (1970); Johnston (1968); Moreau and Dolp (1970); Odum et al. (1961, 1964). **C** *Circle* represents the mean value of reed warblers preparing to cross the desert (Fogden 1972); *arrows* indicate normal water content (*right arrow*) and lower level of tolerance against dehydration (*left arrow*) in resident desert birds (After Skadhauge 1981)

Five "normal" birds (small passerines) of various species captured in the Tanezrouft of the western Sahara in spring had a mean water content ± SD of 63.7% ±1.20; a sixth had a water content of only 53.7% (Haas and Beck 1979). Seven willow warblers trapped in an oasis in the Libyan desert in autumn had a mean water content ± SD of 67.4% ±3.84 (Biebach, unpublished work). The same species in a rocky area in the Libyan desert in autumn had a mean ± SD of 67.5 ±0.89 (n = 13), among which two birds were close to death (fluffed feathers, caught by hand) with a fat content of only 0.5 and 0.2 g and a water content of 68.3 and 67.9% respectively (Biebach, unpublished data). Unfortunately, no data on the water content of birds immediately after the desert crossing are available.

In general, it appears in outlines that the water content of migratory birds resting in the desert is in the range of hydrated birds. Skadhauge (1981) comes to the same conclusion regarding birds after long migratory flights in stating that "investigations have failed to reveal convincing evidence of dehydration". This general conclusion is supported by the suggestion of Haas and Beck (1979) that small birds become exhausted because they ran out of fat, but not water.

Water Budget During Non-Stop Flight. Evidence for a non-stop flight or an intermittent migratory strategy can also come from the knowledge of the water budget during the crossing. Data on the evaporative water loss of free-flying

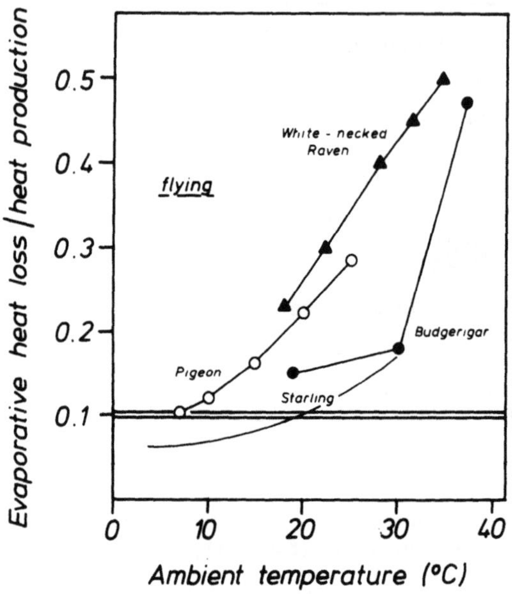

Fig. 7. Ratio of evaporative cooling to metabolic rate during flight in a wind tunnel for the budgerigar (Tucker 1968), the starling (Torre-Bueno 1978), the white-necked raven (Hudson and Bernstein 1981) and the pigeon (Biesel and Nachtigall 1987). A possibly balanced water budget (evaporative waterloss = metabolic water production) for starlings is shown by the *double line*

migrants are not available, but it should be possible to use measurements of flying birds in a wind tunnel to estimate water loss under field conditions (Fig. 7). Wind tunnel data from starlings (Torre-Bueno 1978), budgerigar (Tucker 1968), white-necked raven (Hudson and Bernstein 1981) and pigeons (Biesel and Nachtigall 1987) show that evaporative water loss increases with temperature, slowly at low air temperatures but more steeply at high temperatures. At low air temperatures the heat produced during flight is dissipated mainly by convection. At higher air temperatures the efficiency of this mechanism decreases and heat dissipation is gradually dominated by evaporative heat loss.

Evaporative water loss is to some extent counterbalanced by metabolic water production in the extremely high metabolism during flight. Metabolism of fat or starch results in the production of 26.3 and 31.0 mg of water per kJ respectively. In this context starch is slightly more favourable than fat. At an air temperature between 8° and 12°C metabolic water production (from fat and starch) just balances the evaporative water loss. That means that birds flying at or below about 10°C should have no net water loss even during extremely long flights. In the desert in autumn an air temperature of about 10° should be found during the day at about 3000 m above the ground and during the night at about 1000 m above the ground, if one assumes an air temperature at ground level during day and night of 30° and 18°C respectively, and a decline of 7°C with each 1000-m increase in elevation. Radar studies in Europe and North America (Able 1970; Blokpoel and Burton 1975; Gauthreaux 1972; Hilditch et al. 1973; Richardson 1972; Williams et al. 1974) show that migration of small passerines at these heights is not unusual. Information on the altitude of migration over the Sahara is not available, except for one radar study in autumn by Schäfer, who found that the bulk of migrants were flying at elevations between 1500 and

2500 m, with some as high as 3000 (cited in Moreau 1972, p. 33), details about time of day and size of birds were not provided.

In conclusion I hypothesize that crossing of the Sahara by a non-stop flight should be possible with a balanced water budget, if an altitude with an air temperature of about 10 °C is chosen.

Water Budget During Intermittent Flight. Birds stopping in the desert during the day are confronted with high air temperatures and low humidity. The water loss under these extreme conditions has been measured for a number of passerines (Biebach 1988; Biebach, unpublished data). For example, evaporative water loss of willow warblers resting in normal stopover places in autumn, is 0.056 g/h, or 0.56 g during a 10-h day. At the same time 0.12 g of metabolically formed water is produced if we assume a rate of fat metabolism of 0.84 kJ/h (direct measurements by Pohl, cited in Aschoff and Pohl 1970). Willow warblers are thus left with a total water loss per 10-h day of 0.35 g or 7.2% of total body water. Assuming a balanced water budget during flight in the night, the water content after two nights of flight and one day of stopover is reduced from 67.5 to 62.9%, which is well within the range of the dehydration tolerance. Even if two stopover days and three nights of flight are considered, the water content will be decreased only to 57.9%, which is still above critical values of dehydration (see Fig. 6).

In summary, we expect birds performing a non-stop migration to arrive after the desert crossing with a balanced water budget, whereas birds with an intermittent migratory strategy end up with lower levels of hydration, but not so slow as to be intolerable.

4 Wind Conditions

Are there environmental conditions that make one or the other strategy more favourable? Wind speed and direction generally play an important role during migration (Able 1973; Richardson 1972) and the special situation over the Sahara desert may be the key to understanding the preferred migratory strategy. In the eastern sector of the Sahara, in Egypt and the Sudan, autumn tailwinds (northerly winds) are most likely to be encountered at an altitude of 500 to 1000 m. The reverse is seen with headwinds (southerly winds) with an increasing probability at higher altitudes (Fig. 8; see Moreau 1961). The same tendency is true for the entire Sahara.

The following hypothesis, based on these wind conditions together with constraints set by the energy budget and the water budget, is proposed to explain why both non-stop flight and intermittent flight patterns might be expected.

5 Hypothesis

I consider a simplified wind and temperature situation. Assuming that during the daytime tailwinds occur at 0 to 3000 m, headwinds occur above 3000 m and air

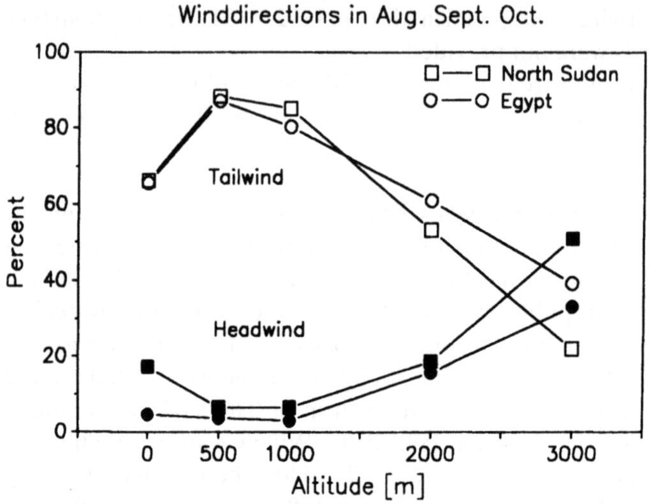

Fig. 8. The probability of meeting head- or tailwinds at different altitudes in the eastern Sahara during the autumn migration period. Tail- and headwinds correspond to northerly and southerly winds respectively. Data from Moreau (1961, Table 8)

temperature decreases with altitude reaching 10°C at 3000 m. Nighttime conditions are the same as the day, except that air temperature reaches 10° at an altitude of 1000 m.

As mentioned earlier, a migrating bird must fly at an air temperature of 10°C or less to maintain a balanced water budget. Such an air layer is encountered during the day at 3000 m, at which elevation the birds will encounter the tailwinds that are necessary from the point of view of the energy budget. At night the same conditions occur at lower altitudes (1000 m). In this way birds can perform a non-stop flight (Fig. 9) flying during both day and night with a balanced water budget and sufficient energy reserves.

However, if wind conditions change and headwinds are encountered at lower altitudes (e.g. as low as 2000 m) during the day, the birds run into the following problems: They can either choose an air layer at 3000 m, where temperatures will be low enough (10°) to fulfill the needs for a balanced water budget, but the headwind will be incompatible with the energy budget, or they can choose an air layer below 2000 m where they will encounter a tailwind, but temperatures will be too high to maintain the water reserve. In both cases the birds have to make a landfall during the day and wait for the night to take off again, when the 10°C temperature zone is lower and is more likely to coincide with the zone of tailwind; the birds perform an intermittent migratory strategy (Fig. 9).

As pointed out earlier (Biebach et al. 1986; Bairlein 1985) a considerable number of other passerine bird species show a similar resting behaviour as willow warblers and are similar with respect to their water and energy budget. We can therefore expect them to perform a similar strategy of desert crossing as hypothesized for willow warblers, namely an intermittent or a non-stop migration depending on weather conditions.

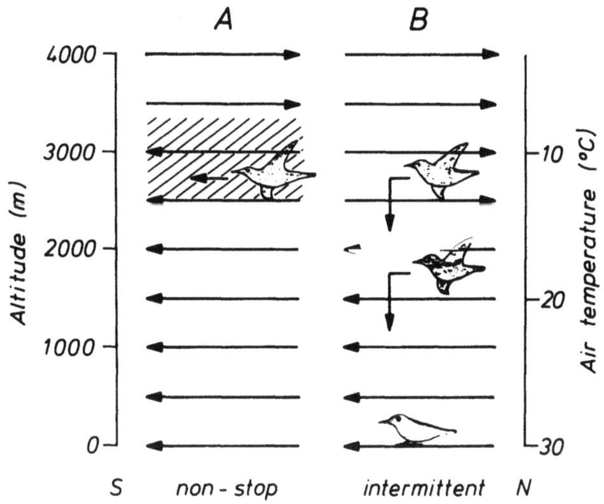

Fig. 9. Model for the effect of wind and temperature conditions on the migratory strategy of desert crossing during the day. *Long arrows* indicate wind direction. To allow a balanced water budget birds have to fly at about 10° C, and to allow the crossing without refueling the birds depend on tailwinds. Both conditions are encountered simultaneously only in the *hatched zone*. Conditions during the night are acceptable with a broad overlap of tailwind and low temperatures. Therefore, conditions in **A** allow flight during the day (non-stop flight during day and night), in **B** only during the night (intermittent, stopover during the day and flight during the night)

On the other hand, bird species with relatively small drag values, e.g. like swifts, swallows or bee-eaters and bird species with relatively greater body mass and higher flight speed, e.g. like cookoos and rollers, are less dependent on wind conditions. They might be able to continue their flight at wind conditions that force the small passerines to make a landfall and they might preferably perform a non-stop flight.

This means that under good weather conditions (tailwind at high altitude) all species make a non-stop flight, whereas under bad weather conditions (head-winds at high altitudes) some species make an intermittent and others a non-stop flight.

These predictions can be tested by measuring the wind direction and speed, temperature, and humidity at different altitudes together with the behaviour of the migrating birds, e.g. altitude and ground speed during flight and the number of stopping birds.

Moreover, the physiological state (energy reserves and water content) of resting birds during the day at intermediate points during the crossing and after the crossing together with the weather conditions during the day will be important to verify the hypothesis. Finally, the relative importance of an intermittent flight to a non-stop flight can be found by direct observations (e.g. radar) of migrating birds during day and night and simultaneous counts of resting birds during the day.

References

Able KP (1970) A radar study of the altitude of nocturnal passerine migration. Bird-Banding 41:282-290

Able KP (1973) The role of weather variables and flight direction in determining the magnitude of nocturnal bird migration. Ecology 54:1031-1041

Aschoff J, Pohl H (1970) Der Ruheumsatz von Vögeln als Funktion der Tageszeit und der Körpergröße. J Ornithol 11:38-47

Bairlein F (1985) Body weights and fat deposition of Palearctic passerine migrants in the central Sahara. Oecologia (Berl) 66:141-146

Bairlein F (1987) The migratory strategy of the garden warbler: a survey of field and laboratory data. Ringing & Migr 8:59-72

Bairlein F, Beck P, Feiler W, Querner U (1984) Herbstbeobachtungen paläarktischer Zugvögel in der algerischen Sahara. Vogelwelt 105:1-9

Berthold P (1975) Migration: control and metabolic physiology. In: Farnes DS, King JR (eds) Avian biology, vol 5. Academic Press, London, pp 77-128

Biebach H (1988) Ecophysiology of resting willow warblers (*Phylloscopus trochilus*) crossing the Sahara. Proc 19th Int Ornithol Congr, pp 2162-2168

Biebach H, Friedrich W, Heine G (1986) Interaction of body mass, fat, foraging and stopover period in trans-Sahara migrating passerine birds. Oecologia (Berl) 69:370-379

Biesel W, Nachtigall W (1987) Pigeon flight in a wind tunnel. J Comp Physiol B Biochem Syst Environ Physiol 157:117-118

Blem CR (1980) The energetics of migration. In: Gauthreaux SA Jr (ed) Animal migration, orientation and navigation. Academic Press, pp 175-224

Blokpoel H, Burton J (1975) Weather and height of nocturnal migration in east-central Alberta: a radar study. Bird-Banding 46:311-328

Brederode van NE, Piersma T (1989) The estimation of fat reserves in coastal waders before their departure from northwest Africa in spring. Ardea (in press)

Calder III WA (1984) Size, function and life history. Cambridge Harvard University Press

Child GI, Marshall SG (1970) A method of estimating carcass fat and fat-free weights in migrant birds from water content of specimens. Condor 72:116-119

Davidson NC (1984) How valid are flight range estimates for waders? Ringing & Migr 5:49-64

Davidson NC, Evans RR (1982) Mortality of redshanks and oystercatchers from starvation during severe weather. Bird Study 29:183-188

Fogden MPL (1972) Premigratory dehydration in the reed warbler *Acrocephalus scirpaceus* and water as a factor limiting migratory range. Ibis 114:548-552

Fry CH, Ash JS, Ferguson-Lees IJ (1970) Spring weights of some palearctic migrants of Lake Chad. Ibis 112:58-82

Gauthreaux SA Jr (1972) Behavioral responses of migrating birds to daylight and darkness: a radar and direct visual study. Wilson Bull 84:136-148

Haas W, Beck P (1979) Zum Frühjahrszug paläarktischer Vögel über die westliche Sahara. J Ornithol 120:237-246

Hilditch CDM, Williams TC, Nisbet ICT (1973) Autumnal bird migration over Antigua, W.I. Bird-Banding 44:171-179

Hudson DM, Bernstein MH (1981) Temperature regulation and heat balance in flying white-necked ravens, *Corvus cryptoleucus*. J Exp Biol 90:207-281

Johnston DW (1968) Body characteristics of palm warblers following an overwater flight. Auk 85:13-18

Masman D, Klaasen M (1987) Energy expenditure during free flight in trained and free living Eurasian kestrels, *Falco tinnunculus*. Auk 104:603-616

Mathiasson S (1963) Visible diurnal migration in the Sudan. Proc 13th Int Ornithol Congr, pp 430-435

Moreau RE (1961) Problems of Mediterranean Sahara migration. Ibis 103:373-421, 580-623

Moreau RE (1969) Comparative weights of some trans-Sahara migrants at intermediate points. Ibis 111:621-624

Moreau RE (1972) The Palearctic-African bird migration systems. Academic Press, London

Moreau RE, Dolp RM (1970) Fat, water, weights and wing-lengths of autumn migrants in transit on the northwest coast of Egypt. Ibis 112:209–228
Odum EP, Connell CE, Stoddard HL (1961) Flight energy and estimated flight ranges of some migratory birds. Auk 78:515–527
Odum EP, Rogers DT, Hicks DL (1964) Homeostatis of the nonfat components of migrating birds. Science 143:1037–1039
Phillips NJ, Round PD (eds) (1975) Crete ringing group report, 1973–1975
Richardson WJ (1972) Timing and amount of bird migration in relation to weather – a review. Oikos 30:224–272
Safriel UN, Lavée D (1988) Weight changes of cross-desert migrants at an oasis – do energetic considerations alone determine the length of stopover? Oecologia (Berl) 76:611–619
Skadhauge E (1974) Renal concentrating ability in selected West Australian birds. J Exp Biol 61:269–276
Skadhauge E (1981) Osmoregulation in birds. Zoophysiology 12. Springer, Berlin Heidelberg New York
Summers RW, Waltner M (1979) Seasonal variations in the mass of waders in southern Africa, with special reference to migration. Ostrich 50:21–37
Torre-Bueno JR (1978) Evaporative cooling and water balance during flight in birds. J Exp Biol 75:231–236
Tucker VA (1968) Respiratory exchange and evaporative water loss in the flying Budgerigar. J Exp Biol 48:67–87
Tucker VA (1974) Energetics of natural avian flight. In: Paynter RA (ed) Avian energetics. Nutall Ornithol Club, Cambridge, Mass, pp 298–333
Valverde JA (1957) Aves del Sahara Espanol (Estudio ecologico del desierto). Instituto de estudios africanos, Consejo superior de investigaciones cientificas, Madrid
Valverde JA (1958) Some observations on the migration through the occidental Sahara. Bull Br Ornithol Club 78:1–5
Vaughan R (1960) Notes on autumn migrants in Morocco. Ibis 102:330–332
Williams JM, Williams TC, Ireland CC (1974) Bird migration over the North Atlantic Ocean. In: Gauthreaux SA (ed) Proc Conf on the Biol Aspects of Bird/Aircraft Collision problem. Air Force Off Sci Res, Clemson, SC
Yapp WB (1956) Two physiological considerations in bird migration. Wilson Bull 68:312–319
Yapp WB (1962) Some physical limitations on migration. Ibis 104:86–89

Bird Migration Across Arid and Mountainous Regions of Middle Asia and Kasakhstan

V.R. Dolnik[1]

This is a brief review of research on migratory birds and the dynamics of migration in the deserts and mountains of central Asia (Dolnik 1985a,b,c, 1987a,b).

1 The Dynamics and Distribution of Migration

Nocturnal migration was studied by a moon-watch method from 13 observation sites during spring and 23 during autumn (Fig. 1). About 1.5 billion birds cross the study area during autumn migration, and about half this number – 750 million – return during spring migration. The number of migrants at night was six to eightfold greater than during the day. About 85% of the migrants are passerines. Daytime migration began after sunrise and ended before sunset at all inland observation sites. Nocturnal migration began during the first hour after sunset and ended before sunrise at all observation sites except the eastern shore of the Caspian Sea. The numbers of migrants increase rapidly after sunset, decrease slowly during the night, and fall to zero before sunrise (Fig. 2). There is a pause in migratory activity between daytime and nighttime passage across deserts and mountains. After nocturnal flights, most birds pause in the open desert during the daytime. Migrants in open deserts begin nocturnal migratory flights within a short span of time after sunset, but individuals land throughout the night on self-determined schedules.

In spring most birds migrate over lowland deserts, and fewer cross the Tien Shan range and the Pamirs. In autumn birds migrate over both landscapes, but mainly over mountains (Fig. 1). Lowland Asiatic deserts are more favourable for migrants in spring than in autumn. The mountains are, on the contrary, very mild during autumn, but very harsh and snow-covered during the spring migration. It appears that north-bound migrants tend to pass to the west of the mountains in spring.

There are two main directions of migration through the region in both seasons (Fig. 1). One is NW-SE (between Siberia and Indostan) and the other is NE-SW (between Siberia and Africa). The migratory vector in autumn shifts from SW in the western part of the region toward S in the eastern part, in

[1]Zoological Institute of the USSR, Academy of Sciences, Leningrad 199034, USSR

E. Gwinner (Ed.) Bird Migration

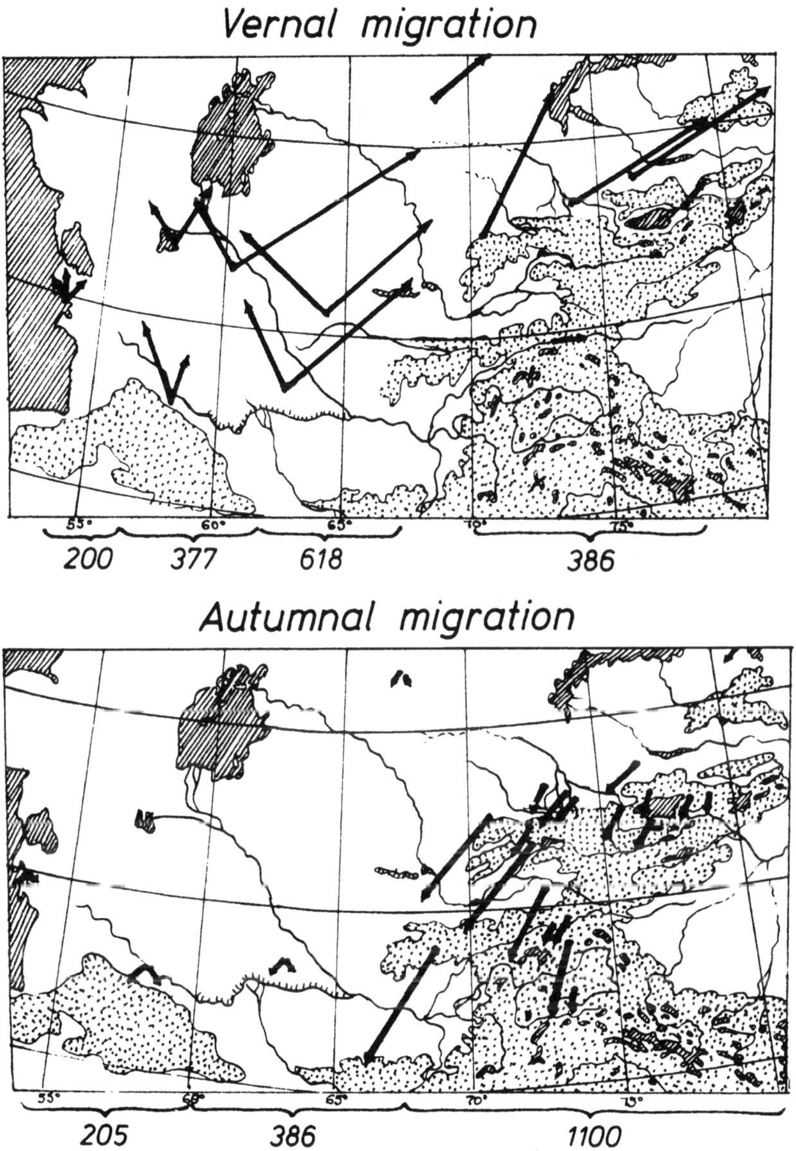

Fig. 1. Density (*length of arrow*) and main direction (*azimuth of arrow*) of noctural bird migration above moon-watch sites indicated by the origins of the arrows. The *numbers* beneath the maps show the total number of migrants (in millions) that passed in each season through the fronts delimited by the *brackets*. Large water bodies are indicated by *diagonal hatching*, the largest being the eastern shore of the Caspian Sea at the *left* and the Aral Sea at longitude 60° E. Mountain ranges are *stippled*; the Elburz range is at the *lower left*; the Hindu Kush, the Pamirs, and the Tien Shan, respectively, are eastward from ca. 65° E

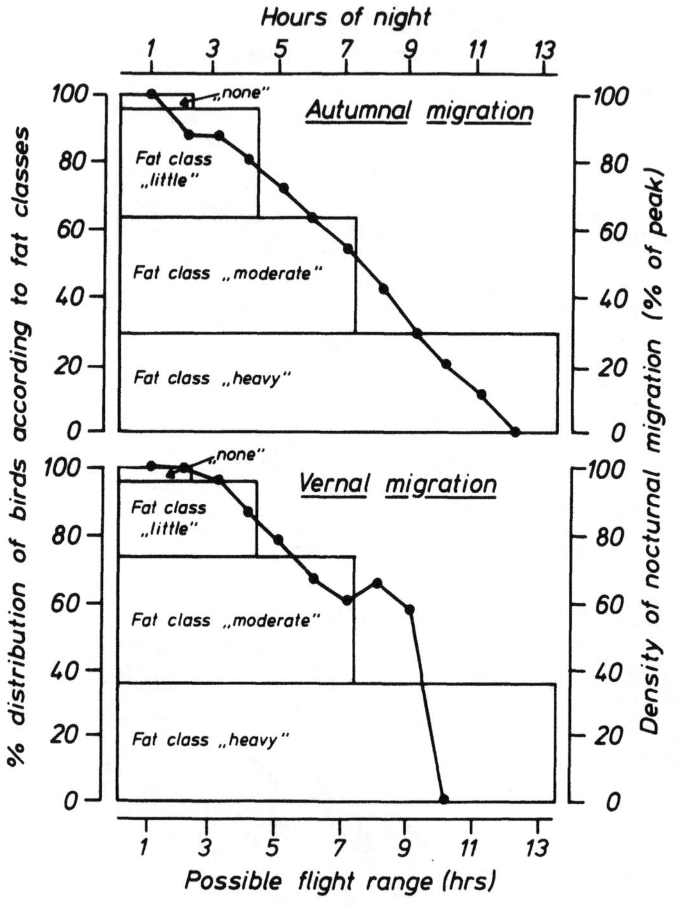

Fig. 2. Frequency distribution of fat classes and corresponding maximum possible hours of flying (histograms with scales at *left* and *bottom*) in relation to the density of nocturnal passage (by moon-watch, *filled circles*, scales at *right* and *top*) across the desert in birds that initiated nocturnal migration from desert sites

accordance to the diminishing importance of Africa as the destination of migrants from W to E along this transect. In the spring, migrants flying N or NW occur only in the western part of the region. Toward the east, the migrants move strongly only toward the NE, reflecting the passage of birds that had skirted the western edge of the mountains but then headed toward eastern destinations.

The median altitude of nocturnal migration in autumn was 1000 m above the lowland deserts, 700 m above forested zones, and 2000–3000 m above large bodies of water (e.g., the Caspian Sea). The median altitudes of nocturnal passage above the Pamirs were 1300 m in passerines, 2100 m in shorebirds, and 3200 m in ducks (5000, 5800, and 6900 m, respectively, above sea level). A few migrants were observed as high as 9700 m above sea level (Fig. 3).

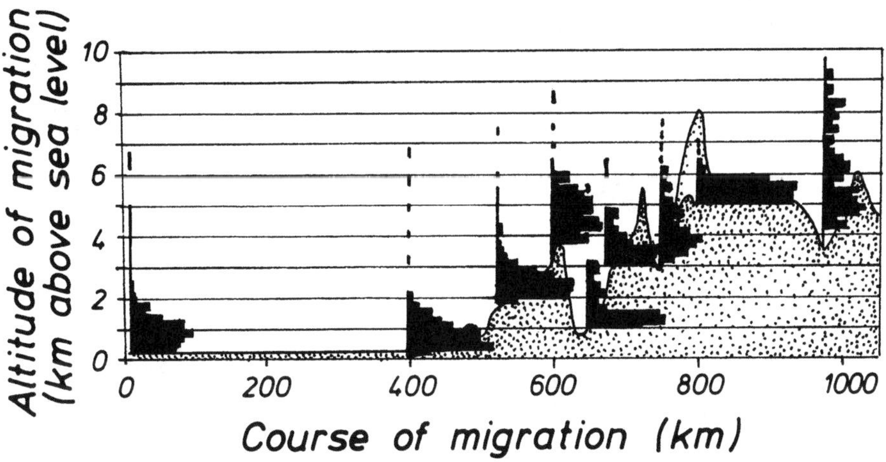

Fig. 3. Frequency distributions (*filled histograms*) of the altitude of nocturnal migration above deserts, the Tien Shan, and the Pamirs in a section along latitude 72°E

2 Patterns of Migratory Flight and Fuel Supplies

Small passerine birds were trapped in mist nets at 10 locations during spring migration and at 17 locations during autumn migration (Fig. 4). One trapping site was on the desert coast of the Caspian Sea (= 0 in Fig. 4), another (A) on a desert island in the Aral Sea, and the rest were at various inland locations. Four were in the subalpine zone (M, N) or high altitude desert (R, T). Six were in lowland sandy or clay deserts (B, C, G, L, Q, and U). Two were in small nonagricultural oases (D, P), two were on the shores of brackish desert lakes (E, F), two were on the shores of freshwater desert reservoirs (K, S), and two were on the shores of desert rivers (H, I). There were no trapping stations located in cultivated areas.

A total of 11 168 birds of 60 species were trapped, banded, and examined. Banded birds were retrapped one to seven times. After banding or upon retrapping, the birds were weighed, classified according to species, sex, age, and fatness (classes I–V), and the wing length was measured. The fat classes were calibrated in terms of the actual fat content of the body (expressed as percent fat-free body mass, or % FFM) by extraction of total body fat (with petroleum ether in a Soxhlet apparatus) from 397 birds of 7 species in which the fat class had been recorded beforehand.

Fat class		Fat content Mean % FFM ± SE
I	None	5.0 ± 0.32
II	Slight	10.4 ± 0.42
III	Moderate	16.6 ± 0.96
IV	Heavy	31.1 ± 1.50
V	Very heavy	37.0 ± 2.10

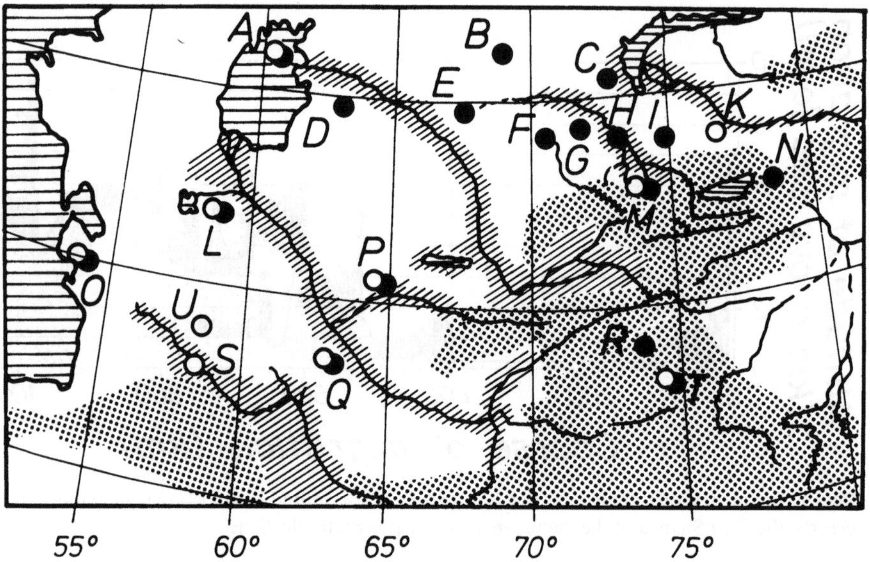

Fig. 4. Locations of trapping stations during the spring (*open circles*) and autumn (*filled circles*) migratory periods. See text for explanation of the *letters*. Elevations above 2000 m are *stippled*; agricultural zones are *hatched*; and deserts are *white*

The measured fat contents were rounded to integers, and the ratio 5:10:17:31:37 was used to estimate the mean fat content of samples of trapped birds from their frequency distribution of fat classes.

The energy density of changes in body mass was estimated by comparison with extractable fat changes. It was 35.7 ± 2.6 kJ/g change of body mass, which did not differ significantly from the energy density of avian at (Yablonkevich et al. 1985). We assumed that the average change of body mass was the same as the change of fat content. The original data for every species and trapping station were published previously. In the present report I have combined data on variously sized species. Fat class and rate of fat deposition are expressed on a per gram body mass basis as they are linearly correlated (slope of 1) with the species-specific body mass. Minimum stopover periods (T) were independent of body mass of the species and were directly pooled. The rate of fat consumption during flight was assumed to be 2.3% loss of body mass per hour ($E_f = 0.05 \times 11.9^{0.69}$ (g/h) for all species; Kendeigh et al. 1977).

2.1 Variations of Body Mass Along the Migratory Route

The body mass of migrants varies as a result of varying fat content in different regions of the migratory route (Fig. 5). The indices M_{max}/FFM and M_h/FFM were larger in the deserts and mountains of Asia than in the forest zone near the

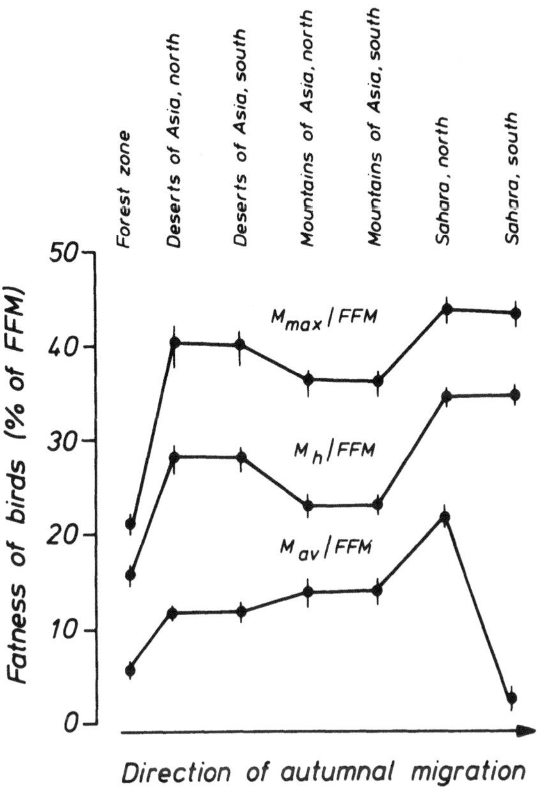

Fig. 5. Average fatness (means ± SE) of all captured birds, (M_{av}/FFM), birds in the highest fat class (M_h/FFM), and maximal fatness (M_{max}/FFM) at selected locations in Asia and Africa during the autumn migration. The data for African birds; presented for comparison, were adapted from Bairlein (1985); Bairlein et al. (1983); Biebach (1985); Biebach et al. (1986); and Lindström (1985)

Baltic Sea, but lower than in the Sahara. These differences may reflect the different foraging possibilities en route with abundant food in the forest zone and almost no food in the Sahara desert. Birds of fat class IV have fuel enough for about 13 h of flight, but this is not enough to enable them to cross the Asian deserts without refueling.

The level of fatness (M_{av}/FFM) in the combined sample of captured birds did not change between the northern and southern borders of the Asian deserts during autumn migration, in contrast to the large decrease across the trans-Saharan region. During vernal migration the fatness of migrants even increases slowly during their passage across the Asian deserts. Migrants traversing these deserts can and do store enough fat at stopover sites to enable them to move in short flights instead of a single nonstop effort.

2.2 Fatness of Migrants in Various Regions

The average fatness in pooled samples and frequency distributions of fat classes varied in different localities (Figs. 6 and 7). The largest fat content was found in birds grounded at Barsakelmes Island (A in Figs. 6 and 7; a desert island) in the Aral Sea during both migratory seasons. Average fatness was less in birds

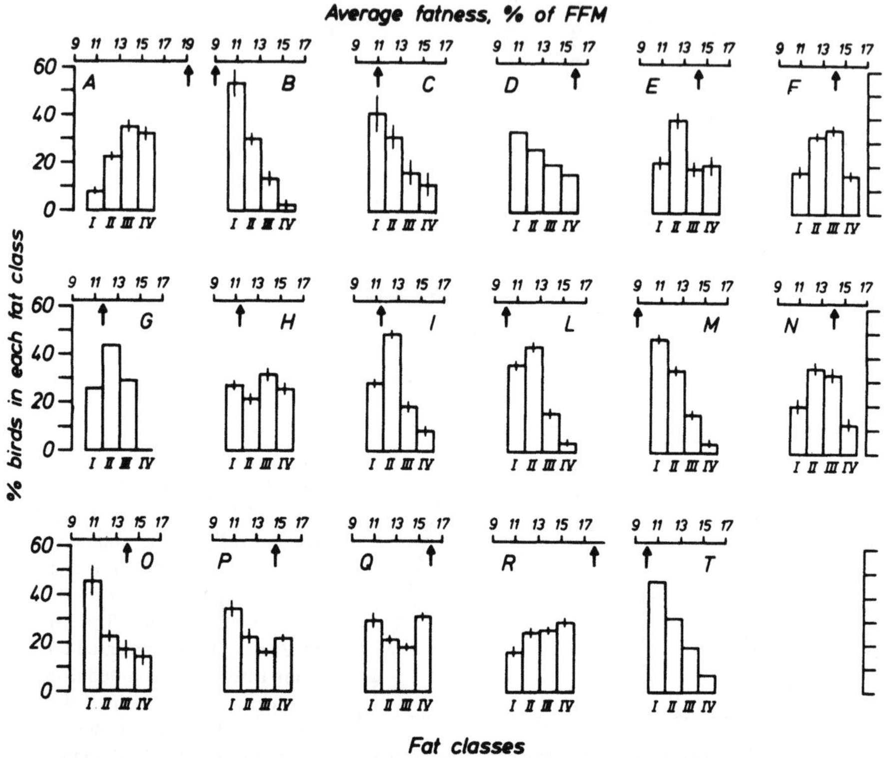

Fig. 6. Frequency distributions of fat classes *I* to *IV* (in %) and average fatness (*arrow* at top axis) in birds captured during spring migration. Means ± SE are given except where data were too scanty to calculate SE. Letters A to T indicate the trapping stations shown on Fig. 4

captured at several points (L and Q in Figs. 6 and 7) in the sandy Karakum deserts where *Haloxylon* sp., *Acacia* sp., *Caligonum* sp., and *Astragalus* sp. were planted, but greater than in birds captured at other localities in sand or clay deserts (Muyunkum, Kysilkum, and Betpakdala deserts; B, C, and G in Fig. 6). In some desert sites most captured migrants were lean.

Migrants trapped at two small oases (D and P in Fig. 6) were moderately fat, as were those trapped at the banks of rivers and of brackish and freshwater lakes in the deserts (E, F, K, S, H, and I in Fig. 6). The fat content of migrants was high in the high-altitude desert of the Alay Valley (3300 m a.s.l.) at the northern border of the Pamirs (R in Fig. 6), but in contrast, was low in both seasons in higher Pamir mountain deserts (3800 m a.s.l.; T in Fig. 6). It is possible that most migrants cross the Pamirs by a nonstop flight. The samples are small from high-mountain systems in the spring because the danger of avalanches makes it difficult to enter the region. We had only one trapping station there in the spring, located in the Tien Shan (2300 m a.s.l.; M in Fig. 6). The number of birds captured was low and all were lean or only with slight fat, in contrast to the abundant captures of fat migrants in the northern foothills of the Tien Shan at the same time (A in Fig. 6).

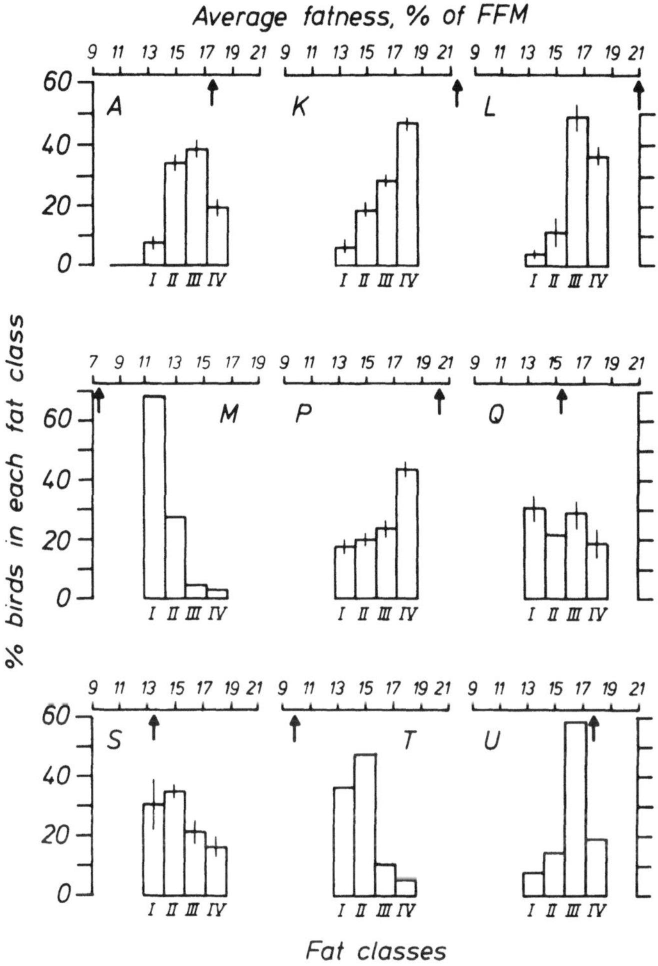

Fig. 7. Frequency distribution of fat classes *I* to *IV* (in %) as in Fig. 6 but during autumn migration. For further explanations see Fig. 6

There were five variants of the frequency distribution of fat classes at the trapping stations (Figs. 6 and 7); (1) monotonic decrease of frequency of fat-classes; (2) monotonic increase; (3) unimodal distribution; (4) nearly uniform; and (5) bimodal frequency, or M-shaped. The first three variants are simple. The more complex forms of the other two results from the mixing of simpler distributions, as suggested by the example in Fig. 8.

2.3 Actual Number of Birds Stopping Over

Some of the birds banded at a trapping station remain there for several days but are not recaptured. Calculation of the proportion of migrants stopping over based

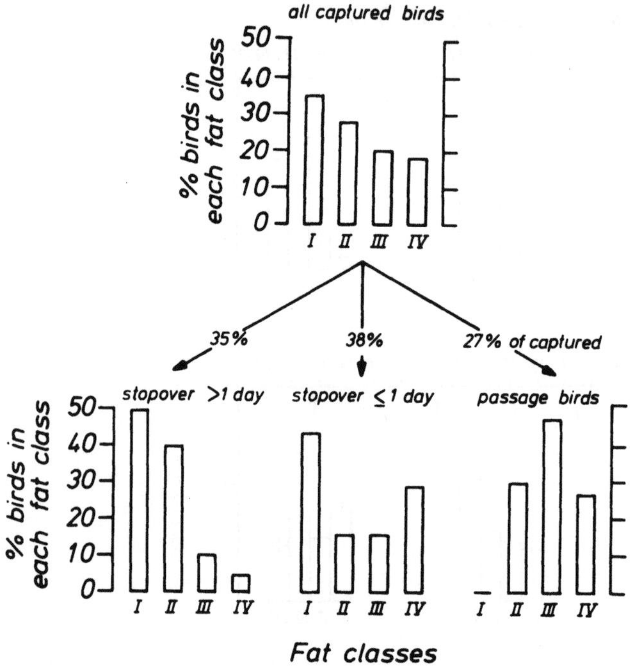

Fig. 8. Frequency distribution of fat classes in autumn at a small oasis in the northern Karakum desert, segregated as recaptures over more than 1 day (*left*), recaptures over 1 day (*middle*), and passage birds that were caught only once (*right*). The numbers and fat classes were corrected according to their different probability of recapture in the three groups (see text)

only on the numbers of recaptured birds therefore underestimates the true proportion. The following procedure was used to estimate the number of birds actually stopping over (N_s) at a trapping station:

$$R = N_3/N_2$$
$$N_s = N_2/R$$
$$\text{or } N_s = (N_2)^2/N_3.$$

N_2 = number of banded birds recaptured twice;
N_3 = number of banded birds recaptured three times;
R = efficiency of recapturing stopover birds.

Birds recaptured within 1 day or less were considered separately from those recaptured at longer intervals because the former group is a mixture of birds actually stopping over and birds pausing between successive days or nights of migration but caught only once. The number of passage migrants (those not stopping over) was calculated by subtracting the number of all recaptured birds from the total number of banded birds. The estimated percentages of birds with a stopover period of more than 1 day was large in the subalpine and alpine zones in the autumn (but nil in the spring), near freshwater lakes in the spring, and at

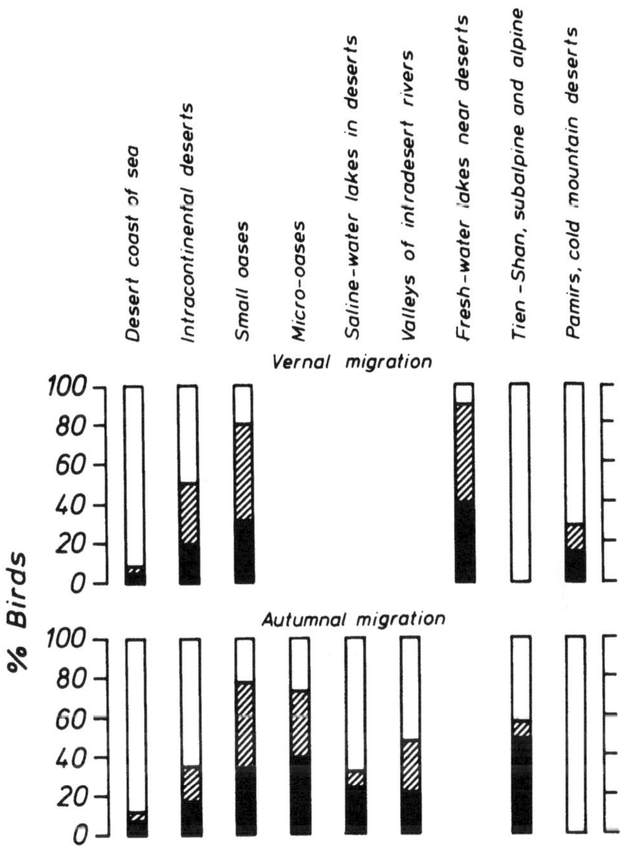

Fig. 9. Proportions of stopover birds (≤ 1 day (hatched) and > 1 day (black) and passage migrants (open) in various biotopes. The percentages were estimated as explained in the text

small oases in both seasons (Fig. 9). The proportion of these birds at desert sites varied from zero to 30% of all banded birds, and was slightly larger in spring than in autumn.

2.4 Fat Reserves of Stopover Birds

At most of the trapping stations the average fatness of migrants was lower in stopover birds than in all banded birds combined (Fig. 10). The difference was smallest in the mountains and largest at small oases. The smaller fat classes predominated in the stopover birds and the larger fat classes predominated in the passage birds. Birds stopping over in deserts were on average fatter than birds stopping at oases, where the proportion of lean birds was higher than at desert sites. Oases appear to attract lean birds.

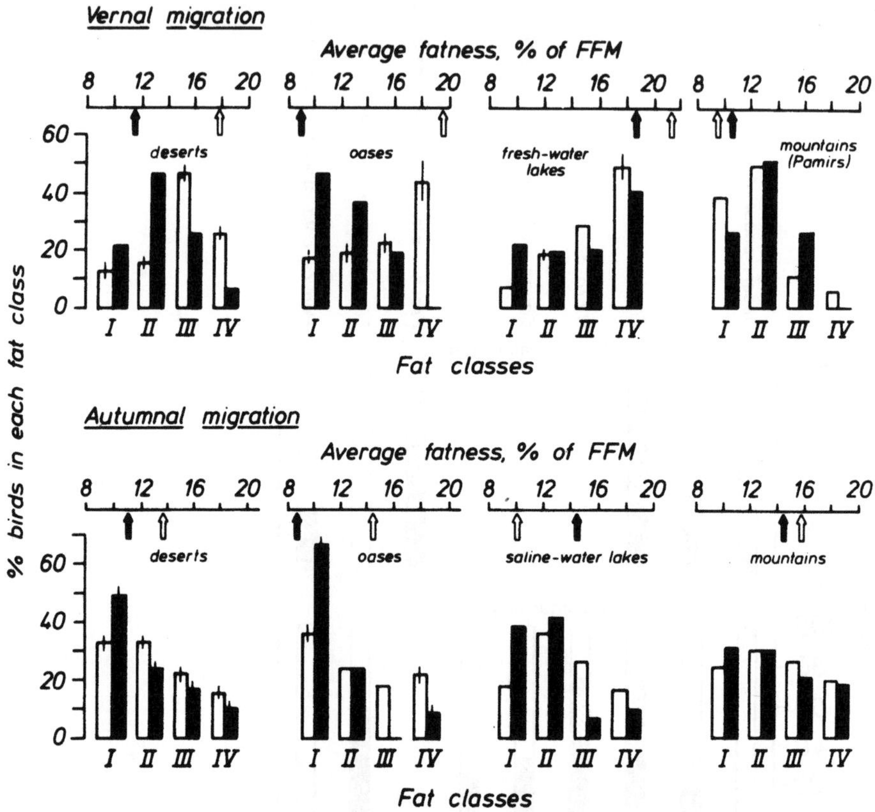

Fig. 10. Average fatness (scale at *top*) and frequency distributions of fat classes in recaptured migrants (*black arrows* and *black columns*) compared with all captured birds (*white arrows* and *white columns*) in various biotopes. For further explanations see Fig. 6

2.5 *Gain or Loss of Body Mass by Stopover Birds*

The body mass of about 8–15% of stopover birds was the same at the time of the final recapture as at the time of initial capture. Body mass in the other stopover birds may either increase or decrease between initial and final captures (Fig. 11). The proportion of birds in which body mass increased tended to be largest in the mountains and near rivers and lakes (40–60%). The proportion was 35–40% in deserts in both seasons, and 40 and 48% at oases in autumn and spring, respectively. The proportion of birds that gained mass was independent of initial body mass in samples from mountains and freshwater lakes, but was correlated inversely with initial fatness in deserts, oases, and brackish lakes (Fig. 12).

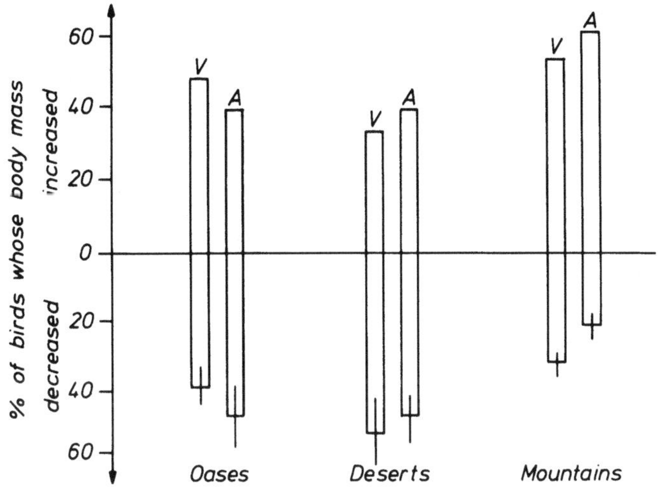

Fig. 11. Percentages of recaptured birds in various biotopes in which body mass increased or decreased between initial and final captures. *V* Vernal; *A* autumnal migration. In all three biotopes and both seasons body mass was unchanged in about 10% of all birds. For further explanations see Fig. 6

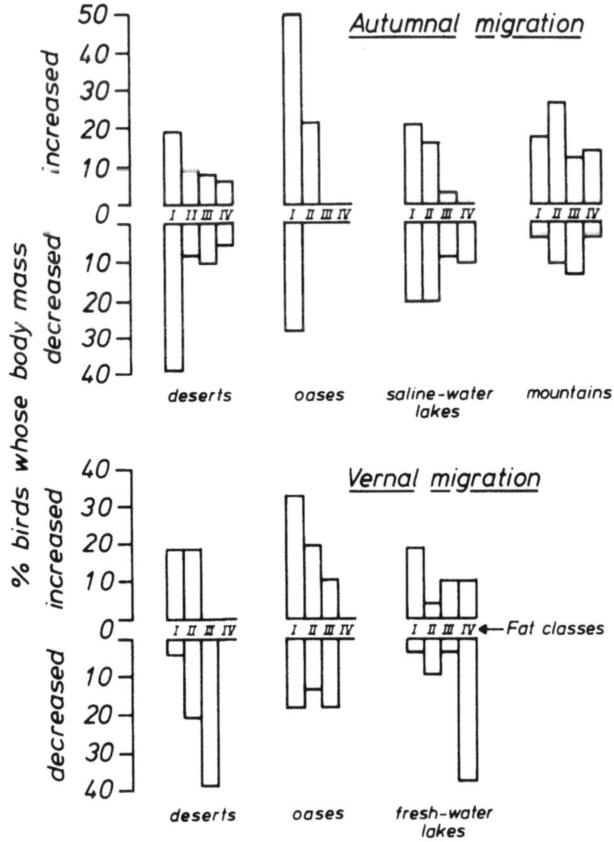

Fig. 12. Percentage changes of body mass in recaptured birds in relation to fat class of the same birds when they were first captured

2.6 Rate of Body-Mass Gain by Stopover Birds

To calculate this rate only data from birds that had increased their body mass between capture and final recapture were used. The rate at which body mass increased during stopover periods was greatest in mountains and desert rivers in autumn and it was low in the Asian deserts and at Pamirs in both seasons (Fig. 13). In Asian deserts, the rate of body-mass increase was 1.4 times higher in spring than in autumn. The average rate of body-mass increase was lower in Asian deserts than in European forest zones, but higher than in Sahara oases. During spring migration body-mass increase was higher in small oases than in deserts but during fall migration there were no differences between these two places. The lack of a difference in autumn presumably was due to the high concentration of birds in oases resulting in high competition. The maximal rates of body-mass gain (also as % FFM/day ± SE) were:

	In spring	In autumn
In deserts	9.0 ± 1.7	6.2 ± 1.3
In oases	—	5.9 ± 0.6
Near desert rivers	7.2[2]	—
In mountains	—	8.4 ± 1.0

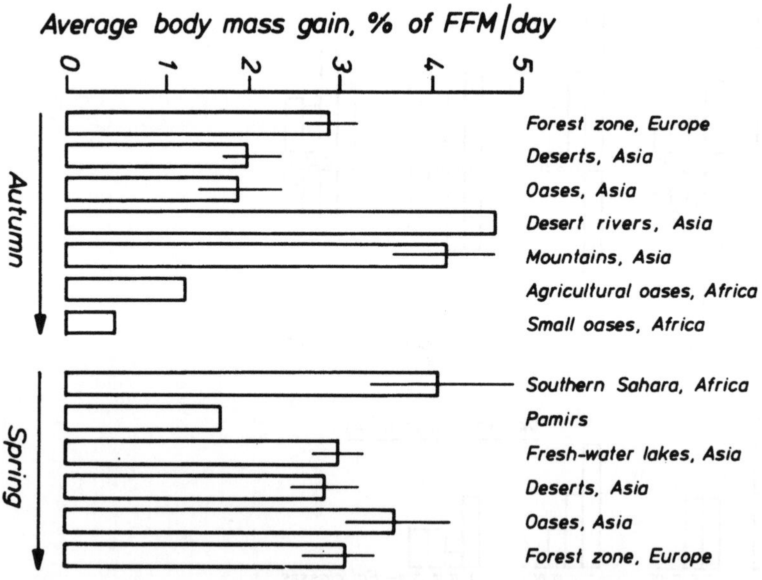

Fig. 13. Average rate (% FFM/day) of body-mass gain in migrants stopping over in selected biotopes in Asia and Africa. The data from Africa were adapted from Bairlein (1985); Bairlein et al. (1983); Biebach et al. (1986); and Lindström (1986). Data for forest-zone birds were obtained near the Baltic Sea. The sequence of biotopes from *left to right* is the order in which migrants encounter them en route. For further explanations see Fig. 6

[2]Data were too scanty to calculate SE.

Note that even in the desert some birds are able to find favourable places at which they achieve very high rates of fat storage. These places in the Asian desert are probably found by short flights during day and night.

One day's fat deposition at the average rate for the particular biotope supplies enough fuel for 1.14 h of flight by a migrant stopping over in the desert in spring, 0.58 h if stopped in the desert in autumn, 0.66 h if stopped at an oasis in autumn, and 1.70 h if stopped in mountains in autumn.

2.7 Average "Minimum" Stopover Period (MSP)

The average time intervals (in h) between first capture and final recapture were:

	In spring	In autumn
In deserts	77 ±20	91 ±12
In oases	98 ±19	97 ±14
At brackish lakes	–	88 ±10
At freshwater lakes	99 ±11	–
In mountains	79 ±17	76 ± 7

The average MSPs of all stopover birds in a particular biotope, as summarized above, provide only limited and potentially misleading information because the MSP is affected by many factors (e.g., locality, initial fatness, proportions gaining or losing body mass, and so forth) whose influence may differ among trapping stations or biotopes (Fig. 14). Theoretically, MSP should be correlated inversely with a migrant's fat reserves when it arrives at a stopover site if the aim of stopping is to restore these reserves. This simple relationship occurred in general only in migrants in which body mass increased during the stopover (Fig. 14). In migrants in which body mass decreased during the stopover the MSP tended to be unimodal in relation to fat class. Lean birds (class I) stopped for a short period, moderately fat birds (class II) stopped for a longer period, and fat birds (classes III and IV) stopped for an intermediate or short period. In general, the class II, III, and IV birds tended to stop over longer in the more attractive vicinity of lakes, rivers, and oases than in the more homogeneous landscapes of open deserts and mountains. I suggest that migrants that are lean upon arrival at a stopover site, and hence confronted by very limited energy (and time) in adjusting to the foraging conditions there, soon move on to another site if food-finding is difficult. If fat birds encounter poor feeding conditions at a stopover site they tend to move on. They have enough fuel reserves for several short flights in search of a good feeding place. Some very fat birds even lost body mass slightly, or failed to gain any, while stopping over for several days at attractive resting sites such as rivers or freshwater lakes. In contrast, the moderately fat (class II) birds lack enough fuel to support several flights in search of more favorable sites, but they may have enough energy and time to allow them to learn, over several days, how to forage effectively in their current locality. Note that most localities in the Asian deserts have an

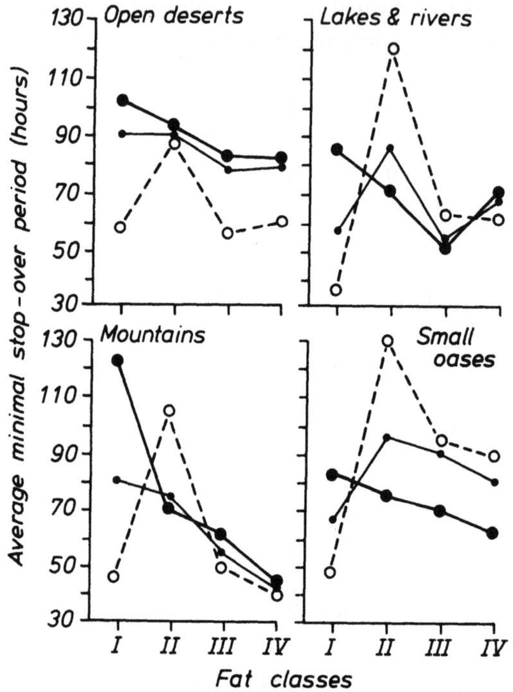

Fig. 14. Average minimum stopover periods in migrants recaptured after more than 1 day in relation to initial fat class and to the direction of body-mass change between initial and final capture. *Small dots* All recaptured birds pooled; *larger filled circles* only birds in which body mass increased; *open circles* only birds in which body mass decreased

adequate potential food supply for migrants, and the birds' principal problem is adjustment to new foods and new methods of foraging during the course of migration.

2.8 The Actual Stopover Period

It is clear that the actual stopover period must be longer than the MSP because a bird arrives at a stopover site at an unknown time before it is captured and departs at an unknown time after its final recapture. The following procedure was used to estimate a more realistic, actual stopover period (T, h). The average intervals (t, h) between the first and second, second and third, third and fourth, and so on, recaptures were 33 ± 4.7, 33 ± 6.1, 32 ± 6.3, 30 ± 7.4, 27[3] and 34[3] h in 44 bird captured three to seven times with an overall average of 31 h. The probability, k, of recapturing a particular bird was estimated as the ratio of the number of birds recaptured a third time to the number recaptured a second time, the number recaptured a fourth time to the number recaptured a third time, and so on. The ratios varied from 0.32 to 0.40 and averaged 0.34. The actual stopover period was then estimated as:

$$T = t + t \cdot k + t \cdot k^2 + t \cdot k^3 + t \cdot k^4 \ldots\ldots, \text{or}$$
$$T = 33 + 33 \cdot 0.34 + 33 \cdot 0.34^2 + 33 \cdot 0.34^3 \ldots\ldots\ldots = 50\ \text{h}.$$

[3]Data were too scanty to calculate SE.

The actual stopover period (50 h) averaged 1.5 times longer than the MSP (33 h).

Using the multiplier 1.5, the data on MSP, and the data on average rates of body-mass increase, it was estimated how long migrants could fly using fat that they had stored during actual stopover periods at representative sites:

Deserts, autumn	5.0 h
Deserts, spring	5.8
Oases, autumn	5.0
Mountains, autumn	8.6

2.9 The Fat Reserves of Passage Migrants

Data on the frequency distributions of fat classes in passage migrants (birds that paused between successive migratory flights but did not stopover) at all desert localities were combined and average distributions were calculated for deserts during spring and autumn migrations. The potential flight ranges (in h) were also estimated for these birds (Fig. 2). The flight ranges estimated from the distribution of fat classes predicts that the density of migrants will decrease during the course of the night, as was observed directly by the moon-watching technique (Fig. 2).

As shown by Fig. 15, passage migrants were much fatter in the Saharan Desert than in Asian deserts both at oases and in the open desert. This reflects different strategies of migration across these desert zones. The strategy in the Sahara, as described by Biebach et al. (1986), is the simpler one, based more on the reliance of large fat reserves and less on effective foraging en route. Some species migrate across both desert zones, but use different strategies in Asia and Africa.

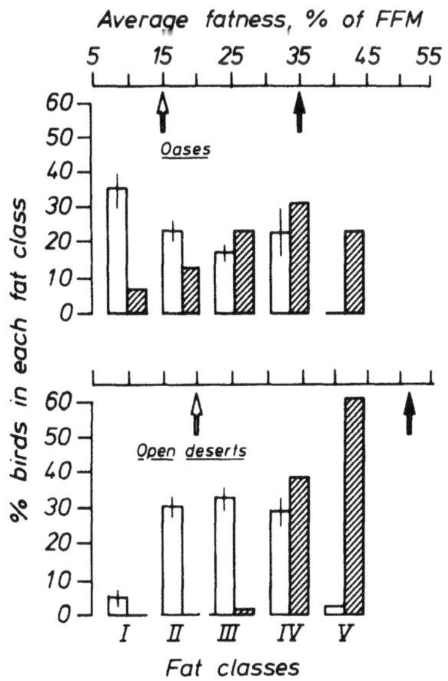

Fig. 15. Average fatness (*top axis* in each panel) and frequency distributions of fat classes in autumn migrants in Asia (*open columns* and *arrow*) and Africa (*hatched column, black arrow*; data adapted from Biebach et al. 1986). For the oasis in Asia (site P in Fig. 4) and the oasis in Africa the pooled data for all captured birds are presented. For the open desert in Asia (sites *A, C, G, L, O,* and *Q* combined) the data shown exclude stopover birds, and hence are for passage migrants only. In the open desert in Africa all the birds captured were passage migrants. For further explanations see Fig. 6

The average fatness of migrants initiating nocturnal flights in Asian deserts was 20.0% FFM in spring and 18.3% FFM in autumn. At some desert stations birds were trapped just at the end of their nocturnal flights. Their average fatness was 10.5% FFM in spring and 9.2% FFM in autumn. The difference between these initial and final overnight averages was 9.5% FFM in spring and 9.1% FFM in autumn, which is equal to enough fuel for about 4 h of nocturnal flight.

3 Comparisons of Independent Results

In a few cases it is possible to examine the reliability of results by comparing estimates derived from independent methods. Using the fuel cost of flight (E_f = 2.3% FFM/h), data on the actual stopover period (T_s), the average observed duration of nocturnal flight (T_f), the average rate of body-mass increase (F_{av}), and the difference of initial and final fat reserves of overnight migrants (F_l), three of four indices of energy balance were calculated during flights across deserts: the expected fat loss during nocturnal flight, the expected fat gain (F_g) during a stopover period, and the expected duration of nocturnal flight:

$$F_g = T_s \cdot F_{av};$$
$$F_l = T_f \cdot E_f;$$
$$F_g = F_l.$$

The concordance between indices measured directly and estimated indirectly is very good in the autumn migration (see above). The slightly greater deviations in spring may result from the small increase of average fatness ($F_g > F_l$) between southern and northern borders of the desert zone in that season, unlike the uniformity observed in the autumn. This may impose a small error on the indirect estimates in the spring.

4 The Main Features of Migration Across the Study Area

Passerine birds cross the desert zone of central Asia on a "wide front" mainly at night. Many birds land at night, and the others land before sunrise. Most of them land in the deserts. In the daytime the migrants forage and feed to restore fat reserves. In general, the loss of fat during migratory flights is balanced by the gain during stopover periods. On average, a migrant accumulates during 1 day of foraging in the desert enough fat to support 1.14 h of migratory flight in spring and 0.58 h in autumn. The average duration of an overnight flight is 4.5 h in both spring and autumn. Many birds begin an overnight flight without enough fuel to fly the entire night.

After landing, many birds move from place to place in search of a favorable stopover site. If a bird discovers a particularly favorable site, it may accumulate fat at up to three times the average rate. The birds diminish the risk of uncertain food supply by storing more fat during the desert crossing than in the mountains, but still store less than their physiological upper limit.

Small oases and desert lakes and rivers do not provide adequate refuges for the 0.75–1.5 billion passerines that migrate across the study area in spring and autumn. The small oases attract too many birds. Because of the intense competition for food, the rate of fat deposition in oases is the same as in the open desert. At the worst, an oasis may be a "black hole" for certain lean birds that do not gain in fat reserves and are not triggered to continue their migration. Remarkable numbers of migrants, mostly lean or with slight fat reserves, stop over in irrigated parts of lowland deserts and foothills.

In the spring, the lowland deserts offer better stopover sites than the snow-covered mountain ranges. Only a small number of migrants cross these ranges by high-altitude nocturnal flight in the spring. Most birds migrating from Indostan pass to the west of the Hindu Kush, Tien Shan, and the Pamirs, across the southern Karakum deserts.

In autumn, in contrast, mountain sites are much more favourable than desert sites for stopping over and refueling, and nocturnal migration is denser over mountains than over deserts in this season. Many birds cross the mountain ranges at very high altitudes, while others fly lower and vary their altitude in accordance to topographic relief. As in the deserts, some migrants in the mountains land at night and the others lands before sunrise, except in the cold high-altitude desert zone of the Pamirs, which most birds cross in both spring and autumn without stopping.

The average fatness of autumn migrants in the mountains is greater than in the desert even though ample food supplies are certain at all stopover sites in this season. These mountain travellers may need extra reserves to diminish the risk of nocturnal flights in the windy atmosphere of this region and to meet the increased cost of thermoregulation during the cold nights.

In general, the basic strategy used by passerine birds migrating across the arid zones of central Asia relies more on effective refueling during stopovers than on maximal fuel storage before starting on the journey. This elicits a moderate rate of travel across deserts, limited by the opportunities for refueling. Such a strategy may be accompanied by selection pressures that favor the evolution of special foraging and feeding adaptations to arid habitats in species that otherwise breed and overwinter in more humid biotopes.

Acknowledgments. I am grateful to Dr. H. Biebach and Dr. J. King for critically reading the manuscript and improving the English.

References

Bairlein F (1985) Body weights and fat deposition of Palaearctic passerine migrants in the central Sahara. Oecologia (Berl) 66:141–146

Bairlein F, Beck P, Feiler W, Querner U (1983) Autumn weights of some Palaearctic migrants in the Sahara. Ibis 125:404–407

Biebach H (1985) Sahara stopover in migratory flycatchers: fat and food affect the time program. Experientia (Basel) 41:695–697

Biebach H, Friedrich W, Heine G (1986) Interaction of body mass, fat, foraging and stopover period in trans-Saharan migrating passerine birds. Oecologia (Berl) 69:370–379
Dolnik VR (ed) (1985a) Spring nocturnal bird passage over arid and mountainous regions of Middle Asia and Kazakhstan. Zool Inst USSR Acad Sci, Leningrad, p 296 (in Russian, English summaries)
Dolnik VR (ed) (1985b) Energy resources of birds migrating across arid and mountainous regions of Middle Asia and Kazakhstan. Proc Zool Inst USSR Acad Sci Leningrad 137:1–195
Dolnik VR (ed) (1985c) Nocturnal bird migration over arid and mountainous regions of Middle Asia and Kazakhstan. Proc Zool Inst USSR Acad Sci Leningrad 169:1–147 (in Russian, English summaries)
Dolnik VR (ed) (1987a) Autumn nocturnal bird passage in southern Middle Asia and Kazakhstan. Proc Zool Inst USSR Acad Sci Leningrad 169:3–7 (in Russian, English summaries)
Dolnik VR (ed) (1987b) Study of bird migration in arid and mountainous regions of Middle Asia and Kazakhstan. Proc Zool Inst USSR Acad Sci Leningrad 173:3–12 (in Russian, English summaries)
Kendeigh SC, Dolnik VR, Gavrilov VM (1977) Avian energetics. In: Pinowski J, Kendeigh SC (ed) Granivorous birds in ecosystems. IBP, vol 12. Cambridge University Press, pp 127–204
Lindström A (1986) Fat deposition in migrating birds. Introductory paper no 44. Reprocentralen Lunds Universitet, Lund, pp 5–26
Yablonkevich ML, Shapoval AP, Dolnik VR (1985) Water and energy resources of passerine birds migrating in autumn across the Middle Asian deserts. Zool Zh 64:877–888 (in Russian, English summary)

Migration Strategies and Tactics of Waders Breeding in Arctic and North Temperate Latitudes

P.R. Evans[1] and N.C. Davidson[2]

1 Introduction

Most waders (often known as shorebirds, particularly in North America) share several characteristics that are of particular relevance to their migration strategies. Most are ground-nesting birds; many lay a fixed clutch of four eggs and few attempt second or even replacement clutches except in the southernmost (temperate) parts of their breeding ranges. Breeding success is highly irregular from year to year, either because of unpredictably severe weather, particularly in the arctic, or because of predation, by arctic foxes in relation to lemming cycles, (Summers and Underhill 1987) or by a much wider spectrum of predators in temperate areas (Pienkowski 1984). The strategies adopted by individual adults of these species to maximize lifetime reproductive success are to maximize their reproductive life spans by minimizing losses on the breeding grounds, on migration and in winter. As adults, they have considerably lower annual mortalities than do passerine migrants of the same size and body mass (Evans and Pienkowski 1984).

Previous reviews of aspects of migration strategies of shorebirds include those by Pienkowski and Evans (1984); Davidson and Evans (1989); Evans and Townshend (1989) and Myers et al. (1985). Descriptions of shorebird migration systems in different parts of the world are given by Morrison (1984) and in books edited by Evans et al. (1984) and Davidson and Pienkowski (1987). This review addresses not only the strategies of migration that have evolved to minimize mortality in an "average" year, but also those (perhaps more appropriately termed "tactics") that are used in direct response to unusually severe conditions imposed by weather, scarcity of food or intraspecific competition, since it is under these conditions that selection may be strongest. Emphasis will be placed on ecophysiological constraints on migration patterns; navigational problems will not be discussed here.

As a group of birds, waders possess several features which make them particularly suitable for the study of problems associated with migration. Several arctic-breeding species undertake some of the longest non-stop flights known amongst land birds; large parts of these flights take place over the oceans. With the exception of a few species such as phalaropes *Phalaropus*, most waders cannot

[1]Department of Biological Sciences, University of Durham, South Road, Durham, DH1 3LE, UK
[2]Nature Conservancy Council, Northminster House, Peterborough PE1 1UA, UK

E. Gwinner (Ed.) Bird Migration

rest on water for long periods. Hence they must have evolved extreme physiological (and navigational) adaptations to undertake these long flights successfully. The feeding habitats they utilize during the non-breeding season, or at staging posts during migration, are highly circumscribed. When in Europe and Africa, the high arctic breeding species tend to use coastal habitats, particularly sand- or mud-flats and river estuaries; the temperate-zone breeding species tend to concentrate on shallow inland wetlands, or flooded agricultural land. (In North America, some populations of high-arctic species, particularly those from central parts of arctic Canada, also use inland wetlands extensively during migration, while those from Alaska and the eastern Canadian arctic tend to follow more coastal routes.) In all these open habitats, waders are relatively easy to observe and count accurately, which assists studies of the time course of their migrations. Most species roost in dense flocks, from which large samples of birds may be caught, by techniques such as cannon-netting, for marking and measurement of body condition indices. Although they are somewhat smaller than wildfowl, waders are large enough to carry unique identification marks, e.g. combinations of colour rings or radio-tags, which enable the behaviour of individuals to be followed in the field at a particular site without the need for recapture.

There are, however, some disadvantages to the use of shorebirds in migration studies. First, birds of a single species from a particular breeding site disperse over a considerable geographical area in the non-breeding season. Unlike geese, therefore, it is difficult to follow more than a few individually marked birds during their complete migrations, once they have left the site of marking. Second, although waders can be kept in captivity successfully, they display migratory restlessness, *Zugunruhe*, in a form which is much less easy to quantify than the wing movements which have proved such a valuable behaviour for laboratory studies of the ecophysiology of songbird migrants (see e.g., Berthold, Gwinner, Wingfield and Schwabl, this Vol.). Finally, most waders feed by night as well as by day, so quantification of changes in total daily food intake before migration can be difficult under field conditions.

2 Migration as a Component of Shorebird Life Cycles

Chicks of most species of waders feed themselves from the moment of hatching. Adults often leave the immediate area of the breeding site just before the juveniles fledge and the young usually migrate independently of the adults, though often in flocks of their own age group. Whilst it is clear that marked year-to-year variations in breeding success occur in the arctic, it is not known to what extent these are compounded by variations in mortality of young during their first migrations to produce the great differences seen from one year to the next in the proportions of juveniles in flocks on the wintering areas. Losses of juveniles on migration may arise from directional errors during flight, as well as from inadequate physiological preparation before departure or choice of inappropriate weather conditions in which to set out. These losses can be high. Amongst records of waders arriving in exhausted condition at staging posts or potential wintering areas, the majority refer to juveniles. Occurrences of waders in unusual habitats in autumn also refer to young birds. However, losses during the subsequent winter appear to be no higher amongst juveniles than amongst adults (Pienkowski and Evans 1985).

Most of the larger species do not breed until they are 3–4 years old and even the smallest species may not breed as one-year-olds if they belong to arctic, rather than temperate zone, populations of a particular species. For example, most British ringed plovers, *Charadrius hiaticula*, which scarcely migrate, breed at 1 year of age, but individuals from arctic populations which winter in northwest Africa usually do not breed until 1 year later. An unknown proportion of such non-breeding 1-year-old birds stay in their "winter" quarters instead of migrating back to the breeding grounds. In turnstone, *Arenaria interpres*, in Scotland and curlew sandpipers, *Calidris ferruginea*, in South Africa, for example, juveniles that overwintered and then stayed into the northern summer did not increase in body mass in April, whereas adults which fed alongside them laid down fat and duly migrated northwards (Elliott et al. 1976; Metcalfe and Furness 1984). In larger species, showing more prolonged delayed maturity, immatures leave the non-breeding areas progressively earlier each year as they grow older, as has been shown in individually marked grey plovers, *Pluvialis squatarola*, in northeast England (Evans, in prep.). Not all such immature migrants travel as far as the breeding grounds, but there is very little information on the precise distances they travel or on the extent and rate of fat deposition before departure from the non-breeding areas. Those immatures that migrate towards the breeding areas often return towards the non-breeding areas in mid-summer, as also do failed breeders, before moulting their plumage, flight as well as body feathers (Chapman 1924). Successful breeding birds of most species return from the arctic to the temperate zone in late summer to moult, whereas several temperate-zone breeding species interrupt their migrations several times to moult parts of their plumage at each staging post, e.g. redshank, *Tringa totanus*, along the coast of western Europe and northwest Africa (Pienkowski et al. 1976). Adults and immatures of most species moult their body plumage again in spring before leaving their wintering areas and/or at a staging post en route to the breeding grounds.

3 Overall Distances Travelled: the Establishment of Migration Patterns

3.1 Interspecific Differences

Many cases are known of differences in the non-breeding areas used by different species breeding in the same country. From Iceland, for example, redshank and black-tailed godwits, *Limosa limosa*, migrate only as far as the British Isles and western France, whereas whimbrel, *Numenius phaeopus*, and ringed plover travel to west and northwest Africa. It is unlikely that physiological constraints lead to this situation. Theoretical arguments suggest that larger species and individuals should be able to overwinter at lower ambient temperatures and therefore higher latitudes than smaller ones (Ketterson and Nolan 1983). However, ecological differences between species, associated with diet and foraging behaviour (as well as the existence in some cases of interspecific competitors), mask any physiological dictates. Pienkowski (1979) pointed out a major differences between two taxa of waders that influences their non-breeding distributions. Plovers, *Charadrii*, with short beaks, forage by sight and depend on the availability and detectability of prey at the ground surface, whereas sandpipers, *Scolopaci*, with longer beaks, often probe deep into the ground and detect prey by touch or chemosensory means; they do not require prey to be active. As temperatures fall, many intertidal crustaceans and polychaetes become less active and so undetectable by plovers while they still remain within reach of probing sandpipers. Thus, in general, plovers need to move to warmer coastal non-breeding areas than sandpipers.

3.2 Intraspecific Differences

Arctic-breeding populations migrate obligatorily to lower latitudes in late summer and early autumn, travelling sufficiently far that they reach sites which, especially if coastal, are normally unfrozen for most of the winter. Individuals of several species are faithful, from year to year, to the area where they settled successfully as juveniles (e.g., Ketterson and Nolan, this Vol.). To reach that area on their first migration they may use several staging posts. Their behaviour is consistent with a model that assumes genetic control of a minimum distance (time) of travel from the breeding grounds, followed in some instances by competition for a place on a potential non-breeding site. Such competition has been demonstrated in juvenile grey plovers by Townshend (1985) on the Tees estuary, northeast England. Larger individuals (as measured by wing length) tended to oust smaller individuals from feeding territories. Some ousted birds migrated further, travelling several 100 km from northeast England to France without staying long, if at all, at any intermediate sites. Juveniles displaced early in the autumn prolonged migration further than those displaced later.

Not all juvenile grey plovers that failed to obtain feeding territories left the estuary; others stayed and fed non-territorially. Conversely, some birds left the estuary early in the autumn without even having attempted to acquire a feeding territory. These birds were heavy for their body size and presumably had used the estuary as a staging post and refuelling area, during the genetically programmed phase of their first migration. All were of smaller body size (wing length) than those that had attempted to stay (Evans and Townshend 1989).

Overtly competitive behaviour is not seen in any juveniles of some wader species, such as sanderling, *Calidris alba*, at potential wintering sites in Britain. Individuals that move on are not significantly smaller than those that stay. In the New World, sanderlings from the Canadian arctic spend the non-breeding seasons chiefly along the coasts of California, Peru and Chile. Those that travel much further encounter better food resources and possibly a decrease in competition with other species (Myers et al. 1985). Whether the energetic costs of long migrations but favourable wintering conditions balance those of short migrations but less favourable winters is not known (see Drent and Piersma, this Vol.).

In addition to the behaviours summarized above, which occurred in every autumn on the Tees estuary, additional grey plovers arrive in late December and early January in winters which are particularly cold. Certain individually marked birds were seen only in the winters of 1978/79, 1981/82, 1984/85 and 1986/87 and presumably had spent milder winters in the Wadden Sea region, as argued by Townshend (1982). Periods of cold weather there later in the winter do not induce westward movements. Birds caught on the Tees estuary in January 1987 included some of these proven cold-winter visitors. The body sizes (wing lengths) were somewhat larger than those that had been successful in the competition for territories as juveniles. This may explain why, in January 1982, one cold-winter visitor succeeded in acquiring a feeding territory by squeezing between and reducing the size of the territories of another two birds that had been on the estuary since late autumn.

If all juvenile grey plovers wintering in western Europe originate from the same part of their Siberian breeding grounds (as seems possible by analogy with the widely scattered winter distributions of individuals of other species ringed on limited parts of their temperate breeding grounds), it may be that larger juveniles are programmed to travel less far than smaller individuals.

The occurrence of extensions to the normal distance of migration, in severe winters only, has been known for many years also in two species of plover that normally feed inland, the lapwing *Vanellus vanellus* and golden plover *Pluvialis apricaria*. "Cold-weather movements" have also been noted in redshank which, unlike most other sandpipers but like the plovers, seems to rely chiefly on sight to detect prey items, at least by day. Furthermore, some redshank feed in winter by freshwater pools and in wet meadows, which may become frozen over.

Not all waders show the same degree of fidelity to the non-breeding sites they used in previous years. In particular, individual knot have been recorded on different western European estuaries in the same month in different years. Since they also move around within each winter (Dugan 1981), the behavioural mechanisms that lead to changes between years are not clear. Nor are the mechanisms known which lead to a large degree of segregation in the wintering areas used by the two sexes of certain waders. In the ruff, *Philomachus pugnax*, for example, those present in the non-breeding season in Britain are chiefly males, whereas those in Africa are chiefly females (Pienkowski and Evans 1984).

4 The Locations, Separation and Timing of Use of Migration Staging Posts

4.1 Interspecific Differences

Several instances are known of different patterns of use of staging posts by shorebirds of different species during journeys from similar breeding areas to common wintering areas (e.g., Piersma 1987). A particularly remarkable example refers to the migrations of Hudsonian godwits *Limosa haemastica* and knot

Calidris canutus from arctic Canada to South America. Both species use James Bay in northern Ontario as a refuelling area. The godwits then fly at least 4500 km non-stop to a (presently unknown) staging post in northern South America, where they refuel sufficiently to fly to southern Argentina, their main non-breeding quarters. The smaller knot reach a similar "wintering" area by a series of shorter flights, with known staging posts on the eastern coast of America and in the Guianas and a suspected staging post in northern Brazil (Morrison 1984).

If physiological limitations controlled the lengths of the successive flight stages, smaller species might be expected to fly longer stages, because maximum fat loads transportable (expressed as *percentages* of total body mass) should be larger in such species (Pennycuick 1969). This does not accord with all interspecific differences found in the field, and the reasons for these differences are not clear. On the eastern Atlantic seaboard, while migrating along the major shorebird flyways from northwest Africa through western Europe to either Greenland or Siberia, knot display some of the longest non-stop flights between staging posts of all the wader species involved (Davidson et al. 1986; Dick et al. 1987) and apparently longer than those flown by the same species in the New World. Possibly their reliance on particular size classes of small bivalves as food restricts the number of sites at which suitably abundant food resources can be guaranteed to occur. Certainly, in winter, knot occur on a much smaller proportion of estuaries in Britain than do most other waders.

Competition between species might also affect the suitability of particular intertidal areas as staging posts. Recher (1966) suggested that during migration shorebird densities on feeding grounds were higher than at other periods of the annual cycle, and that segregation amongst different species occurred by microhabitat and differential timings of passage. Schneider and Harrington (1981) also implied that competition could occur between species on staging areas. They found that 60% of the standing crop of macroinvertebrates was removed by shorebirds at one staging post during the autumn migration period; but it is not known whether such a large reduction in prey density depressed food intake rates. Insofar as competition usually favours larger species, it is these that might have evolved the most favourable pattern of use of staging sites, both in relation to timing and places visited. On this hypothesis, smaller species would have had to adjust their migratory programmes to avoid the times and places used by larger species taking the same type of prey; this might lead to use of a series of shorter flight stages, with less demand on resources for refuelling at each staging post. Yet not all small species use such a strategy. Any benefits from migration by a few long flight stages may be counterbalanced by increasing reliance on the predictability of food resources at the few staging posts and at the destination. Arrival in the arctic in spring after a long flight, if this meant exhausting fat reserves, would be hazardous, since weather conditions can restrict feeding opportunities unexpectedly.

Piersma (1987) has argued that a series of short flights is always energetically cheaper than one long flight to cover the same distance. This is true of level flight but the costs of climbing to the appropriate altitude at the start of each flight stage are also important. It is also possible that reducing overall flight time may be important. Birds carrying heavy fat deposits, capable therefore of prolonged flight, have higher optimal speeds in the early part of their migration than birds carrying little fat, capable of only short flights (Pennycuick 1969). Thus the overall flight time for the total journey should be shorter if migration proceeds as a single long flight. Departure with large fat reserves and high air speeds will also be advantageous if adverse winds are likely to be encountered.

4.2 Intraspecific Differences

One of the most thoroughly studied migratory pathways is that of the knot. There is a marked difference in timing of use of a major staging area in the Wadden Sea by Nearctic and Siberian breeding populations, the former departing to the arctic in late April and early May, the latter in late May and early June. The Nearctic birds gather in the Wadden Sea after spending the winter on estuaries and coastal flats chiefly around the North and Irish Seas; the Siberian birds winter chiefly in north-west Africa and migrate to the Wadden Sea in early May, scarcely overlapping in time of occurrence with the Nearctic birds (Dick et al. 1987). There is also some segregation of the preferred major feeding areas (Prokosch 1988). The timing difference could be interpreted as avoidance of competition, but may be related simply to the generally later onset of favourable conditions on the Siberian than on the Greenland/arctic Canadian breeding grounds. The difference in feeding sites used within the refuelling area could, however, be related to depletion of prey early in the spring by Nearctic knots in their preferred feeding areas, so that the Siberian birds find better feeding elsewhere, but this has not been examined in detail.

The same species, indeed the same populations, may use different staging posts, and therefore distances flown non-stop between them, at different times of year and in different parts of their migrations. In autumn, adult dunlin *Calidris alpina* from breeding sites in northern European Russia migrate via staging areas around the southern coasts of the Baltic to moulting areas in the Wadden Sea and The Wash (southeastern England). After moult, many move varying distances further to wintering sites around the North and Irish Sea coasts. In spring, migration from these areas back to the arctic probably takes place in a single flight, to judge from the high lipid levels of birds departing from Britain and the absence of staging areas around the Baltic (Davidson and Evans, 1989).

Dunlin migrating from breeding quarters in Iceland to north-west Africa have a long (1000 km) overwater flight to and from the British Isles but, thereafter, in autumn they are seen at many staging posts along the coastline of western Europe and northwestern Africa. It is not clear whether individual birds use many sites or whether this is a population phenomenon. In spring, individual migrants return by reasonably long flights, as has been proven by recoveries in eastern England of birds marked only a few days previously in western France and by studies of fat deposition by migrants in Morocco (Piersma 1987).

Some species use different routes to and from their breeding areas. Although most of the Nearctic breeding population of knot that winters in western Europe migrates in spring via staging areas in Iceland, a proportion (perhaps 20%) travel by way of a staging area in north Norway. (This area includes two fjords some 250 km apart. In any one spring an individual bird probably uses only one or other fjord, but the exact site used may differ between years.) None of the migrants that travel via Norway in May return by that route or staging area in autumn. Some may travel directly to western Europe, most stage in western Iceland.

5 The Ecophysiology of Shorebirds at Migration Staging Posts

Before departure from non-breeding areas or staging posts, most shorebirds increase in body mass, a result chiefly of deposition of fat. The increased daily food intake needed for this may be achieved by increasing intake rate and decreasing vigilance, as in the turnstone (Metcalfe and Furness 1984) or lengthening the period of feeding. Several species also increase the size of their flight muscles at certain staging posts, before and during both spring and autumn migrations (Davidson and Evans 1989, in prep.). Increases in muscle mass might be required to provide extra power, to enable migrants to fly at optimum speeds while carrying substantial masses of fat at the start of each flight stage; or they might be used as a protein store, to enhance survival at the destination or accelerate egg production at the breeding grounds. If flight muscles are to be used as a protein store, their mass should not be allowed to decrease appreciably during long flights. If, however, the premigratory increase is related merely to power requirements, then muscle mass could decrease as total body mass decreases during flight, since progressively less power is required to fly at maximum range speeds.

Studies of knot in northern Norway in late spring indicate that their flight muscle masses immediately after arrival are considerably higher than on the wintering grounds (Table 1) but that further hypertrophy takes place at the staging post. At Ellesmere Island (82°N), one of the highest arctic breeding sites of any wader, knot were collected just after arrival from either Norway or Iceland. Their flight muscle masses were lower than on arrival in Norway but considerably higher than predicted from aerodynamic arguments (Morrison and Davidson 1990, in press). In parallel with these findings, levels of fat in migrants arriving on both the staging site in Norway and the breeding grounds in Ellesmere Island were relatively high, over 10% of total body mass. Thus these knot carry a considerable safety margin of both fat and muscle mass from the wintering area to a staging post, as well as from the final staging post to the breeding grounds. This provides an insurance against poor feeding conditions both at the staging post and in the arctic and probably shortens the journey times. Such reserves appear to be vital for survival in the arctic in years with severe weather in late spring (Morrison and Davidson 1990, in press).

Dates of arrival and duration of stay of knot in northern Norway varied very little between 1985 and 1986 and the fat and muscle levels at arrival and departure in the 2 years were also closely similar (Table 1). This indicates that any differences in food availability between the very cold spring of 1985 and the

Table 1. Body condition of Nearctic knots *Calidris canutus islandica* during their migrations

Site/date	Days after main arrival (before main departure)	Lipid index[a] (%)	Flight muscle index[b]	Sample size
Balsfjord, northern Norway				
12.5.1985	2	11	0.308	19
26.5.1985	(2)	30.5	0.326	16
11.5.1986	1	14	0.308	24
27.5.1986	(1)	33.5	0.327	27
Ellesmere Island, Canada				
1/2.6.1986	2/3	16	0.288	10
Teesmouth, northeastern England				
1.8.1985	2	4.5	0.224	7
10.12.1984	–	8	0.287	10

[a] Lipid index = mass of fat/total body mass.
[b] Flight muscle index: see Evans and Smith (1975).

normal temperatures of 1986 did not affect the rate of muscle hypertrophy or fat deposition. It appears that birds departed either according to a strict time schedule or only when they had reached some threshold level of body composition. In support of the latter interpretation, a sample of knot caught on 28 May 1986 from amongst a few hundred that stayed on the staging post after the main departure had significantly lower body masses than on 27 May.

If the density and availability of food is low enough, it might affect rates of fat deposition and time spent on staging grounds. Consistent with this idea, Piersma (1987) showed that dunlin stayed on a Moroccan estuary for an average of 16 days when food was scarce but only 11 days in the following spring when food was abundant.

Long overwater flights form the first stages of the journey of adult knot from Greenland to the British Isles. Normally they fly directly to the moulting sites (The Wash and the Dutch parts of the Wadden Sea) but in some years, parties of adults in breeding plumage settle briefly on other British estuaries, leaving again before commencing moult. Catches of such birds have been made within a few days of their arrival on the Tees estuary. Total body masses, lipid levels and flight muscle masses (Table 1) were well below those of birds arriving at the Norwegian staging post in spring or at other times of year. This suggests that birds interrupt migration to use an extra staging post when body mass and condition fall below a threshold. Siberian knot sometimes arrive in northwest Africa in autumn with total body masses even lower than those of the Greenland knot described above (Dick and Pienkowski 1979). Suitable refuelling sites along the northwest African coast are few and separated by many hundreds of kilometres, forcing birds to prolong migration even when body reserves fall below the supposed threshold.

On their return migration from northwest Africa in spring, most Siberian knot fly directly to a staging post in the German Wadden Sea, but some birds stage also in the Vendee, western France (Dick et al. 1987). Such birds, caught in May soon after arrival, had body masses similar to those knot stopping on the Tees, short of their normal destination, in autumn. Birds pausing at the Vendée departed after a few days, once their body masses exceeded about 130 g, i.e. with fat reserves sufficient only to complete their journey to their normal refuelling site in the Wattenmeer. Possibly such birds failed to reach there directly because they met headwinds en route or had departed from northwest Africa with inadequate fat reserves, the result perhaps of a strictly controlled time programme dictating their date of departure. Indeed, they may have been individuals that had spent the northern winter in South Africa and had arrived late at their northwest African staging post. This idea is developed further by Piersma (1987).

At the moment, most information on body condition of waders at both regular and additional staging posts refers to total body mass and, by inference, to fat reserves. Information on muscle hypertrophy is sparse and only one study has been made so far of the accompanying changes in the structure of shorebird flight muscles. In knot at the north Norwegian staging post and in dunlin and sanderling before departure from wintering sites in northeast England, about two-thirds of the increases in flight muscle mass resulted from increases in myofibril mass (associated with a proportionate increase in cross-sectional area) and about one-quarter from additional mitochondrial mass. Sarcoplasm increased very little (Davidson and Evans, in prep.).

6 Behaviour of Migrant Waders at Departure and in Flight

Radar observations indicate that in autumn many shorebirds depart from staging posts in the hour before sunset, in contrast to passerines that leave somewhat later. It is now clear that migratory departures may also take place just before high water, perhaps an hour after feeding has stopped. Possibly birds clear their guts before take-off. Flocks often call as they climb and circle, before finally heading off on a somewhat zig-zag course, not always in the appropriate migratory direction. The rate of climb varies between species (T. Piersma pers. comm.) but it may be that all species eventually select a similar flight altitude, dependent upon favourability of the wind.

Waders tend to fly considerably faster (50–80 kph) than passerines (30-60 kph) (Lack and Parslow 1962; Rayner 1985), and perhaps for this reason are more ready to migrate in light opposed winds as well as with tail winds (reviewed by Richardson 1978, and this Vol.). At present, it is not known whether waders fly at maximum range speeds during migration. The changes in muscle mass recorded in knot cast some doubt on this (Davidson and Evans 1989).

7 Outstanding Ecophysiological Problems in the Migrations of Waders

As already indicated, the development of migration patterns in individuals of wader species that do not breed until they are 3–4 years old is poorly understood. Mechanisms of control of the timing of their migrations (progressively earlier in spring), the distances travelled (progressively further each spring) and of the extent of fat deposition and muscle hypertrophy in different years await detailed study. So also, remarkably little is known about physiological preparation for the first migration of juveniles, in particular how much variation in body condition occurs amongst individuals leaving the same breeding grounds, variation that may lead to mortality before wintering areas are reached. Understanding of the reasons for the differing types of migratory strategies, e.g. short or long flights between staging posts; levels of fat deposition and muscle hypertrophy, needs further development. Finally, it is not known whether the biochemical mechanisms involved in preparation for migration in waders are identical to those in passerines.

References

Chapman A (1924) The borders and beyond – arctic, Cheviot, tropic. Gurney & Jackson, London, Edinburgh

Davidson NC, Evans PR (1989) Prebreeding accumulation of fat and muscle protein by Arctic-breeding shorebirds. Proc 19th Int Ornithol Congr, Ottawa, pp 342–352

Davidson NC, Pienkowski MW (eds) (1987) The conservation of international flyway populations of waders. Wader Study Group Bull 49, Suppl/IWRB Special Publ 7

Davidson NC, Strann K-B, Crockford NJ et al. (1986) The origins of knots in arctic Norway in spring. Ornis Scand 17:175–179

Dick WJA, Pienkowski MW (1979) Autumn and early winter weights of waders in north-west Africa. Ornis Scand 10:117–123

Dick WJA, Piersma T, Prokosch P (1987) Spring migration of the Siberian knot *Calidris canutus canutus*: results of a co-operative Wader Study Group project. Ornis Scand 18:5–16

Dugan PJ (1981) Seasonal movements of shorebirds in relation to spacing behaviour and prey availability. PhD Thesis, Durham University, UK

Elliot CCH, Waltner M, Underhill LG, Pringle JS, Dick WJA (1976) The migration system of the curlew sandpiper *Calidris ferruginea* in Africa. Ostrich 47:191–213

Evans PR, Pienkowski MW (1984) Population dynamics of shorebirds. In: Burger J, Olla BL (eds) Shorebirds: breeding behavior and populations. Plenum, New York (Behavior of marine animals, vol 5)

Evans PR, Smith PC (1975) Fat and pectoral muscle as indicators of body condition in the bar-tailed godwit. Wildfowl 26:64–76

Evans PR, Townshend DJ (1989) Site faithfulness of waders away from the breeding grounds: how individual migration patterns are established. Proc 19th Int Ornithol Congr, Ottawa, pp 594–603

Evans PR, Goss-Custard JD, Hale WG (eds) (1984) Coastal waders and wildfowl in winter. Cambridge University Press, Cambridge

Ketterson ED, Nolan V Jr (1983) The evolution of differential bird migration. In: Johnston RF (ed) Current Ornithology, vol 1. Plenum, New York, pp 357–402

Lack D, Parslow JLF (1962) Falls of night migrants on the English east coast in autumn 1960 and 1961. Bird Migr 2:187–201

Metcalfe NB, Furness RW (1984) Changing priorities: the effect of pre-migratory fattening on the trade-off between foraging and vigilance. Behav Ecol Sociobiol 15:203–206

Morrison RIG (1984) Migration systems of some New World shorebirds. In: Burger J, Olla BL (eds) Shorebirds: migratory and foraging behavior. Plenum, New York (Behavior of marine animals, vol 6)

Morrison RIG, Davidson NC (1990) Migration and arrival patterns of shorebirds at Alert, Ellesmere Island, NWT, Syllogeus (Bull Natl Mus Nat Hist Ottawa) (in press)

Myers JP, Maron JL, Sallaberry M (1985) Going to extremes: why do sanderlings migrate to the Neotropics. In: Buckley PA, Foster MS, Morton ES (eds) Neotropical ornithology. Ornithol Monogr 36:520–535

Pennycuick CJ (1969) The mechanics of bird migration. Ibis 111:525–556
Pienkowski MW (1979) Differences in habitat requirements and distribution patterns of plovers and sandpipers as investigated by studies of feeding behaviour. Verh Ornithol Ges Bayern 23:105–124
Pienkowski MW (1984) Breeding biology and population dynamics of ringed plovers *Charadrius hiaticula* in Britain and Greenland: nest-predation as a possible factor limiting distribution and timing of breeding. J Zool (Lond) 202:83–114
Pienkowski MW, Evans PR (1984) Migratory behavior of shorebirds in the western Palearctic. In: Burger J, Olla BL (eds) Shorebirds: migratory and foraging behavior. Plenum, New York (Behavior of marine animals, vol 6)
Pienkowski MW, Evans PR (1985) The role of migration in the population dynamics of birds. In: Sibly RM, Smith RH (eds) Behavioural ecology: the ecological consequences of adaptive behaviour. Blackwells, Oxford (Symposium of Brit Ecol Soc)
Pienkowski MW, Knight PJ, Stanyard DJ, Argyle FB (1976) The primary moult of waders on the Atlantic coast of Morocco. Ibis 118:347–365
Piersma T (1987) Hink, stap of strong? Reisbeperkingen van arctische steltlopers door voedselzoeken, vetopbouw en vliegsnelheid. Limosa 60:185–194
Prokosch P (1988) Arktische Watvögel in Wattenmeer. Corax 12:273–442
Rayner JMV (1985) Article 'Flight, speeds of'. In: Campbell B, Lack E (eds) A dictionary of birds. Poyser, Calton
Recher H (1966) Some aspects of the ecology of migrant shorebirds. Ecology 47:393–407
Richardson WJ (1978) Timing and amount of bird migration in relation to weather: a review. Oikos 30:224–272
Schneider D, Harrington BA (1981) Timing of shorebird migration in relation to prey depletion. Auk 98:801–811
Summers RW, Underhill L (1987) Factors related to breeding production of brent geese *Branta b. bernicla* and waders (*Charadrii*) on the Taimyr Peninsula. Bird Study 34:161–171
Townshend DJ (1982) The Lazarus syndrome in grey plovers. Wader Study Group Bull 34:11–12
Townshend DJ (1985) Decisions for a lifetime: establishment of spatial defence and movement patterns by juvenile grey plovers *Pluvialis squatarola*. J Anim Ecol 54:267–274

An Exploration of the Energetics of Leap-Frog Migration in Arctic Breeding Waders*

R. DRENT[1] and T. PIERSMA[1,2]

1 Introduction

Of the many problems posed by the phenomenon of bird migration, the interpretation of current migratory patterns in an evolutionary perspective remains a very thorny one. A widespread feature is for populations of the same species occupying distinctly disjunct breeding ranges to show differences in their wintering range as well. In some cases the more northerly breeding forms pass over the other populations to winter beyond the more southerly breeding forms. This migratory pattern, termed leap-frog migration, was first recognized more than 50 years ago and has been extensively discussed by Salomonsen (1955a,b), who in particular extended the list to include a number of waders. All authors who have grappled with this problem are in agreement that the displacement of the populations in winter must be attributed to the avoidance of competition (for review see Greenberg 1986). We accept this essential ingredient, and from this basic premise we will explore the implications of the spectacular differences in wintering area for the energy budgets of populations so segregated. In essence, we believe that pinpointing the essential differences in the energetic balance of the leap-frogging component may help to identify the selective advantages associated with the evolution of this migratory pattern.

2 An Example: Bar-tailed Godwits

Waders have provided classical examples of leap-frogging migrant birds (Salomonson 1955a,b): ringed plover *Charadrius hiaticula* and redshank *Tringa totanus*. A few other species, e.g. dunlin *Calidris alpina* (Alerstam and Högstedt 1980, 1985; Pienkowski et al. 1985) and snipe *Gallinago gallinago* (Gudmundsson 1988), also display some of the characteristics of leap-frog migrants. Our involvement with studies on the feeding ecology, energetics and migrations of bar-tailed godwits *Limosa lapponica* (Piersma 1982; Piersma and Jukema 1990; Smit

*Dedicated to Prof. Dr. J. Aschoff in recognition of his long-term support for studies of both animal energetics and migration

[1]University of Groningen, Zoological Laboratory, P.O. Box 14, 9750 AA Haren, The Netherlands
[2]Netherlands Institute for Sea Research, P.O. Box 59, 1790 AB Den Burg, Texel, The Netherlands

E. Gwinner (Ed.) Bird Migration

and Piersma 1989) has recently led us to recognize that this species represents a particularly neat example of a leap-frog migrant.

Bar-tailed godwits are a sexually dimorphic (females are larger and have a paler summer plumage than males) wader of fairly large size in the weight range 200–500 g. A Palearctic species, they breed predominantly on low arctic tundra, ranging from northern Norway in the west, to westernmost Alaska in the east (Cramp and Simmons 1983). Bar-tailed godwits winter in coastal areas at temperate, subtropical and tropical latitudes (Fig. 1). Although the overall wintering range is large, the great majority of bar-tailed godwits along the East

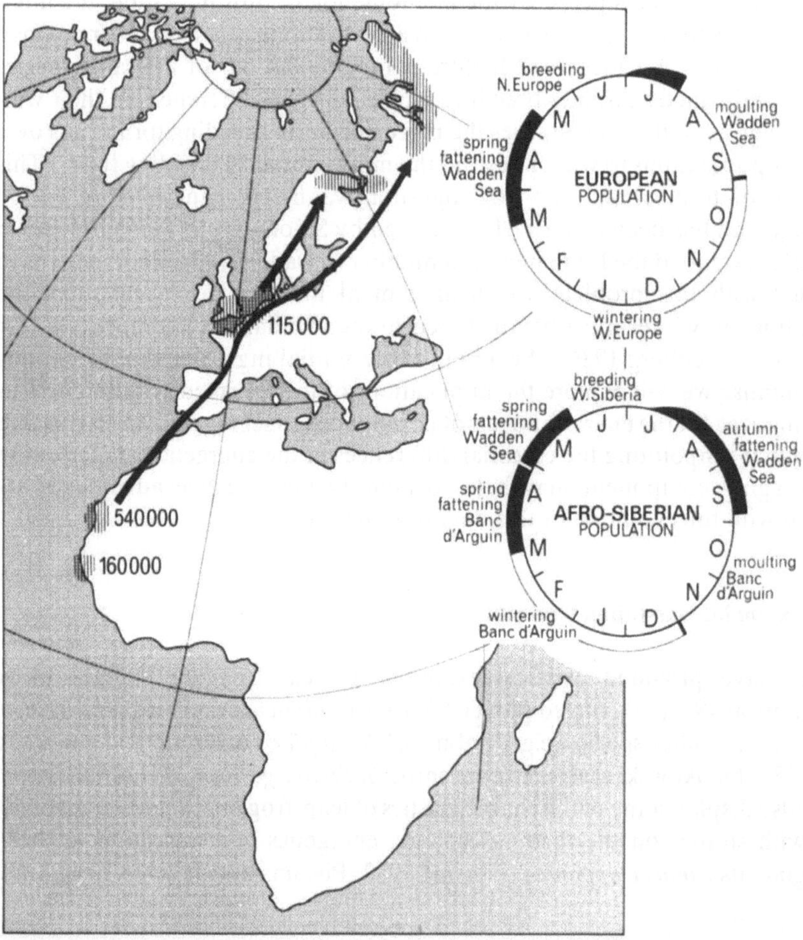

Fig. 1. Diagram of the main breeding and wintering areas of the two populations (European and Afro-Siberian) of bar-tailed godwits using the East Atlantic flyway (breeding areas generalized after Cramp and Simmons 1983; M. Engelmoer in prep.; wintering areas after Smit and Piersma 1989). The annual *clocks* indicate the sequence of phases experienced by birds of the two populations in the course of the year (compiled from Glutz von Blotzheim et al. 1977; Cramp and Simmons 1983; M. Engelmoer in prep.)

Atlantic winter at only three locations (Smit and Piersma 1989). About 115 000 birds winter in the estuaries around the southern North Sea and Irish Sea. The largest wintering concentration (540 000 birds) is found on the Banc d'Arguin in Mauritania, while another 160 000 godwits occur 1000 km more to the south, around the Archipélago dos Bijagos in Guinea-Bissau.

Recent evaluations of information on morphometrics, recoveries and re-sightings, and the timing of migration (Prokosch 1988; Piersma and Jukema 1990; M. Engelmoer, in prep.) have shown that birds wintering in W Europe are likely to all breed in northern Europe, around the basin of the White Sea (Fig. 1; Cramp and Simmons 1983; Mjelstad and Saetersdal 1986). The W African birds move further east and to higher latitudes in summer, their breeding area stretching from the Pechora river to eastern Taymyr Peninsula in Siberia (Fig. 1). The W African birds use two or more long-distance flights to reach their breeding grounds, stopping over at some of the same estuaries in Europe where the north European breeders winter (Piersma 1987; Prokosch 1988; Piersma and Jukema 1990). We shall distinguish the long-distance migrants by calling them the *Afro-Siberian* population, the others as the *European* population. Together they provide a perfect leap-frog migration system.

The annual cycles of the two populations of bar-tailed godwits are summarized in Fig. 1. According to the available information (Glutz von Blotzheim et al. 1977; Cramp and Simmons 1983), the two populations spend about the same amount of time on the breeding grounds. The Afro-Siberian birds, who have to cover 8300 km one-way, spend much more time on migration than the birds of the European population (2500 km one-way). Both populations moult their flight feathers upon arrival on the wintering grounds (Dick 1975; Boere 1976). The moult of contour feathers in spring occurs in the wintering area in European birds, and in both wintering and spring staging areas in Afro-Siberian godwits (J. Jukema and T. Piersma in prep.).

3 Energetics of Leap-Frog Migration

As we have seen, the Afro-Siberian population of the bar-tailed godwit leap-frogs over the European population in the course of its migration. It will be our task here to construct provisional energy budgets for the two populations to assess the repercussions. Our goal is to provide estimates of the daily energy requirement for the various phases of the annual cycle, treated here in the succession wintering-migration-breeding-migration-moult. For simplicity, all budgets will be constructed for *males* only (see note below on sexual segregation in winter). The basis for the calculations can be presented briefly as follows:

Wintering. In view of what is known about lower critical temperatures of waders (Kersten and Piersma 1987: 9 °C in the Oystercatcher *Haematopus ostralegus*, at 500 g much heavier than our godwits in winter), costs for temperature regulation cannot be ignored for the population wintering in European estuaries (January air temperatures average 5–9 °C). Extrapolation from food requirements of caged waders in winter (Kersten and Piersma 1987) indicate a level of three times basal metabolic rate (BMR) as a reasonable approximation, and indeed Smith (1975: p. 19) managed to obtain food intake observations on bar-tailed godwits for the Lindisfarne wintering area (October-April: ca. 600 kJ/day metabolizable energy, ME), equivalent to about 3 BMR. Requirements for birds wintering in Africa are expected to be much lower, and a minimal estimate can be obtained from caged birds under thermoneutral conditions (existence energy or existence metabolism, ME_{cage}). Cage existence metabolism in waders can be estimated from the allometric relation:

$$ME_{cage} = 912\ M^{0.704}\ (\text{ME in kJ/day, M = body mass in kg}),$$

based on measurements of nine species (Kersten and Piersma 1987). This is almost exactly equivalent to 2 BMR as determined empirically in this group:

$$BMR = 437\ M^{0.729}\ (\text{BMR in kJ/day, M = body mass in kg}),$$

according to data on six species assembled by Kersten and Piersma (1987).

There is some justification for these approximations of energy requirements (ME of 2 BMR in Africa, 3 BMR in Europe; taking differences in mass into account, this translates into 300 and 600 kJ/day, respectively), from a recent study on the sanderling *Calidris alba* (Fig. 2). Utilizing the doubly-labeled water method, Castro (1988) managed to make direct determinations of energy expenditure of sanderlings in winter, spanning the climatic extremes encountered by the Nearctic population, from New Jersey down to Peru. The minimal level (observed in Panama) is equivalent to 2 BMR (BMR = 48 kJ/day as measured in Peruvian birds; Castro 1987) and the peak values (New Jersey) are roughly double this. Castro (1988) interprets these differences as reflecting primarily differences in thermoregulatory costs between the sites, although he was unable

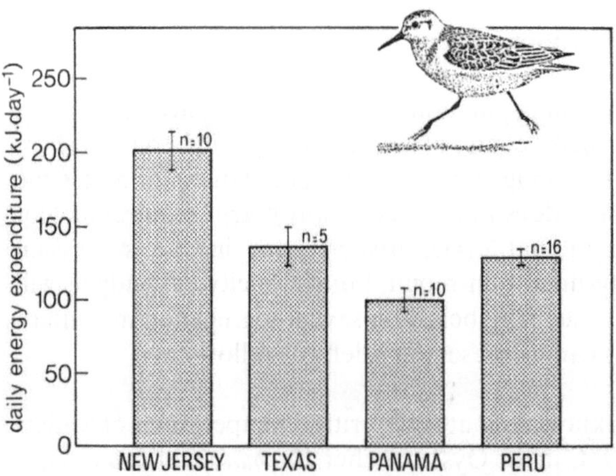

Fig. 2. Daily energy expenditure (ME, kJ) for free-living sanderlings in winter (doubly-labeled water method, means with SE and sample size indicated, from Castro 1988) under the field conditions prevailing in the winter range (arranged from north to south)

to completely disentangle the effects of ambient temperature, wind speed and body mass. These empirical data on energy expenditure confirm the large savings in daily maintenance cost enjoyed by individuals migrating to the tropics, deduced from activity budgets and body mass (Myers et al. 1985).

Migration. Estimates of energy requirements for migration may best be based on a consideration of the daily rate of accumulation of energy reserves, and converting this into a daily cost factor. For the spring migration, intensive netting of bar-tailed godwits allows precise reconstruction of the population-average trajectories of body mass, both for the Banc d'Arguin and the two populations staging in the Wadden Sea (Fig. 3). Perhaps surprisingly, the very extensive data demonstrate linear increments over long periods. There is insufficient data to allow detailed reconstruction of the time course of accumulation of fat and protein reserves, but we do have sufficient carcass analyses to substantiate that in both areas fat and protein are deposited in the proportion 48:52 (fresh mass basis) in males (Piersma and Jukema 1990). Trials with caged waders (six species, but not including the bar-tailed godwit) in both Africa and Europe provide empirical estimates of the daily energy requirements to lay down reserves of this composition. The most extensive data refer to the knot, *Calidris canutus*, where 67 trials are available relating ME to daily change in mass. The slope of this relation (gain and loss of mass fitting the same line) estimates the energetic input at 25.6 kJ/g body reserves (Klaassen et al. 1990). Depending on the efficiency of synthesis assumed to apply under these conditions (75–85%), this cost factor would be associated with the deposition of 40–45% fat by weight, the remainder being protein (see Klaassen et al. 1990, for details of computation). This mix of

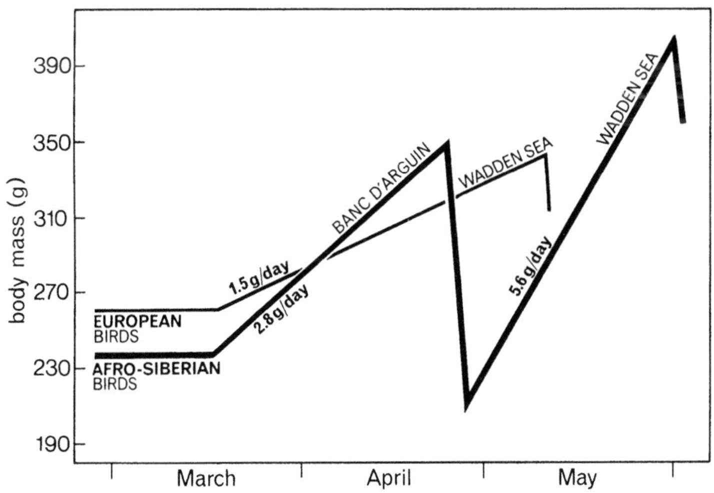

Fig. 3. Body-mass changes during spring migration of males of two populations of bar-tailed godwits. Data for European birds from Prokosch (1988: Table 26 and Fig. 28), and for Afro-Siberian birds from Piersma and Jukema (1990: Fig. 9)

substrates resembles closely the 48:52 proportion determined for the male godwits, and we prefer using the empirically determined cost factor of 25.6 kJ/g rather than making a further slight adjustment which would lend a spurious precision to our budgets. Daily increments (from Fig. 3) and the duration concerned are then estimated as follows:

Afro-Siberian pop., Africa: 2.8 g/d = 72 kJ ME for 45 days;
Afro-Siberian pop., Europe: 5.6 g/d = 143 kJ ME for 32 days;
European pop., Wadden Sea: 1.5 g/d = 38 kJ ME for 53 days.

The cost involved in making the return journey cannot be estimated with the same confidence, since body mass changes at that time of the year are not yet known. A first approximation has been derived as follows. During the spring movement from Mauritania to the Wadden Sea, the entire energy reserve accumulated is expended (see Fig. 3). This is equivalent to a cost of 3163 kJ to traverse 4300 km (assuming no feeding en route, see Piersma and Jukema 1990 for details), i.e. 0.74 kJ/km. The return trip (2500 km for the European population, 4000 km for the Afro-Siberian population, to reach the Wadden Sea again after breeding) probably involves shorter hops with an energetic saving (less excess mass to transport, see also the chapter by Alerstam and Lindström, this Vol.). As a first approximation, we have reduced the cost factor to two-thirds of the spring value (hence 0.5 kJ/km) and for convenience the resulting total cost has been averaged over the month during which the journey is performed. Although it is likely that the actual fattening period is shorter, this computation procedure does not affect the monthly means which we require.

Incidentally, the cost of flight deduced from these mass changes works out at 7–9 BMR, depending on the degree of tailwind assistance the godwits enjoy during the spring flight from W Africa (see Piersma and Jukema 1990). This cost level seems a reasonable approximation for a long-distance migrant with unmistakable wing-shape adaptations to reduce flight cost (Masman and Klaassen 1987).

Breeding and Moult. Only the most sketchy knowledge of energy expenditure of waders during breeding is currently available. We base our estimate of 3.5 BMR for the high arctic situation on preliminary results from doubly-labeled water trials on free-living purple sandpipers, *Calidris maritima*, studied on Spitsbergen (Pierce 1989; 237 kJ/day, mean for ten birds with average body mass of 63.6 g). This value is somewhat lower than the provisional estimate deduced from time budget data for the semipalmated sandpiper *Calidris pusilla*, studied at Barrow by Ashkenazie and Safriel (1979), but is in substantial agreement with provisional estimates for incubating ringed plovers studied in northern England (3.5–4 BMR, P.R. Evans *in litt.*). For the European tundra breeding bar-tailed godwits daily ME is estimated at 475 kJ.

For moult in the Wadden Sea area we have added 15% to the ME estimates prevailing at that season (an admittedly crude approximation based on the only extensive studies on non-passerine birds currently available; Wijnandts 1984; Masman 1986).

Resumé of the Annual Cycle. Since assimilation values have not been determined empirically for the range of prey types commonly encountered in the field (a thorny topic recently reviewed by Speakman 1987 and by Castro et al. 1989) and since there is evidence for a complicated temperature-dependent relationship in waders (Klaassen et al. 1990), we base our overview on estimated metabolizable energy (ME) instead of estimated gross intake rates. ME can be taken as indicative of energy expenditure if summed over long periods when body mass increments are effectively balanced by losses. Over the short term, whenever energy is being accumulated in the form of body reserves, intake will of course exceed expenditure.

When our provisional estimates are plotted (Fig. 4) the Afro-Siberian population appears to experience peak demands during pre-migratory hyperphagia preceding the final lap of the migratory journey (the 4000 km separating the breeding area from the Wadden Sea), whereas the European population is predicted to sustain a high cost throughout its non-breeding phase, the costs of premigratory fattening in the spring compensating for the reduced costs of thermoregulation during that season.

When these monthly values are assigned to the various phases of the annual cycle, a crude breakdown of the total annual ME can be given (see insets, Fig. 4). The summed values do not differ inordinately when the two populations are compared (bearing in mind the considerable difference in mean mass, see Fig. 3). Total annual ME is of the same order as the far more refined figures presented for the Kestrel *Falco tinnunculus*, a non-passerine of similar mass as the Afro-Siberian male godwits (Masman et al. 1988).

4 Discussion

The primary objective of this exercise has been to highlight differences in the patterns of energy intake through the year for populations experiencing a wide latitudinal range of conditions in the course of the annual cycle. The Afro-Siberian population of the bar-tailed godwit enjoys the advantage of a low-cost wintering area, but faces the very considerable cost of crossing the gap to the Wadden Sea area of NW Europe in spring. Here, energetic supplies can be accumulated to cover both the final lap of the journey as well as provide an energy reserve to bridge the period of food scarcity encountered upon arrival on the high-arctic breeding area. Feeding conditions on the wintering range in W Africa may conceivably place a constraint on the calendar period in which an increase of intake rate, sufficient to meet the cost of accumulation of body reserves, can be achieved. The lower rate of body mass gain experienced in Africa when compared to the rate achieved by the same population in the second period of hyperphagia in the Wadden Sea may be interpreted in two ways. Our view is that environmental constraints acting in Africa (such as prey availability, assimilation efficiency and cost of foraging in relation to ambient climatological conditions) prevent the godwits from achieving a higher rate of energy gain at this time. It is, however, conceivable that a lower daily rate of energy gain has been selected for

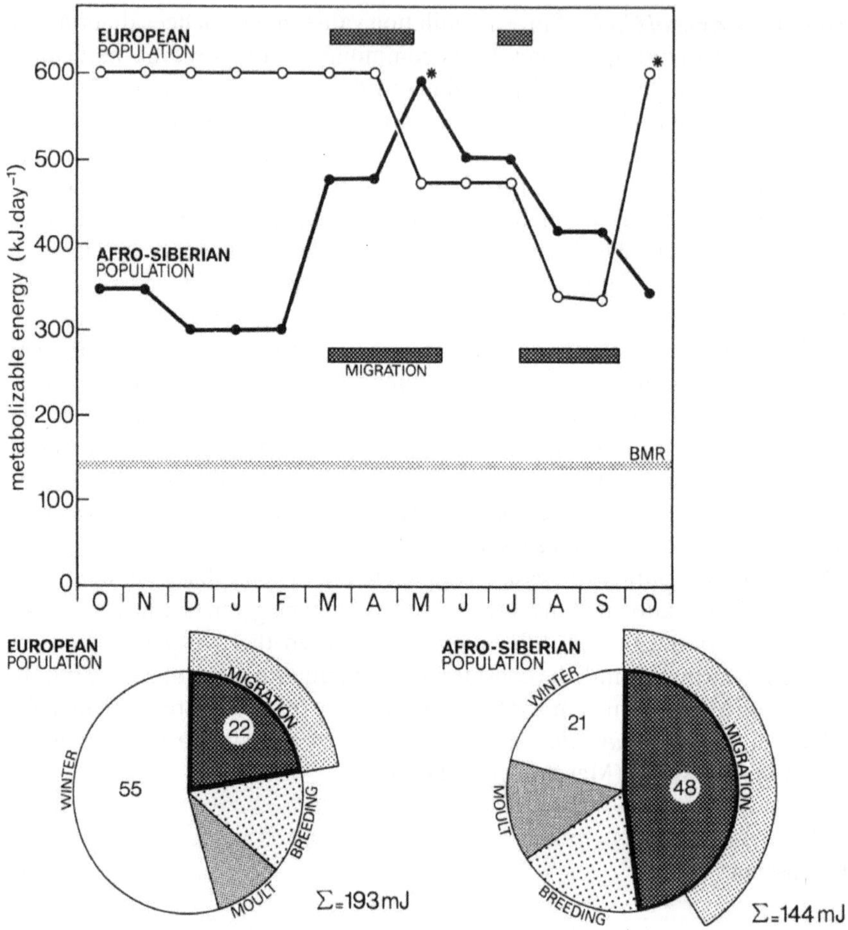

Fig. 4. Provisional energy budgets (ME = metabolizable energy) in the course of the annual cycle for the European (*open circles*) and Afro-Siberian (*filled circles*) population of the bar-tailed godwit. *BMR* scaling refers to the average empirical value for two individuals of the Afro-Siberian population (Piersma and Jukema 1990; European birds have not yet been studied in the laboratory). The two monthly means marked by * indicate periods for which independent direct observations confirm the estimated level (May = Petersen 1981, assuming a mean prey ash-free dry mass of 10 mg; October through April = Smith 1975). The annual totals are collected in the pie diagrams below, where the percentages of annual ME are shown for the calendar periods concerned with migration, breeding, moult and winter. For migration the proportion of the annual time budget is indicated on the *outer circle*

in order to maximize the net efficiency of a process involving a complicated balance between gain and cost (whereby costs of tissue maintenance must also be considered). A case in point is the enhanced cost of locomotion as the birds become heavier. The shape of these cost curves has not yet been established empirically with sufficient precision to follow this up.

Similarly, we interpret the difference in the rate of mass gain of the two godwit populations in the Wadden Sea area (1.5 g/day in April and 5.6 g/day in May), as reflecting primarily enhanced feeding conditions later in spring (perhaps a temperature-mediated increase in prey availability; cf. Pienkowski 1983). Accepting this line of thought, we argue that the godwits wintering in Africa are prevented from returning to Europe earlier in the season by the W African environment, but that this loss of time is compensated for by a much higher intake rate once the Wadden Sea has been reached. Two questions arise. First, what places a ceiling on the Wadden Sea figure for ME? The rate of energy intake required to build up the body reserves in May (in agreement with Petersen's 1981 direct observations) is close to the empirical limit postulated by Kirkwood (1983) to apply to birds and mammals undergoing maximum hyperphagia. This finding hints at a physiological constraint in the bird, rather than an environmentally induced limitation. Secondly, the narrow margins in departure reserves of the Afro-Siberian population departing from the African staging sites (our calculations indicating that the migratory jump to the Wadden Sea area can only be achieved if the flocks are wind assisted; see Piersma and Jukema 1990) hints at a stringent limitation. Either the transportation cost rises prohibitively with increasing mass or the pressures of time favour an early shift to the higher profitability site.

Indirectly it is possible to allocate the reserves accumulated during this final staging period in the Wadden Sea to migration (transport) and reserves required for breeding. Assuming that the cost deduced from the first leg of the journey applies here as well, 2900 kJ will be expended in reaching the breeding grounds, leaving approximately 800 kJ as reserve upon arrival (equivalent to 14–17% of body mass). That waders upon arrival on high-arctic breeding grounds indeed carry with them reserves of such magnitude has been confirmed in recent work on Ellesmere Island (Davidson and Evans 1988; Morrison and Davidson 1988), concerning carcass analyses of knots *Calidris canutus* and turnstones, *Arenaria interpres*, recently arrived from northern Europe.

The assumption that the underlying selective advantage of leap-frog migration is to avoid what Salomonsen (1955a) so aptly described as the "devastating competition" inevitably associated with sharing the same winter quarters, is of course impossible to test directly under present-day conditions. If we attempt to visualize some incipient stage of the differentiation in winter goal, it is well to bear in mind that the effects of food competition in the Wadden Sea area will be compounded as winter draws near, by the sharp increase in daily ration needed to balance the budget. An individual escaping to more southerly areas might well experience such a relaxation of demand that the prey collection rate could be halved, and it is necessary to invoke advantages in wintering as close to the breeding area as possible (Alerstam and Högstedt 1980) to explain why any godwits remain behind in the seemingly inhospitable coastal region of Europe. Although the modeling exercises undertaken seem reassuring (Sutherland and Parker 1985; Lundberg and Alerstam 1986), there is a real need for observational data on survival rates of populations within the framework of migration strategies, where the mortality penalty associated with more northerly wintering can be balanced against the hazards incurred during the more extended travels of the leap-frogging component (Pienkowski and Evans 1985). This is a rather tall order, since differences in reproductive output, and more particularly the predictability of reproductive performance are likely to prevail between the two populations (the tropical wintering segment facing less consistent breeding results).

A curious but perhaps significant corollary of this variability in output is that in the "bumper years" of the high-arctic breeders (Summers and Underhill 1987; de Boer and Drent 1989), a large proportion of the population reaching the wintering grounds will consist of inexperienced individuals, highly susceptible to the effects of competition (see Evans and Davidson, this Vol.). Townshend (1985) working with grey plover, *Pluvialis squatarola*, investigated competitive exclusion at a northern wintering site, and demonstrated that long-winged juveniles were more likely to acquire territories and remain. A direct impact of physical attributes of the individual is implicated, perhaps correlated with earliness of hatch. Displaced juveniles were reported almost 1000 km further south. The partial segregation of the sexes during winter (in the Afro-Siberian population of bar-tailed godwits the females winter further south) is in line with other wader species where brood care is the sole responsibility of the male (Myers 1981), and there is the merest shred of evidence from the breeding area that this may be true for bar-tailed godwits in the high arctic (Cramp and Simmons 1983). This pattern of partial overlap provides opportunities for observational tests of the reality of intraspecific competition, inspired by the work on juveniles mentioned earlier.

A critical comparison of the energy budgets of social categories defined in terms of dominance has yet to be undertaken for the field situation, but would appear a promising avenue in view of the tantalizing results from semi-captive great tit *Parus major* flocks (Järvi 1983). The relevance of dominance hierarchies as a factor impinging on intake rates has emerged from recent work on the oystercatcher (Ens and Goss-Custard 1984; Goss-Custard and Durell 1987).

5 Future Research Needs

Several points have emerged from the present analysis that identify pressing needs for further research:

1. Comparison of carcass analysis of godwits shortly before departure from the African wintering grounds and directly after arrival in the Wadden Sea area 4300 km away (Piersma and Jukema 1990) indicates that fuel consumption during this migratory flight involves nearly equal amounts of protein and fat on a weight basis. Fat thus provides some 90% of the energy required, in substantial agreement with the situation during moderate starvation (Cheryl et al. 1988). The finding that during a virtual non-stop flight fat and protein are drawn upon in the same proportions as during starvation points to some underlying biochemical principle. The rule-of-thumb in common use among field workers handling migratory birds, by which mass change is equated to change in fat content alone, is obviously no longer tenable. There is scope here for application of non-destructive techniques of assessing body components in studies of free-living migrants and in longitudinal studies of captive birds (such as the plethysmometric

determination of total body volume to estimate total body density, reviewed by Gundlach and Visscher 1986, or total body electrical conductivity, described by Walsberg 1988). That the mix of substrates deposited by waders during migration stopovers may apply to a much wider taxonomic array of birds is implied by Dolnik and Gavrilov's (1973) finding that the caloric equivalent of the variation in mass of chaffinches, *Fringilla coelebs*, amounted to 6.10 kcal/g (=25.5 kJ/g) virtually identical to the value obtained for knots in Africa (25.6 kJ/g, Klaassen et al. 1990). Recently, laboratory work on the garden warbler, *Sylvia borin*, has provided elegant confirmation of this constant during the period of pre-migratory hyperphagia (Biebach, this Vol.).

2. The distinction between a high-cost but high-yield European wintering zone and a low-cost, low-yield situation in the W African wintering area raises the question of specific metabolic adaptation in the godwit populations concerned. To date only individuals of the African wintering population have been subjected to respirometry (Piersma and Jukema 1990) and basal metabolic rate (BMR) was within 5% of prediction by the six-species relation adduced for waders by Kersten and Piersma (1987). Further work is urgently needed to define the metabolic response to ambient temperature change in both populations in the search for energy-saving mechanisms.

3. The magnitude of the energy requirement for thermoregulation during winter for godwits inhabiting the wind-swept mudflats of temperate Europe underlines the relevance of a thermal modelling approach for estimating energy budgets in these birds (Bakken et al. 1981; recently applied in the field situation by Masman et al. 1988). The empirical data for free-living sanderlings studied by Castro (1988) confirm the reality of the high costs associated with wintering at temperate latitudes. Employing heated mounts to investigate the thermal advantage of utilizing sheltered areas for roosting, desisting from feeding under certain conditions, and other related problems would appear a promising avenue, as already explored for chick waders by Chappell (1980).

4. The energetic cost of migration as assessed from observed rates of accumulation of body reserves looms large in the annual energy budget of the long-distance migrant, the Afro-Siberian population (48% of annual expenditure), compared to the European population (22%). On account of the lower wintering costs, the annual total for the two populations is not greatly dissimilar, if dissimilar at all. The salient feature of the comparison is to highlight the time constraints imposed by the lengthy process of accumulating the reserves required for the migratory journeys. For example, the Afro-Siberian population expends the energy accumulated during 24 h of spring hyperphagia on the Banc d'Arguin in just over 1 h of migratory flight. In our interpretation, Africa is available as a wintering site only for populations nesting in the high arctic where the relatively late advent of conditions favourable for breeding provides the leeway necessary for fitting in the 3 months of preparatory staging. Conversely, the potential of the vast Siberian breeding range can only be realized because of the extensive capacity of the wintering quarters in Africa, which in turn can only be reached thanks to the Wadden Sea situated as a stepping stone between the two.

5. The evolutionary interpretation of migratory patterns will not be solved by recourse to energetic considerations alone, but must include new approaches towards quantifying the importance of competition in defining winter quarters. Here lies a challenge for innovative research in eco-ethology, and it is the message of this contribution to emphasize the need for inclusion of field metabolism measurements in any such attempt.

Acknowledgements. Explorers need beacons to guide them, and we acknowledge the generosity of Gonzalo Castro, Eilin Pierce and Peter R. Evans for sharing their DLW determinations on energy metabolism of free-living waders before publication. Bruno Ens provided some insightful criticisms on a draft. Without mentioning each of them personally, we would like to acknowledge that we have been instructed, critically guided and stimulated by many different colleagues: we hope that the co-operative spirit pervading the Wader Study Group will long continue.

References

Alerstam T, Högstedt G (1980) Spring predictability and leap-frog migration. Ornis Scand 11:196–200

Alerstam T, Högstedt G (1985) Leap-frog arguments: reply to Pienkowski, Evans and Townshend. Ornis Scand 16:71–74

Ashkenazie S, Safriel UN (1979) Time-energy budget of the semipalmated sandpiper *Calidris pusilla* at Barrow, Alaska. Ecology 60:783–799

Bakken GS, Buttemer WA, Dawson WR, Gates DM (1981) Heated taxidermic mounts: a means of measuring the standard operative temperature affecting small animals. Ecology 62:311–318

Boer F de, Drent RH (1989) Breeding success of high arctic breeding geese and waders: eating or being eaten? Wader Study Group Bull 55:11–16

Boere GC (1976) The significance of the Dutch Waddenzee in the annual life cycle of arctic, subarctic and boreal waders, Part 1. The function as a moulting area. Ardea 64:210–291

Castro G (1987) High basal metabolic rate in sanderlings (*Calidris alba*). Wilson Bull 99:267–268

Castro G (1988) Ecophysiology of sanderlings migrating to four different latitudes. Thesis, University of Pennsylvania, Philadelphia

Castro G, Stoyan N, Myers JP (1989) Assimilation efficiency in birds: a function of taxon or food type? Comp Biochem Physiol A Comp Physiol 92:271–278

Chappell MA (1980) Thermal energetics of chicks of Arctic-breeding shorebirds. Comp Biochem Physiol A Comp Physiol 65:311–317

Cherel Y, Robin J-P, Maho Y Le (1988) Physiology and biochemistry of long-term fasting in birds. Can J Zool 66:159–166

Cramp S, Simmons KEL (eds) (1983) The birds of the western Palearctic, vol 3. Oxford University Press Oxford

Davidson NC, Evans PR (1988) Pre-breeding accumulation of fat and protein by arctic-breeding shorebirds. Proc XIX Int Congr Ornithol Ottawa, pp 342–352

Dick WJA (ed) (1975) Oxford and Cambridge Mauritanian Expedition 1973. Report Cambridge

Dolnik VR, Gavrilov VM (1973) Caloric equivalent of body weight variations in chaffinches (*Fringilla coelebs*) In: Bykhovskii BE (ed) Bird migrations, ecological and physiological factors. Wiley, New York, pp 273–287

Ens BJ, Goss-Custard JD (1984) Interference among oystercatchers, *Haematopus ostralegus* L., feeding on mussels, *Mytilus edulis* L., on the Exe estuary. J Anim Ecol 53:217–232

Glutz von Blotzheim UN, Bauer KM, Bezzel E (1977) Handbuch der Vögel Mitteleuropas, Band 7. Akademische Verlagsgesellschaft, Wiesbaden

Goss-Custard JD, Durell SEA le V dit (1987) Age-related effects in oystercatchers, *Haematopus ostralegus*, feeding on mussels, *Mytilus edulis*. J Anim Ecol 56:521–558

Greenberg R (1986) Competition in migrant birds in the nonbreeding season. Curr Ornithol 3:281–307

Gudmundsson GA (1988) Intraspecific variation in bird migration patterns. Introductory paper no 49. Department of Animal Ecology, University of Lund

Gundlach BL, Visscher GJW (1986) The plethysmometric measurement of total body volume. Human Biol 58:783–799

Järvi T (1983) On the evolution of inter- and intraspecific communication through natural and sexual selection. Thesis, University of Stockholm

Kersten M, Piersma T (1987) High levels of energy expenditure in shorebirds; metabolic adaptations to an energetically expensive way of life. Ardea 75:175–187

Kirkwood JK (1983) A limit to metabolizable energy intake in mammals and birds. Comp Biochem Physiol A Comp Physiol 75:1–3

Klaassen M, Kersten M, Ens BJ (1990) Energetic requirements for maintenance and premigratory body mass gain of waders wintering in Africa. Ardea 78: (in press)

Lundberg S, Alerstam T (1986) Bird migration patterns: conditions for stable geographical population segregation. J Theor Biol 123:403–414

Masman D (1986) The annual cycle of the Kestrel *Falco tinnunculus*: a study in behavioural energetics. Thesis, University of Groningen

Masman D, Klaassen M (1987) Energy expenditure during free flight in trained and free-living Eurasian kestrels (*Falco tinnunculus*). Auk 104:603–616

Masman D, Daan S, Beldhuis HJA (1988) Ecological energetics of the kestrel: daily energetic expenditure throughout the year based on time-energy budget, food intake and doubly labeled water methods. Ardea 76:64–81

Mjelstad H, Saetersdal M (1986) Density, population size and breeding distribution of spotted redshank *Tringa erythropus*, bar-tailed godwit *Limosa lapponica* and jack snipe *Lymnocryptes minimus* in Norway. Fauna Norv Ser C, Cinclus 9:13–16

Morrison RIG, Davidson NC (1988) Migration, body condition and behaviour of shorebirds during spring migration at Alert, Ellesmere Island, N.W.T. Syllogeus (in press)

Myers JP (1981) Cross-seasonal interactions in the evolution of sandpiper social systems. Behav Ecol Sociobiol 8:195–202

Myers JP, Maron JL, Sallaberry M (1985) Going to extremes: why do sanderlings migrate to the Neotropics. Ornithol Monogr 36:520–535

Petersen BD (1981) Vadefuglenes fouragering og predation paå bundfaunaen paå Tipperne. Dansk Ornithol Foren Tidsskr 75:7–22

Pienkowski MW (1983) Surface activity of some intertidal invertebrates in relation to temperature and the foraging behaviour of their shorebird predators. Mar Ecol Prog Ser 11:141–150

Pienkowski MW, Evans PR (1985) The role of migration in the population dynamics of birds. In: Sibly RM, Smith RH (eds) Behavioural ecology: ecological consequences of adaptive behaviour. Oxford Blackwell Scientific, pp 331–352

Pienkowski MW, Evans PR, Townshend DJ (1985) Leap-frog and other migration patterns of waders: a critique of the Alerstam and Högstedt hypothesis, and some alternatives. Ornis Scand 16:61–70

Pierce E (1989) Breeding biology of the purple sandpiper on Spitsbergen. Thesis, University of Bergen

Piersma T (1982) Foraging of bar-tailed godwits. In: Altenburg W, Engelmoer M, Mes R, Piersma T (eds) Wintering waders on the Banc d'Arguin, Mauritania. Stichting Veth tot steun aan Waddenonderzoek, Leiden, pp 135–148

Piersma T (1987) Hink, stap of sprong? Reisbeperkingen van arctische steltlopers door voedselzoeken, vetopbouw en vliegsnelheid. Limosa 60:185–194

Piersma T, Jukema J (1990) Budgeting the flight of a long-distance migrant: changes in nutrient reserve levels of bar-tailed godwits at successive spring staging sites. Ardea 78 (in press)

Prokosch P (1988) Das Schleswig-Holsteinische Wattenmeer als Frühjahrs-Aufenthaltsgebiet arktischer Watvogel-Populationen am Beispiel von Kiebitzregenpfeifer (*Pluvialis squatarola*, L. 1758), Knutt (*Calidris canutus*, L. 1758) und Pfuhlschnepfe (*Limosa lapponica*, L. 1758). Corax 12:274–442

Salomonsen F (1955a) Evolution and bird-migration. Acta XI Congr Int Ornithol, pp 337–339

Salomonsen F (1955b) The evolutionary significance of bird-migration. Dan Biol Medded 22(6):1–61

Smit CJ, Piersma T (1989) Numbers, mid-winter distribution and migrations of wader populations using the East Atlantic flyway. In: Boyd H, Pirot J-Y (eds) Flyways and reserve networks for waterbirds. IWRB Spec Publ No 9. Canadian Wildlife Service, Ottawa, pp 24–63

Smith PC (1975) A study of the winter feeding ecology and behaviour of the bar-tailed godwit *Limosa lapponica*. Thesis, University of Durham

Speakman JR (1987) Apparent absorption efficiencies for redshank (*Tringa totanus* L.) and oystercatcher (*Haematopus ostralegus* L.): implications for the predictions of optimal foraging models. Am Nat 130:677–691

Summers RW, Underhill LG (1987) Factors related to breeding production of brent geese *Branta b. bernicla* and waders (Charadrii) on the Taimyr Peninsula. Bird Study 34:161–172

Sutherland WJ, Parker GR (1985) Distribution of unequal competitors. In: Sibly RM, Smith RH (eds) Behavioural ecology: ecological consequences of adaptive behaviour. Blackwell Scientific, Oxford, pp 255–273

Townshend DJ (1985) Decisions for a lifetime: establishment of spatial defence and movement patterns of juvenile grey plovers (*Pluvialis squatarola*). J Anim Ecol 54:267–274

Walsberg GE (1988) Evaluation of a nondestructive method for determining fat stores in small birds and mammals. Physiol Zool 61:153–159

Wijnandts H (1984) Ecological energetics of the long-earned owl (*Asio otus*). Ardea 72:1–92

Problems Inhibiting Energetic Analyses of Migration

G.E. Walsberg[1]

1 Introduction

A major goal of students of avian migration is to understand the suite of factors that has produced the impressive diversity of migratory patterns observed in birds and the significance of such variation for the organism's fitness. Prudence generally dictates suspicion of single-factor explanations for complex biological problems. In the case of avian migration, however, it seems likely that dissecting the energetic bases and consequences of variable migratory patterns will be a particularly fruitful approach. The behavior of birds in migration is largely directed toward the single task of moving long distances and the resource of overwhelming importance for this activity clearly is energy. This contrasts with other portions of the annual cycle such as reproduction, during which tactics for energy acquisition may be significantly complicated by simultaneous demands for nutrients such as protein (e.g., Drobney 1980). Such considerations strongly suggest that analyses of migration energetics are likely to yield substantial insights.

A particular illuminating device for such analyses is construction of an energy budget quantifying avenues of energy acquisition and allocation, and allowing analyses of important determinants of variation in this budget. This approach allows the costs and benefits of varying behavioral and physiological patterns, such as fat storage, migratory routes, and stopover behavior, to be set in an appropriate comparative context. In recent years, energy budgeting techniques have been significantly refined and used successfully in many studies of non-migrating birds (e.g., Ettinger and King 1980; Mugaas and King 1981). These analyses have become increasingly accurate as workers have identified and reduced important sources of error (e.g., Weathers et al. 1984; Williams and Nagy 1984; Masman 1986).

Major difficulties arise, however, when applying such techniques to migrating birds. Most obvious, the transient nature of the animal provides major difficulties. In addition, allocation of a major fraction of the animal's total energy expenditure to flight substantially reduces the probable accuracy of energy budgets. For birds not in migration, improving estimates of resting and thermostatic expenditure increases substantially the accuracy of the total budget

[1]Department of Zoology, Arizona State University, Tempe, Arizona 85287-1501, USA

E. Gwinner (Ed.) Bird Migration

(Weathers et al. 1984). This is unlikely to be true of migrants, however. Time budgets of nonmigrating birds commonly reveal that only a small percentage of time is spent in flight (Walsberg 1983). This has been fortunate, because power consumption during flight is intense as well as being poorly understood and difficult to quantify compared to other avian activities.

Though construction of such energy budgets is highly desirable, current data do not allow more than a coarse examination of modes of energy allocation and acquisition during migration. Traditionally, workers have focused on two aspects of migration energetics, premigratory acquisition of energy stores and expenditure of those stores during flight. In limited cases this view may be realistic, as in the case of individuals whose migration is essentially limited to movement between two sites separated by a region unsuitable for foraging. For most migrants, however, a third critically important aspect exists. This is replenishing energy stores in the course of migration. A minimal data base necessary to begin quantifying the ecological energetics of migrant birds must include all three of these components. Development of even an extensive data base describing any single component will be inadequate until it can be placed in the context of a complete, or nearly complete, energy budget.

Of these three major aspects of migration energetics, the physiology of premigratory fat storage has been most extensively analyzed (Blem 1980; King 1972; Farner et al. 1968; Berthold 1975), though the ecology and ecological energetics of birds in this phase is essentially unknown. The energetics of flight also has received extensive attention, but the extant data base remains unsatisfactory. The third component, replenishing fat stores in the course of migration, is the least studied of these three major aspects. Thus, there is for no single species the information necessary to begin construction of an adequate energy budget for migration. This suggests that major emphasis be devoted to developing our understanding of the energetics of flight and the energetics of foraging by birds in migratory transit.

2 En Route Foraging

Long-distance migrants typically punctuate periods of movement with "stopovers" in which they rest and have the opportunity to forage and replenish fat stores in preparation for the next phase of migratory flight. Such restoration of fat stores clearly is critical for successful completion of long-distance migrations, but is likely to be seriously complicated by the necessity for the animal to forage in an unfamiliar, perhaps inhospitable, habitat in which it may experience substantial competition for resources from both resident and migrant birds. Though few data are available describing the energetics of these periods, analyses of the ecology of such stopovers in two systems illustrate the complexity of the factors affecting migratory patterns.

For example, Rappole and Warner (1976) analyzed the ecology of migrants (particularly the northern waterthrush, *Seiurus noveboracensis*) interrupting their northward migration in coastal Texas after crossing the Gulf of Mexico. Only a minority of .the species observed exhibited average weight gains during their stopover. Indeed, one-half of the species studies actually lost weight during the stopover, suggesting that many individuals do not profit energetically from this interruption of migration. One basis for this variable profitability is suggested by data for the northern waterthrush. Individuals known to be territorial gained weight on average during their stopover period, while other individuals lost weight. Even individuals that eventually acquired territories typically declined in body mass, and presumably fat stores, during the initial few days of the stopover period that typically are required for territorial establishment. Thus, Rappole and Warner's analysis demonstrates that the ability to replenish fat stores during a particular interruption of migration is a complex function of variables that include, for example, the animal's capacity to defend resources in an intensely competitive environment.

Passerine birds migrating across the Sahara Desert provide a second example of the apparent interactions of energy supply and demand with migratory patterns. Contrary to earlier expectations (Moreau 1972), many birds do not cross this region in a single flight but rather interrupt their migratory flights with stopovers devoted to rest and perhaps foraging (Barlein 1985; Biebach 1985 and this Vol.; Biebach et al. 1986). In nearly vegetationless regions of the eastern Sahara Desert, Biebach et al. (1986) found that some individuals occupied oases characterized by an abundant food supply. Unexpectedly, however, birds at these few and isolated oases probably account for only a minor fraction of migrants on the ground. Many more birds apparently rest in the open, nearly vegetationless desert that composes most of this region. In such sites, no water and essentially no food is available and shade is presented only by scattered shrubs, large rocks, or holes in the ground. Also unexpectedly, birds found in these regions clearly have not been forced to land because of expended fuel reserves. Individuals found resting in the open desert have fat stores apparently sufficient to allow them to fly continuously over the remaining portion of the desert. In contrast, birds stopping in oases tend to be lower in body mass and fat stores than those found in the open desert. From these observations, Biebach et al. (1986) suggest that patterns of migratory flight and stopover depend upon a series of decision-making processes based upon resource availability and the animal's physiological status. These authors propose that if migrants retain ample fat stores at the end of a nocturnal flight, the birds will land and rest if shade is available. This landing apparently is not, however, dependent on availability of food or water. Migrants that do not retain adequate fat stores at the end of a nocturnal flight apparently refrain from landing to rest unless both shade and a rich foraging site are available, such as is presented by an oasis. The length of the stopover period in these oases is probably dependent upon the animal's success in replenishing its fat stores. If such a rich foraging site cannot be found, these workers suggest that the animal continues on its migratory flight during daylight hours. This would be advantageous because the additional energy expended during a daylight rest period would only decrease their probability of successfully crossing the desert.

These analyses demonstrate the importance of such interruptions of migratory flights for understanding general patterns of migration, and also clearly suggest that energetics are of major importance. Though energy budgets for this phase of migration are unavailable, such analyses clearly are an important first step.

In addition to their insights into the behavior and energetic status of birds interrupting migratory flights in the Sahara Desert, these studies also illustrate a puzzling problem common to many migrants. This is use of a migratory route that appears unnecessarily dangerous. For example, passerine birds cross the extremely inhospitable eastern Sahara Desert and apparently do not take particular advantage of the Nile River Valley (Moreau 1972; Biebach et al. 1986). Use of this valley would essentially eliminate a major ecological barrier, as this north-south corridor is abundantly supplied with water and foraging sites. Failure to concentrate along such routes is curious and could result from, for example, functional limitations on the navigational system used by migrants. Alternatively, such behavior may be a historical artifact of milder conditions that prevailed a few 1000 years ago in what is now the Sahara Desert. In either case, such behavior provides a cautionary note against analyzing biological patterns by uncritically expecting that organisms are performing "optimally".

3 Flight

Two contrasting approaches have been used to analyze the energy requirements of flight. These are empirical analyses in which metabolic power input is measured in flying birds and theoretical analyses in which mechanical power output is estimated based on aerodynamic theory.

Determining metabolic power input during flight is technically challenging. Such measurements have been made both for animals in free flight and flying within wind tunnels, with these two techniques providing a contrasting set of advantages and disadvantages (see also Rayner, this Vol.). Measurements within wind tunnels allow precise control of the conditions to which the animal is exposed (e.g., wind speed, air temperature). However, such requirements require extensive selection and training of test subjects that is tedious and can importantly affect results (e.g., Tucker 1968; Rothe and Nachtigall 1987; Nachtigall, this Vol.). In addition, flight within wind tunnels has the potential to produce important artifacts in resultant data. The flight environment is noisy and unnatural in its appearance, unusual flight patterns commonly are noted (e.g., Tucker 1972; Rothe and Nachtigall 1987), and workers often have been unable to maintain the birds in flights sustained beyond a fraction of an hour. Though some workers have measured the metabolic rate of unemcumbered birds flying in closed-circuit wind tunnels (e.g., Torre-Bueno and Larochelle 1978), most analyses have entailed fitting the animal with a facial mask to collect expired respiratory gases (e.g., Tucker 1968, 1972; Rothe et al. 1987). The aerodynamic drag produced by this apparatus has been estimated as requiring an additional power consumption of about 12–30% (Tucker 1972; Rothe et al. 1987).

In contrast, measurements made using doubly labeled water of power consumption by unencumbered birds in free flight avoid these complications but produce uncertainties of their own. For example, the exact distance and speed of flight often are unknown (LeFebvre 1964; Gessaman and Nagy 1988).

The result of these technical limitations is that, in spite of a series of careful analyses, substantial uncertainties exist in extant data that suggest substantial caution in their application to problems of migration. A salient example is flight power consumption in the pigeon, *Columba livia*, the species for which the most extensive data set is available. Values for power consumption vary up to 2.3-fold, with the maximum and minimum both occurring at the same flight speed of 16 m s^{-1} (Fig. 1). In addition, major effects of attaching drag-inducing apparatus to birds are apparent in these data. Gessaman and Nagy (1988) concluded that attaching a harness and radio transmitter weighing only 2.5–5% of the bird's body weight produced increases in power consumption averaging 41–52% and decreased flight speed more than 25% (Fig. 1). These increases combine to elevate by 85–100% the total energy expenditure required to fly a given distance. That similar experimental artifacts may occur broadly in other studies is suggested by Masman and Klaassen (1986), who analyzed data for flight metabolism in 58

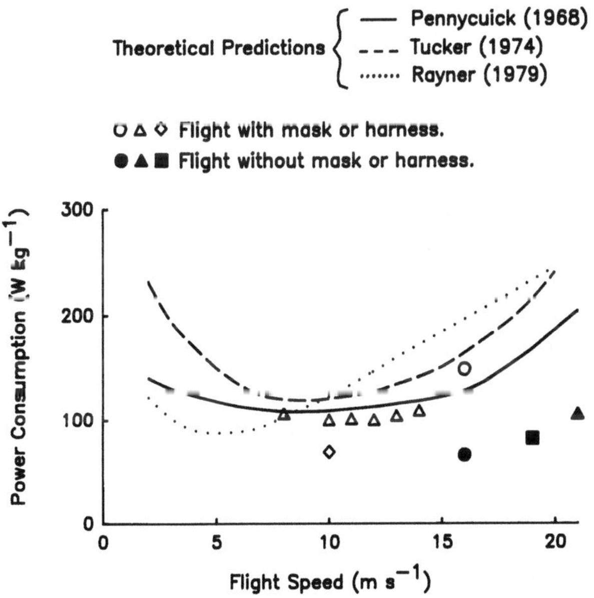

Fig. 1. Measured power consumption during flight in *Columba livia* compared to theoretical predictions. *Open diamonds* (Butler et al. 1977) and *open triangles* (Rothe et al. 1987) are values measured for birds flying in a wind tunnel and wearing a face mask to collect respiratory gases. *Open circle* is value for free flight with bird carrying a harness and transmitter weighing 2.5–5% of body mass (Gessaman and Nagy 1988). *Filled symbols* are values for free flight with animal unencumbered by harnesses or masks. Sources of latter data: *circle* = LeFebvre (1964), *square* = Polus (1985, cited in Rothe et al. 1987), *triangle* = Gessaman and Nagy (1988). Theoretical predictions from Pennycuick (1968) and Rayner (1979) are those author's calculations of mechanical power output converted to metabolic power input by assuming that gross exercise efficiency = 0.2. Theoretical prediction from Tucker (1974) was calculated using this author's theory, the data of Pennycuick (1968) for *Columba livia*, and assuming efficiency = 0.2

species and concluded that measurements made in wind tunnels tend to be elevated about 50% over those measured for free-flying animals.

Finally, it is particularly notable that empirical studies have not yet focused on aspects of flight that are of prime interest to students of migration. For example, no adequate data set is available describing the flight energetics of an insectivorous passerine (the typical long-distance migrant) or have explored the consequences for power consumption and flight speed of varying morphology (e.g., wing length and fat load).

Analyses of flight energetics based on aerodynamic theory potentially can address such problems and have greatly stimulated general concepts of avian flight performance. Over the past 20 years, a series of complex and potentially valuable theoretical analyses of bird flight have been generated (e.g., Pennycuick 1968; Tucker 1974; Greenewalt 1975; Rayner 1979)[2]. Such models have been used by workers to estimate power consumption during migratory flights and to account for a host of effects such as flight speed (e.g., Pennycuick 1975; Baggott 1986). It is striking, however, that there have been only limited attempts to empirically validate any of these theoretical models (e.g., Tucker 1974). To a substantial degree, this is because biophysical models predict mechanical power *output* by a flying bird while physiological measurements quantify metabolic power *input*. Workers relying on aerodynamic models to estimate power consumption of migrating birds must therefore estimate the gross efficiency of exercise (*sensu* Stainsby et al. 1980), which is the ratio of external work accomplished (output) to energy expenditure (input). This is a poorly known parameter that may be subject to much variation, both between individuals and within individuals as the form or intensity of exercise varies (Rayner 1979; Stainsby et al. 1980). Estimates of power input relying on these models commonly estimate gross efficiency as about 20%, but with little empirical basis. Useful insights are provided by reversing this normal conversion and estimating gross exercise efficiency by dividing theoretical predictions of power output by experimentally determined metabolic power input (Table 1). Using the data presented in Fig. 1 for *Columba livia*, various aerodynamic theories predict exercise efficiencies of 17–59% (Table 1). Variation in these estimates could be due to real variation in efficiency, errors in model predictions, and errors in measurements of metabolic power input. The relative contribution of these sources cannot yet be quantified. However, it is notable that maximum gross muscle efficiency probably does not exceed 30% (Stainsby et al. 1980) and that this value does not include components of total body power consumption such as energy expenditure by nonexercising portions of the body. Inclusion of such components further reduces the maximum expected value of whole-body gross efficiency of exercise. For model predictions of mechanical power output to be accurate, empirical measurements of metabolic power input therefore must be in error up to approximately 100%. That is,

[2]During the symposium, I was alerted by Dr. Rayner to a revision of his model that generally predicts lower values for power consumption at high flight speeds (Rayner 1986). Unfortunately, I was unable to obtain a copy of this paper within the time necessary to adequately evaluate these revised predictions.

Table 1. Gross energy efficiency during flight in *Columba livia* predicted by theoretically predicted values for mechanical power output and measured metabolic power input. Values are percentages calculated as 100 × (output)/(input). Input and output data are those in Fig. 1

Source of measured power input	Source of predicted power output		
	Pennycuick (1968)	Tucker (1974)	Rayner (1979)
Butler et al. (1977)	32	36	35
Gessaman and Nagy (1988)[a]	38	51	49
Gessaman and Nagy (1988)[b]	17	22	26
LeFebvre (1964)	39	50	59
Polus (1985; cited in Rothe et al. 1987)	41	52	57
Rothe et al. (1987)	20–23	23–26	19–32

[a] Value for bird flying without harness or transmitter.
[b] Value for bird flying with harness and transmitter weighing 2.5–5% of body weight.

reported values in several studies must equal about one-half the actual value. This error would be in the direction opposite to that which should be produced by the attachment of drag-inducing apparatus to the animal. Errors in empirical measurements undoubtedly occur, but are unlikely to be this large. Such considerations strongly suggest that at least one important component of the disparity between measured and predicted values is model error.

4 Concluding Comments

A primary factor restricting analyses of the energetics of migrating birds is the difficulty of confidently estimating power consumption in flight. Unfortunately, energetic predictions from extant models are, in an important sense, untestable at present. Such tests would require knowledge of whole-body exercise efficiency, a property that cannot as yet be reliably estimated and which might vary widely between species or within an individual as form or intensity of exercise changes.

Beyond theoretical predictions, few recourses are available for workers requiring estimates of flight energy expenditure by species not yet subjected to extensive experimental measurements. One approach is to estimate power consumption using an empirically based estimating equation, such as allometric equations describing the average relation between power consumption and body mass (e.g. Butler 1980). Interspecific variation around such allometric averages is impressive, however, and may well vitiate the utility of resulting energy budgets. Closer approximations may be produced by equations such as that of Masman and Klaassen (1987), which is based upon empirical data for 22 species not confined in wind tunnels and uses Greenewalt's (1975) model as a template to incorporate data on wing span, wing area, and body mass. It impressively accounts for 84% of variance in measured flight power consumption. Even this equation, however, is not designed to account for effects of variables of critical importance to migrating birds, such as flight speed and fat load.

Resolving these difficulties could follow two possible courses. One would be to improve theoretical approaches. This is attractive because of the potential power of such aerodynamic analyses. Fundamentally new approaches may be required, however, to deal with the critical obstacle of estimating exercise efficiency – a long-standing problem that has frustrated exercise physiologists in general (Stainsby et al. 1980). A fruitful alternative may be to substantially expand the empirical data base and significantly refine analyses of it. The type of analysis exemplified by Masman and Klaassen (1986), which combines empirical data and theoretically generated insights, may well serve as a useful model.

References

Baggott GK (1986) The fat contents and flight ranges of four warbler species on migration in North Wales. Ringing & Migr 7:25–36

Barlein F (1985) Body weights and fat deposition of Palearctic passerine migrants in the central Sahara. Oecologia (Berl) 66:141–146

Berthold P (1975) Migration: control and metabolic physiology. In: Farner DS, King JR (eds) Avian biology, vol 5. Academic Press, London

Biebach H (1985) Sahara stopover in migratory flycatchers: fat and food affect the time program. Experientia (Basel) 41:695–697

Biebach H, Friedrich B, Heine G (1986) Interaction of body mass, fat, foraging and stopover period in trans-Sahara migrating passerine birds. Oecologia (Berl) 69:370–379

Blem CR (1980) The energetics of migration. In: Gauthreaux SA (ed) Animal migration, orientation, and navigation. Academic Press, London, pp 175–224

Butler PJ (1980) Respiration during flight. Adv Physiol Sci 10:155–164

Butler PJ, West NH, Jones DR (1977) Respiratory and cardiovascular responses of the pigeon to sustained, level flight in a wind-tunnel. J Exp Biol 71:7–26

Drobney RD (1980) Reproductive bioenergetics of wood ducks. Auk 97:480–490

Ettinger AO, King JR (1980) Time and energy budgets of the willow flycatcher (*Empidonax traillii*) during the breeding season. Auk 97:533–546

Farner DS, King JR, Stetson MH (1968) The control of fat metabolism in migratory birds. Proc 3rd Int Congr Endocrinol, pp 152–157

Gessaman JA, Nagy KA (1988) Transmitter loads affect the flight speed and metabolism of homing pigeons. Condor 90:662–668

Greenwalt CH (1975) The flight of birds. Transactions of the American Philosophical Society, New Series 65:1–67

King JR (1972) Adaptive periodic fat storage by birds. Proc 15th Int Ornithol Congr, pp 200–217

LeFebvre EA (1964) The use of D_2O^{18} for measuring energy metabolism in *Columba livia* at rest and in flight. Auk 81:403–416

Masman D (1986) The annual cycle of the kestrel *Falco tinnunculus*. A study in behavioral energetics. Thesis, University of Groningen, Groningen

Masman D, Klaassen M (1987) Energy expenditure for free flight in trained and wild kestrels, *Falco tinnunculus*. Auk 104:603–616

Moreau RE (1972) The Palearctic-African bird migration systems. Academic Press, London

Mugaas JN, King JR (1981) Annual variation of daily energy expenditure by the black-billed magpie: a study of thermal and behavioral energetics. Stud Avian Biol 5:1–78

Pennycuick CJ (1968) Power requirements for horizontal flight in the pigeon *Columba livia*. J Exp Biol 49:527–555

Pennycuick CJ (1975) Mechanics of flight. In: Farner DS, King JR (eds) Avian biology, vol 5. Academic Press, New York, pp 1–75

Polus M (1985) Quantitative and qualitative respiratory measurements on unrestrained free flying pigeons by AMACS (Airborne measuring and control systems). In: Nachtigall W (ed) BIONA-report 3. Akad Wiss Mainz, Fischer, Stuttgart, pp 293–301

Rappole JD, Warner DW (1976) Relationships between behavior, physiology, and weather in avian transients at a migration stopover. Oecologia 26:193–212

Rayner JMV (1979) A new approach to animal flight mechanics. J Exp Biol 88:17–54

Rayner JMV (1986) Vertebrate flapping flight mechanics and aerodynamics, and the evolution of flight in bats. In: Nachtigall W (ed) Bat flight – Fledermausflug. Fischer, Stuttgart, pp 27–74 (Biona Report 5)

Rothe HJ, Nachtigall W (1987) Pigeon flight in a wind tunnel. I. Aspects of wind tunnel design, training methods and flight behaviour of different pigeon races. J Comp Physiol B Biochem Syst Environ Physiol 157:91–98

Rothe HJ, Biesel W, Nachtigall W (1987) Pigeon flight in a wind tunnel. II. Gas exchange and power requirement. J Comp Physiol B Biochem Syst Environ Physiol 157:99–109

Stainsby WN, Gladden LB, Barclay JK, Wilson BA (1980) Exercise efficiency: validity of base-line subtractions. J Appl Physiol 48:518–522

Torre-Bueno JR, Larochelle J (1987) The metabolic cost of flight in unrestrained birds. J Exp Biol 75:223–229

Tucker VA (1968) Respiratory exchange and evaporative water loss in the flying budgerigar. J Exp Biol 48:67–87

Tucker VA (1972) Metabolism during flight in the laughing gull, *Larus atricilla*. Am J Physiol 222:237–245

Tucker VA (1974) Bird metabolism during flight: evaluation of a theory. J Exp Biol 58:689–709

Walsberg GE (1983) Avian ecological energetics. In: Farner DS, King JR, Parkes KC (eds) Avian biology, vol 7. Academic Press, London, pp 161–220

Weathers WW, Buttemer WA, Hayworth AM, Nagy KA (1984) An evaluation of time-budget estimates of daily energy expenditure in birds. Auk 101:459–472

Williams JB, Nagy KA (1984) Daily energy expenditure of savannah sparrows: comparison of time-energy budget and doubly labelled water estimates. Auk 101:221–229

Species Index

Subject Index

MIX
Papier aus verantwortungsvollen Quellen
Paper from responsible sources
FSC® C105338

If you have any concerns about our products,
you can contact us on
ProductSafety@springernature.com

In case Publisher is established outside the EU,
the EU authorized representative is:
Springer Nature Customer Service Center GmbH
Europaplatz 3, 69115 Heidelberg, Germany

Printed by Libri Plureos GmbH
in Hamburg, Germany